Studies in Systems, Decision and Control

Volume 408

The series "Studies in Systems, Decision and Control" (SSDC) covers both new developments and advances, as well as the state of the art, in the various areas of broadly perceived systems, decision making and control–quickly, up to date and with a high quality. The intent is to cover the theory, applications, and perspectives on the state of the art and future developments relevant to systems, decision making, control, complex processes and related areas, as embedded in the fields of engineering, computer science, physics, economics, social and life sciences, as well as the paradigms and methodologies behind them. The series contains monographs, textbooks, lecture notes and edited volumes in systems, decision making and control spanning the areas of Cyber-Physical Systems, Autonomous Systems, Sensor Networks, Control Systems, Energy Systems, Automotive Systems, Biological Systems, Vehicular Networking and Connected Vehicles, Aerospace Systems, Automation, Manufacturing, Smart Grids, Nonlinear Systems, Power Systems, Robotics, Social Systems, Economic Systems and other. Of particular value to both the contributors and the readership are the short publication timeframe and the world-wide distribution and exposure which enable both a wide and rapid dissemination of research output.

Indexed by SCOPUS, DBLP, WTI Frankfurt eG, zbMATH, SCImago.

All books published in the series are submitted for consideration in Web of Science.

More information about this series at https://link.springer.com/bookseries/13304

Miguel Bernal · Antonio Sala · Zsófia Lendek ·
Thierry Marie Guerra

Analysis and Synthesis of Nonlinear Control Systems

A Convex Optimisation Approach

Springer

Miguel Bernal
Department of Electrical and Electronics Engineering
Sonora Institute of Technology
Obregon, Sonora, Mexico

Zsófia Lendek
Department of Automation
Technical University of Cluj-Napoca
Cluj-Napoca, Romania

Antonio Sala
Instituto Universitario de Automática e Informática Industrial
Universitat Politècnica de València
Valencia, Spain

Thierry Marie Guerra
Department of Automation and Control, LAMI HUMR CNRS 8201
Université Polytechnique Hauts-de-France
Valenciennes, France

ISSN 2198-4182 ISSN 2198-4190 (electronic)
Studies in Systems, Decision and Control
ISBN 978-3-030-90772-3 ISBN 978-3-030-90773-0 (eBook)
https://doi.org/10.1007/978-3-030-90773-0

MATLAB is a registered trademark of The MathWorks, Inc. See https://www.mathworks.com/trademarks for a list of additional trademarks.

Mathematics Subject Classification: 34H05, 37N35, 93B36, 93B51, 93C10, 93C35, 93D05, 93D09, 93D20, 93D23, 93D25, 93D30, 52A41, 90C25, 90C22, 90C23

This Springer imprint is published by the registered company Springer Nature Switzerland AG
The registered company address is: Gewerbestrasse 11, 6330 Cham, Switzerland

Contents

Notation and Abbreviations

Conventions

The following conventions are used throughout the book:

- The standard control-theoretic conventions are used. For instance, the state is denoted by $\boldsymbol{x}$, the control action by $\boldsymbol{u}$, the process dynamics by $\boldsymbol{f}$, and the measurements by $\boldsymbol{y}$.
- All the vectors are column vectors. The transpose of a vector is denoted by the superscript T. For instance, the transpose of $\boldsymbol{x}$ is $\boldsymbol{x}^T$.
- Boldface notation is used for vector or vector functions, e.g. $\boldsymbol{f}$ is a vector function.

Symbols and Notation

General Notation

I	Identity matrix.
0	Zero matrix.
$(*)$	Hermitian of a matrix $A + (*) = A + A^T$.
$A > 0$	A is a positive-definite matrix.
$\widehat{\boldsymbol{x}}$	Estimated value of the signal x.
$\dot{\boldsymbol{x}}$	Derivative of the signal x.
$\mathcal{X}$	Domain where the variable x is defined.
$\mathcal{C}$	Domain where the variable z is defined.
$\lVert \cdot \rVert$	Euclidean norm of a vector or induced norm of a matrix.
γ, μ	Positive constants used as bounds.
i, j, k, l	Indices.

Dynamic Systems

$\boldsymbol{x}$	State vector.
$\boldsymbol{u}$	Input vector.
$\boldsymbol{y}$	Output vector.
$\boldsymbol{d}$	Disturbance/unknown input.
$\boldsymbol{e}$	Error vector (observation/tracking).
n	Dimension of the state vector.
m	Dimension of the input vector.
q	Dimension of the output vector.
$\boldsymbol{f}$	State transition function; general nonlinear vector function.
$\boldsymbol{h}$	Measurement function; general nonlinear vector function.
A	State transition matrix (linear systems).
B	Input matrix (linear systems).
C	Measurement matrix (linear systems).
a	Affine term in the state equation (linear systems).
c	Affine term in the output equation (linear systems).
t	Time (real for continuous-time systems, integer for discrete-time ones).

TS Models

z_i	ith scheduling variable.
z	Scheduling vector.
p	Dimension of the scheduling vector.
$w_{i_j}^j$	Weighting function of z_j.
h_i	Membership function of z.
i, j	Indices for vertex models.
r	Number of vertex models.
A_i	State matrix of the ith vertex model.
B_i	Input matrix of the ith vertex model.
C_i	Measurement matrix of the ith vertex model.
a_i	Affine term in the ith vertex model.
c_i	Affine term in the ith vertex measurement model.
P	Lyapunov's matrix.
V	Lyapunov's function.
K_i	Standard observer gain of the ith vertex model.
F_i	Standard controller gain of the ith vertex model.
$A(\boldsymbol{x})$	Matrix function.

Chapter 1
Introduction

Since the advent of linear matrix inequality (LMI) theory and toolboxes in the early nineties, many problems in control which did not have an explicit analytical solution can be solved (numerically) via convex optimisation. A reinterpretation of classical linear controller and observer setups (LQR, Kalman Filter, $\mathcal{H}_\infty$ Riccati equations) was made in terms of LMIs; generalisations to *polytopic* uncertain linear systems were quickly addressed. Since bounded nonlinearities in a region of the state space can be fitted inside "boxes", many nonlinear systems can be embedded into time-varying polytopic linear ones, so the above-mentioned techniques can apply to them. Based on that fact, fuzzy rules, which were used since their inception as a sort of smooth interpolators between several behaviours (instead of abrupt switching from ordinary control rules in, say, ladder diagrams and relays) acquired a new meaning and modern fuzzy control, jointly with its close cousin *linear parameter varying* (LPV) control (gain scheduling) arose in the late 90's as a successful approach to control nonlinear systems (modelled in the so-called Takagi–Sugeno (TS) framework via linear time-varying ones). Since then, the older heuristic approach to mimicking human operators by setting up some expert rules has been displaced from the research arena of nonlinear control.

The seminal book on LMIs in control by [1] and the widely cited book of [2] on LMI-based analysis and design of nonlinear systems in the form of TS models, summarised the main ideas of the then incipient approach and made them available in high-quality sources. Since then, a vast amount of literature on TS-based control has been generated. Drawing conclusions on nonlinear systems via exact convex representations resembling TS models became increasingly common and less conservative; it helped this area to further the distance from its fuzzy origins and LPV interpretations while consolidating as a legitimate tool for analysis and design of nonlinear control systems.

This book aims to present the main modelling and control ideas behind TS control with a more current perspective, a comprehensive approach which includes other models with natural convex structures such as linear fractional (LFT) or polyhedral,

M. Bernal et al., *Analysis and Synthesis of Nonlinear Control Systems*, Studies in Systems, Decision and Control 408, https://doi.org/10.1007/978-3-030-90773-0_1

or induced convex schemes such as polynomial or descriptors. The reader should keep in mind that nowadays the concept of TS model may as well be rid of fuzzy rules and that the state- or parameter-dependency of the interpolation functions plays a crucial role in the most recent refinements.

Structure of the Book

Chapter 2 sets off a discussion on the problems this book is concerned about. The following chapters covering modelling (Chap. 3) and stability (Chap. 4) are similarly organised to discuss their respective topics for TS models, descriptors, polynomial models, and linear fractional transformation structures. Once this has been made, state feedback design (Chap. 5) as well as more complex control problems (Chap. 6) are considered: the sort of convex models and design choices will progressively shrink with respect to the simpler and more general problem of stability.

This book is mainly concerned with exact modelling; therefore, a lot of attention is devoted to the sector nonlinearity approach and its Taylor-based polynomial generalisation in Chap. 3, though approximate models based on linearisation are also briefly described. The use of linear fractional transformation structures for modelling of uncertain systems is also discussed.

Chapter 4 begins by considering quadratic stability of TS models; then, it smoothly moves to non-quadratic approaches such as those based on piecewise and parameter-dependent Lyapunov functions. Once these subjects are covered, stability analysis of polynomial and descriptor models can be generalised from them. Quadratic and non-quadratic stability of linear fractional transformation systems is finally considered. In all cases, it is understood that stability results apply to nonlinear control systems which have been modelled in any of the considered convex structures; all of the results can be stated as convex optimisation problems in the form of linear matrix inequalities or sum of squares.

State feedback is the subject of Chap. 5. Naturally, the number of models and Lyapunov functions that have reached the maturity required for such task is less than that for stability analysis since, as it will be discussed, many problems arise. Controller design based on the assumption of measurable premises and combined with performance specifications is covered, with unified descriptions when available (for instance, for the parameter-dependent discrete-time case for TS models), sometimes only pointed out (for certain non-quadratic schemes in the continuous-time context, for example). The research in this area is still ongoing to tackle several issues (such as conditions in terms of bilinear matrix inequalities or local versus global results): they are, nevertheless, briefly discussed. Performance specifications (input/output constraints, $\mathcal{H}_\infty$ disturbance rejection) and robustness is discussed before a real-time application for automated wheelchair swing-up control is presented.

The last part of this book, Chap. 6, is concerned with more complex control problems than the previous one, mainly observer design (both measured and unmeasured premises) and output feedback. An even more reduced set of structures and Lyapunov function choices is presented as a natural consequence of the difficulties researchers have found to deal with these advanced problems. Nevertheless, the reader is referred to up-to-date reports on these matters in order to pursue her/his investigation into still not-closed research problems. Two real-time applications are presented: sitting control for people living with spinal cord injury and observer-based trajectory tracking for power-assisted wheelchairs.

Finally, Chap. 7 gathers some conclusions and perspectives by considering the most recent works and ramifications in the field at bird's eye.

References

1. Boyd, S., El Ghaoui, L., Féron, E., Balakrishnan, V.: Linear Matrix Inequalities in System and Control Theory. Studies in Applied Mathematics. Society for Industrial and Applied Mathematics, Philadelphia, PA (1994)
2. Tanaka, K., Wang, H.O.: Fuzzy Control Systems Design and Analysis, analysis Wiley, New York (2001)

Chapter 2
Problems to Be Solved and Scope of the Book

This chapter intends to motivate the reader to go further in the quest for nonlinear control based on models with convex structure. Besides some preliminary definitions and control objectives which are common to most control techniques, the wide variety of models with convex structures is briefly visited, their advantages and drawbacks described. It will become apparent that convex structures lead to numerical rather than just formal design, something valuable for practical engineers since most formal nonlinear techniques cannot handle systems with order greater than 2 or 3, or without a special form.

2.1 Some Useful Classes of Dynamical Systems

This book is focussed on convex models of nonlinear systems for which a mathematical model is available, for instance, by first principles. In general, such models can be arranged in the state-space form that follows:

Definition 2.1 [1] A *nonlinear system* is a set of coupled first-order ordinary differential equations (ODEs) gathered in a state equation of the form

$$\dot{\boldsymbol{x}}(t) = \boldsymbol{f}(\boldsymbol{x}, \boldsymbol{u}, t), \quad \boldsymbol{x}(t+1) = \boldsymbol{f}(\boldsymbol{x}, \boldsymbol{u}, t), \tag{2.1}$$

where $\boldsymbol{x} \in \mathbb{R}^n$ denotes the system state, $\boldsymbol{u} \in \mathbb{R}^m$ is the system input, $\boldsymbol{f}(\boldsymbol{x}, \boldsymbol{u}, t) : \mathbb{R}^n \times \mathbb{R}^m \times \mathbb{R} \mapsto \mathbb{R}^n$, $t \in \mathbb{R}$ for continuous-time systems (left) and $t \in \mathbb{Z}$ for discrete-time systems (right), $n, m \in \mathbb{N}$.

For convenience, the single notation $\delta \boldsymbol{x}(t) = \boldsymbol{f}(\boldsymbol{x}, \boldsymbol{u}, t)$ will be used in the sequel, where $\delta \boldsymbol{x}(t)$ should be understood as $\delta \boldsymbol{x}(t) \equiv \dot{\boldsymbol{x}}(t)$ for continuous-time models, and $\delta \boldsymbol{x}(t) \equiv \boldsymbol{x}(t+1)$ for discrete-time models. The vector field $\boldsymbol{f}(\cdot)$ is usually piecewise continuous in t and locally Lipschitz in $(\boldsymbol{x}, \boldsymbol{u})$ over a compact set of interest $\mathcal{C}$;

M. Bernal et al., *Analysis and Synthesis of Nonlinear Control Systems*, Studies in Systems, Decision and Control 408, https://doi.org/10.1007/978-3-030-90773-0_2

unless stated otherwise, these assumptions will be extended to vector fields including disturbances, parameters, or any other argument. Apart from the dynamic equation above, the models may include some output equation

$$\boldsymbol{y}(t) = \boldsymbol{h}(\boldsymbol{x}, \boldsymbol{u}, t), \tag{2.2}$$

describing a set of variables of interest gathered in $\boldsymbol{y}(t) \in \mathbb{R}^q$ which might be controlled or measured, $\boldsymbol{h}(\boldsymbol{x}, \boldsymbol{u}, t) : \mathbb{R}^n \times \mathbb{R}^m \times \mathbb{R} \mapsto \mathbb{R}^q$.

General Differential-Algebraic Models

In some cases, models do not come in the state-space ODE form (2.1), but in terms of equations involving other describing variables whose derivatives (future values, in discrete cases) cannot be trivially solved for and moved to the left-hand side. Such models, in the most general form, will be expressions of the sort $\boldsymbol{f}(\boldsymbol{x}, \dot{\boldsymbol{x}}, \boldsymbol{u}, t) = 0$, where some of the rows could be actually only algebraic equations, i.e. $f_i(\boldsymbol{x}, \boldsymbol{u}, t) = 0$ for some row i; they are called *differential-algebraic equations* (DAEs) [2], being the ODE (2.1) a particular case. In some cases, such models will give rise to the so-called *descriptor* representations. For brevity, details will be left out of this motivating chapter and discussed later on.

There are quite a few different classes of dynamical systems reported in the literature. In this book, the following definitions will be used.

Linear Dynamic Systems

Definition 2.2 [3] A *linear time-varying* system (LTV) is a dynamical system whose evolution is defined by equations of the form

$$\begin{aligned} \delta\boldsymbol{x}(t) &= A(t)\boldsymbol{x}(t) + B(t)\boldsymbol{u}(t), \\ \boldsymbol{y}(t) &= C(t)\boldsymbol{x}(t) + D(t)\boldsymbol{u}(t), \end{aligned} \tag{2.3}$$

where $A(t)$, $B(t)$, $C(t)$, $D(t)$ are matrices of suitable dimensions; as before, $\boldsymbol{x}$, $\boldsymbol{u}$, and $\boldsymbol{y}$ stand for the system state, input, and output, respectively.

In LTV systems the expressions for the time dependence in A, B, C, D are assumed to be known beforehand. A particular case of LTV systems are the *linear time-invariant* (LTI) ones, in which A, B, C, D are constant matrices.

Definition 2.3 [4] A *linear parameter-varying* (LPV) system is defined as a linear system whose system matrices depend on an exogenous time-varying parameter vector $\boldsymbol{\theta}(t)$:

$$\begin{aligned} \delta \boldsymbol{x}(t) &= A(\boldsymbol{\theta})\boldsymbol{x}(t) + B(\boldsymbol{\theta})\boldsymbol{u}(t), \\ \boldsymbol{y}(t) &= C(\boldsymbol{\theta})\boldsymbol{x}(t) + D(\boldsymbol{\theta})\boldsymbol{u}(t). \end{aligned} \tag{2.4}$$

The scheduling parameters $\boldsymbol{\theta}$ take values in some prescribed set $\boldsymbol{\theta} \in \Theta$. In some cases, a set Θ_d is also available in the model description, constraining the derivatives of the parameters as $\dot{\boldsymbol{\theta}} \in \Theta_d$. The difference with LTV systems lies in the fact that only the sets Θ, Θ_d are assumed known in the stability analysis or control design phase, instead of the actual expression of $\boldsymbol{\theta}(t)$.

In real-time operation of a candidate LPV design, some components of the scheduling parameters $\boldsymbol{\theta}$ may be measurable. Of course, in the measurable case, better performance can be achieved. Sometimes, the values of $\boldsymbol{\theta}$ range in a finite set (equivalently, $A(\boldsymbol{\theta})$ and $B(\boldsymbol{\theta})$ are piecewise constant); in that case, such systems are denoted as *jump linear systems*.

Definition 2.4 If the scheduling parameters are actually endogenous to the state dynamics (e.g. $\boldsymbol{\theta} \supset \boldsymbol{x}$), the system is named *quasi-LPV* [5].

Quasi-LPV systems usually arise from (conservative) nonlinear system manipulations: for instance, $\dot{x} = -x^3$ can be expressed as $\dot{x} = -\theta \cdot x, \theta = x^2$. Disregarding the latter equality, an LPV expression $\dot{x} = -\theta \cdot x$ is obtained (i.e. a quasi-LPV one). Some authors denote these manipulations as *LPV embedding* [6]. This idea is, in fact, at the root of many of the models to be discussed later in this work allowing handling nonlinear systems with linear tools.

If the values of $\boldsymbol{\theta}(t)$ are generated by discrete-event dynamics whose behaviour can depend on the continuous-time variables, the systems are denoted as *switched* linear systems.

Nonlinear Dynamic Systems

Regarding nonlinear dynamic systems, a particular class to be studied in this book roots in the polynomial ones:

Definition 2.5 A *polynomial* system is a dynamical system (2.1) where $\boldsymbol{f}$ is a polynomial vector field in $\boldsymbol{x}$ and $\boldsymbol{u}$.

Polynomial time-varying and *polynomial parameter-varying* systems can be defined by adapting those in the linear case above.

Definition 2.6 An *affine in control* system is a dynamical system (2.1) where $\boldsymbol{f}$ is affine in $\boldsymbol{u}$, i.e. redefining $\boldsymbol{f}$, $\delta \boldsymbol{x}(t) = \boldsymbol{f}(\boldsymbol{x}, t) + \boldsymbol{g}(\boldsymbol{x}, t)\boldsymbol{u}(t)$.

Evidently, all linear systems (LTI, LPV, LTV) are affine in control. Non-affine in control systems are much harder to deal with and, usually, linearisation or extended (linear) dynamics are employed in order to get an affine in control model; see Section 2.3.1.

Discretisation and Sampling

Note that discrete-time models arising from sampling continuous-time first-principle ones can be discretised in the linear+zero-order hold case (matrix exponential formulae), but not so in a general nonlinear case. In that case, approximations (Euler, mid-point, etc. [7]) need to be made: hence, there will possibly exist a discretisation error (which will increase with the sampling period T—for instance, the error is in the order of T^2 in the Euler discretisation) so that the subsequent discrete-time design must be robust enough to accommodate it (and T should be reasonably small).

2.2 Control Objectives

When discussing models for control, the system inputs may be split into manipulated inputs and non-manipulated ones; moreover, possibly time-varying parameters may be included. A notation incorporating the nuances of the above definitions may be set up denoting now with $\boldsymbol{u}$ the manipulated inputs, with $\boldsymbol{d}$ the non-manipulated ones, and with $\boldsymbol{\theta}$ the possibly time-varying parameters. Instead of Definition 2.1, dynamics are expressed as follows:

Definition 2.7 A *parameter-dependent time-varying nonlinear system* subject to external disturbances is a set of first-order ordinary differential (difference) equations gathered in the state equation

$$\delta \boldsymbol{x}(t) = \boldsymbol{f}(\boldsymbol{x}, \boldsymbol{u}, \boldsymbol{d}, \boldsymbol{\theta}, t), \tag{2.5}$$

where the particular expressions of $\boldsymbol{d}(t)$—disturbances—and $\boldsymbol{\theta}(t)$—uncertain parameters—are assumed to be unknown beforehand.

If the dynamics is in the form

$$\delta \boldsymbol{x}(t) = \boldsymbol{f}(\boldsymbol{x}, \boldsymbol{u}, \boldsymbol{d}), \tag{2.6}$$

the system will be called *time-invariant*.

If the dynamics is in the form

$$\delta \boldsymbol{x}(t) = \boldsymbol{f}(\boldsymbol{x}, \boldsymbol{u}, \boldsymbol{d}, \boldsymbol{\theta}), \tag{2.7}$$

the system will be called *parameter-varying*.

Systems of the latter class where the parametric uncertainties $\boldsymbol{\theta}$ are *unknown, but fixed* (i.e. constant), are usually referred to simply as *uncertain*.. Keep in mind that there are many classes of uncertainty, unknown-but-constant parameter uncertainty being just one of them: we might have neglected dynamics, linearisation error, etc.

Remark 2.1 The distinction between disturbances and parameter variations is subtle (well, actually, non-existing: both are unknown time-dependent signals in a general nonlinear setting). In LPV models, parameters $\boldsymbol{\theta}$ do *not* change the equilibrium point, whereas disturbances *do* change it. Something may change *both* things (linearised model matrices *and* equilibrium point): then we need to do the quasi-LPV trick: split in two the effects for a Takagi–Sugeno model (TS, to be defined later). See, for instance, Example 3.16 on resistor variation in RC circuit: it changes both the equilibrium point and the time constant. A TS model as well as a linear fractional transformation (LFT, to be defined later) are shown therein.

Stabilisation

The objectives of most developments in this book will be stabilising a particular operation point or trajectory of the generic nonlinear system (2.5).

If all expressions of time are known beforehand (i.e. there is no $\boldsymbol{d}$ and $\boldsymbol{\theta}$), in some cases it is possible to compute a desired trajectory $(\boldsymbol{x}_r(t), \boldsymbol{u}_r(t))$ such that it verifies:

$$\delta \boldsymbol{x}_r(t) = \boldsymbol{f}(\boldsymbol{x}_r, \boldsymbol{u}_r, t).$$

In that case, the *local stabilisation* problem will be that of computing a *feedback* control law $\boldsymbol{u}(\boldsymbol{x}, t)$ such that $\boldsymbol{x}(t) \to \boldsymbol{x}_r(t)$ as $t \to \infty$ for all $\boldsymbol{x}(0) \in \mathcal{D}$ where $\mathcal{D}$ is a known initial condition set. State feedback might require some states to be recovered from the output via an observer, thus leading to *output feedback*, which can be also dynamic; see Chap. 6.

The most common case is $\boldsymbol{x}_r$ being a constant operating point and $\boldsymbol{u}_r$ being a feed-forward control action cancelling the known time variation. In case of undisturbed time-invariant models (i.e. constant $\boldsymbol{d}$ in (2.6)), the resulting $\boldsymbol{u}_r$ is also constant. Time-varying trajectories are mostly used in robotics [8–10] or batch processes [11, 12]. Synchronisation [13–15] is also a related issue.

Disturbance Rejection

If there are unknown beforehand disturbances or parameter variations in time, then the expression for the feedforward $\boldsymbol{u}_r$ cannot be obtained in a general case. Of course, it can be obtained online in special cases when $\boldsymbol{d}$ or $\boldsymbol{\theta}$ are *measurable*. The computation of such feedforward laws for nonlinear system is out of the scope of this work; the interested reader is referred to [16].

Assuming such exact input and trajectory cannot (or *will* not, if too complex) be computed *a priori* or as a function of measurements, the *disturbance rejection* objective is usually cast as obtaining a control law which minimises the distance in some signal norm (see the Appendix A.7) between a reference trajectory $\boldsymbol{\psi}_r(t) = (\boldsymbol{x}_r(t), \boldsymbol{u}_r(t))$ and the actual $\boldsymbol{\psi}(t) = (\boldsymbol{x}(t), \boldsymbol{u}(t))$.

In most practical cases, $\delta \boldsymbol{x}_r(t) = \boldsymbol{f}(\boldsymbol{x}_r, \boldsymbol{u}_r, \boldsymbol{d}_r, \boldsymbol{\theta}_r, t)$ for some nominal reference values of disturbances and parameters, but of course, it does *not* fulfil $\delta \boldsymbol{x}_r(t) = \boldsymbol{f}(\boldsymbol{x}_r, \boldsymbol{u}_r, \boldsymbol{d}, \boldsymbol{\theta}, t)$.

In a general nonlinear control setup, solutions for the disturbance-rejection problem are, too, local: a bound on $\|\boldsymbol{\psi}(t) - \boldsymbol{\psi}_r(t)\|$ may only be guaranteed for some initial states, reference trajectories, disturbances and parameter values fulfiling some constraints on their values and their time derivatives. The locality of the disturbance-rejection results is usually overlooked in literature; it will be discussed in Chap. 4.

Robust Control

In the particular case that there are parameters which do not change the equilibrium point (or trajectory), the *robust stabilisation* problem of a time-varying (known variation) system around a trajectory which fulfils:

$$\delta \boldsymbol{x}_r(t) = \boldsymbol{f}(\boldsymbol{x}_r, \boldsymbol{u}_r, \boldsymbol{\theta}, t), \quad \forall \boldsymbol{\theta}(t),$$

will be that of computing a *feedback* control law $\boldsymbol{u}(\boldsymbol{x}, t)$ such that $\boldsymbol{x}(t) \to \boldsymbol{x}_r$ as $t \to \infty$ for all $\boldsymbol{x}(0) \in \mathcal{D}$ where $\mathcal{D}$ is a known initial condition set and for all values of $\boldsymbol{\theta}$ in a prescribed set. As for stabilisation, robust control may also be performed via output instead of state feedback.

A *robust disturbance rejection* problem may be posed analogously.

State Estimation

In many practical cases, a model from a physical system is available, incorporating some *output equations*:

$$\delta \boldsymbol{x}(t) = \boldsymbol{f}(\boldsymbol{x}, \boldsymbol{u}, \boldsymbol{d}, t), \tag{2.8}$$

$$\boldsymbol{y}(t) = \boldsymbol{h}(\boldsymbol{x}, \boldsymbol{u}, \boldsymbol{d}, t), \tag{2.9}$$

and experimental input–output data $(\boldsymbol{u}(t), \boldsymbol{y}(t))$ for t in a given interval must be fit to such a model. Usually, the model assumes that the statistical average of $\boldsymbol{d}(t)$ is zero.

As future outputs of dynamical systems depend on initial conditions, the optimal value of such initial conditions so that the best fit to the data is achieved is a relevant problem in many applications. The result of this is a smoothed best-fit state and output trajectory, and this problem is usually called as the *smoothing* problem. For instance, for linear models with Gaussian noise, the optimal solution is the so-called RTS smoother [17–19].

Also, in an online situation, knowing the best estimate of $\boldsymbol{x}(t)$ given a past record of input-output data is, too, a problem of interest both for its own sake (the so-called *filtering* problem, in which a "clean" internal state must be constructed from noisy measurements [20]), in order to *diagnose* the system (in models in which some state variables are assumed to be fault states or fault parameters [21]) or, in order to *control* the system [22]. The state estimator for this problem is usually called *state observer*, introduced in the linear case by Kalman and Luenberger [23, 24].

The difference between a smoother and a filter is the past-only information assumed in the latter, whereas a smoother has a finite window of past and future information in order to obtain a state estimate at a particular instant (except at the extreme initial and final data points).

An observer is said to be *stable* if the estimated value of $\boldsymbol{x}(t)$ converges to the actual one as the length of the data set tends to infinity, in the case $\boldsymbol{d}(t) = 0, \forall t$.

In some cases, there is the so-called *separation principle* in which a stable observer jointly with a stable state-feedback controller $\boldsymbol{u}(\boldsymbol{x})$, provides a stable *output-feedback* controller (in which $\boldsymbol{u}(t)$ is the output of a dynamic system whose inputs are the measurements in $\boldsymbol{y}(t)$). The linear case is the most paradigmatic example [25]; in the nonlinear case, such separation does not hold in general; some examples will be discussed later in the book. Nevertheless, separation is used and relied upon in nonlinear control practice [26]. This book will discuss the particular nuances of the state estimation approach for the convex-structure models which are the main *leitmotifs* of the following chapters.

Parameter Scheduling

If some of the values of $\boldsymbol{\theta}$ are measurable, denoted as $\boldsymbol{\theta}_m(t)$, the above robust control laws may depend on them $\boldsymbol{u}(\boldsymbol{x}, \boldsymbol{\theta}_m, t)$.

2.3 Models

Generic nonlinear systems may be difficult to control in the terms above discussed. Hence, some simplifications, approximations or transformations of the original nonlinear models are usually carried out in the literature in order to pose a control problem in a form amenable for further analysis and design following a particular technique.

Let us discuss general ideas on the topic, and let us defer further discussion for Chap. 3.

2.3.1 Model Approximations

In literature, approximations valid in the neighbourhood of an operation point are usually set up in order to get simplified representations.

The first approximation to think of is converting a nonlinear system into an affine in control one. Indeed, there are very specific cases in which the conversion is straightforward: for instance, if an invertible transformation $\boldsymbol{v} = \boldsymbol{n}(\boldsymbol{u})$ of the input such that

$$\delta \boldsymbol{x}(t) = \boldsymbol{f}(\boldsymbol{x}, \boldsymbol{u}, t) = \bar{\boldsymbol{f}}(\boldsymbol{x}, t) + \boldsymbol{g}(\boldsymbol{x}, t)\boldsymbol{n}(\boldsymbol{u})$$

exists, then $\boldsymbol{u} = \boldsymbol{n}^{-1}(\boldsymbol{v})$ is easy to compute. However, non-invertible nonlinearities in the input are hard to deal with.

So, a Taylor series in $\boldsymbol{u}$ for a virtually fixed $\boldsymbol{x}$ is usually taken and the model (2.1) gets written as

$$\delta \boldsymbol{x}(t) = \boldsymbol{f}(\boldsymbol{x}, \boldsymbol{u}, t) \approx \boldsymbol{f}(\boldsymbol{x}, \boldsymbol{u}_r, t) + \left.\frac{\partial \boldsymbol{f}}{\partial \boldsymbol{u}}\right|_{(\boldsymbol{x}, \boldsymbol{u}_r, t)} \cdot (\boldsymbol{u} - \boldsymbol{u}_r), \tag{2.10}$$

where a state- and time-dependent Jacobian appears. Note that disturbance inputs can also be approximated by affine forms, when convenient.

In the sequel, unless otherwise stated, we will assume models are affine in control, as many nonlinear control literature results do. Nevertheless, it should be kept in mind that terms depending on control or disturbance signals may be treated as bounded nonlinearities, at least for stability and disturbance analysis [27].

When linearising both in state and input around a trajectory for a time-varying system (i.e. with *known* expressions of the variation in time), we get

$$\begin{aligned}\delta \boldsymbol{x}(t) = \boldsymbol{f}(\boldsymbol{x}, \boldsymbol{u}, t) \approx & \boldsymbol{f}(\boldsymbol{x}_r, \boldsymbol{u}_r, t) + \left.\frac{\partial \boldsymbol{f}}{\partial \boldsymbol{x}}\right|_{(\boldsymbol{x}_r, \boldsymbol{u}_r, t)} \cdot (\boldsymbol{x} - \boldsymbol{x}_r) \\ & + \left.\frac{\partial \boldsymbol{f}}{\partial \boldsymbol{u}}\right|_{(\boldsymbol{x}_r, \boldsymbol{u}_r, t)} \cdot (\boldsymbol{u} - \boldsymbol{u}_r),\end{aligned}$$

and, assuming $\delta \boldsymbol{x}_r(t) = \boldsymbol{f}(\boldsymbol{x}_r, \boldsymbol{u}_r, t)$ (i.e. the reference trajectory is compatible with the model equations) and subtracting on both sides, the usual increment notation $\bar{\boldsymbol{x}} = \boldsymbol{x} - \boldsymbol{x}_r$, $\bar{\boldsymbol{u}} = \boldsymbol{u} - \boldsymbol{u}_r$ yields the LTV system with zero equilibrium point in the new coordinates

$$\delta \bar{\boldsymbol{x}}(t) \approx \left.\frac{\partial \boldsymbol{f}}{\partial \boldsymbol{x}}\right|_{(\boldsymbol{x}_r, \boldsymbol{u}_r, t)} \cdot \bar{\boldsymbol{x}} + \left.\frac{\partial \boldsymbol{f}}{\partial \boldsymbol{u}}\right|_{(\boldsymbol{x}_r, \boldsymbol{u}_r, t)} \cdot \bar{\boldsymbol{u}} = A(t)\bar{\boldsymbol{x}} + B(t)\bar{\boldsymbol{u}}. \tag{2.11}$$

The best known application of the above idea is the particular case of linearisation of a time-invariant system (2.6) around a constant equilibrium point $\boldsymbol{x}_r$, $\boldsymbol{u}_r$, $\boldsymbol{d}_r = 0$, $\delta \boldsymbol{x}_r = 0$. Indeed, removing the time variation above, we get the well-known LTI system with zero equilibrium point

$$\delta \bar{x}(t) \approx \left.\frac{\partial f}{\partial x}\right|_{(x_r, u_r)} \cdot \bar{x} + \left.\frac{\partial f}{\partial u}\right|_{(x_r, u_r)} \cdot \bar{u} = A\bar{x} + B\bar{u}. \tag{2.12}$$

For brevity, the bar notation in increments will be omitted and all linear systems will be assumed in the sequel to be linearisations in incremental coordinates of the original nonlinear system. Also, in systems with disturbances and parameter variation, linearisation points or trajectories can be stated for some nominal values (or functions of time, i.e. trajectories) for both parameters and disturbances, in order to set up suitable disturbance-rejection problems.

If further terms in the Taylor series are taken, the result is a polynomial approximation of the system dynamics [28].

The basic issue arising from the above techniques is the fact that the model is an *approximation*, only valid in a restricted neighbourhood of the origin; in fact, the neighbourhood size is, in theory, *infinitesimal*. In practice, linearisation errors might be comparably lower than other modelling errors in a non-infinitesimal region but assessing its actual size is, of course, a problem. In the next section, we will address an alternative modelling framework without linearisation errors.

2.3.2 Models with Convex Structure

The so-called models with convex structure will be those which can be expressed as a (nonlinear) interpolation between simpler models. Let us describe the model classes to be analysed in later chapters of this book.

Vertex Interpolation Models

The main developments in this book discuss models expressed as a blending of some simpler models via nonlinear functions known as membership functions (MFs):

Definition 2.8 A *convex model* is a set of first-order ODEs whose right-hand side can be expressed as a convex sum of vector fields (models), namely

$$\delta x(t) = \sum_{i=1}^{r} h_i(z) f_i(x, u), \tag{2.13}$$

where $r \in \mathbb{N}$ represents the number of models in the convex sum, $z \in \mathbb{R}^p$ may depend on the state x, input u, time t, disturbances or exogenous parameters, and the nonlinear functions $h_i(z)$, $i \in \{1, 2, \ldots, r\}$ hold the *convex sum property* in a compact set $\mathcal{C} \subseteq \mathbb{R}^p$, i.e.:

$$\sum_{i=1}^{r} h_i(z) = 1, \quad 0 \le h_i(z) \le 1, \quad \forall z \in \mathcal{C}. \tag{2.14}$$

Definition 2.9 The vector z in a convex model (2.13) is known as the *premise* or *scheduling vector*; it is assumed to be bounded and continuously differentiable in a compact set $\mathcal{C} \subseteq \mathbb{R}^p$ of the scheduling/premise space.

Definition 2.10 The nonlinear functions $h_i(z)$, $i \in \{1, 2, \ldots, r\}$ in a convex model (2.13) are known as *membership functions* (MFs); they hold the convex sum property in a compact set $\mathcal{C} \subseteq \mathbb{R}^p$ of the scheduling space.

Convex models may include nonlinear outputs which, in turn, share the convex structure of the differential/difference equations

$$y(t) = \sum_{i=1}^{r} h_i(z)\boldsymbol{h}_i(\boldsymbol{x}, \boldsymbol{u}).$$

Definition 2.11 If $\boldsymbol{f}_i(\boldsymbol{x}, \boldsymbol{u})$ are linear, i.e. $\boldsymbol{f}_i(\boldsymbol{x}, \boldsymbol{u}) = A_i\boldsymbol{x} + B_i\boldsymbol{u}$, (2.13) is called a *Takagi–Sugeno* (TS) model [29, 30]; if, in addition, z does not include the state $\boldsymbol{x}$ it is called a *linear polytopic model* [31, 32]; if the TS model comes from fuzzy modelling techniques it is also referred to as *TS fuzzy model.*

Definition 2.12 If $\boldsymbol{f}_i$ are polynomial, the result is known as *fuzzy polynomial* [33, 34]; in the sequel we will refer to them as *polynomial representations* or *convex polynomial models* as we do not pursue fuzzy modelling, but exact rewriting of nonlinear expressions.

Normally, when this sort of model arises from the *sector nonlinearity approach*, to be described later, the number of models is related to the dimension of the premise vector by $r = 2^p$.

Box Models

An alternative representation of polytopic models in literature is

$$\delta\boldsymbol{x}(t) = \boldsymbol{f}_0(\boldsymbol{x}, \boldsymbol{u}) + \sum_{i=1}^{r} p_i(z)\boldsymbol{f}_i(\boldsymbol{x}, \boldsymbol{u}), \tag{2.15}$$

where $p_i(z) \in [p_i^0, p_i^1]$ and $\boldsymbol{f}_i(\boldsymbol{x}, \boldsymbol{u})$ are given (and usually linear); they are called *affine models* in [32]. These box models can be written, using the bounds on the box parameters p_i, as

$$\delta x(t) = f_0(x, u) + \sum_{i=1}^{r} \sum_{j=0}^{1} h_{ij}(z) p_i^j f_i(x, u). \tag{2.16}$$

Equation (2.13) is more general, because (2.15) can be expressed as (2.13) as the vertex interpolation can represent arbitrary convex polyhedra, not only rectangles). Hence, these parameter box models will not be specifically considered in the theoretical developments of this work as they are a particular case of the representation (2.13).

Linear Fractional Models

Yet a third representation of models having convex structure is in the form of a feedback connection between a time-invariant model M and a static (memoryless) function block Δ, where Δ is assumed to be amenable to expression in convex structure of one of the two options described above. So Δ can be nonlinear, uncertain, and time-varying. This class of models can be expressed as

$$M := \left\{ \begin{aligned} \delta x(t) &= f_1(x, u, w_N) \\ y(t) &= f_2(x, u, w_N) \\ z_N(t) &= f_3(x, u, w_N) \end{aligned} \right\}, \tag{2.17}$$

$$w_N = \Delta(z_N). \tag{2.18}$$

For simplicity, and given the fact that most current literature only considers the linear case for M, i.e. LTI f_1, f_2, f_3, this is why only the so-called Linear Fractional Models [5, 35–38] will be considered in the sequel.

In some cases, the application of the small-gain theorem (Theorem A.1 in Appendix A.8) does allow Δ to be a dynamic system.

This class of models is very general, including the two above as particular cases. In fact, the resulting feedback interconnection is *rational* in Δ. For further details; see Sect. 3.5 in Chap. 3. We refer to these models as having convex structure, because, as above said, (2.18) will be bounded by a vertex interpolation or box model in most cases. Nevertheless, the representation is able to represent non-convex functions of the elements of Δ (the mentioned rational cases) and, hence, using convex optimisation techniques on these cases will entail accepting significant conservatism (less conservative options, say μ-synthesis [39], require non-convex optimisation without guaranteed convergence or global optimality in a general case).

Tensor-Product Models

In some (quite common, as later discussed) cases, the system model (for instance, a state equation) can be expressed as

$$\delta x(t) = \sum_{i_1=1}^{r_1} \sum_{i_2=1}^{r_2} \cdots \sum_{i_q=1}^{r_q} h_{i_1}^{[1]}(z) h_{i_2}^{[2]}(z) \cdots h_{i_q}^{[q]}(z) f_{i_1 i_2 \dots i_q}(x, u), \tag{2.19}$$

in which

$$\sum_{i_s=1}^{r_s} h_{i_s}^{[s]}(z) = 1, \quad h_{i_s}^{[s]}(z) \in [0, 1], \quad s = 1, 2, \dots, q,$$

i.e. there are q independent convex interpolations which are later mixed together in the above tensor-product formulation. In fact, as discussed in Sect. 3.2.1, the number of vertices in each interpolation is $r_s = 2$ in most cases of practical interest. This class of models is called *tensor-product* (TP) models.

Of course, a single flattened Takagi–Sugeno model (2.13) with a total of $r_1 \times r_2 \times \cdots \times r_q$ vertices can be set up, as it is trivial to see that

$$\sum_{i_1=1}^{r_1} \sum_{i_2=1}^{r_2} \cdots \sum_{i_q=1}^{r_q} h_{i_1}^{[1]}(z) h_{i_2}^{[2]}(z) \cdots h_{i_q}^{[q]}(z) = 1.$$

However, in some analysis steps, such mixing may be quite conservative and keeping the tensor-product separation would be beneficial. Similarly, the tensor-product version of the box models can also be set up.

Homogeneous Multiple-Sum Models

If system (2.19) verifies that $h_i^{[j]} \equiv h_i^{[k]}$ for all j, k (i.e. all independent interpolations are actually identical, so they can be simply written as $h_i^{[j]} \equiv h_i$), then the TP expression can be written as

$$\delta x(t) = \sum_{i_1=1}^{r} \sum_{i_2=1}^{r} \cdots \sum_{i_q=1}^{r} h_{i_1}(z) h_{i_2}(z) \cdots h_{i_q}(z) f_{i_1 i_2 \dots i_q}(x, u), \tag{2.20}$$

which is denoted as homogenous q-dimensional fuzzy summation. For instance, when system (2.13) is controlled with a TS controller whose membership functions are, intentionally, set to be identical to those of the controlled plant (the so-called *parallel distributed compensation*, PDC [40]), if f_i are affine in control the result is a two-dimensional ($q = 2$) homogeneous summation; see (5.3). As the above-discussed flattening in this widely-used case is overly conservative, special developments for it have been developed in literature. The whole Appendix D is devoted to the analysis of homogeneous multidimensional sums.

Remark 2.2 An intermediate case in which some of the TP memberships are equal but others not, say $h_{i_1} \equiv h_{i_2}$, $h_{i_3} \equiv h_{i_4}$, but $h_{i_1} \neq h_{i_3}$ may also be possible (if a TP controller (PDC) is used to control a TP process). Details are omitted here for brevity.

Exact Representation of Nonlinear Systems with Convex Structure

One of the basic leitmotifs in this book is the fact that many nonlinear systems can be expressed in *exact* form in the convex-structure settings under discussion. For instance, a model

$$\dot{x}(t) = -\sin(1 + x)x(t) + u(t) \tag{2.21}$$

can be written, interpolating the sinusoid between its minimum (-1) and maximum $(+1)$, i.e. writing $\sin(1 + x) = h_1(x) \cdot 1 + h_2(x) \cdot (-1)$, as

$$\dot{x}(t) = h_1(x) \cdot (-x(t)) + h_2(x) \cdot x(t) + u(t), \tag{2.22}$$

where $h_1(x) = ((\sin(x + 1) + 1)/2)$ and $h_2(x) = 1 - h_1(x)$, Note that (2.22) is *not* an approximation, but an equality.

If we do not take into account that $h_1(x)$ is actually a sinusoidal, we might design controllers for whatever shape h_1 might have. This idea motivates the search for such design techniques, as discussed in the next section and expanded in later chapters.

Uncertain Models

Uncertainty naturally occurs in many realistic applications. We might have:

- uncertain model parameters;
- neglected dynamics;
- unmeasurable premise variables as will be defined later;
- uncertain membership functions.

Dealing with uncertainty is the subject of robust control, a field that goes back to the very foundations of control systems: first in the frequency domain [41, 42], later based on state-space representations [39, 43]. Naturally, within the context of convex models and linear matrix inequalities, it has also been developed as bounded uncertainties can, of course, be written as convex sums of their bounds, then grouped in nested convex sums of identical nature of the state- or time-dependent ones from a polytopic point of view [44, 45].

Other Models

The ideas of jump/switching systems can be extended to nonlinear systems and the above convex-structure models; introducing delay also gives rise to a fruitful line of research. For brevity and simplicity, they will not be covered in the present monograph. The interested reader is referred to [46–50].

2.4 The Convex-Structure Approach to Control: Motivation

Consider system (2.21) and the equivalent form (2.22). If we can prove some stability or performance statement for $\dot{x}(t) = h_1 \cdot (-x(t)) + h_2 \cdot x(t) + u(t)$ for any arbitrarily time-varying $h_1(t)$ and $h_2(t)$ with the only constraints being $h_1 \geq 0$, $h_2 \geq 0$, $h_1 + h_2 = 1$, then we have proved such statement for the general case of the nonlinear model.

This is the so-called *membership-shape independent* approach to nonlinear analysis: setting up convex optimisation setups to provide solutions to control problems for all h.

Of course, in the above example, forgetting that h_1 is sinusoidal might introduce some conservativeness; this is inherent to the membership-shape independent choice to deal with stability analysis and control. For instance, the linearisation around $x = 0$ of (2.21) yields $\dot{x}(t) \approx -0.54x(t) + u(t)$ so even in open-loop $u = 0$ we know that the origin is (locally) stable. However, we cannot ensure that the origin of (2.22) is open-loop stable for an arbitrary h_1 as, indeed, for $h_1(x) = 0$, $h_2(x) = 1$ we get $\dot{x}(t) = x(t) + u(t)$. So we have an exact description of the system (2.22) which can be unstable (if shape is disregarded) and an approximate one which is stable.

Fortunately, anyway, the origin of (2.22) is *stabilisable* so shape-independent controllers can be designed (for instance, a linear one $u = -2x$). Indeed, with the proposed controller the closed-loop can be written as $\dot{x} = h_1(x) \cdot (-3x) + h_2(x) \cdot (-x)$ and, considering a Lyapunov function $V(x) = x^2$ we have $\dot{V}(x) = h_1(x) \cdot (-6x^2) + h_2(x) \cdot (-2x^2)$ which, being an interpolation between $-6x^2$ and $-2x^2$ is, of course, strictly negative for $x \neq 0$, for any h_1 and h_2 as long as they are positive and add up to unity.

Of course, even if the approach is conservative, it can provide tractable and systematic solutions, whereas approaches keeping the nonlinearity might fail or be complicated to set up in a general case. Generalising the idea to higher-order systems can be dealt with reasonable ease; this is the main idea of the book.

2.5 Summary and Conclusions

This chapter has discussed general definitions of linear and nonlinear dynamic systems, the objectives relating to their control, as well as standard (linearisation or Taylor series) approximations around the origin (or around a trajectory). Models with convex structure have been presented and a very simple motivating case has been addressed in which an exact representation of a nonlinear system is put forward, even if a conservative shape-independent step could not prove stability (but it was successful in stabilisation). Discussing a systematic modelling procedure for high-order systems and generalising the Lyapunov analysis as well as addressing the sources of conservativeness, in the different control-related problems earlier discussed, is the objective of the remainder of the book.

References

1. Khalil, H.K.: Nonlinear Systems. Prentice-Hall, Upper Saddle River, NJ (2002)
2. Dai, L.: Singular Control Systems. Lecture Notes in Control and Information Sciences, Springer, Berlin (1989)
3. Chen, C.-T.: Linear System Theory and Design. The Oxford Series in Electrical and Computer Engineering, Oxford University Press, Oxford, UK (1999)
4. Shamma, J., Athans, M.: Analysis of nonlinear gain-scheduled control systems. IEEE Trans. Autom. Control **35**(1), 898–907 (1990)
5. Apkarian, P., Gahinet, P.: A convex characterization of gain-scheduled h_∞ controllers. IEEE Trans. Autom. Control **40**(5), 853–864 (1995)
6. Bruzelius, F.: Linear Parameter-Varying Systems. Departmente of Signlas and Systems, Chalmers University of Technology, Goteborg, Sweden (2004)
7. Yee, H., Sweby, P., Griffiths, D.: Dynamical approach study of spurious steady-state numerical solutions of nonlinear differential equations. I. The dynamics of time discretization and its implications for algorithm development in computational fluid dynamics. J. Comput. Phys. **97**(2), 249–310 (1991)
8. Kim, Y.H., Lewis, F.L., Abdallah, C.T.: A dynamic recurrent neural-network-based adaptive observer for a class of nonlinear systems. Automatica **33**, 1539–1543 (1997)
9. Kim, Y., Lewis, F.: Neural network output feedback control of robot manipulators. IEEE Trans. Robot. Autom. **15**, 301–309 (1999)
10. Lewis, F., Dawson, D., Abdallah, C.: Robot Manipulator Control: Theory and Practice. Marcel Dekker Inc, New York (2004)
11. Chen, T., Morris, J., Martin, E.: Particle filters for state and parameter estimation in batch processes. J. Process Control **15**, 665–673 (2005)
12. Herrera, E., Castillo, B., Ramirez, J., Ferreira, E.: Exact fuzzy observer for a baker's yeast fed-batch fermentation process. In: Proceedings of the IEEE International Conference on Fuzzy Systems. IEEE, New York (2007)
13. Pecora, L.M., Carroll, T.L.: Synchronization in chaotic systems. Phys. Rev. Lett. **64**(8), 821–824 (1990)
14. Loría, A., Panteley, E., Zavala, A.: Adaptive observers with persistency of excitation for synchronization of chaotic systems. IEEE Trans. Circuits Syst. I Regul. Pap. **56**(12), 2703–2716 (2009)
15. Aguiar, B., Gonzalez, T., Bernal, M.: A way to exploit the fractional stability domain for robust chaos suppression and synchronization via LMIs. IEEE Trans. Autom. Control **61**(10), 2796–2807 (2016)

16. Fliess, M., Levine, J., Rouchon, P.: Flatness and defect of nonlinear systems -introductory theory and examples. Int. J. Control **61**(6), 1327–1361 (1995)
17. PooGyeon, P., Kailath, T.: New square-root smoothing algorithms. IEEE Trans. Autom. Control **41**(5), 727–732 (1996)
18. Asif, A.: Fast Rauch-Tung-Striebel smoother-based image restoration for noncausal images. IEEE Trans. Signal Process. **11**(3), 371–374 (2004)
19. Sarkka, S.: Unscented Rauch-Tung-Striebel smoother. IEEE Trans. Autom. Control **53**(3), 845–849 (2008)
20. Arulampalam, S., Maskell, S., Gordon, N.J., Clapp, T.: A tutorial on particle filters for on-line nonlinear/non-Gaussian Bayesian tracking. IEEE Trans. Signal Process. **50**(2), 174–188 (2002)
21. Frank, P.M.: Fault diagnosis in dynamic systems using analytical and knowledge-based redundancy: A survey and some new results. Automatica **26**(3), 459–474 (1990)
22. Schneider, H., Frank, P.: Observer-based supervision and fault detection in robots using nonlinear and fuzzy logic residual evaluation. IEEE Trans. Control Syst. Technol. **4**, 274–282 (1996)
23. Kalman, R.E.: A new approach to linear filtering and prediction problems. Transactions of the ASME-Journal of Basic Engineering **82**(1), 35–45 (1960)
24. Luenberger, D.G.: Observers for multivariable systems. IEEE Trans. Autom. Control **11**(2), 190–197 (1966)
25. Kailath, T.: Linear Systems. Prentice Hall, Upper Saddle River, NJ, USA (1980)
26. Åström, K.J., Wittenmark, B.: Computer Controlled Systems. Prentice-Hall, Upper Saddle River, NJ (1990)
27. Guerra, T., Guerra, J., Bernal, M.: LMI-based dynamic control of non-affine nonlinear systems via Takagi–Sugeno models. In: 2018 IEEE International Conference on Fuzzy Systems (FUZZ-IEEE). IEEE, New York, pp. 1–6 (2018)
28. Sala, A., Ariño, C.: Polynomial fuzzy models for nonlinear control: A Taylor series approach. IEEE Trans. Fuzzy Syst. **17**(6), 1284–1295 (2009)
29. Takagi, T., Sugeno, M.: Fuzzy identification of systems and its applications to modeling and control. IEEE Trans. Syst. Man Cybern. **15**(1), 116–132 (1985)
30. Tanaka, K., Wang, H.O.: Fuzzy Control System Design and Analysis: A Linear Matrix Inequality Approach. Wiley, New York (2001)
31. Boyd, S., El Ghaoui, L., Féron, E., Balakrishnan, V.: Linear Matrix Inequalities in System and Control Theory: Studies in Applied Mathematics. Society for Industrial and Applied Mathematics, Philadelphia, PA (1994)
32. Scherer, C., Weiland, S.: Linear Matrix Inequalities in Control. Delft University, Delft, The Netherlands (2005)
33. Tanaka, K., Yoshida, H., Ohtake, H., Wang, H.: A sum of squares approach to modeling and control of nonlinear dynamical systems with polynomial fuzzy systems. IEEE Trans. Fuzzy Syst. **17**(4), 911–922 (2009)
34. Bernal, M., Sala, A., Jaadari, A., Guerra, T.-M.: Stability analysis of polynomial fuzzy models via polynomial fuzzy Lyapunov functions. Fuzzy Sets Syst. **185**, 5–14 (2011)
35. Apkarian, P., Adams, R.J.: Advanced gain-scheduling techniques for uncertain systems. IEEE Trans. Control Syst. Technol. **6**(1), 21–32 (1998)
36. Scherer, C.W.: LPV control and full block multipliers. Automatica **37**(3), 361–375 (2001)
37. Wu, F., Dong, K.: Gain-scheduling control of LFT systems using parameter-dependent Lyapunov functions. Automatica **42**(1), 39–50 (2006)
38. Rugh, W.J., Shamma, J.S.: Research on gain scheduling. Automatica **36**(10), 1401–1425 (2000)
39. Zhou, K., Doyle, J.C., Glover, K., et al.: Robust and Optimal Control, vol. 272. Prentice Hall, Englewood Cliffs, NJ (1996)
40. Wang, H., Tanaka, K., Griffin, M.: An approach to fuzzy control of nonlinear systems: stability and design issues. IEEE Trans. Fuzzy Syst. **4**(1), 14–23 (1996)
41. Safonov, M.G.: Stability and Robustness of Multivariable Feedback Systems. The MIT Press, Cambridge, MA (1980)

42. Bhattacharyya, S., Keel, L., Chapellat, H.: Robust Control: The Parametric Approach. Prentice Hall, New York (1995)
43. Ioannou, P.A., Sun, J.: Robust Adaptive Control. Prentice Hall, Upper Saddle River, NJ (1996)
44. Feng, G.: Robust H_∞ filtering of fuzzy dynamic systems. IEEE Trans. Aerosp. Electron. Syst. **51**(2), 658–670 (2005)
45. Sun, Y.-J.: Robust stability of uncertain T–S fuzzy time-varying systems. Chaos, Solitons Fractals **39**(4), 1588–1594 (2009)
46. Cao, Y., Frank, P.: Analysis and synthesis of nonlinear time-delay systems via fuzzy control approach. IEEE Trans. Fuzzy Syst. **8**(2), 200–211 (2000)
47. Wu, L., Su, X., Shi, P., Qiu, J.: A new approach to stability analysis and stabilization of discrete-time T–S fuzzy time-varying delay systems. IEEE Trans. Syst. Man Cybern. B Cybern. **41**(1), 273–286 (2011)
48. Lin, C., Wang, G., Lee, T.H., He, Y.: LMI approach to analysis and control of Takagi–Sugeno fuzzy systems with time delay, vol. 351. Springer Science & Business Media, Berlin (2007)
49. Benzaouia, A., Hajjaji, A., Tadeo, F.: Stabilization of switching Takagi–Sugeno systems by switched Lyapunov function. Int. J. Adapt. Control Signal Process. **25**(12), 1039–1049 (2011)
50. de Souza, W.A., Teixeira, M.C., Cardim, R., Assunçao, E.: On switched regulator design of uncertain nonlinear systems using Takagi–Sugeno fuzzy models. IEEE Trans. Fuzzy Syst. **22**(6), 1720–1727 (2014)

Chapter 3
Modelling via Convex Structures

This chapter presents a variety of nonlinear models related to the Takagi–Sugeno form (TS): some of them are directly inspired from the fuzzy approach in which they first appeared; others are generalisations or derivative structures of the original one. Their common characteristic is the presence of convex structures of the state, time, or exogenous parameters. Since this book is focussed on model-based approaches, different methods are presented to construct the convex models from a given nonlinear one.

3.1 Introduction

The physical laws that describe the behaviour of a system are the natural option for constructing a mathematical model of the plant; these descriptions differ in complexity, accuracy, and usefulness with respect to some purpose. In the control field, linear time-invariant (LTI) structures have been preferred as mathematical models because they allow control and observer design to be easily and comprehensively implemented [1–3]. Nevertheless, the nature of a system (highly nonlinear behaviour, for example) or the required precision of the control task (accomplish a performance which necessarily has to take small effects into account) may cancel the validity of an LTI representation, thus compelling to consider nonlinear modelling. Among the latter, few structures are available that preserve their advantage for control purposes despite their increased complexity. The different classes of model structures outlined in the previous chapter will be now discussed in more depth, indicating how such models can be obtained from the original nonlinear models.

M. Bernal et al., *Analysis and Synthesis of Nonlinear Control Systems*, Studies in Systems, Decision and Control 408, https://doi.org/10.1007/978-3-030-90773-0_3

3.2 Takagi–Sugeno Models

The so-called Takagi–Sugeno (TS) form will be widely used in the sequel due to its convex structure. TS models can be combined with the direct Lyapunov method to obtain conditions in the form of linear matrix inequalities (LMIs), which are efficiently solved via convex optimisation techniques [4].

Originally, TS models were conceived within the fuzzy rule-based structure inherited from the model-free fuzzy approach [5]. In its origins, fuzzy logic in control was conceived as a softening of the hard switching between different outputs that ordinary binary-logic circuits (originated in relay-based ladder diagrams in the 1920s) produced in industrial processes. In this way, e.g. a smooth control part of traditional PIDs could be set up in order to control nonlinear systems by interpolating the different modes of control set up by an expert. They reflected an effort to connect traditional control techniques with the heuristic fuzzy logic-based control area; they were primarily viewed as an application of artificial intelligence in control (in a reduced interpretation of mimicking human operators).

Later on, a model-based approach emerged whose techniques were more appealing to the control community due to the disappearance of heuristics and their strong resemblance of linear parameter-varying (LPV) models. This book is concerned with the latter approach; therefore, the heuristic rule-based structure will be obviated, though it should be kept in mind that there is a straightforward underlying linguistic interpretation that may be useful in some applications. The interested reader is referred to [6] for details.

Definition 3.1 A *Takagi–Sugeno model* is a set of first-order ODEs whose right-hand side can be expressed as a convex sum of linear models (known as *consequents* or *vertex* models) via nonlinear functions (known as membership functions, MFs) that hold the convex sum property [7], i.e.

$$
\begin{aligned}
\delta \boldsymbol{x}(t) &= \sum_{i=1}^{r} h_i(z)(A_i \boldsymbol{x}(t) + B_i \boldsymbol{u}(t)), \\
\boldsymbol{y}(t) &= \sum_{i=1}^{r} h_i(z)\left(C_i \boldsymbol{x}(t) + D_i \boldsymbol{u}(t)\right),
\end{aligned}
\tag{3.1}
$$

where $\delta \boldsymbol{x}(t) = \dot{\boldsymbol{x}}(t)$ and $t \in \mathbb{R}$ for continuous-time models, $\delta \boldsymbol{x}(t) = \boldsymbol{x}(t+1)$ and $t \in \mathbb{Z}$ for discrete-time models, $r \in \mathbb{N}$ is the number of vertex models, $\boldsymbol{x} \in \mathbb{R}^n$ is the state vector, $\boldsymbol{u} \in \mathbb{R}^m$ is the input vector, $\boldsymbol{y} \in \mathbb{R}^q$ is the output vector, $z \in \mathbb{R}^p$ is the premise vector (which may depend on the state, time, or exogenous parameters and is assumed to be bounded and smooth in a compact set $\mathcal{C} \subseteq \mathbb{R}^p$), $A_i,\ B_i,\ C_i,\ D_i,\ i \in \{1, 2, \ldots, r\}$ are matrices of proper dimensions, and the MFs $h_i(z),\ i \in \{1, 2, \ldots, r\}$ (MFs) hold the *convex sum property* in $\mathcal{C}$, i.e.

$$\sum_{i=1}^{r} h_i(z) = 1, \quad 0 \leq h_i(z) \leq 1. \tag{3.2}$$

Some important remarks follow regarding the nature of TS models. We begin this discussion by introducing the following concept:

Definition 3.2 [4] A *linear differential inclusion* (LDI) is described by

$$\dot{\boldsymbol{x}}(t) \in \Omega \boldsymbol{x}(t), \ \boldsymbol{x}(t_0) = \boldsymbol{x}_0, \tag{3.3}$$

where $\Omega \subseteq \mathbb{R}^{n \times n}$ is a (known) set of matrices.

For a given initial condition $\boldsymbol{x}(t_0) = \boldsymbol{x}_0$, there is a family of trajectories satisfying (3.3); this family belongs to $\Omega \boldsymbol{x}(t)$. A particular case, which gives rise to convenient computational methods, is the following:

Definition 3.3 A *polytopic LDI* is an LDI where $\Omega = \text{co}\{A_1, A_2, \ldots, A_r\}$, where $\text{co}\{\cdot\}$ stands for the convex hull of the respective set. Matrices A_i can be understood as *vertex* model matrices.

This representation is of particular interest since many results on the stability of nonlinear models in the TS form are derived from the stronger condition of stability of an LDI containing the trajectories of the nonlinear model.

A linear parameter-varying (LPV) model

$$\dot{\boldsymbol{x}}(t) = A(\boldsymbol{\theta})\boldsymbol{x}(t), \ \boldsymbol{x}(t_0) = \boldsymbol{x}_0 \tag{3.4}$$

can be considered a class of LDI; depending on the nature of the parameters (whether they are just bounded and uncertain or governed by a subsequent dependence on the states or time), they can be used to represent the following:

- linear uncertain systems (LUS) which have *unknown but fixed* (constant in time) parameters;
- linear time-varying (LTV) models whose parameters are *explicitly known* functions of time;
- linear parameter-varying (LPV) models whose parameters do vary with time, but whose explicit expression is assumed unknown and only LDI bounds $A(\boldsymbol{\theta}) \in \Omega$ are used in the analysis.

As can be seen from (3.1), TS models are a special class of LPV models also known as quasi-LPV ones, whose parameters z can depend on the state $\boldsymbol{x}$, time t, or exogenous parameters (uncertainties) $\boldsymbol{\theta}$.

Example 3.1 In order to exemplify these categories, consider the LPV model:

$$\dot{\boldsymbol{x}}(t) = \begin{bmatrix} -2 + \theta_1(t) & 1 \\ 0 & -3 - \theta_2(t) \end{bmatrix} \boldsymbol{x}(t) \tag{3.5}$$

Fig. 3.1 TS models: where are we?

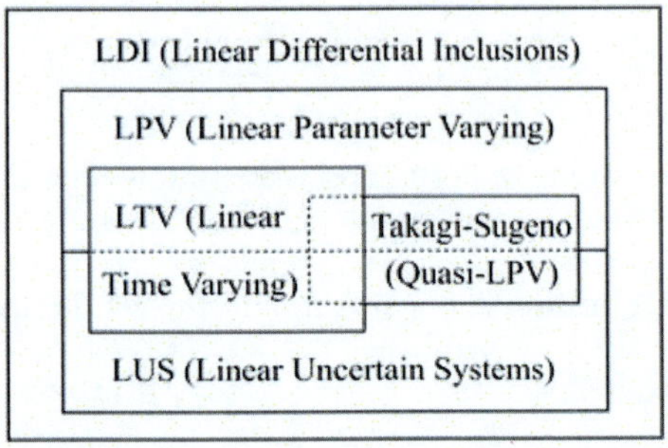

with *varying* parameters $\theta_1(t) \in [-1, 1]$ and $\theta_2(t) \in [0, 1]$ (if the parameters were constant, but unknown, lying in the range, we would be faced to a LUS), as well as the following LTV one:

$$\dot{\boldsymbol{x}}(t) = \begin{bmatrix} -2 + \sin t & 1 \\ 0 & -3 - \cos^2 t \end{bmatrix} \boldsymbol{x}(t), \tag{3.6}$$

and finally the nonlinear model:

$$\dot{\boldsymbol{x}}(t) = \begin{bmatrix} -2 + \sin x_1 & 1 \\ 0 & -3 - \cos^2 x_2 \end{bmatrix} \boldsymbol{x}(t). \tag{3.7}$$

Due to the nonlinear functions involved, the LTV model (3.6) and the nonlinear one (3.7) are particular cases of the LPV (3.5). All these systems as well as any parameter-, time-, or state-dependent combination of them can be exactly represented (using the sector nonlinearity approach, which will be explained in detail in Sect. 3.2.1) by a TS model with an appropriate definition of the premise variables z. This explains the fact that TS models are also known as quasi-LPV ones; it also highlights the interest of the TS approach in contrast with the LPV point of view, since the former allows more specific design than the latter, therefore relaxing sufficient conditions and approaching them to necessity. This discussion is illustrated in Fig.3.1. □

Historically, TS models in their original fuzzy form have not considered any further dependence of MFs h_i, but treated them as *already given parameters holding the convex sum property*; hence, the global character of the results thus derived [8]. In the discrete-time context, an interesting discussion on the topic can be found in [9].

The next sections present some methodologies to obtain a TS model from a nonlinear one as the first step towards stability analysis and controller design. Two main approaches can be used to obtain TS fuzzy models: 1) identifying the model using measured or simulated data and 2) analytic construction of a TS model that exactly represents or approximates a given nonlinear dynamical model. Here, we consider the second approach, i.e. constructing a TS model from the nonlinear dynamical one. For identification-based TS systems, the interested reader is referred to [10–16].

Several methods exist that construct a TS representation or an approximation of a given model. The so-called sector nonlinearity approach [17] privileges model accuracy over simplicity while the linearisation approach allows simpler representations at the price of being approximate. In applications, variables are in general defined over a compact set; in such a case, TS models have been proven to be able of approximating any nonlinear function to an arbitrary degree of accuracy (if linearisation is pursued) or exactly representing any nonlinear function as a convex rewriting (if the sector nonlinearity approach is employed) [18–21]. A TS model can be constructed using the method described in Chapter 14 of [7] such that both the nonlinear system and its derivative are approximated. Other methods that approximate a given nonlinear system are dynamic linearisation [22], which is in fact a Taylor series expansion in several operating points, or the substitution method developed by [23]. Some of these methods will be presented in what follows.

3.2.1 The Sector Nonlinearity Approach

Presented by, for instance, [24] and [17], the sector nonlinearity approach is a methodology that allows rewriting a nonlinear model into the TS form . As implied, in this approach, the TS model is not an approximation, but an *exact representation* of the nonlinear one. The approach requires that nonlinearities are explicitly known in closed form or in terms of their bounds inside the modelling region, usually a compact set containing the origin $\boldsymbol{x} = 0$; the latter induces a compact in the premise space, namely $\mathcal{C} \subseteq \mathbb{R}^p$. The nonlinearities are captured into MFs which hold the convex sum property in the compact set (whether the induced or the original one) and interpolate among linear models that can be viewed as vertices of a polytopic LDI.

To introduce the technique in its simplest case, consider a nonlinear function $f : [a, b] \to \mathbb{R}$, bounded on $[a, b]$, and denote $\overline{\mathrm{nl}} = \max_{x \in [a, b]} f(x)$ and $\underline{\mathrm{nl}} = \min_{x \in [a, b]} f(x)$. Then f can be written as

$$f(x) = w_1(x)\underline{\mathrm{nl}} + w_2(x)\overline{\mathrm{nl}},$$

with

$$\begin{aligned} w_1(x) &= \frac{\overline{\mathrm{nl}} - f(x)}{\overline{\mathrm{nl}} - \underline{\mathrm{nl}}}, \\ w_2(x) &= \frac{f(x) - \underline{\mathrm{nl}}}{\overline{\mathrm{nl}} - \underline{\mathrm{nl}}}, \\ w_1, w_2 &\in [0,\ 1], \end{aligned}$$

i.e. it can be expressed as the convex combination of two constants. Note that $w_1(x) + w_2(x) = 1$.

Similarly, let us consider a nonlinearity such that $f(0) = 0$, i.e. functions of the form $f : \mathbb{R} \to \mathbb{R}$ with

$$f(x) = \overline{f}(x)x,$$

where $\overline{f}(x)$ is bounded. Then, one can denote

$$\overline{\text{nl}} = \max_{x \in [a,\, b]} \overline{f}(x),$$
$$\underline{\text{nl}} = \min_{x \in [a,\, b]} \overline{f}(x)$$

and express the function f as

$$f(x) = w_1(x)\underline{\text{nl}} \cdot x + w_2(x)\overline{\text{nl}} \cdot x,$$

with

$$w_1(x) = \frac{\overline{\text{nl}} - \overline{f}(x)}{\overline{\text{nl}} - \underline{\text{nl}}},$$
$$w_2(x) = \frac{\overline{f}(x) - \underline{\text{nl}}}{\overline{\text{nl}} - \underline{\text{nl}}},$$
$$w_1,\, w_2 \in [0,\ 1],$$

i.e. it can be expressed as the convex combination of two linear functions. Again, $w_1(x) + w_2(x) = 1$.

Example 3.2 Consider the function $f : [-\pi/4,\ \pi/4] \to \mathbb{R},\ f(x) = x \sin x$. The function $\sin x$ on the interval $[-\pi/4,\ \pi/4]$ is bounded, with

$$\overline{\text{nl}} = \max_{x \in [-\pi/4,\, \pi/4]} \sin x = \frac{\sqrt{2}}{2},$$
$$\underline{\text{nl}} = \min_{x \in [-\pi/4,\, \pi/4]} \sin x = -\frac{\sqrt{2}}{2}.$$

Therefore, $f(x)$ can be written as

$$f(x) = w_1(x)\left(-\frac{\sqrt{2}}{2}x\right) + w_2(x)\left(\frac{\sqrt{2}}{2}x\right),$$

with

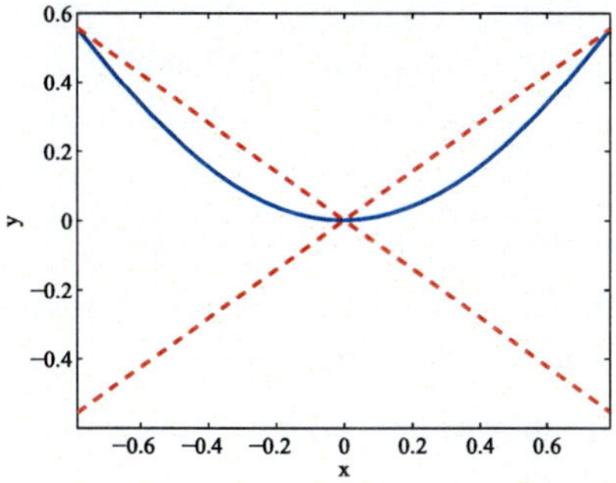

Fig. 3.2 The function $f(x) = x \sin x$ and the two linear bounds

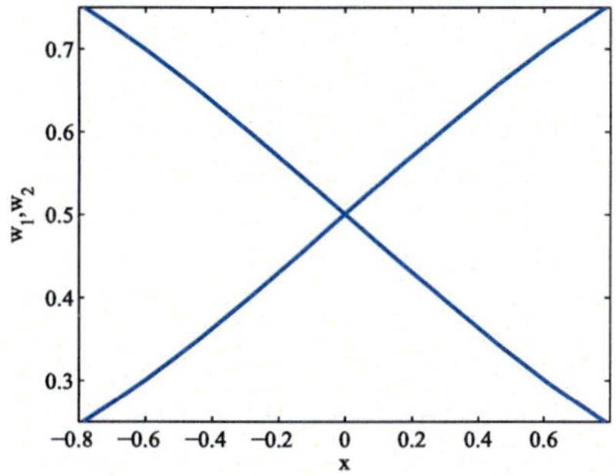

Fig. 3.3 Membership functions for Example 3.2

$$
\begin{aligned}
w_1(x) &= \frac{\sqrt{2}/2 - \sin x}{2(\sqrt{2}/2)},\\
w_2(x) &= \frac{\sin x + \sqrt{2}/2}{2(\sqrt{2}/2)},\\
w_1(x),\ w_2(x) &\in [0,\ 1].
\end{aligned}
$$

The function $f(x) = x \sin x$ and the two linear bounds ($-\sqrt{2}/2x$ and $\sqrt{2}/2x$) are illustrated in Fig. 3.2, and the membership functions $w_1(x)$ and $w_2(x)$ are given in Fig. 3.3. As can be seen, both $w_1(x)$ and $w_2(x)$ are positive and their sum is 1. □

These examples illustrate the fact that any bounded expression can be written as a convex sum of its bounds [24]; moreover, they also point out that convex sums of the sort shown here can be stacked together at the leftmost side of expressions as they amount to 1; such expressions can of course be of matrix nature, but also vector fields, polynomials, etc [25].

We are now ready to generalise the sector nonlinearity technique to nonlinear dynamical systems in continuous or discrete time, i.e. systems of the form:

$$
\begin{aligned}
\delta \boldsymbol{x}(t) &= \boldsymbol{f}(\boldsymbol{x}, \boldsymbol{u}, \boldsymbol{d}, \boldsymbol{\theta}, t),\\
\boldsymbol{y}(t) &= \boldsymbol{h}(\boldsymbol{x}, \boldsymbol{u}, \boldsymbol{d}, \boldsymbol{\theta}, t),
\end{aligned}
\tag{3.8}
$$

where $\delta\boldsymbol{x}(t) = \dot{\boldsymbol{x}}(t)$ and $t \in \mathbb{R}$ for continuous-time models; $\delta\boldsymbol{x}(t) = \boldsymbol{x}(t+1)$ and $t \in \mathbb{Z}$ for discrete-time models; $\boldsymbol{f}$ and $\boldsymbol{h}$ are smooth nonlinear functions that stand for the state and the measurement model, respectively; $\boldsymbol{x} \in \mathbb{R}^n$ is the state vector; $\boldsymbol{u} \in \mathbb{R}^m$ is the input vector; $\boldsymbol{y} \in \mathbb{R}^q$ is the measurement (output) vector; $\boldsymbol{\theta}$ gathers time-varying parameters; and $\boldsymbol{d}$ are disturbances acting on the system.

Constructing a TS model that exactly represents the nonlinear system (3.8) requires identifying a set of expressions $z_j(\cdot)$, $j \in \{1, 2, \ldots, p\}$, in $\boldsymbol{f}(\boldsymbol{x}, \boldsymbol{u}, \boldsymbol{d}, \boldsymbol{\theta}, t)$ and $\boldsymbol{h}(\boldsymbol{x}, \boldsymbol{u}, \boldsymbol{d}, \boldsymbol{\theta}, t)$, such that

1. they might depend on the state $\boldsymbol{x}$, input $\boldsymbol{u}$, time t, parameters $\boldsymbol{\theta}$, and disturbances $\boldsymbol{d}$;
2. they are bounded by z_j^0 and z_j^1, i.e. $z_j^0 \leq z_j(\boldsymbol{x}, \boldsymbol{u}, \boldsymbol{d}, \boldsymbol{\theta}, t) \leq z_j^1$, $j \in \{1, 2, \ldots, p\}$;
3. they are sufficiently smooth in a compact set $\mathcal{C} \subseteq \mathbb{R}^p$;
4. they allow writing $\boldsymbol{f}(\boldsymbol{x}, \boldsymbol{u}, \boldsymbol{d}, \boldsymbol{\theta}, t)$ and $\boldsymbol{h}(\boldsymbol{x}, \boldsymbol{u}, \boldsymbol{d}, \boldsymbol{\theta}, t)$ as polynomials in $z_j(\cdot)$ whose coefficients may depend on $\boldsymbol{x}$, $\boldsymbol{u}$, $\boldsymbol{d}$, $\boldsymbol{\theta}$, and t. Rational expressions of z_j may also be considered in so-called linear fractional models, to be later reviewed.

The expressions $z_j(\cdot)$, $j \in \{1, 2, \ldots, p\}$ constitute what in the TS model context is called the *premise* or *scheduling vector* $\mathbf{z}(\cdot) = \begin{bmatrix} z_1(\cdot) & z_2(\cdot) & \cdots & z_p(\cdot) \end{bmatrix}^T$. Notice that the boundedness of these expressions may arise from the above continuity assumption and its arguments lying on a compact set, i.e. from the states, time, or exogenous variables being, in turn, bounded. Since most convex modelling is done in order to perform stability analysis and design, the modelling region in the state space usually involves the origin $\boldsymbol{x} = 0$, as it is assumed to be an equilibrium point of (3.8) whose stability properties are going to be either analysed or imposed.

Once this is made, the following p pairs of weighting functions can be constructed:

$$w_0^j(\cdot) = \frac{z_j^1 - z_j(\cdot)}{z_j^1 - z_j^0}, \quad w_1^j(\cdot) = 1 - w_0^j(\cdot), \quad j \in \{1, 2, \ldots, p\}. \tag{3.9}$$

Each pair of weighting functions holds the convex sum property in the compact set $\mathcal{C} \subseteq \mathbb{R}^p$. Since convex sums in additive/multiplicative terms can be gathered at the leftmost side of expressions, a convex model can be readily obtained as follows:

$$\begin{aligned} \delta\boldsymbol{x}(t) &= \sum_{i_1=0}^{1}\sum_{i_2=0}^{1}\cdots\sum_{i_p=0}^{1} w_{i_1}^1(z_1)w_{i_2}^2(z_2)\cdots w_{i_p}^p(z_p)\widehat{\boldsymbol{f}}_{(i_1,i_2,\ldots,i_p)}, \\ \boldsymbol{y}(t) &= \sum_{i_1=0}^{1}\sum_{i_2=0}^{1}\cdots\sum_{i_p=0}^{1} w_{i_1}^1(z_1)w_{i_2}^2(z_2)\cdots w_{i_p}^p(z_p)\widehat{\boldsymbol{h}}_{(i_1,i_2,\ldots,i_p)}, \end{aligned} \tag{3.10}$$

where

$$\begin{aligned} \widehat{\boldsymbol{f}}_{(i_1,i_2,\ldots,i_p)} &= \boldsymbol{f}|_{w_{i_1}^1=w_{i_2}^2=\cdots=w_{i_p}^p=1}, \\ \widehat{\boldsymbol{h}}_{(i_1,i_2,\ldots,i_p)} &= \boldsymbol{h}|_{w_{i_1}^1=w_{i_2}^2=\cdots=w_{i_p}^p=1}. \end{aligned}$$

This model is known as a tensor-product (TP) one.

More commonly, a convex model (a "flattened" single-dimensional summation) is obtained using the weights above to define the *membership functions* (MFs) $h_i(z)$, $i \in \{1, 2, \ldots, r\}$, $r = 2^p$ as follows:

$$h_i(z) = h_{d_1^i \times 2^{p-1} + \cdots + d_{p-1}^i \times 2 + d_p^i + 1} = \prod_{j=1}^{p} w^j_{d_j^i}(z_j), \tag{3.11}$$

with $i = d_1^i \times 2^{p-1} + \cdots + d_{p-1}^i \times 2 + d_p^i + 1 \in \{1, \ldots, 2^p\}$ and $d_j^i \in \{0, 1\}$ the jth bit of the p-bit binary representation of $i - 1$. Inherited from the weight pairs, MFs also satisfy the convex sum property (3.2) in $\mathcal{C}$.

Notice that each of the r MFs h_i represents a combination of extreme values of the nonlinearities $z_j(\cdot)$; to complete the TS description, state and measurement functions $\boldsymbol{f}$ and $\boldsymbol{h}$ are evaluated in each of these combinations, i.e. $\widehat{\boldsymbol{f}}_i = \boldsymbol{f}|_{h_i=1}$ and $\widehat{\boldsymbol{h}}_i = \boldsymbol{h}|_{h_i=1}$. Then, a TS model of (3.8) can be written as follows:

$$\begin{aligned} \delta \boldsymbol{x}(t) &= \sum_{i=1}^{r} h_i(z) \widehat{\boldsymbol{f}}_i, \\ \boldsymbol{y}(t) &= \sum_{i=1}^{r} h_i(z) \widehat{\boldsymbol{h}}_i. \end{aligned} \tag{3.12}$$

The functions $\widehat{\boldsymbol{f}}_i$ and $\widehat{\boldsymbol{h}}_i$ are usually less complex than the original nonlinear functions $\boldsymbol{f}$ and $\boldsymbol{h}$ and are often chosen as constant, linear, or affine functions of the state; they are referred to as *local models* in fuzzy contexts because the whole system would reduce to their dynamics if $h_i(z) = 1$; in our context, these are referred to as *vertex models*.

Remark 3.1 Even if we will usually express the models in the flattened expression (3.12), with $r = 2^p$, the tensor-product structure (3.10) must be kept in mind because, as later results will show, it leads to less-conservative results than the flattened version, [26].

In this book, we use TS systems with linear or affine vertex models: the first class usually comes from affine-in-control nonlinear systems while the second one often appears under linearisation. They are both particular instances of the general case described above: $\widehat{\boldsymbol{f}}_i(t) = A_i \boldsymbol{x}(t) + B_i \boldsymbol{u}(t)$, $\widehat{\boldsymbol{h}}_i(t) = C_i \boldsymbol{x}(t)$ for the linear case; $\widehat{\boldsymbol{f}}_i(t) = A_i \boldsymbol{x}(t) + B_i \boldsymbol{u}(t) + \boldsymbol{a}_i$, $\widehat{\boldsymbol{h}}_i(t) = C_i \boldsymbol{x}(t) + \boldsymbol{c}_i$ for the affine case. Why do we move from (3.8) to such instances? The answer lies on the convenience of a matrix form to apply the direct Lyapunov method for analysis and synthesis purposes: it requires a system of the form (3.8) to be written as

$$\begin{aligned} \delta \boldsymbol{x}(t) &= A(\boldsymbol{x}, \boldsymbol{u}, \boldsymbol{d}, \boldsymbol{\theta}, t)\boldsymbol{x}(t) + B(\boldsymbol{x}, \boldsymbol{u}, \boldsymbol{d}, \boldsymbol{\theta}, t)\boldsymbol{u}(t), \\ \boldsymbol{y}(t) &= C(\boldsymbol{x}, \boldsymbol{u}, \boldsymbol{d}, \boldsymbol{\theta}, t)\boldsymbol{x}(t) + D(\boldsymbol{x}, \boldsymbol{u}, \boldsymbol{d}, \boldsymbol{\theta}, t)\boldsymbol{u}(t), \end{aligned}$$

with nonlinear matrix expressions $A(\cdot)$, $B(\cdot)$, $C(\cdot)$, and $D(\cdot)$ clearly mimicking the state-space representation of a linear system. While the sector nonlinearity methodology can be directly applied in this context, classical results exclude the appearance of the control input $\boldsymbol{u}$ within the nonlinear matrix expressions $A(\cdot)$, $B(\cdot)$, $C(\cdot)$, and $D(\cdot)$ as this may lead to algebraic loops [27]; similarly, disturbances $\boldsymbol{d}$ are only considered for disturbance rejection approaches (see Chapter 6). Usually disturbances are also factored out as inputs; unknown signals that do not change the equilibrium point are usually denoted as θ, and if they do change the equilibrium point, they are usually disturbances; therefore, a more common factorisation of (3.8) is the particular case

$$\begin{aligned} \delta \boldsymbol{x}(t) &= A(\boldsymbol{x}, \boldsymbol{\theta}, t)\boldsymbol{x}(t) + B(\boldsymbol{x}, \boldsymbol{\theta}, t)\boldsymbol{u}(t), \\ \boldsymbol{y}(t) &= C(\boldsymbol{x}, \boldsymbol{\theta}, t)\boldsymbol{x}(t) + D(\boldsymbol{x}, \boldsymbol{\theta}, t)\boldsymbol{u}(t), \end{aligned} \tag{3.13}$$

which, once the non-constant terms in $A(\cdot)$, $B(\cdot)$, $C(\cdot)$, and $D(\cdot)$ are grouped in the scheduling vector $z(\boldsymbol{x}, \boldsymbol{\theta}, t)$, can be seen as

$$\begin{aligned} \delta \boldsymbol{x}(t) &= A(z)\boldsymbol{x}(t) + B(z)\boldsymbol{u}(t), \\ \boldsymbol{y}(t) &= C(z)\boldsymbol{x}(t) + D(z)\boldsymbol{u}(t). \end{aligned} \tag{3.14}$$

Clearly, this factorisation is not unique, which implies a first source of conservativeness for methods based on TS models. For instance, the equation $\dot{x} = x_1 x_2$ may be conceived as $\dot{x} = z \cdot x_2$ with $z \equiv x_1$, or $\dot{x} = z \cdot x_1$ with $z \equiv x_2$. Subsequent stability analysis or control design conditions might give different results for each choice of premise variable z. A recent quest on appropriately modelling a nonlinear system to attain the best value for an optimisation problem can be found in [28, 29].

A proper TS model of the form (3.1) can, thus, be found, i.e.

$$\begin{aligned} \delta \boldsymbol{x}(t) &= \sum_{i=1}^{r} h_i(z) \left(A_i \boldsymbol{x}(t) + B_i \boldsymbol{u}(t)\right), \\ \boldsymbol{y}(t) &= \sum_{i=1}^{r} h_i(z) \left(C_i \boldsymbol{x}(t) + D_i \boldsymbol{u}(t)\right), \end{aligned} \tag{3.15}$$

where $A_i = A(z)|_{h_i=1} \in \mathbb{R}^{n \times n}$, $B_i = B(z)|_{h_i=1} \in \mathbb{R}^{n \times m}$, $C_i = C(z)|_{h_i=1} \in \mathbb{R}^{q \times n}$, $D_i = D(z)|_{h_i=1} \in \mathbb{R}^{q \times m}$, and $h_i(z)$ are defined as in (3.11), $i \in \{1, 2, \ldots, r\}$. TS models with affine vertex models as consequents will be studied in more detail in Sect. 3.2.2.

Thanks to the normalised membership functions, the linear (affine) dynamical TS model is in fact a convex combination of local linear (affine) models. This property facilitates the stability analysis of the fuzzy system (see Chapter 4).

Notation for Convex Sums

Handling convex sums of expressions Υ_i (usually matrices) may be cumbersome; the following shorthand notation has been commonly used and will be adopted in this book:

$$\Upsilon_h = \sum_{i=1}^{r} h_i(z)\Upsilon_i. \tag{3.16}$$

This notation not only underlines the fact that the convex sum depends on the MFs (something that might be expressed as $\Upsilon(h)$ with h appropriately defined as a vector gathering the different MFs), but also that the MFs hold the convex sum property for some compact set $\mathcal{C} \subseteq \mathbb{R}^p$. This notation intentionally resembles linear models as convexity allows nonlinear systems in such form to be treated *almost* as linear systems; nevertheless, the reader should keep in mind that these are nonlinear expressions if the explicit "shape" of $h_i(z)$ was considered.

Using this notation, the TS model in (3.1) or (3.15) can be rewritten as follows:

$$\begin{aligned} \delta x(t) &= A_h x(t) + B_h u(t), \\ y(t) &= C_h x(t) + D_h u(t). \end{aligned} \tag{3.17}$$

For a TP model (3.10), a slightly different notation is adopted: let $\mathbf{i} = (i_1, i_2, \ldots, i_p)$, $\mathbf{w}(z) = w^1_{i_1}(z)w^2_{i_2}(z)\cdots w^p_{i_p}(z)$, then

$$\Upsilon_{\mathbf{w}} = \sum_{\mathbf{i}\in\mathbb{B}^p} \mathbf{w_i}(z)\Upsilon_{\mathbf{i}} = \sum_{i_1=0}^{1}\sum_{i_2=0}^{1}\cdots\sum_{i_p=0}^{1} w^1_{i_1}(z_1)w^2_{i_2}(z_2)\cdots w^p_{i_p}(z_p)\Upsilon_{(i_1,i_2,\ldots,i_p)} \tag{3.18}$$

for a set of (usually matrix) expressions $\Upsilon_{(i_1,i_2,\ldots,i_p)}$.

Thus, assuming that the weights yielding the MFs are used, the TS model (3.1) or (3.15) can be rewritten as follows:

$$\begin{aligned} \delta x(t) &= A_{\mathbf{w}} x(t) + B_{\mathbf{w}} u(t), \\ y(t) &= C_{\mathbf{w}} x(t) + D_{\mathbf{w}} u(t), \end{aligned} \tag{3.19}$$

where the corresponding matrices $A_{\mathbf{i}}$, $B_{\mathbf{i}}$, $C_{\mathbf{i}}$, and $D_{\mathbf{i}}$, are A_i, B_i, C_i, and D_i, respectively, provided $\mathbf{i}$ is the p-bit binary representation of $i-1$.

The following examples illustrate how the sector nonlinearity approach can be employed to construct an exact TS representation of a given nonlinear dynamical system. The first one details the methodology on an arbitrary nonlinear model; the second one underlines its applicability and non-uniqueness for physical models; and the third one points out how this approach can be used for LPV and LTV models as well.

Example 3.3 Consider the nonlinear model with two states, x_1 and x_2, one input u, and one measurement y given as

$$\begin{aligned}\dot{x}_1(t) &= x_1x_2^2 + 3x_1^2 + x_2 + x_1^2u,\\ \dot{x}_2(t) &= 3x_1 + x_1x_2^2 + u,\\ y(t) &= 2x_1x_2 + 2x_2,\end{aligned} \tag{3.20}$$

with the variables defined on the compact set $\mathcal{X} = \{x : |x_i| \le 1, i = 1, 2\}$. This system can be rewritten in the form (3.14) as follows:

$$\begin{aligned}\dot{x}(t) &= \begin{bmatrix} x_2^2 + 3x_1 & 1 \\ 3 & x_1x_2 \end{bmatrix} x(t) + \begin{bmatrix} x_1^2 \\ 1 \end{bmatrix} u(t),\\ y(t) &= \begin{bmatrix} 2x_2 & 2 \end{bmatrix} x(t).\end{aligned} \tag{3.21}$$

The scheduling (premise) variables, i.e. the non-constant elements in the matrix functions

$$A(x) = \begin{bmatrix} x_2^2 + 3x_1 & 1 \\ 3 & x_1x_2 \end{bmatrix}, \quad B(x) = \begin{bmatrix} x_1^2 \\ 1 \end{bmatrix}, \quad C(x) = \begin{bmatrix} 2x_2 & 2 \end{bmatrix} \tag{3.22}$$

may be chosen as $z_1(x) = x_2^2 + 3x_1$, $z_2(x) = x_1x_2$, $z_3(x) = x_1^2$, and $z_4(x) = 2x_2$, so

$$A(x) = \begin{bmatrix} z_1 & 1 \\ 3 & z_2 \end{bmatrix}, \quad B(x) = \begin{bmatrix} z_3 \\ 1 \end{bmatrix}, \quad C(x) = \begin{bmatrix} z_4 & 2 \end{bmatrix}.$$

Clearly, the bounds in $\mathcal{X}$ induce the compact set

$$\mathcal{C} = \{z : -3 \le z_1 \le 4,\ -1 \le z_2 \le 1,\ 0 \le z_3 \le 1,\ -2 \le z_4 \le 2\}.$$

For each of these four terms, the two weighting functions and the corresponding matrix elements are computed as follows:

1. $z_1(x) = x_2^2 + 3x_1 \in [-3, 4]$, i.e. $z_1^0 = -3$ and $z_1^1 = 4$. The first weighting function is

$$w_0^1(z) = \frac{z_1^1 - z_1(x)}{z_1^1 - z_1^0} = \frac{4 - (x_2^2 + 3x_1)}{4 - (-3)} = \frac{4 - x_2^2 - 3x_1}{7}.$$

   The second weighting function is

$$w_1^1(z) = 1 - w_0^1(z) = \frac{x_2^2 + 3x_1 + 3}{7}.$$

   Then, the scheduling variable $z_1(x)$ is represented as the weighted sum

$$z_1(\boldsymbol{x}) = -3w_0^1(z) + 4w_1^1(z) = \sum_{i_1=0}^{1} w_{i_1}^1(z) z_1^{i_1}.$$

2. $z_2(\boldsymbol{x}) = x_1 x_2 \in [-1,\ 1]$, i.e. $z_2^0 = -1$ and $z_2^1 = 1$. The first weighting function is

$$w_0^2(z) = \frac{z_2^1 - z_2(\boldsymbol{x})}{z_2^1 - z_2^0} = \frac{1 - x_1 x_2}{1 - (-1)} = \frac{1 - x_1 x_2}{2},$$

while the second one is

$$w_1^2(z) = 1 - w_0^2(z) = \frac{1 + x_1 x_2}{2}.$$

The scheduling variable $z_2(\boldsymbol{x})$ is represented as

$$z_2(\boldsymbol{x}) = -1w_0^2(z) + 1w_1^2(z) = \sum_{i_2=0}^{1} w_{i_2}^2(z) z_2^{i_2}.$$

3. $z_3(\boldsymbol{x}) = x_1^2 \in [0,\ 1]$, i.e. $z_3^0 = 0$ and $z_3^1 = 1$. The corresponding weights are

$$w_0^3(z) = \frac{z_3^1 - z_3(\boldsymbol{x})}{1 - 0} = 1 - x_1^2$$

and

$$w_1^3(z) = 1 - w_0^3(z) = x_1^2,$$

which means the scheduling variable $z_3(\boldsymbol{x})$ can be written as

$$z_3(\boldsymbol{x}) = 0w_0^3(z) + 1w_1^3(z) = \sum_{i_3=0}^{1} w_{i_3}^3(z) z_3^{i_3}.$$

4. $z_4(\boldsymbol{x}) = 2x_2 \in [-2,\ 2]$, i.e. $z_4^0 = -2$ and $z_4^1 = 2$. Defining

$$w_0^4(z) = \frac{z_4^1 - z_4(\boldsymbol{x})}{z_4^1 - z_4^0} = \frac{2 - 2x_2}{4} = \frac{1 - x_2}{2}$$

and

$$w_1^4(z) = 1 - w_0^4(z) = \frac{1 + x_2}{2}$$

as the fourth set of weights, the corresponding scheduling variable $z_4(\boldsymbol{x})$ is written as

$$z_4(\boldsymbol{x}) = -2w_0^4(z) + 2w_1^4(z) = \sum_{i_4=0}^{1} w_{i_4}^4(z) z_4^{i_4}.$$

Since there are 4 pairs of weights, the number of membership functions $h_i(z)$ in (3.11) (which is the same number of local linear models) is $2^4 = 16$. They correspond to all the min-max combinations of $z_1(\boldsymbol{x})$, $z_2(\boldsymbol{x})$, $z_3(\boldsymbol{x})$, and $z_4(\boldsymbol{x})$. For illustration purposes, some of these vertex models and MFs are shown below:

Vertex model 1:

$$\begin{aligned} \dot{\boldsymbol{x}}(t) &= A_1\boldsymbol{x}(t) + B_1\boldsymbol{u}(t), \\ \boldsymbol{y}(t) &= C_1\boldsymbol{x}(t), \end{aligned}$$

with

$$\begin{aligned} A_1 &= \begin{bmatrix} z_1^0 & 1 \\ 3 & z_2^0 \end{bmatrix} = \begin{bmatrix} -3 & 1 \\ 3 & -1 \end{bmatrix}, \\ B_1 &= \begin{bmatrix} z_3^0 \\ 1 \end{bmatrix} = \begin{bmatrix} 0 \\ 1 \end{bmatrix}, \\ C_1 &= \begin{bmatrix} z_4^0 & 2 \end{bmatrix} = \begin{bmatrix} -2 & 2 \end{bmatrix}, \end{aligned}$$

and the corresponding MF computed as $h_1(z) = w_0^1(z)w_0^2(z)w_0^3(z)w_0^4(z)$.

Vertex model 2:

$$\begin{aligned} \dot{\boldsymbol{x}}(t) &= A_2\boldsymbol{x}(t) + B_2\boldsymbol{u}(t), \\ \boldsymbol{y}(t) &= C_2\boldsymbol{x}(t), \end{aligned}$$

with

$$\begin{aligned} A_2 &= \begin{bmatrix} z_1^0 & 1 \\ 3 & z_2^0 \end{bmatrix} = \begin{bmatrix} -3 & 1 \\ 3 & -1 \end{bmatrix}, \\ B_2 &= \begin{bmatrix} z_3^0 \\ 1 \end{bmatrix} = \begin{bmatrix} 0 \\ 1 \end{bmatrix}, \\ C_2 &= \begin{bmatrix} z_4^1 & 2 \end{bmatrix} = \begin{bmatrix} 2 & 2 \end{bmatrix}, \end{aligned}$$

and the MF of the vertex computed as $h_2(z) = w_0^1(z)w_0^2(z)w_0^3(z)w_1^4(z)$.

Vertex model 7:

$$\begin{aligned} \dot{\boldsymbol{x}}(t) &= A_7\boldsymbol{x}(t) + B_7\boldsymbol{u}(t), \\ \boldsymbol{y}(t) &= C_7\boldsymbol{x}(t), \end{aligned}$$

with

$$\begin{aligned} A_7 &= \begin{bmatrix} z_1^0 & 1 \\ 3 & z_2^1 \end{bmatrix} = \begin{bmatrix} -3 & 1 \\ 3 & 1 \end{bmatrix}, \\ B_7 &= \begin{bmatrix} z_3^1 \\ 1 \end{bmatrix} = \begin{bmatrix} 1 \\ 1 \end{bmatrix}, \\ C_7 &= \begin{bmatrix} z_4^0 & 2 \end{bmatrix} = \begin{bmatrix} -2 & 2 \end{bmatrix}, \end{aligned}$$

and the MF of the vertex computed as $h_7(z) = w_0^1(z)w_1^2(z)w_1^3(z)w_0^4(z)$.

Vertex model 16:

$$\begin{aligned} \dot{\boldsymbol{x}}(t) &= A_{16}\boldsymbol{x}(t) + B_{16}\boldsymbol{u}(t), \\ \boldsymbol{y}(t) &= C_{16}\boldsymbol{x}(t), \end{aligned}$$

with

$$\begin{aligned} A_{16} &= \begin{bmatrix} z_1^1 & 1 \\ 3 & z_2^1 \end{bmatrix} = \begin{bmatrix} 4 & 1 \\ 3 & 1 \end{bmatrix}, \\ B_{16} &= \begin{bmatrix} z_3^1 \\ 1 \end{bmatrix} = \begin{bmatrix} 1 \\ 1 \end{bmatrix}, \\ C_{16} &= \begin{bmatrix} z_4^1 & 2 \end{bmatrix} = \begin{bmatrix} 2 & 2 \end{bmatrix}, \end{aligned}$$

and the MF of the vertex computed as $h_{16}(z) = w_1^1(z)w_1^2(z)w_1^3(z)w_1^4(z)$.

The remaining vertices are defined in a similar manner, corresponding to all $r = 16$ combinations; they all allow rewriting (3.21) as a 16-vertex TS model with the convex structure given in (3.15).

Clearly, (3.21) can be rewritten as

$$\begin{aligned} \dot{\boldsymbol{x}}(t) &= \begin{bmatrix} \sum_{i_1=0}^{1} w_{i_1}^1(z)z_1^{i_1} & 1 \\ 3 & \sum_{i_2=0}^{1} w_{i_2}^2(z)z_2^{i_2} \end{bmatrix} \boldsymbol{x}(t) + \begin{bmatrix} \sum_{i_3=0}^{1} w_{i_3}^3(z)z_3^{i_3} \\ 1 \end{bmatrix} \boldsymbol{u}(t), \\ y(t) &= \begin{bmatrix} \sum_{i_4=0}^{1} w_{i_4}^4(z)z_4^{i_4} & 2 \end{bmatrix} \boldsymbol{x}(t), \end{aligned}$$

or, equivalently,

$$\begin{aligned} \dot{\boldsymbol{x}}(t) &= \sum_{i_1=0}^{1}\sum_{i_2=0}^{1}\sum_{i_3=0}^{1}\sum_{i_4=0}^{1} w_{i_1}^1(z)w_{i_2}^2(z)w_{i_3}^3(z)w_{i_4}^4(z)\left(\begin{bmatrix} z_1^{i_1} & 1 \\ 3 & z_2^{i_2} \end{bmatrix}\boldsymbol{x}(t) + \begin{bmatrix} z_3^{i_3} \\ 1 \end{bmatrix}\boldsymbol{u}(t)\right) \\ &= \sum_{\mathbf{i}\in\mathbb{B}^4} \mathbf{w_i}(z)\left(A_{\mathbf{i}}\boldsymbol{x}(t) + B_{\mathbf{i}}\boldsymbol{u}(t)\right) = A_{\mathbf{w}}\boldsymbol{x}(t) + B_{\mathbf{w}}\boldsymbol{u}(t), \end{aligned}$$

$$\begin{aligned} y(t) &= \sum_{i_1=0}^{1}\sum_{i_2=0}^{1}\sum_{i_3=0}^{1}\sum_{i_4=0}^{1} w_{i_1}^1(z)w_{i_2}^2(z)w_{i_3}^3(z)w_{i_4}^4(z)\begin{bmatrix} z_4^{i_4} & 2 \end{bmatrix}\boldsymbol{x}(t) \\ &= \sum_{\mathbf{i}\in\mathbb{B}^4} \mathbf{w_i}(z)C_{\mathbf{i}}\boldsymbol{x}(t) = C_{\mathbf{w}}\boldsymbol{x}(t), \end{aligned}$$

where notation in (3.18) has been used with

$$\begin{aligned} A_{(i_1,i_2,i_3,i_4)} &= A_{\mathbf{i}} = \begin{bmatrix} z_1^{i_1} & 1 \\ 3 & z_2^{i_2} \end{bmatrix}, \quad B_{(i_1,i_2,i_3,i_4)} = B_{\mathbf{i}} = \begin{bmatrix} z_3^{i_3} \\ 1 \end{bmatrix}, \\ C_{(i_1,i_2,i_3,i_4)} &= C_{\mathbf{i}} = \begin{bmatrix} z_4^{i_4} & 2 \end{bmatrix}. \end{aligned}$$

The convex sum property allows gathering the weights on the leftmost side of the expressions and even allows them to appear multiplied with terms that are not involved with them as in that case the convex sum amounts to 1. This is a TP model.

A classical TS model requires further grouping the weights in MFs $h_i(z)$ and grouping the nested convex sums into a single one:

$$\dot{x}(t) = \sum_{i=1}^{16} h_i(z) \left(\begin{bmatrix} z_1^{i_1} & 1 \\ 3 & z_2^{i_2} \end{bmatrix} x(t) + \begin{bmatrix} z_3^{i_3} \\ 1 \end{bmatrix} u(t) \right) = A_h x(t) + B_h u(t),$$

$$y(t) = \sum_{i=1}^{16} h_i(z) \begin{bmatrix} z_4^{i_4} & 2 \end{bmatrix} x(t) = C_h x(t),$$

where notation in (3.16) has been used with

$$A_i = \begin{bmatrix} z_1^{i_1} & 1 \\ 3 & z_2^{i_2} \end{bmatrix}, \quad B_i = \begin{bmatrix} z_3^{i_3} \\ 1 \end{bmatrix}, \quad C_i = \begin{bmatrix} z_4^{i_4} & 2 \end{bmatrix},$$

index i being the decimal equivalence of the binary number $[i_4 i_3 i_2 i_1]$ plus 1. Importantly, the tensor-product and the classical TS models are algebraically equivalent, though a more appropriate selection of nonlinearities may produce associations which in turn may create different vertexes in the tensor-product one. Such associations will prove key for relaxing results as they will get closer to the actual nonlinear model.

In general, since nonlinearities in the model can be grouped in different ways, the resulting TS models are not unique. For instance, the nonlinear system (3.20) can be equivalently written, instead of (3.21), as

$$\begin{aligned} \dot{x}(t) &= \begin{bmatrix} 3x_1 & 1 + x_1 x_2 \\ 3 & x_1 x_2 \end{bmatrix} x(t) + \begin{bmatrix} x_1^2 \\ 1 \end{bmatrix} u(t), \\ y(t) &= \begin{bmatrix} 0 & 2x_1 + 2 \end{bmatrix} x(t). \end{aligned} \tag{3.23}$$

Although the non-constant terms in the above matrices are $3x_1$, $1 + x_1 x_2$, $x_1 x_2$, x_1^2, and $2x_1 + 2$, i.e. there are 5 terms, actually there are only three different weighting functions, corresponding to x_1, x_1^2, and $x_1 x_2$. An affine transformation of a nonlinearity will affect only the maximum and the minimum, but not the weighting functions.

For instance, consider $z_1 = x_1 \in [-1, \ 1]$, the weighting functions are

$$w_0^1(z) = \frac{1 - z_1}{2}, \quad w_1^1(z) = 1 - w_0^1(z) = \frac{z_1 + 1}{2}.$$

Consider now the affine transformation $z_1 = a x_1 + b$, with a and b constants and $x_1 \in [-1, \ 1]$: $a x_1 + b \in [b - a, \ a + b]$, and the weighting functions are

$$w_0^1(z) = \frac{a+b-(ax_1+b)}{a+b-(b-a)} = \frac{a-ax_1}{2a} = \frac{1-x_1}{2},$$
$$w_1^1(z) = 1 - w_0^1(z) = \frac{x_1+1}{2}.$$

With the considerations above, system (3.20) can be exactly represented using the sector nonlinearity approach by 8 vertices, by choosing as scheduling variables $z_1(\boldsymbol{x}) = x_1$, $z_2(\boldsymbol{x}) = x_1^2$, and $z_3(\boldsymbol{x}) = x_1x_2$. Details are left to the reader.

Using the same decoupling as in (3.23), let us now consider only the two pairs of weights $(w_0^1(z), w_1^1(z))$ and $(w_0^2(z), w_1^2(z))$ ultimately depending on x_1 and x_2, respectively, with $x_i \in [-1, 1]$, $z_i^0 = -1$, $z_i^1 = 1$ $i \in \{1, 2\}$, in order to illustrate how tensor-product models may lead to new vertex models with only two scheduling variables:

$$\dot{\boldsymbol{x}}(t) = \begin{bmatrix} 3\sum_{i_1=0}^{1} w_{i_1}^1(z)z_1^{i_1} & 1+\left(\sum_{i_1=0}^{1} w_{i_1}^1(z)z_1^{i_1}\right)\left(\sum_{i_2=0}^{1} w_{i_2}^2(z)z_2^{i_2}\right) \\ 3 & \left(\sum_{i_1=0}^{1} w_{i_1}^1(z)z_1^{i_1}\right)\left(\sum_{i_2=0}^{1} w_{i_2}^2(z)z_2^{i_2}\right) \end{bmatrix}\boldsymbol{x}(t) \tag{3.24}$$
$$+ \begin{bmatrix} \left(\sum_{i_1=0}^{1} w_{i_1}^1(z)z_1^{i_1}\right)^2 \\ 1 \end{bmatrix}\boldsymbol{u}(t)$$

$$= \sum_{i_1=0}^{1}\sum_{i_2=0}^{1} w_{i_1}^1(z)w_{i_2}^2(z)\begin{bmatrix} 3z_1^{i_1} & 1+z_1^{i_1}z_2^{i_2} \\ 3 & z_1^{i_1}z_2^{i_2} \end{bmatrix}\boldsymbol{x}(t) + \begin{bmatrix} \sum_{i_3=0}^{1} w_{i_3}^1(z)z_1^{i_3} \\ 1 \end{bmatrix}\boldsymbol{u}(t)$$
$$= \sum_{i_1=0}^{1}\sum_{i_2=0}^{1}\sum_{i_3=0}^{1} w_{i_1}^1(z)w_{i_2}^2(z)w_{i_3}^1(z)\left(\begin{bmatrix} 3z_1^{i_1} & 1+z_1^{i_1}z_2^{i_2} \\ 3 & z_1^{i_1}z_2^{i_2} \end{bmatrix}\boldsymbol{x}(t) + \begin{bmatrix} z_1^{i_1}z_1^{i_3} \\ 1 \end{bmatrix}\boldsymbol{u}(t)\right),$$

$$y(t) = \begin{bmatrix} 0 & 2\sum_{i_1=0}^{1} w_{i_1}^1(z)z_1^{i_1}+2 \end{bmatrix}\boldsymbol{x} = \sum_{i_1=0}^{1} w_{i_1}^1(z)\begin{bmatrix} 0 & 2z_1^{i_1}+2 \end{bmatrix}\boldsymbol{x}(t),$$

where a third convex sum appears without using another weighting function $w_{i_3}^1(z)$; it corresponds to the term x_1^2. It is obvious that despite the fact that the extended sum has 8 terms, some of them will have the same w-decomposition leading to association of the corresponding matrices, i.e. new vertexes. □

Remark 3.2 Note that the last of the models, (3.24), in the above example depends only on "two" nonlinear functions w_1^1 and w_1^2, thus being more efficient for later results: although the diverse modelling options are "equivalent" ways of writing the original nonlinearity, the results in the following chapters will be independent of the "shape" of the actual w_j^i, so we can (informally) say that the less of them we

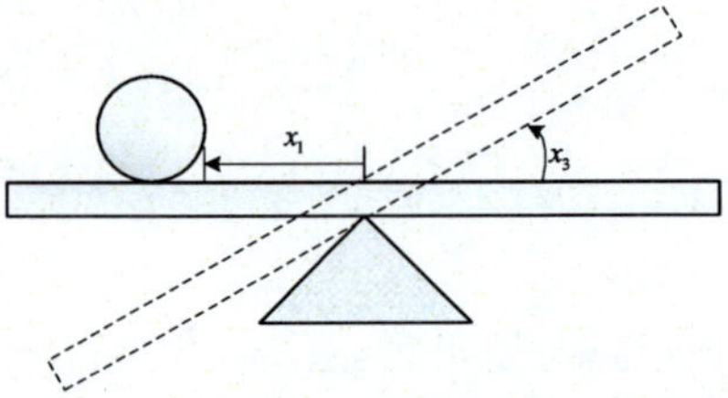

Fig. 3.4 The ball and beam system

consider, the lower the conservatism. The last option in the said example results in a multiple-summation model which would need sum relaxations in Appendix D, even in stability analysis (for single-summation models $\sum_{i=1}^{r} h_i(z) \cdot (A_i \boldsymbol{x} + B_i \boldsymbol{u})$ these kinds of relaxations are needed only in controller design), as discussed in later chapters of this work.

Example 3.4 Consider the following nonlinear model of the ball and beam system shown schematically in Fig. 3.4:

$$\dot{\boldsymbol{x}}(t) = \begin{bmatrix} x_2 \\ a x_1 x_4^2 - b \sin x_3 \\ x_4 \\ u(t) \end{bmatrix}, \tag{3.25}$$

with $\boldsymbol{x} = [x_1\, x_2\, x_3\, x_4]^T$, $x_1 \in [-0.5,\ 0.5]$ being the distance of the ball from the beam centre, $x_2 \in [-1,\ 1]$ being the ball linear velocity, $x_3 \in [-0.25\pi,\ 0.25\pi]$ being the beam angle with respect to the horizontal line, $x_4 \in [-1,\ 1]$ being the angular velocity, u the control input, and a and b system parameters.

The nonlinear model (3.25) can be rewritten as $\dot{\boldsymbol{x}}(t) = A(\boldsymbol{x})\boldsymbol{x}(t) + B(\boldsymbol{x})\boldsymbol{u}(t)$, i.e.

$$\dot{\boldsymbol{x}}(t) = \begin{bmatrix} 0 & 1 & 0 & 0 \\ a x_4^2 & 0 & -b \sin x_3 / x_3 & 0 \\ 0 & 0 & 0 & 1 \\ 0 & 0 & 0 & 0 \end{bmatrix} \boldsymbol{x}(t) + \begin{bmatrix} 0 \\ 0 \\ 0 \\ 1 \end{bmatrix} u(t),$$

from which the non-constant terms $z_1(\boldsymbol{x}) = a x_4^2$ and $z_2(\boldsymbol{x}) = -b \sin x_3 / x_3$ are identified as to induce the following compact set:

$$\mathcal{C} = \{z : 0 \leq z_1 \leq a,\ -b \leq z_2 \leq -0.9b\}$$

once bounds on x_3 and x_4 are taken into account.

According to the notation, $z_1^0 = 0$, $z_1^1 = a$, $z_2^0 = -b$, and $z_2^1 = -0.9b$. The weighting functions for z_1 are

$$w_0^1(z) = \frac{a - z_1}{a - 0} = 1 - x_4^2,$$
$$w_1^1(z) = 1 - w_0^1(z) = x_4^2,$$

and for z_2 are

$$w_0^2(z) = \frac{-0.9b - z_2}{-0.9b - (-b)} = -9 + 10\frac{\sin x_3}{x_3},$$
$$w_1^2(z) = 1 - w_0^2(z) = 10 - 10\frac{\sin x_3}{x_3}.$$

Using these weights, the following MFs can be defined according to (3.11):

$$h_1(z) = w_0^1(z)w_0^2(z),\ h_2(z) = w_0^1(z)w_1^2(z),$$
$$h_3(z) = w_1^1(z)w_0^2(z),\ h_4(z) = w_1^1(z)w_1^2(z).$$

The corresponding matrices of the vertex models are as follows:

$$A_1 = \begin{bmatrix} 0 & 1 & 0 & 0 \\ z_1^0 & 0 & z_2^0 & 0 \\ 0 & 0 & 0 & 1 \\ 0 & 0 & 0 & 0 \end{bmatrix} = \begin{bmatrix} 0 & 1 & 0 & 0 \\ 0 & 0 & -b & 0 \\ 0 & 0 & 0 & 1 \\ 0 & 0 & 0 & 0 \end{bmatrix},\ A_2 = \begin{bmatrix} 0 & 1 & 0 & 0 \\ z_1^0 & 0 & z_2^1 & 0 \\ 0 & 0 & 0 & 1 \\ 0 & 0 & 0 & 0 \end{bmatrix} = \begin{bmatrix} 0 & 1 & 0 & 0 \\ 0 & 0 & -0.9b & 0 \\ 0 & 0 & 0 & 1 \\ 0 & 0 & 0 & 0 \end{bmatrix},$$
$$A_3 = \begin{bmatrix} 0 & 1 & 0 & 0 \\ z_1^1 & 0 & z_2^0 & 0 \\ 0 & 0 & 0 & 1 \\ 0 & 0 & 0 & 0 \end{bmatrix} = \begin{bmatrix} 0 & 1 & 0 & 0 \\ a & 0 & -b & 0 \\ 0 & 0 & 0 & 1 \\ 0 & 0 & 0 & 0 \end{bmatrix},\ A_4 = \begin{bmatrix} 0 & 1 & 0 & 0 \\ z_1^1 & 0 & z_2^1 & 0 \\ 0 & 0 & 0 & 1 \\ 0 & 0 & 0 & 0 \end{bmatrix} = \begin{bmatrix} 0 & 1 & 0 & 0 \\ a & 0 & -0.9b & 0 \\ 0 & 0 & 0 & 1 \\ 0 & 0 & 0 & 0 \end{bmatrix},$$

$$B_i = \begin{bmatrix} 0 & 0 & 0 & 1 \end{bmatrix}^T, \quad i \in \{1, 2, 3, 4\}.$$

Taking into account these definitions, the nonlinear model (3.25) is equivalent to the following 4-vertex TS model

$$\dot{\boldsymbol{x}}(t) = \sum_{i=1}^{4} h_i(z)\left(A_i\boldsymbol{x}(t) + B_i\boldsymbol{u}(t)\right), \tag{3.26}$$

within $\mathcal{C}$. It is worth noting that the equivalence holds even outside the compact set $\mathcal{C}$ (the TS model is a *rewriting* of the original nonlinear model), but the TS structure cannot be called such if the convex sum property does not hold, which is what happens when some MFs $h_i(z)$ become negative or supraunitary outside $\mathcal{C}$.

As implied by the method, the sector nonlinearity approach does not yield a unique representation of the nonlinear model (3.25); another way to rewrite it is the following:

$$\dot{\boldsymbol{x}}(t) = \begin{bmatrix} 0 & 1 & 0 & 0 \\ 0.5ax_4^2 & 0 & -b\frac{\sin x_3}{x_3} & 0.5ax_1x_4 \\ 0 & 0 & 0 & 1 \\ 0 & 0 & 0 & 0 \end{bmatrix}\boldsymbol{x}(t) + \begin{bmatrix} 0 \\ 0 \\ 0 \\ 1 \end{bmatrix}u(t).$$

This form allows defining three non-constant terms instead of two as in the previous example: $z_1(\boldsymbol{x}) = ax_4^2$, $z_2(\boldsymbol{x}) = -b\sin x_3/x_3$, and $z_3(\boldsymbol{x}) = ax_1x_4$. Once bounds on x_1, x_3, and x_4 are taken into account, the following induced compact set is defined:

$$\mathcal{C} = \{z : 0 \leq z_1 \leq a,\ -b \leq z_2 \leq -0.9b,\ -0.25a \leq z_3 \leq 0.25a\}.$$

Thus, three pairs of weighting functions arise:

$$\begin{aligned} w_0^1(z) &= \frac{a - z_1}{a - 0} = 1 - x_4^2,\ w_1^1(z) = 1 - w_0^1(z) = x_4^2, \\ w_0^2(z) &= \frac{-0.9b - z_2}{0.9b - (-b)} = -9 + 10\frac{\sin x_3}{x_3},\ w_1^2(z) = 1 - w_0^2(z) = 10 - 10\frac{\sin x_3}{x_3}, \\ w_0^3(z) &= \frac{0.5a - z_3}{0.5a - (-0.5a)} = 0.5 - x_1x_4,\ w_1^3(z) = 1 - w_0^3(z) = 0.5 + x_1x_4. \end{aligned}$$

Each of the corresponding 2^3 MFs have the form $h_i(z) = w_{i_1}^1(z)w_{i_2}^2(z)w_{i_3}^3(z)$, from which eight corresponding matrices emerge with the following structure:

$$A_i = \begin{bmatrix} 0 & 1 & 0 & 0 \\ z_1^{i_1} & 0 & z_2^{i_2} & z_3^{i_3} \\ 0 & 0 & 0 & 1 \\ 0 & 0 & 0 & 0 \end{bmatrix}.$$

As before, nonlinearities $z_1(\boldsymbol{x})$, $z_2(\boldsymbol{x})$, and $z_3(\boldsymbol{x})$ in the vertex model matrices switch according to the values they acquire when their respective MF $h_i(z) = 1$. The whole TS model is, thus, defined as (3.26) with $r = 8$. □

Example 3.5 Besides the sector nonlinearity approach applied to state-dependent nonlinear models as those in Examples 3.3 and 3.4, bounding can be used to exactly represent LPV, LTV, and every combination of state-, time-, or parameter-dependent nonlinear model in the TS form.

Effectively, following the sector nonlinearity methodology, it can be checked that the LPV model (3.5) as well as the LTV one (3.6) have the same vertex models of the TS representation of the state-dependent nonlinear model (3.7), i.e.

$$\begin{aligned} A_1 &= \begin{bmatrix} -2 + z_1^0 & 1 \\ 0 & -3 - z_2^0 \end{bmatrix} = \begin{bmatrix} -3 & 1 \\ 0 & -3 \end{bmatrix}, \\ A_2 &= \begin{bmatrix} -2 + z_1^0 & 1 \\ 0 & -3 - z_2^1 \end{bmatrix} = \begin{bmatrix} -3 & 1 \\ 0 & -4 \end{bmatrix}, \\ A_3 &= \begin{bmatrix} -2 + z_1^1 & 1 \\ 0 & -3 - z_2^0 \end{bmatrix} = \begin{bmatrix} -1 & 1 \\ 0 & -3 \end{bmatrix}, \\ A_4 &= \begin{bmatrix} -2 + z_1^1 & 1 \\ 0 & -3 - z_2^1 \end{bmatrix} = \begin{bmatrix} -1 & 1 \\ 0 & -4 \end{bmatrix}, \end{aligned}$$

with $z_1(\cdot) \in [z_1^0, z_1^1] = [-1, 1]$ and $z_2(\cdot) \in [z_2^0, z_2^1] = [0, 1]$. This implies that the associated polytope of these TS models is the same, despite the fact that their MFs differ since they reflect the parameter, time, or state dependence of their original models, i.e.

1. LPV model (3.5):

$$w_0^1(z) = \frac{z_1^1 - \theta_1}{z_1^1 - z_1^0},\ w_0^2(z) = \frac{z_2^1 - \theta_2}{z_2^1 - z_2^0}\ ;$$

2. LTV model (3.6):

$$w_0^1(z) = \frac{z_1^1 - \sin t}{z_1^1 - z_1^0},\ w_0^2(z) = \frac{z_2^1 - \cos^2 t}{z_2^1 - z_2^0}\ ;$$

3. State-dependent model (3.7):

$$w_0^1(z) = \frac{z_1^1 - \sin x_1}{z_1^1 - z_1^0},\ w_0^2(z) = \frac{z_2^1 - \cos^2 x_2}{z_2^1 - z_2^0}$$

with $h_1(z) = w_0^1(z)w_0^2(z)$, $h_2(z) = w_0^1(z)w_1^2(z)$, $h_3(z) = w_1^1(z)w_0^2(z)$, and $h_4(z) = w_1^1(z)w_1^2(z)$. □

Of course, if terms in $\boldsymbol{f}(\boldsymbol{x}, \boldsymbol{u}, \boldsymbol{d}, \boldsymbol{\theta}, t)$ and $\boldsymbol{h}(\boldsymbol{x}, \boldsymbol{u}, \boldsymbol{d}, \boldsymbol{\theta}, t)$ in (3.8) were bounded in all of $\mathbb{R}^\xi$, ξ being the number of their arguments, then we would obtain *globally* valid TS models. This happens with, for instance, sine or cosine functions.

The main advantage of the sector nonlinearity approach is that the obtained TS model is an exact representation of the nonlinear system in the considered compact set. However, the number of vertex models may be exponential in the number of nonlinearities, which may lead to intractable design problems. Linear fractional models in Sect. 3.5 may help in reducing the number of vertices, as premise variables may be the arguments of more complex rational nonlinearities on them.

Therefore, unless instability, unobservability, or uncontrollability of the vertex models is an issue, a representation with a minimum number of vertices should be chosen. A second shortcoming is that the obtained consequent models are not guaranteed to be stable, observable, or controllable. Available methods to analyse the stability of a TS model require that the linear vertex models are stable. Likewise, the methods for controller/observer design require that the vertex models are stabilizable/detectable. Depending on the nonlinear system considered, another representation of the nonlinear system may be chosen to have the desired properties. Otherwise, methods that obtain an approximate fuzzy model, whose vertex models have the same properties as the nonlinear system, can be used, such as those presented in the next section [28, 29].

3.2.2 *Approximate Models*

When the models obtained by the sector nonlinearity approach are not suitable for analysis or design, e.g. due to the very large number of vertex models or loss of useful properties, approximate models may be derived. For instance, linearising a nonlinear model is a usual step for control purposes since it allows obtaining a control law via linear methods. Extensions to these and other techniques are discussed below; very recently, a variety of works have also appeared that intend to optimise the modelling task with respect to a performance measure; the interested reader is referred to [28–30] and the references therein.

Note, however, that these approximate models are an *ad hoc* solution to be used only as a last resort option, given that the sector nonlinearity is exact (given a nonlinear model), so later stability and control design techniques will guarantee performance for the original nonlinear model; on the other hand, using approximate models in this section will lose the aforementioned guarantees even if, however, the approximate approach might prove helpful in experimental applications with simple nonlinearities. Hence, the ideas in this section are not recommended as a first option.

Linearisation in Multiple Points

A TS model of a nonlinear one can be constructed from a set of linearised models interconnected by MFs; these MFs are usually defined to attain their maximum values in the linearisation points as well as to cover the region of interest of the state space in a uniform manner. As it will turn out, the possibility of linearising in non-equilibrium points obliges us to consider affine terms in the linearised models.

Consider the nonlinear dynamical model

$$\begin{aligned}\delta\boldsymbol{x}(t) &= \boldsymbol{f}(\boldsymbol{x},\boldsymbol{u}),\\ \boldsymbol{y}(t) &= \boldsymbol{h}(\boldsymbol{x}),\end{aligned} \tag{3.27}$$

where, as before, $\delta\boldsymbol{x}(t)=\dot{\boldsymbol{x}}(t)$ and $t\in\mathbb{R}$ for continuous-time models, $\delta\boldsymbol{x}(t)=\boldsymbol{x}(t+1)$ and $t\in\mathbb{Z}$ for discrete-time models, $\boldsymbol{x}\in\mathbb{R}^n$ is the vector of state variables, $\boldsymbol{y}\in\mathbb{R}^q$ is the vector of measurements, $\boldsymbol{u}\in\mathbb{R}^m$ is the input vector, and $\boldsymbol{f}$ and $\boldsymbol{h}$ are smooth nonlinear vector functions.

The goal is to obtain an approximation of the nonlinear model (3.27) as a nonlinear interpolation of r linear models, i.e.

$$\delta x_{app}(t) = \sum_{i=1}^{r} h_i(z)(A_i x(t) + B_i u(t) + a_i),$$
$$y(t) = \sum_{i=1}^{r} h_i(z)(C_i x(t) + c_i), \tag{3.28}$$

where A_i, B_i, C_i and a_i, c_i are the matrices and vector biases of the local linear models, respectively; $z \in \mathbb{R}^p$ is the scheduling vector that determines which of the linear models are active at a certain moment; and $h_i(z)$, $i \in \{1, 2, \ldots, r\}$ are the MFs according to which the linearisation is made. Each of the r linearisation points (x_i, u_i), $i \in \{1, 2, \ldots, r\}$ has p scheduling variables z_j, $j \in \{1, 2, \ldots, p\}$, which describe the model nonlinearities, i.e. a selection of inputs, states, and measurements. For each scheduling variable z_j, a certain number of weighting functions $w_j^k(z_j)$, centred on the values where the involved linearisation points attain their maximum, are defined; they are usually chosen with triangular or trapezoidal shapes. Each MF is a product of p weighting functions, so the resulting TS structure is similar to that obtained via the sector nonlinearity approach.

For instance, a TS fuzzy approximation of a given nonlinear model can be obtained by a Taylor series expansion in different representative points [22], which may or may not be equilibria. However, this approximation will in general result in affine consequents.

The consequent matrices are obtained as

$$A_i = \left.\frac{\partial f}{\partial x}\right|_{(x_i, u_i)}, \qquad B_i = \left.\frac{\partial f}{\partial u}\right|_{(x_i, u_i)}, \qquad C_i = \left.\frac{\partial h}{\partial x}\right|_{(x_i, u_i)},$$

where $|_{(x_i, u_i)}$ denotes the evaluation of the expression on the left in the value corresponding to (x_i, u_i) for those state and input variables that are scheduling variables and 0 for those states and inputs that are not in z.

Since generally the linearisation is not done in equilibria, to obtain an accurate approximation of the nonlinear model, affine terms (biases) expressed as

$$a_i = f(x, u)|_{(x_i, u_i)} - (A_i x)|_{(x_i, u_i)} - (B_i u)|_{(x_i, u_i)},$$
$$c_i = h(x)|_{(x_i, u_i)} - (C_i x)|_{(x_i, u_i)}$$

have to be added. It is not Jacobian at (x_i, u_i), but p-dimensional grid point.

In contrast with the sector nonlinearity approach, linearisation leads to TS models that approximate the original nonlinear one. As the number of well-chosen linearisation points increases, a better TS approximation of the nonlinear model can be found at the price of increasing the number of affine models; it will turn out that this augments the computational burden of the subsequent analysis or controller synthesis.

The method is illustrated in the following examples.

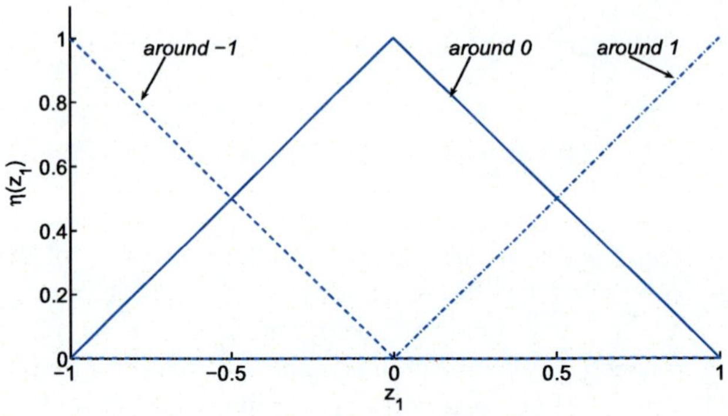

Fig. 3.5 Membership functions for the scheduling variable z_1

Example 3.6 Consider the discrete-time nonlinear system

$$\begin{aligned} \boldsymbol{x}(t+1) &= \begin{bmatrix} 2x_1x_2 - 3x_1^2 + 2x_2 + u \\ x_1 + 3x_2 + x_2^2 + x_1u \end{bmatrix}, \\ y(t) &= 2x_1^2 + x_2, \end{aligned}$$

where $\boldsymbol{x} = [x_1\, x_2]^T$ and the variables defined on $\mathcal{C} = \{\boldsymbol{x},\, y,\, u\,|u,\, y \in \mathbb{R},\, |x_i| \leq 1,\, i = 1,\, 2\}$.

Since the state transition function $\boldsymbol{f} = \begin{bmatrix} 2x_1x_2 - 3x_1^2 + 2x_2 + u \\ x_1 + 3x_2 + x_2^2 + x_1u \end{bmatrix}$ is affine in the input u and nonlinear in the state variables x_1 and x_2, the scheduling vector is chosen as $z = \left[x_1\; x_2\right]^T$.

For this example, we choose nine linearisation points as $\{(x_1,\, x_2, u) : x_1,\, x_2 \in \{-1, 0, 1\}, u = 0\}$. A common choice of weighting functions is normalised triangular or trapezoidal ones that attain their maximum in the linearisation points, for each scheduling variable. These functions are easily represented, and for each variable, at most two are activated. In this example, the membership functions chosen are shown in Fig. 3.5.

The membership functions for z_1 are defined as

$$\begin{aligned} w_1^1(z_1) &= \begin{cases} -z_1, & \text{if } z_1 \leq 0 \\ 0, & \text{otherwise} \end{cases}; \\ w_1^2(z_1) &= 1 - |z_1|; \\ w_1^3(z_1) &= \begin{cases} z_1, & \text{if } z_1 \geq 0 \\ 0, & \text{otherwise.} \end{cases} \end{aligned}$$

Similar functions are defined as membership functions for z_2

$$w_2^1(z_2) = \begin{cases} -z_2, & \text{if } z_2 \leq 0 \\ 0, & \text{otherwise} \end{cases};$$
$$w_2^2(z_2) = 1 - |z_2|;$$
$$w_2^3(z_2) = \begin{cases} z_2, & \text{if } z_2 \geq 0 \\ 0, & \text{otherwise.} \end{cases}$$

The partial derivatives of the state and measurement equations with respect to the scheduling vector are

$$\frac{\partial \boldsymbol{f}}{\partial \boldsymbol{x}} = \begin{bmatrix} 2x_2 - 6x_1 & 2x_1 + 2 \\ 1 + u & 3 + 2x_2 \end{bmatrix}, \quad \frac{\partial \boldsymbol{f}}{\partial u} = \begin{bmatrix} 1 \\ x_1 \end{bmatrix}, \quad \frac{\partial \boldsymbol{h}}{\partial \boldsymbol{x}} = \begin{bmatrix} 4x_1 & 1 \end{bmatrix}.$$

These matrices are evaluated in the selected linearisation points to obtain the matrices of the vertex models. The affine terms are obtained from the evaluation in the linearisation points of the expressions

$$\boldsymbol{f}(\boldsymbol{x}, u) - \frac{\partial \boldsymbol{f}}{\partial \boldsymbol{x}}\boldsymbol{x} - \frac{\partial \boldsymbol{f}}{\partial u}u = \begin{bmatrix} 3x_1^2 - 2x_1x_2 \\ -x_2^2 - x_1u \end{bmatrix},$$
$$\boldsymbol{h}(\boldsymbol{x}) - \frac{\partial \boldsymbol{h}}{\partial \boldsymbol{x}}\boldsymbol{x} = -2x_1^2.$$

Then, the approximate TS fuzzy model will have the form (3.28) with $\delta\boldsymbol{x}(t) = \boldsymbol{x}(t+1)$ and $r = 9$ vertex models; some of them are shown below for illustration:

Vertex model 1:

$$A_1 = \begin{bmatrix} 2x_2 - 6x_1 & 2x_1 + 2 \\ 1 + u & 3 + 2x_2 \end{bmatrix}\Bigg|_{(z_{0,1},0)} = \begin{bmatrix} 4 & 0 \\ 1 & 1 \end{bmatrix},$$
$$B_1 = \begin{bmatrix} 1 \\ x_1 \end{bmatrix}\Bigg|_{(\boldsymbol{x}_1,0)} = \begin{bmatrix} 1 \\ -1 \end{bmatrix},$$
$$\boldsymbol{a}_1 = \begin{bmatrix} 3x_1^2 - 2x_1x_2 \\ -x_2^2 - x_1u \end{bmatrix}\Bigg|_{(\boldsymbol{x}_1,0)} = \begin{bmatrix} 1 \\ -1 \end{bmatrix},$$
$$C_1 = \begin{bmatrix} 4x_1 & 1 \end{bmatrix}\Big|_{(\boldsymbol{x}_1,0)} = \begin{bmatrix} -4 & 1 \end{bmatrix},$$
$$c_1 = -2x_1^2\Big|_{(\boldsymbol{x}_1,0)} = -2,$$

where $\boldsymbol{x}_1 = \begin{bmatrix} -1 & -1 \end{bmatrix}^T$, and the membership function of the vertex is computed as $h_1(\boldsymbol{z}) = w_1^1(\boldsymbol{z})w_2^1(\boldsymbol{z})$.

Vertex model 2:

$$
\begin{aligned}
A_2 &= \begin{bmatrix} 2x_2 - 6x_1 & 2x_1 + 2 \\ 1 + u & 3 + 2x_2 \end{bmatrix}\bigg|_{(\boldsymbol{x}_2,0)} = \begin{bmatrix} 6 & 0 \\ 1 & 3 \end{bmatrix}, \\
B_2 &= \begin{bmatrix} 1 \\ x_1 \end{bmatrix}\bigg|_{(\boldsymbol{x}_2,0)} = \begin{bmatrix} 1 \\ -1 \end{bmatrix}, \\
\boldsymbol{a}_2 &= \begin{bmatrix} 3x_1^2 - 2x_1x_2 \\ -x_2^2 - x_1u \end{bmatrix}\bigg|_{(\boldsymbol{x}_2,0)} = \begin{bmatrix} 3 \\ 0 \end{bmatrix}, \\
C_2 &= \begin{bmatrix} 4x_1 & 1 \end{bmatrix}\Big|_{(\boldsymbol{x}_2,0)} = \begin{bmatrix} -4 & 1 \end{bmatrix}, \\
c_2 &= -2x_1^2\Big|_{(\boldsymbol{x}_2,0)} = -2,
\end{aligned}
$$

where $\boldsymbol{x}_2 = \begin{bmatrix} -1 & 0 \end{bmatrix}^T$, and the membership function of the vertex is computed as $h_2(z) = w_1^1(z)w_2^2(z)$.

Vertex model 3:

$$
\begin{aligned}
A_3 &= \begin{bmatrix} 2x_2 - 6x_1 & 2x_1 + 2 \\ 1 + u & 3 + 2x_2 \end{bmatrix}\bigg|_{(\boldsymbol{x}_3,0)} = \begin{bmatrix} 8 & 0 \\ 1 & 5 \end{bmatrix}, \\
B_3 &= \begin{bmatrix} 1 \\ x_1 \end{bmatrix}\bigg|_{(\boldsymbol{x}_3,0)} = \begin{bmatrix} 1 \\ -1 \end{bmatrix}, \\
\boldsymbol{a}_3 &= \begin{bmatrix} 3x_1^2 - 2x_1x_2 \\ -x_2^2 - x_1u \end{bmatrix}\bigg|_{(\boldsymbol{x}_3,0)} = \begin{bmatrix} 5 \\ -1 \end{bmatrix}, \\
C_3 &= \begin{bmatrix} 4x_1 & 1 \end{bmatrix}\Big|_{(\boldsymbol{x}_3,0)} = \begin{bmatrix} -4 & 1 \end{bmatrix}, \\
c_3 &= -2x_1^2\Big|_{(\boldsymbol{x}_3,0)} = -2,
\end{aligned}
$$

where $\boldsymbol{x}_3 = \begin{bmatrix} -1 & 1 \end{bmatrix}^T$, and the membership function of the vertex is computed as $h_3(z) = w_1^1(z)w_2^3(z)$.

Vertex model 4:

$$
\begin{aligned}
A_4 &= \begin{bmatrix} 2x_2 - 6x_1 & 2x_1 + 2 \\ 1 + u & 3 + 2x_2 \end{bmatrix}\bigg|_{(\boldsymbol{x}_4,0)} = \begin{bmatrix} -2 & 2 \\ 1 & 1 \end{bmatrix}, \\
B_4 &= \begin{bmatrix} 1 \\ x_1 \end{bmatrix}\bigg|_{(\boldsymbol{x}_4,0)} = \begin{bmatrix} 1 \\ 0 \end{bmatrix}, \\
\boldsymbol{a}_4 &= \begin{bmatrix} 3x_1^2 - 2x_1x_2 \\ -x_2^2 - x_1u \end{bmatrix}\bigg|_{(\boldsymbol{x}_4,0)} = \begin{bmatrix} 0 \\ -1 \end{bmatrix}, \\
C_4 &= \begin{bmatrix} 4x_1 & 1 \end{bmatrix}\Big|_{(\boldsymbol{x}_4,0)} \begin{bmatrix} 0 & 1 \end{bmatrix}, \\
c_4 &= -2x_1^2\Big|_{(\boldsymbol{x}_4,0)} = 0,
\end{aligned}
$$

where $x_4 = \left[0\ -1\right]^T$, and the membership function of the vertex is computed as $h_4(z) = w_1^2(z)w_2^1(z)$.

The remaining models are constructed in a similar manner.

Note that since the individual membership functions for z_1 and z_2, respectively, are normal, the obtained combined membership functions h_i are also normalised. □

Example 3.7 Consider again the nonlinear model of the ball and beam system in (3.25). Linearisation around a certain point x_0 will require the following computations:

$$A = \left.\frac{\partial f}{\partial x}\right|_{x=x_0,u=u_0} = \begin{bmatrix} 0 & 1 & 0 & 0 \\ 0.7143x_4^2 & 0 & -7.0073\cos x_3 & 1.4286x_1x_4 \\ 0 & 0 & 0 & 1 \\ 0 & 0 & 0 & 0 \end{bmatrix},$$

$$B = \left.\frac{\partial g}{\partial u}\right|_{x=x_0,u=u_0} = \begin{bmatrix} 0 \\ 0 \\ 0 \\ 1 \end{bmatrix},$$

$$a = f(x_i, 0) - \left.\frac{\partial f}{\partial x}x_i\right|_{(x_i,0)} - \left.\frac{\partial f}{\partial u}u_0\right|_{(x_i,0)}$$

$$= \begin{bmatrix} 0 \\ -1.4284x_1x_4^2 - 7.0073(\sin x_3 + x_3\cos x_3) \\ 0 \\ 0 \end{bmatrix}.$$

If linearisation is made choosing $z = x_3$ as the scheduling variable for the linearisation points $x_1 = [0\ 0\ 0\ 0]^T$ and $x_2 = [0\ 0\ \pm\pi/4\ 0]^T$, the following vertex models arise:

Vertex model 1, linearisation point $(x_1, x_2, x_3, x_4) = (0, 0, 0, 0)$:

$$\dot{x}(t) = A_1x(t) + Bu(t) + a_1,$$

with

$$A_1 = \begin{bmatrix} 0 & 1 & 0 & 0 \\ 0 & 0 & -7.0073 & 0 \\ 0 & 0 & 0 & 1 \\ 0 & 0 & 0 & 0 \end{bmatrix},\quad B = \begin{bmatrix} 0 \\ 0 \\ 0 \\ 1 \end{bmatrix},\quad a_1 = \begin{bmatrix} 0 \\ 0 \\ 0 \\ 0 \end{bmatrix}.$$

Vertex model 2, linearisation point $(x_1, x_2, x_3, x_4) = (0, 0, \pm\pi/4, 0)$:

$$\dot{x}(t) = A_2x(t) + Bu(t) + a_2,$$

with

$$A_2 = \begin{bmatrix} 0 & 1 & 0 & 0 \\ 0 & 0 & -4.9549 & 0 \\ 0 & 0 & 0 & 1 \\ 0 & 0 & 0 & 0 \end{bmatrix}, \quad B = \begin{bmatrix} 0 \\ 0 \\ 0 \\ 1 \end{bmatrix}, \quad \boldsymbol{a}_2 = \begin{bmatrix} 0 \\ -1.0633 \\ 0 \\ 0 \end{bmatrix}.$$

MFs for interpolating between these models can be defined as follows (recall $z_1 = x_3$):

$$h_1(z) = \begin{cases} 0, & z_1 < -\pi/4 \\ 1 + \frac{4}{\pi} z_1, & -\pi/4 \le z_1 < 0 \\ 1 - \frac{4}{\pi} z_1, & 0 \le z_1 < \pi/4 \\ 0, & z_1 \ge \pi/4 \end{cases};$$

$$h_2(z) = \begin{cases} 1, & z_1 < -\pi/4 \\ -\frac{4}{\pi} z_1, & -\pi/4 \le z_1 < 0 \\ \frac{4}{\pi} z_1, & 0 \le z_1 < \pi/4 \\ 1, & z_1 \ge \pi/4. \end{cases}$$

The TS model is then written as (3.28) with $\delta \boldsymbol{x}(t) = \dot{\boldsymbol{x}}$ and $r = 2$, which constitutes a remarkable reduction in the number of vertex models with respect to Example 3.4, where sector nonlinearity was used; the shortcoming is, of course, that the reduced model is only an approximation. □

The Substitution Method

The advantage of constructing TS models using linearisation is that although the fuzzy system is only an approximation of the original nonlinear system, the consequents retain important properties of the nonlinear system in the linearisation points [31]. For instance, if the nonlinear system is locally observable in a neighbourhood of the linearisation point, then the corresponding vertex model is observable. An important disadvantage, in particular when performing stability analysis or controller design, is that, since the linearisation will generally not be performed in equilibrium points, affine TS models will be obtained, for which standard methods cannot be applied. This is why, if piecewise methods are not employed [32, 33], in some cases, it is preferable to use the substitution method of [23], explained next.

Consider again the system (3.27), with the assumption that it can be written as (3.28), similarl to the representation used for the sector nonlinearity approach. However, instead of representing each nonlinearity as the convex sum of two linear function, the values are simply substituted for the chosen points. If the nonlinear system has the form (3.28), a TS model with linear consequents that approximates the original nonlinear model is obtained. The method is illustrated on the following example.

Example 3.8 Consider the discrete-time nonlinear system from Example 3.6, i.e.

$$\begin{aligned} \boldsymbol{x}(t+1) &= \begin{bmatrix} 2x_1x_2 - 3x_1^2 + 2x_2 + u \\ x_1 + 3x_2 + x_2^2 + x_1 u \end{bmatrix}, \\ y(t) &= 2x_1^2 + x_2, \end{aligned}$$

with $\boldsymbol{x} = [x_1\, x_2]^T$ and the variables defined on $\mathcal{X} = \{\boldsymbol{x},\ y,\ u\,|u,\ y \in \mathbb{R},\ |x_i| \leq 1,\ i = 1,\ 2\}$.

This system can be rewritten as

$$\begin{aligned} \boldsymbol{x}(t+1) &= \begin{bmatrix} 2x_2 - 3x_1 & 2 \\ 1 & 3 + x_2 \end{bmatrix} \boldsymbol{x}(t) + \begin{bmatrix} 1 \\ x_1 \end{bmatrix} u(t), \\ y(t) &= \begin{bmatrix} 2x_1 & 1 \end{bmatrix} \boldsymbol{x}(t). \end{aligned}$$

The scheduling vector is again chosen as $z(\boldsymbol{x}) = \begin{bmatrix} x_1 & x_2 \end{bmatrix}^T$, the substitution points as $\{(x_1,\ x_2,\ u)\,|x_1,\ x_2 \in \{-1, 0, 1\},\ u = 0\}$, and the triangular membership functions represented in Fig. 3.5. The approximate TS fuzzy model is given by substituting the chosen points into the state, input, and output matrices above. The models are as follows:

Vertex model 1:

$$\begin{aligned} \boldsymbol{x}(t+1) &= A_1\boldsymbol{x}(t) + B_1\boldsymbol{u}(t), \\ \boldsymbol{y}(t) &= C_1\boldsymbol{x}(t), \end{aligned}$$

with

$$\begin{aligned} A_1 &= \begin{bmatrix} 2x_2 - 3x_1 & 2 \\ 1 & 3 + x_2 \end{bmatrix}\Bigg|_{(\boldsymbol{x}_1, 0)} = \begin{bmatrix} 1 & 2 \\ 1 & 2 \end{bmatrix}, \\ B_1 &= \begin{bmatrix} 1 \\ x_1 \end{bmatrix}\Bigg|_{(\boldsymbol{x}_1, 0)} = \begin{bmatrix} 1 \\ -1 \end{bmatrix}, \\ C_1 &= \begin{bmatrix} 2x_1 & 1 \end{bmatrix}\Big|_{(\boldsymbol{x}_1, 0)} = \begin{bmatrix} -2 & 1 \end{bmatrix}, \end{aligned}$$

where $\boldsymbol{x}_1 = (-1,\ -1)$, and the membership function of the vertex is computed as $h_1(z) = w_1^1(z)w_2^1(z)$.

Vertex model 2:

$$\begin{aligned} \boldsymbol{x}(t+1) &= A_2\boldsymbol{x}(t) + B_2\boldsymbol{u}(t), \\ \boldsymbol{y}(t) &= C_2\boldsymbol{x}(t), \end{aligned}$$

with

$$A_2 = \begin{bmatrix} 2x_2 - 3x_1 & 2 \\ 1 & 3 + x_2 \end{bmatrix}\Big|_{(\boldsymbol{x}_2,0)} = \begin{bmatrix} 3 & 2 \\ 1 & 3 \end{bmatrix},$$
$$B_2 = \begin{bmatrix} 1 \\ x_1 \end{bmatrix}\Big|_{(\boldsymbol{x}_2,0)} = \begin{bmatrix} 1 \\ -1 \end{bmatrix},$$
$$C_2 = \begin{bmatrix} 2x_1 & 1 \end{bmatrix}\Big|_{(\boldsymbol{x}_2,0)} = \begin{bmatrix} -2 & 1 \end{bmatrix},$$

where $\boldsymbol{x}_2 = (-1,\ 0)$, and the membership function of the vertex is computed as $h_2(z) = w_1^1(z)w_2^2(z)$.

Vertex model 3:

$$\boldsymbol{x}(t+1) = A_3\boldsymbol{x}(t) + B_3\boldsymbol{u}(t),$$
$$\boldsymbol{y}(t) = C_3\boldsymbol{x}(t),$$

with

$$A_3 = \begin{bmatrix} 2x_2 - 3x_1 & 2 \\ 1 & 3 + x_2 \end{bmatrix}\Big|_{(\boldsymbol{x}_3,0)} = \begin{bmatrix} 5 & 2 \\ 1 & 4 \end{bmatrix},$$
$$B_3 = \begin{bmatrix} 1 \\ x_1 \end{bmatrix}\Big|_{(\boldsymbol{x}_3,0)} = \begin{bmatrix} 1 \\ -1 \end{bmatrix},$$
$$C_3 = \begin{bmatrix} 2x_1 & 1 \end{bmatrix}\Big|_{(\boldsymbol{x}_3,0)} = \begin{bmatrix} -2 & 1 \end{bmatrix},$$

where $\boldsymbol{x}_3 = (-1,\ 1)$, and the membership function of the vertex is computed as $h_3(z) = w_1^1(z)w_2^3(z)$.

Vertex model 4:

$$\boldsymbol{x}(t+1) = A_4\boldsymbol{x}(t) + B_4\boldsymbol{u}(t),$$
$$\boldsymbol{y}(t) = C_4\boldsymbol{x}(t),$$

with

$$A_4 = \begin{bmatrix} 2x_2 - 3x_1 & 2 \\ 1 & 3 + x_2 \end{bmatrix}\Big|_{(\boldsymbol{x}_4,0)} = \begin{bmatrix} -2 & 2 \\ 1 & 2 \end{bmatrix},$$
$$B_4 = \begin{bmatrix} 1 \\ x_1 \end{bmatrix}\Big|_{(\boldsymbol{x}_4,0)} = \begin{bmatrix} 1 \\ 0 \end{bmatrix},$$
$$C_4 = \begin{bmatrix} 2x_1 & 1 \end{bmatrix}\Big|_{(\boldsymbol{x}_4,0)} = \begin{bmatrix} 0 & 2 \end{bmatrix},$$

where $\boldsymbol{x}_4 = (0,\ -1)$, and the membership function of the vertex is computed as $h_4(z) = w_1^2(z)w_2^1(z)$.

The remaining vertices are constructed in a similar manner. In fact, as $A(\boldsymbol{x})$ is linear in $\boldsymbol{x}$ in this case, we would only need to consider the vertices of the modelling region (as $A(\boldsymbol{x})$ in intermediate points will be a convex combination of the value of $A(\boldsymbol{x})$ at the chosen vertices); thus, a literal application of this method is only exact in quadratic nonlinearities so $A(\boldsymbol{x})\boldsymbol{x}$ can be written with linear $A(\boldsymbol{x})$.

The above exactness is not the case in general, where a dense-enough cloud of points might be needed in order to have a reduced approximation error, motivating its inclusion in the approximate model section.

In complex high-order problems, we are exponential in the dimension to generate these clouds of points and approximation quality heavily suffers. An improvement of this method, keeping exactness, is the Kwiatkowski method [34]. □

Least-Squares Fit

Yet another possibility of getting models in the considered classes in this chapter is resorting to system-identification techniques and carrying out, for instance, a conventional least-squares fit on some available data sets.

For instance, given a sufficiently large data set of points $\{(x_k, y_k) \in \mathbb{R}^2,\ k = 1, \ldots, N\}$, a fit to obtain a Takagi–Sugeno model would entail fixing arbitrary antecedent regressors, for instance, a set of r arbitrary functions $w_1(\boldsymbol{x}), w_2(\boldsymbol{x}), \ldots, w_r(\boldsymbol{x})$, and obtaining the membership functions via normalisation:

$$h_i(\boldsymbol{x}) = \frac{w_i(\boldsymbol{x})}{\sum_{j=1}^{r} w_j(\boldsymbol{x})}. \tag{3.29}$$

Once this step is done, a candidate model structure could be stated as follows:

$$y(\boldsymbol{x}) = \sum_{i=1}^{r} h_i(\boldsymbol{x})(a_i \boldsymbol{x} + b_i) + \epsilon, \tag{3.30}$$

and the regression problem below

$$\begin{bmatrix} y_1 \\ \vdots \\ y_N \end{bmatrix} = \begin{bmatrix} h_1(x_1)x_1 & \ldots & h_r(x_1)x_1 & h_1(x_1) & \ldots & h_r(x_1) \\ \vdots & & \vdots & \vdots & & \vdots \\ h_1(x_N)x_N & \ldots & h_r(x_N)x_N & h_1(x_N) & \ldots & h_r(x_N) \end{bmatrix} \begin{bmatrix} a_1 \\ \vdots \\ a_r \\ b_1 \\ \vdots \\ b_r \end{bmatrix} + \begin{bmatrix} \epsilon_1 \\ \vdots \\ \epsilon_N \end{bmatrix} \tag{3.31}$$

can be considered. As most stability and control results in later chapters preclude affine terms, b_i may be omitted, if so wished, at the expense of a larger fit error ϵ, of course.

Introducing the shorthand notation $\boldsymbol{y} = \Xi \boldsymbol{\theta} + \boldsymbol{\epsilon}$, $\boldsymbol{y}$ being the output data vector, Ξ the "membership×states" matrix, and $\boldsymbol{\theta} = [a_1\, a_2\, \cdots\, a_r\, b_1\, b_2\, \cdots\, b_r]^T$ the vector of coefficients, minimising the 2-norm of the residual vector $\boldsymbol{\epsilon}$ can be solved in a straightforward manner via a pseudo-inverse:

$$\hat{\boldsymbol{\theta}} = \Xi^{+} \boldsymbol{y}, \tag{3.32}$$

where Ξ^{+} denotes the left pseudo-inverse of Ξ.

Note that, if the input data $\boldsymbol{x}$ and output data $\boldsymbol{y}$ were a vector, trivial changes would also pose the problem as a standard least-squares regression. As a last remark, the membership themselves might have adjustable parameters $h_i(\boldsymbol{x}, \boldsymbol{\theta})$. In that case, the problem would not be linear in parameter and, hence, pseudo-inverse one-step computation of the optimal parameters would not apply, requiring other optimisation approaches such as the quasi-Newton [35], the Levenberg–Marquardt [36], or directed-random search [37]; these issues are not in the scope of this book.

The basic idea inspiring the above approach is the so-called *universal function approximation* paradigm (UFA) which ensures that many commonsense choices of h_i (see below) allow reducing the maximum error to any desired value by increasing the number of regressors r (vertices) when approximating any continuous function $\boldsymbol{y}(\boldsymbol{x})$ in a compact set, even if the number of data points tends to infinity.

Standard choices for w_i are polynomials $w_i = x^i$, the Gaussian radial-basis regressors $w_i = e^{-(x_i-\zeta_i)^2/\sigma_i^2}$, or interpolation-based schemes (linear interpolation, splines, etc.). All of them can be shown to have the UFA property (details omitted for brevity), so a large-enough r results in a reduced $\|\epsilon\|$. Once normalised by (3.29), they verify the convex sum property; actually, note that, as polynomials may be negative, adding some constants to w_i may be needed before normalisation (without loss of generality).

Let us discuss now in which conditions such models can be used in the framework to be developed in later chapters, which requires exact models to prove stability and also requires zero to be an equilibrium point.

The basic idea is that the above models (3.30) can be understood as exact TS models plus a *disturbance* ϵ, as a bound for ϵ can be extracted from the data used in the fitting process. The presence of this non-zero disturbance excludes proving standard stability (convergence to the origin) and only convergence to a small-enough set around the origin can be actually proven or, alternative, invariance of a large-enough set (when ϵ is negligible compared to $\boldsymbol{x}$), for instance, proving that $\dot{V}(\boldsymbol{x}) < 0$ *outside* some level set $V_\gamma = \{V(\boldsymbol{x}) \leq \gamma\}$ ensures V_γ is invariant.

If piecewise approaches are excluded [32, 33], affine models ($b_i \neq 0$) are very limited when discussing LMIs to guarantee the above issues, but data-based TS models ($b_i = 0$) could be reasonably used with disturbance-related LMI problems (or, if the number of vertices is high-enough, neglecting such disturbance ϵ). However, the collection of a rich data set in order to adjust such models is an issue because a large region of the state space should be explored, and that step has practical difficulties.

As a last remark, the above classes of models (in a general setting, the so-called linearly parameterised approximators $\boldsymbol{y} = \Psi(\boldsymbol{x})\boldsymbol{\theta} + \epsilon$) are widely used in adaptive control under different frameworks (nonlinear [38], neural [39], fuzzy [40]). All of these frameworks share common traits; the reader is referred to the cited sources and references therein for details, out of the scope of the present monograph. The basic idea of the adaptive approach is avoiding the state-space exploration in the identification step mentioned in the previous paragraph, by adjusting the parameters with the controller in place along a closed-loop trajectory; ensuring stability (understood only as boundedness—in a particular sense—because of the non-zero error term) is the key problem addressed in such literature.

Vertex Reduction (Simplification of Sector Nonlinearity Models)

Another source of approximate models comes from the fact that, in a system with a lot of nonlinearities, the sector nonlinearity procedure, even if exact in theory, would produce an intractable high power-of-two number of vertex models and, if some of them were neglected, they might be considered as approximation uncertainty.

For instance, we might have a 12-dimensional box model:

$$\dot{x} = \sum_{i=1}^{12} \sum_{j=1}^{2} h_{ij}(z) A_{ij} x \tag{3.33}$$

but consider that dealing with 2^{12} vertices is intractable for later LMI conditions. Hence, we may decide to consider only three important components (possibly these with larger singular values of A_{ij}), and consider the remaining 9 sums as uncertainty. If the neglected terms have a maximum singular value below some bound $\|A_{ij}\|_\infty \leq \delta$, for all $i > 3$, we would rewrite as follows:

$$\dot{x} = \left(\sum_{i=1}^{3} \sum_{j=1}^{2} h_{ij}(z) A_{ij} + \Delta \right) x \tag{3.34}$$

with $\|\Delta\|_\infty \leq 2 \cdot 9 \cdot \delta$; infinity norm is understood as induced 2-norm (maximum singular value).

If such uncertainty were not considered in later stability/control design conditions, then such procedure would be in the realm of the approximate models considered in this section. If some bounds of the uncertainty can be provided, then such bounds could be considered in the stability conditions.

More detail on these issues appears in Sect. 3.5.4, where a principal component model reduction approach is pursued, starting from a large number of samples of $A(z)$ from the previously discussed substitution method; indeed the above singular-value bounding and principal component analysis of a batch of data are closely related.

3.2.3 Piecewise and Affine Models

Akin to the LS fit options, other literature [41] proposes approximate modelling of nonlinear systems via piecewise affine systems in which a single affine model $\dot{\boldsymbol{x}}(t) = A\boldsymbol{x}(t) + \boldsymbol{a}$ is fit by least squares in a region of the partition. Details are omitted for brevity, as generalisation from the prior least-squares developments is obvious. Nevertheless, exact piecewise modelling is briefly reviewed as it lies in the spirit of the present work whose focus is on nonlinear models exactly rewritten as convex ones via the sector nonlinearity approach. Piecewise affine TS systems lower the separation among the vertex models via a partition of the modelling region which naturally introduces affine terms [42].

Consider an autonomous continuous-time nonlinear model

$$\dot{\boldsymbol{x}}(t) = \boldsymbol{f}(\boldsymbol{x}), \tag{3.35}$$

with $\boldsymbol{x} \in \mathbb{R}^n$ as the state vector and $\boldsymbol{f}(\cdot) : \mathbb{R}^n \to \mathbb{R}^n$ being a $\mathscr{C}^2$ possibly nonlinear vector field. Assume the origin is such that $\boldsymbol{f}(0) = 0$. As $\boldsymbol{f}$ is linearisable at the origin, denoting as A its Jacobian evaluated at the origin, we can rewrite $\boldsymbol{f}$ in (3.35) as

$$\boldsymbol{f}(\boldsymbol{x}) = A\boldsymbol{x} + \sum_{j=1}^{p} \boldsymbol{c}_j z_j(\boldsymbol{x}), \tag{3.36}$$

with $z_j : \mathbb{R}^n \mapsto \mathbb{R}$, for $j = \{1, 2, \dots, p\}$, being a group of nonlinearities whose linearisation is zero and $\boldsymbol{c}_j$ being column vectors indicating how nonlinearity z_j enters in each of the equations of (3.35).

Consider a compact Ω with the origin $\boldsymbol{x} = 0$ belonging to it; since $\boldsymbol{f}$ is $\mathscr{C}^2$, each z_j can be bounded in Ω by two affine functions:

$$z_j^0(\boldsymbol{x}) \leq z_j(\boldsymbol{x}) \leq z_j^1(\boldsymbol{x}), \tag{3.37}$$

where

$$z_j^1(\boldsymbol{x}) = a_1^j \boldsymbol{v}_j^T \boldsymbol{x} + b_1^j, \quad z_j^0(\boldsymbol{x}) = a_0^j \boldsymbol{v}_j^T \boldsymbol{x} + b_0^j, \tag{3.38}$$

being a_i^j, b_i^j scalars, and $\boldsymbol{v}_j^T$ row vectors which constitute arbitrarily tight linear bounds on $z_j(\boldsymbol{x})$. With bounds (3.37) available, we have

$$z_j(\boldsymbol{x}) = \sum_{i=0}^{1} w_i^j(\boldsymbol{x}) \left(a_i^j \boldsymbol{v}_j^T \boldsymbol{x} + b_i^j \right), \tag{3.39}$$

with weights given by the well-known interpolation expression:

$$w_0^j(z_j) := \frac{z_j^1(\boldsymbol{x}) - z_j(\boldsymbol{x})}{z_j^1(\boldsymbol{x}) - z_j^0(\boldsymbol{x})}, \quad w_1^j(z_j) := 1 - w_0^j(z_j). \tag{3.40}$$

Operating with all z_j, for $j \in \{1, 2, \ldots, p\}$, then $r = 2^p$ membership functions can be defined as

$$h_i(z) := \prod_{j=1}^{p} w^j_{i_j}(z), \tag{3.41}$$

with $i \in \{1, 2, \ldots, r\}$ building a binary-digit expression of i as $i = i_p \times 2^{p-1} + \ldots + i_2 \times 2 + i_1 + 1$, $i_j \in \{0, 1\}$, $z = [z_1\, z_2\, \cdots\, z_p]^T$. Obviously, the MFs hold the convex sum property, i.e. $\sum_{i=1}^{r} h_i(z) = 1$, $h_i(z) \geq 0$. Using such memberships, (3.35) can be expressed as follows:

$$\dot{\boldsymbol{x}}(t) = \sum_{i=1}^{r} h_i(z) \left(A\boldsymbol{x}(t) + \sum_{j=1}^{p} \boldsymbol{c}_j \left(a^j_{i_j} \boldsymbol{v}_j^T \boldsymbol{x} + b^j_{i_j} \right) \right). \tag{3.42}$$

If the standard shorthand notation $\Upsilon_h := \sum_{i=1}^{r} h_i(z)\Upsilon_i$ is adopted, from (3.42), denoting $A_i := A + \sum_{j=1}^{p} \boldsymbol{c}_j a^j_{i_j} \boldsymbol{v}_j^T$ and $b_i := \sum_{j=1}^{p} \boldsymbol{c}_j b^j_{i_j}$, the nonlinear model (3.35) in Ω can be compactly written as the following affine TS model:

$$\dot{\boldsymbol{x}}(t) = A_h \boldsymbol{x}(t) + \boldsymbol{b}_h, \quad \boldsymbol{x} \in \Omega. \tag{3.43}$$

A non-affine piecewise generalisation of the sector nonlinearity approach can be found in [43]. Several options for affine piecewise TS modelling are available; the examples in [42] used the minimum-weighted area approach which has been further generalised in [44]. The interested reader is referred to these works for details; some of them are briefly discussed below.

Instead of bounding a single-variable nonlinearity $\rho(x)$ between two sectors defined by lines crossing the origin $y = a_1 x$ and $y = a_2 x$, piecewise models can affinely bound regions in which the origin is *not* in their interior, i.e. nonlinearities are bounded between hyperplanes that do not necessarily pass by the origin. In the scalar case, affine modelling will require bounding the nonlinearity as

$$a_1 x + b_1 \leq z(x) \leq a_2 x + b_2 \tag{3.44}$$

for some a_1, a_2, b_1, b_2, for all $x \in \Omega_k$.

Clearly, such bounds (hyperplanes, lines) are not unique; in [44], optimisation criteria which choose the two lines in which the covered area is smallest are presented; this is done via LMIs. Once the optimal parameters are found, the piecewise affine Takagi–Sugeno (PWATS) model is, in the scalar case, given by

$$z(x) = w_1(z)\bar{a}_1(\boldsymbol{x}) + (1 - w_1(z))\bar{a}_2(\boldsymbol{x}), \tag{3.45}$$

where

$$w_1(z) = \frac{a_2 x + b_2 - z(\boldsymbol{x})}{a_2 x + b_2 - (a_1 x + b_1)}, \quad w_2(z) = 1 - w_1(z),$$

$$\bar{a}_1(x) = a_1 x + b_1, \quad \bar{a}_2(x) = a_2 x + b_2.$$

Generalising this procedure for a modelling compact Ω, which is partitioned into q subregions with disjoint interiors, Ω_k, $k \in \{1, 2, \ldots, q\}$, i.e.

$$\bigcup_{k=1}^{q} \Omega_k = \Omega, \qquad \text{int}(\Omega_k) \cap \text{int}(\Omega_l) = \emptyset,$$

the original nonlinear dynamics can be expressed as a piecewise affine TS model (PWATS) [32] in the form:

$$\begin{aligned} \dot{x}(t) &= A_h^k x(t), \quad x(t) \in \Omega_k, \quad k \in K_0, \\ \dot{x}(t) &= A_h^k x(t) + b_h^k, \quad x(t) \in \Omega_k, \quad k \in K_1, \end{aligned} \tag{3.46}$$

where $K_0 := \{k : 0 \in \Omega_k\}$ is the set of indices of those regions Ω_k that include the origin and $K_1 := \{k : 0 \notin \Omega_k\}$ is the set of indices of the remaining ones (not containing the origin).

Each region Ω_k can be described by a set of constraints $\Omega_k := \{\sigma_j^k(x) \geq 0, \ j \in \{1, 2, \ldots, n_k\}\}$ which can be included into the numerical analysis. If $\sigma_j^k(x)$ are affine functions of x, the partition of Ω is a *polyhedral* one, which are similar to those in [32]; if they have circular boundaries, some results can be found in [45]. Polyhedral partitions can be described as $\sigma_k(x) := \bar{E}_k \bar{x} \succeq 0$, where $\bar{E}_k = \begin{bmatrix} E_k & e_k \end{bmatrix}$, $x \in \Omega_k$, $k \in \{1, 2, \ldots, q\}$; they can be systematically (but not uniquely) constructed as in [32, 46].

For each region Ω_k, all constraints can be joined in a vector of functions $\sigma_k(\cdot) := \left[\sigma_1^k(\cdot) \ \ldots \ \sigma_{n_k}^k(\cdot)\right]^T$; thus, we could define $\Omega_k = \{x : \sigma_k(x) \succeq 0\}$, where "$\succeq 0$" stands for element-wise "greater than 0". Such descriptions, as it will be shown later, can be exploited along with the piecewise Lyapunov functions to relax the conservativeness of a single-region single-Lyapunov function TS/LMI stability analysis. Unfortunately, affine models will not render convex LMI conditions when discussing controller design.

3.3 Polynomial Representations

The sector nonlinearity approach presented in Sect. 3.2.1 as well as the approximation method described in Sect. 3.2.2 produce TS models that can be viewed as convex nonlinear interpolations of linear models: nonlinearities are bounded between constants (sector nonlinearity) or approximated by the first-order Taylor terms (linearisation). Nevertheless, polynomials can also be used to bound nonlinearities via high-order approximations; in this way, the aforementioned methodologies can be generalised as polynomial convex representations [47, 48]. These models have attracted the interest of the control community due to the fact that polynomial positivity can be tested via efficient semidefinite-programming tools such as sum-of-squares (SOS) [49], which offers a computational framework for systematic stability analysis and controller synthesis of nonlinear models just in the same way as the LMI framework does for TS models.

Using high-order approximations instead of linear ones is a straightforward extension of the results showed in the previous section; it is, therefore, omitted. On the other hand, two generalisations of the sector nonlinearity approach have emerged: the extension to TS modelling proposed by [48] which is based on bounding the non-polynomial nonlinearities by constants while preserving the polynomial expressions for further use under SOS, and the generalisation of [47] which bounds the nonlinearities by polynomials of arbitrary order at the expense of the membership functions; since the latter includes the former, this methodology is detailed in the sequel.

Nonlinear dynamical systems as (3.8) will be considered, i.e.

$$\begin{aligned} \delta \boldsymbol{x}(t) &= \boldsymbol{f}(\boldsymbol{x}, \boldsymbol{u}, \boldsymbol{d}, \boldsymbol{\theta}, t), \\ \boldsymbol{y}(t) &= \boldsymbol{h}(\boldsymbol{x}, \boldsymbol{u}, \boldsymbol{d}, \boldsymbol{\theta}, t), \end{aligned} \tag{3.47}$$

where $\delta \boldsymbol{x}(t)$, $\boldsymbol{f}, \boldsymbol{h}, \boldsymbol{x}, \boldsymbol{u}, \boldsymbol{y}, \boldsymbol{d}$, and $\boldsymbol{\theta}$ have the same meanings as before. The construction of an exact polynomial convex representation from a nonlinear model mimics the sector nonlinearity methodology as it rewrites each non-polynomial expression (formerly nonlinearity) as a convex sum of polynomials (formerly linear terms). Thus, once all the non-polynomial functions $\xi_1(\cdot)$, $\xi_2(\cdot)$,..., $\xi_p(\cdot)$ in (3.47) have been identified, they will be rewritten as convex sums of polynomials of arbitrary order, following a Taylor series approach first described in [47] and detailed below. As with the premise variables in the TS context, these non-polynomial terms may depend on the state $\boldsymbol{x}$, the input $\boldsymbol{u}$, time t, disturbances $\boldsymbol{d}$, or parameters $\boldsymbol{\theta}$ and will be assumed to be bounded and smooth in a compact set $\mathcal{C}$. Thus, the vector $\boldsymbol{\xi}(\boldsymbol{z}) = [\xi_1(z_1)\ \xi_2(z_2)\ \cdots\ \xi_p(z_p)]^T$ may be viewed as a generalisation of the *premise* or *scheduling vector* $\boldsymbol{z}(\cdot)$ from the TS context.

In order to write each non-polynomial expression $\xi_j(z_j)$, $j \in \{1, 2, \ldots, p\}$ as a convex sum of polynomials, consider its Taylor approximation of order N around the origin (of course, any operating point for $\boldsymbol{z}$ may be chosen instead of the origin, if convenient):

$$\xi_j^N(z_j) = \sum_{i=0}^{N-1} \frac{\xi_j^{[i]}(0)}{i!} z_j^i,$$

with $\xi_j^0(z_j) = 0$, and let

$$T_j^N(z_j) = \frac{\xi_j(z_j) - \xi_j^N(z_j)}{z_j^N}.$$

In the compact set of interest $\mathcal{C}$, $T_j^N(z_j)$ is bounded; therefore, the following bounds are well defined:

$$\psi_j^1 := \sup_{z_j \in \mathcal{C}} T_j^N(z_j), \ \psi_j^2 := \inf_{z_j \in \mathcal{C}} T_j^N(z_j),$$

based on which the following convex rewriting of $T_j^N(z_j)$ arises:

$$T_j^N(z_j) = w_0^j \psi_1^1 + w_1^j \psi_1^2,$$

with

$$w_0^j(z_j) = \frac{T_j^N(z_j) - \psi_1^2}{\psi_1^1 - \psi_1^2}, \ w_1^j(z_j) = 1 - w_0^j(z_j). \tag{3.48}$$

Hence, as $\xi_j(z_j) = \xi_j^N(z_j) + T_j^N(z_j) z_j^N$, it can be expressed as

$$\xi_j(z_j) = w_0^j(z_j) p_0^j(z_j) + w_1^j(z_j) p_1^j(z_j),$$

where $p_0^j(z_j) = \xi_j^N(z_j) + \psi_1^1 z_j^N$ and $p_1^j(z_j) = \xi_j^N(z_j) + \psi_1^2 z_j^N$ are polynomials of degree N, and $w_0^j(z_j)$, $w_1^j(z_j)$ are weighting functions which hold the convex sum property in the compact $\mathcal{C}$.

Since this is done for every non-polynomial expression in (3.47), the MFs $h_i(\boldsymbol{\xi})$, $i \in \{1, 2, \ldots, r\}$, $r = 2^p$, can be defined as in (3.11) based on (3.48), i.e.

$$h_i(\boldsymbol{\xi}) = h_{d_1^i \times 2^{p-1} + \cdots + d_{p-1}^i \times 2 + d_p^i + 1} = \prod_{j=1}^{p} w_{d_j^i}^j(z_j), \tag{3.49}$$

with $i = d_1^i \times 2^{p-1} + \cdots + d_{p-1}^i \times 2 + d_p^i + 1 \in \{1, \ldots, 2^p\}$ and $d_j^i \in \{0, 1\}$ the jth bit of the p-bit binary representation of $i - 1$. These MFs inherit the convex sum property (3.2) in $\mathcal{C}$ from their constituting weights (3.48).

As in the TS case, each of the r MFs $h_i(\boldsymbol{\xi})$ represents a combination instance of extreme values of the non-polynomial expressions $\xi_j(z_j)$; the full polynomial convex model stems from evaluating the state and measurement functions $\boldsymbol{f}$ and $\boldsymbol{h}$ in each of these combinations, i.e. $\widehat{\boldsymbol{f}}_i = \boldsymbol{f}|_{h_i=1}$ and $\widehat{\boldsymbol{h}}_i = \boldsymbol{h}|_{h_i=1}$. Then, a polynomial convex representation of (3.47) is given by

$$\begin{aligned} \delta \boldsymbol{x}(t) &= \sum_{i=1}^{r} h_i(\boldsymbol{\xi}) \widehat{\boldsymbol{f}}_i, \\ \boldsymbol{y}(t) &= \sum_{i=1}^{r} h_i(\boldsymbol{\xi}) \widehat{\boldsymbol{h}}_i. \end{aligned} \tag{3.50}$$

Due to the way this model has been constructed, the functions $\widehat{\boldsymbol{f}}_i$ and $\widehat{\boldsymbol{h}}_i$ are vectors of polynomials resulting from polynomials already present in $\boldsymbol{f}$ and $\boldsymbol{h}$ as well as from products of polynomials $p_0^j(z_j)$, $p_1^j(z_j)$, $j \in \{1, 2, \ldots, p\}$ produced by the convex rewriting of non-polynomial terms. On the other hand, MFs $h_i(\boldsymbol{\xi})$ capture the non-polynomial expressions of the model into convex shapes, a key property for the Lyapunov-based stability analysis and design.

Note that setting $N = 1$ with $f(0) = 0$, we obtain the usual sector nonlinearity methodology that bounds a function between two first-degree polynomials.

Example 3.9 The discrete-time nonlinear system from Example 3.6, i.e.

$$\boldsymbol{x}(t+1) = \begin{bmatrix} 2x_1x_2 - 3x_1^2 + 2x_2 + u \\ x_1 + 3x_2 + x_2^2 + x_1 u \end{bmatrix},$$
$$y(t) = 2x_1^2 + x_2,$$

is already in the polynomial convex representation (3.50) with $r = 1$, $h_1 = 1$, $\widehat{\boldsymbol{f}}_1 = \begin{bmatrix} 2x_1x_2 - 3x_1^2 + 2x_2 + u \\ x_1 + 3x_2 + x_2^2 + x_1 u \end{bmatrix}$, and $\widehat{\boldsymbol{h}}_1 = 2x_1^2 + x_2$. □

Example 3.10 Consider again the ball and beam system in Example 3.4 whose model is reproduced here for convenience:

$$\dot{\boldsymbol{x}}(t) = \begin{bmatrix} x_2 \\ a x_1 x_4^2 - b \sin x_3 \\ x_4 \\ u \end{bmatrix}, \tag{3.51}$$

with the variables in $\mathcal{C} = \{\boldsymbol{x}, y, u : u, y \in \mathbb{R}, |x_i| \leq 1, i = 1, \ldots, 4\}$, and a and b constant parameters.

There is one non-polynomial nonlinearity in (3.51): $\sin x_3$, whose Taylor series expansion around $x_3 = 0$ is given by

$$\sin x_3 = x_3 - \frac{x_3^3}{3!} + \frac{x_3^5}{5!} - \frac{x_3^7}{7!} + \cdots$$

If only the first-degree term is used to rewrite $\sin x_3$ as a convex expression, it yields the same outcome as if sector nonlinearity was used, i.e. the terms

$$\xi_1^1(z_1) = 0, \quad T_1^1(z_1) = \frac{\sin z_1 - 0}{z_1},$$

where $N = 1$, $z_1 = x_3$, allow bounding $T_1^1(z_1)$ by $\psi_1^1 = 1$ and $\psi_1^2 = 0.8415$ in $|x_3| \leq 1$. Thus, the corresponding weights are defined as follows:

$$w_0^1(z_1) = \frac{T_1^1(z_1) - 0.8415}{1 - 0.8415}, \quad w_1^1(z_1) = 1 - w_0^1(z_1).$$

Since there is only one non-polynomial term in (3.51), these weights are also the MFs $h_i(\boldsymbol{\xi})$, $i = 1, 2$. The corresponding polynomials are

$$p_0^1(z_1) = 0 + 1z_1, \quad p_1^1(z_1) = 0 + 0.8415z_1.$$

Should third-degree polynomials be used instead to rewrite $\sin x_3$, the following expressions arise by the same procedure with $z_1 = x_3$:

$$\xi_1(z_1) = w_0^1(z_1)p_0^1(z_1) + w_1^1(z_1)p_1^1(z_1),$$

with

$$h_1(\xi_1) = w_0^j(z_1) = \frac{T_1^3(z_1) + 0.1667}{-0.1586 + 0.1667}, \; h_2(\xi_1) = w_1^1(z_1) = 1 - w_0^1(z_1),$$
$$p_0^1(z_1) = z_1 - 0.1586z_1^3, \; p_1^1(z_1) = x_3 - 0.1667z_1^3,$$

provided that

$$\xi_1^3(z_1) = 0 + z_1, \quad T_1^3(z_1) = \frac{\sin z_1 - z_1}{z_1^3}, \quad -0.1667 \leq T_1^3(z_1) \leq -0.1586.$$

In any case, the ball and beam nonlinear model (3.51) can be exactly rewritten as (3.50) with

$$\widehat{\boldsymbol{f}}_1 = \begin{bmatrix} x_2 \\ ax_1x_4^2 - bp_0^1 \\ x_4 \\ u \end{bmatrix}, \; \widehat{\boldsymbol{f}}_2 = \begin{bmatrix} x_2 \\ ax_1x_4^2 - bp_1^1 \\ x_4 \\ u \end{bmatrix},$$

and $\delta\boldsymbol{x}(t) = \dot{\boldsymbol{x}}(t)$, where the polynomials $p_0^1(z_1)$, $p_1^1(z_1)$ and the MFs $h_1(\xi_1)$, $h_2(\xi_1)$ are defined accordingly to the desired polynomial degree. □

Example 3.11 Consider the phase-locked-loop model [31]:

$$\dot{\boldsymbol{x}}(t) = \begin{bmatrix} x_2 \\ -c\sin x_1 - (a + b\cos x_1)\,x_2 \end{bmatrix} \tag{3.52}$$

within the compact set $\mathcal{C} = \{\boldsymbol{x} : |x_i| \leq 1, \; i = 1, 2\}$, with a, b, and c constant parameters.

We seek a polynomial convex representation of the form (3.50) for the model (3.52), where the two non-polynomial terms $\sin x_1$ and $\cos x_1$ are rewritten as convex sums of polynomials of fifth degree, i.e. $N = 5$. As before, the following terms are defined from the Taylor series expansion of the non-polynomial terms with $z_1 = x_1$, $z_2 = x_1$, $\xi_1(z_1) = \sin z_1$, and $\xi_2(z_2) = \cos z_2$:

$$\xi_1^5(z_1) = z_1 - \frac{z_1^3}{3!}, \quad \xi_2^5(z_2) = 1 - \frac{z_2^2}{2!} + \frac{z_2^4}{4!}.$$

Based on these terms, the following expressions and their bounds are found:

$$T_1^5(z_1) = \frac{\sin z_1 - z_1 + z_1^3/(3!)}{z_1^5},$$
$$T_2^5(z_2) = \frac{\cos z_2 - 1 + z_2^2/(2!) - z_2^4/(4!)}{z_2^5},$$

$$0.0081376 = \psi_1^1 \le T_1^5(z_1) \le \psi_1^2 = 0.0083336,$$
$$-0.0013643 = \psi_2^1 \le T_2^5(z_2) \le \psi_2^2 = 0.0013643.$$

The corresponding weights are

$$w_0^1(z_1) = \frac{T_1^5(z_1) - 0.0081376}{0.0083336 - 0.0081376}, \quad w_1^1(z_1) = 1 - w_0^1(z_1),$$
$$w_0^2(z_2) = \frac{T_2^5(z_2) + 0.0013643}{0.0013643 + 0.0013643}, \quad w_1^2(z_2) = 1 - w_0^2(z_2),$$

from which the MFs can be computed as

$$h_1(\boldsymbol{\xi}) = w_0^1(z_1)w_0^2(z_2),\ h_2(\boldsymbol{\xi}) = w_0^1(z_1)w_1^2(z_2),$$
$$h_3(\boldsymbol{\xi}) = w_1^1(z_1)w_0^2(z_2),\ h_4(\boldsymbol{\xi}) = w_1^1(z_1)w_1^2(z_2),$$

with $\boldsymbol{\xi} = [\xi_1\ \xi_2]^T$. The corresponding polynomials are

$$p_0^1(z_1) = z_1 - \frac{z_1^3}{3!} + 0.0083336z_1^5, \quad p_1^1(z_1) = z_1 - \frac{z_1^3}{3!} + 0.0081376z_1^5,$$

$$p_0^2(z_2) = 1 - \frac{z_2^2}{2!} + \frac{z_2^4}{4!} + 0.0013643z_2^5, \quad p_1^2(z_2) = 1 - \frac{z_2^2}{2!} + \frac{z_2^4}{4!} - 0.0013643z_2^5.$$

Therefore, under the previous definitions, an exact fifth-degree polynomial convex rewriting of (3.52) comes as

$$\dot{x}(t) = \sum_{i=1}^{4} h_i(\boldsymbol{\xi})\widehat{f}_i, \tag{3.53}$$

with

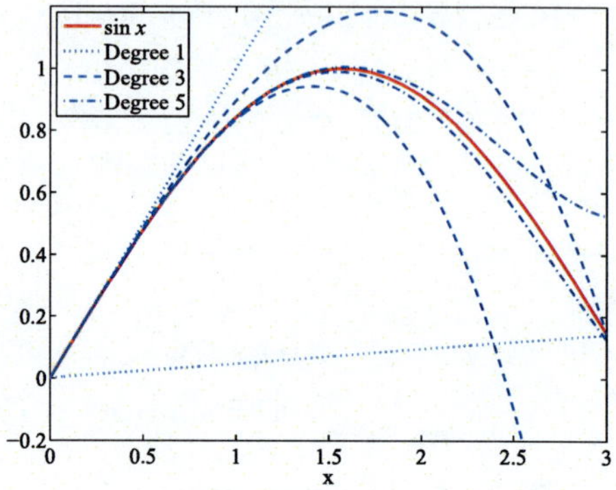

Fig. 3.6 Polynomial bounds of $\sin x$ for $x \in [0, 3]$

$$\widehat{\boldsymbol{f}}_1 = \begin{bmatrix} x_2 \\ -cp_0^1(z_1) - \left(a + bp_0^2(z_2)\right) x_2 \end{bmatrix},$$
$$\widehat{\boldsymbol{f}}_2 = \begin{bmatrix} x_2 \\ -cp_0^1(z_1) - \left(a + bp_1^2(z_2)\right) x_2 \end{bmatrix},$$
$$\widehat{\boldsymbol{f}}_3 = \begin{bmatrix} x_2 \\ -cp_1^1(z_1) - \left(a + bp_0^2(z_2)\right) x_2 \end{bmatrix},$$
$$\widehat{\boldsymbol{f}}_4 = \begin{bmatrix} x_2 \\ -cp_1^1(z_1) - \left(a + bp_1^2(z_2)\right) x_2 \end{bmatrix}.$$

Note that, as expected, $\widehat{\boldsymbol{f}}_i$, $i \in \{1, 2, 3, 4\}$ are vectors of polynomials. □

The function $\sin x$ has been bounded by first-, third- and fifth-degree polynomials through Examples 3.10 and 3.11 in the interval $x \in [-1, 1]$. Figure 3.6 illustrates the fact that these bounds get progressively closer to the actual function; new bounds for the interval $x \in [0, 3]$ have been used.

3.4 Descriptor Representations

Some dynamical systems arising in applications are better described by differential/difference equations along with algebraic restrictions [50, 51].

Definition 3.4 [52] A set of *differential/difference-algebraic equations* (DAEs) is a set of differential/difference equations that hold under a set of algebraic restrictions, namely

$$\boldsymbol{f}(\boldsymbol{x}, \delta\boldsymbol{x}, \boldsymbol{u}, \boldsymbol{d}, \boldsymbol{\theta}, t) = 0, \quad \boldsymbol{g}(\boldsymbol{x}, \boldsymbol{u}, \boldsymbol{d}, \boldsymbol{\theta}, t) = 0, \tag{3.54}$$

where, as before, $\delta\boldsymbol{x} = \dot{\boldsymbol{x}}(t)$ with $t \in \mathbb{R}$ for continuous-time systems (differential equations); $\delta\boldsymbol{x} = \boldsymbol{x}(t+1)$ with $t \in \mathbb{Z}$ for discrete-time systems (difference equations); $\boldsymbol{f}$ and $\boldsymbol{g}$ are smooth nonlinear vector fields that stand for the dynamical and algebraic parts of the DAE, respectively; $\boldsymbol{x} \in \mathbb{R}^n$ and $\boldsymbol{u} \in \mathbb{R}^m$ are state and input vectors, respectively; and $\boldsymbol{\theta}$ and $\boldsymbol{d}$ group parameters and exogenous signals acting on the system, respectively.

As the governing variables are not just state variables, but reaction forces, balances, subject to holonomic or non-holonomic constraints, etc., these variables were historically called *descriptor variables* and their equations called *descriptor systems* [53, 54]; nevertheless, in a broad sense, the terminology "singular system", "descriptor systems", or DAE should be understood as referring to the same thing in this manuscript.

DAE models are the most general way to describe "lumped-parameter" systems, being ordinary differential equations a particular case (although at first sight these representations may seem different from ordinary differential equations (ODEs), most of them are amenable to the latter class via the *Pantelides algorithm* [55] as, actually, in a general case, only ODEs can actually be simulated by computers). Some examples of descriptor modelling of electrical circuits [56], chemical reactors [57], mechanical setups subject to kinematic constraints [58], Markovian processes [59, 60], etc. appear in literature.

There is no loss of generality in expressing the dynamical part as an integrator, as we may express, if so wished $\delta \boldsymbol{x} = y$ and actually describe the rest of the system model with only algebraic equations:

$$\bar{\boldsymbol{f}}(\boldsymbol{x}, y, \boldsymbol{u}, \boldsymbol{d}, \boldsymbol{\theta}, t) = 0, \tag{3.55}$$

where $\bar{\boldsymbol{f}}$ encompasses $\boldsymbol{f}(\cdot)$ and $\boldsymbol{g}(\cdot)$.

The above shows the large degree of redundancy of DAE representations, and they are not unique not even up to invertible transformations (as in the case of ODE state-space representations), but different descriptions with different numbers of variables and equations may refer to the same underlying system, and manipulations of the equations may involve not only algebraic operations but derivatives and integrals.

The DAE representation is so general that, well, they can model meaningless systems with impulsive responses, non-unique solutions, non-holonomic setups, etc. Some "regularity" conditions are needed, which, however, are difficult to state in a "global", nonlinear setup. The linear case is well studied and summarised below:

Definition 3.5 A *linear time-invariant DAE* is one whose model is written as

$$E\dot{\boldsymbol{x}}(t) = A\boldsymbol{x}(t) \tag{3.56}$$

with possibly singular E. There is no loss in generality in assuming E being square as $\boldsymbol{x}$ can be extended and E and A suitably padded with zeros. These systems are also called *LTI descriptor systems* or *singular linear systems.*

In the case of invertible E, we would have the ODE $\dot{\boldsymbol{x}} = E^{-1}A\boldsymbol{x}$.

Definition 3.6 A singular linear system (3.56) is written in the *dynamic decomposition form* (DDF) if it has the following model:

$$\begin{bmatrix} I & 0 \\ 0 & 0 \end{bmatrix} \begin{bmatrix} \dot{x}_1 \\ \dot{x}_2 \end{bmatrix} = \begin{bmatrix} A_{11} & A_{12} \\ A_{21} & A_{22} \end{bmatrix} \begin{bmatrix} x_1 \\ x_2 \end{bmatrix},$$

with x_1 and x_2 being the partition of the new state coordinates.

A singular linear system is always amenable to the DDF, and in fact, (3.55) results in a similar form for a nonlinear case.

Definition 3.7 [61] The pair (E, A) is *regular* if $\det(sE - A) \neq 0$ for some $s \in \mathbb{C}$.

Informally, it means that the Laplace transform can be used to obtain the associated time response for given initial conditions. Regularity implies the existence and uniqueness of solutions; these solutions may, however, have some impulsive terms if initial conditions are arbitrary and they do not obey the algebraic constraints. These situations must, of course, be avoided in actual applications. Again, particularising to the linear case, we may state the following:

Definition 3.8 [62] The singular linear system (3.56) has *non-impulsive* or *impulse-free* behaviour if $\deg(\det(sE - A)) = \text{rank}(E)$, which is equivalent to condition $\det A_{22} \neq 0$ in the DDF form of Definition 3.6.

The fact that impulsive behaviour may arise in singular linear systems can be seen from its general solution, which is based on the Kronecker form:

Definition 3.9 [61] A singular linear system (3.56) is written in the *Kronecker form* if it has the following model:

$$\begin{bmatrix} I & 0 \\ 0 & N \end{bmatrix} \begin{bmatrix} \dot{x}_1 \\ \dot{x}_2 \end{bmatrix} = \begin{bmatrix} A_1 & 0 \\ 0 & I \end{bmatrix} \begin{bmatrix} x_1 \\ x_2 \end{bmatrix},$$

with x_1 and x_2 new state vector partitions.

A singular linear system is always amenable to the Kronecker form.

The general solution of a singular linear system (3.56) is given by

$$x(t, x_0) = P \begin{bmatrix} e^{A_1 t} \begin{bmatrix} I & 0 \end{bmatrix} P^{-1} x_0 \\ -\sum_{i=1}^{h-1} \delta^{(i-1)}(t) N^i \begin{bmatrix} 0 & I \end{bmatrix} P^{-1} x_0 \end{bmatrix},$$

which, if $N \neq 0$, may present impulsive behaviour.

When no input is present, consistent initialisation eliminates impulsive behaviour even if $N \neq 0$.

The general solution of a singular linear system (3.56) is given by

$$x(t, x_0) = P \begin{bmatrix} e^{A_1 t} \begin{bmatrix} I & 0 \end{bmatrix} P^{-1} x_0 \\ -\sum_{i=1}^{h-1} \delta^{(i-1)}(t) N^i \begin{bmatrix} 0 & I \end{bmatrix} P^{-1} x_0 \end{bmatrix},$$

which, if $N \neq 0$, may present impulsive behaviour.

When no input is present, consistent initialisation eliminates impulsive behaviour even if $N \neq 0$. These considerations are important as time-response properties should be redefined in terms of the subset of the state space induced by algebraic restrictions.

The need to constrain variables to the algebraic restrictions gives an alternate definition (or perhaps just an interpretation) of a DAE, a.k.a. singular system:

Definition 3.10 [52] A *singular system* is a set of DAEs (3.54) whose algebraic equations restrict the solutions of the differential/difference equations to a subset of the original solution space.

The concept of (local) regularity of a nonlinear DAE can be informally related to the regularity of its linearisation; also, the idea about a system having or not "impulsive" response can be somehow extended to the nonlinear case with the definition below:

Definition 3.11 [57] The *differential/difference index* of a set of DAEs (3.54) is the number of differentiation/difference steps on the algebraic restriction required to determine the missing dynamics in the differential/difference equations.

Thus, the Pantelides algorithm consists in successive differentiation/forward difference of the algebraic restrictions up to the DAE index, in order to determine a set of ODEs with a subset of solutions that behave exactly as those of the DAE. This process cannot be always applied as many "consistency requisites" intervene [55]. The resulting ODEs usually have more dynamical equations than the original representation due to the introduction of spurious states or the completion of missing dynamics. This apparent growth in dimension is, nevertheless, subject to a *consistent initialisation* of the states as they must hold the algebraic restrictions (those explicitly stated in the original DAE as well as those derived during the application of the Pantelides algorithm). Once a DAE has been transformed into an ODE, it can be correctly simulated via conventional tools; otherwise, recently appeared toolboxes are able to simulate a DAE from scratch [63].

In their most general (nonlinear) form, DAEs are very hard to analyse and are still under study; they present some phenomena which do not exist in ODEs, e.g. indeterminacy, impasse points, and generalised impulse functions in the solution [61]. These phenomena are usually associated with the singular nature of the DAE:

Definition 3.12 A *semilinear descriptor model* is a state-space representation of a DAE with the following form:

$$E(\boldsymbol{x}, \boldsymbol{u}, \boldsymbol{d}, \boldsymbol{\theta}, t)\delta\boldsymbol{x}(t) = \boldsymbol{f}(\boldsymbol{x}, \boldsymbol{u}, \boldsymbol{d}, \boldsymbol{\theta}, t), \tag{3.57}$$

where $\delta\boldsymbol{x}(t)$, $\boldsymbol{x}$, $\boldsymbol{u}$, $\boldsymbol{d}$, $\boldsymbol{\theta}$, and $\boldsymbol{f} \in \mathbb{R}^n$ have the usual meaning, and $E(\cdot) \in \mathbb{R}^{n\times n}$ is a continuously differentiable nonlinear matrix function of its arguments, sometimes denoted as *mass matrix*.

For instance, $(\dot{x})^3 = u - x^2$ is a generic descriptor system, but not semilinear; $(1 + x)\dot{x} = u$ is a semilinear one; its mass matrix becomes singular at $x = -1$.

In general, descriptor models may have *rectangular* $E(\cdot)$, but via the Pantelides algorithm, they are amenable to a *square* form as the one above.

The null space of E implicitly defines algebraic constraints in a generic DAE (3.54). If, as a particular case, $E(\cdot)$ has full rank, the semilinear descriptor form can be reduced to an ODE just by multiplying both sides by the inverse of the mass matrix, i.e.

$$\delta \boldsymbol{x} = E^{-1}(\boldsymbol{x}, \boldsymbol{u}, \boldsymbol{d}, \boldsymbol{\theta}, t)\boldsymbol{f}(\boldsymbol{x}, \boldsymbol{u}, \boldsymbol{d}, \boldsymbol{\theta}, t). \tag{3.58}$$

In the sequel, we will specialise on the modelling, stability, and control design of the above semilinear DAEs, obtaining convex bounding models for both E and $\boldsymbol{f}$. Thus, the term "semilinear" will be omitted most of the times, for brevity. We will, too, assume in the sequel that the modelling region is such that the mass matrix does not change rank in it (if it does, later LMI conditions will not be feasible); actual invertibility of E will, too, be needed for the feasibility of some of the results in next chapters, whereas some others will be able to yield the Lyapunov functions for non-invertible E.

Even in the case *of invertible mass matrix*, the semilinear descriptor form may still preserve some advantages over its ODE counterpart because it avoids matrix inversion; this translates into a lesser number of vertex models when applying the convex modelling techniques in Sects. 3.2 and 3.3 [64, 65] to nonlinear/uncertain systems (of course, no advantage arises if E were a known constant matrix). A non-singular E appears in, for instance, the case of models arising from the Euler–Lagrange or the Newton–Euler descriptions of robotic manipulators [66–68].

Alternatively, the so-called linear fractional (LFT) representations, to be discussed in Sect. 3.5.3, might be a viable alternative for handling such rational expressions of uncertain elements without an excessive number of vertices, also avoiding introducing the descriptor representation in regular cases; in fact, LFT and descriptors are in a way equivalent in some common cases, as discussed later on in this chapter.

Obtaining an exact or approximate, TS or polynomial model of a nonlinear descriptor one (3.57), preserving the right- and left-hand side contents, does not require anything else than the techniques described in the previous sections: the result will be what is known as a *TS descriptor model* or *polynomial descriptor model*. Nevertheless, in contrast with the standard state-space convex representations, the descriptor ones usually consist of two sets of convex models: one with r_e vertices for the left-hand side and the other with r vertices for the right-hand side of the differential/difference equations, i.e.

$$\sum_{i=1}^{r_e} v_i(z) E_i \delta \boldsymbol{x}(t) = \sum_{i=1}^{r} h_i(z) \widehat{\boldsymbol{f}}_i, \tag{3.59}$$

where E_i are matrices of proper size, and $\widehat{\boldsymbol{f}}_i$ are constant, linear, or polynomial functions depending upon the premise vector z, which in turn may depend on states, time,

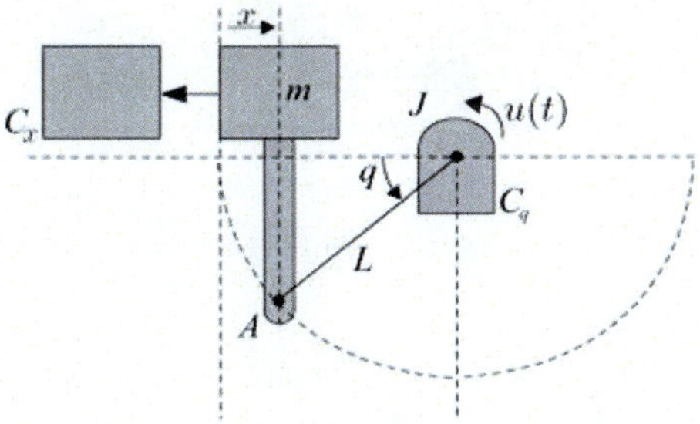

Fig. 3.7 Scotch yoke mechanism

or exogenous variables. As before, each of the r MFs $h_i(z)$ represents a combination instance of extreme values of the nonlinearities on the right-hand side, while the r_e MFs $v_i(z)$ have the same meaning for the left-hand side: they hold the convex sum property for every $z \in \mathcal{C} \subseteq \mathbb{R}^p$. Of course, due to convexity, these two sets of models can be combined into a single one, if convenient, by defining a new set of MFs in terms of the two original sets, i.e. $\bar{h}_k(z) = h_i(z)v_j(z)$, $k \in \{1, 2, \ldots, r \times r_e\}$. The ensuing examples illustrate this discussion.

Example 3.12 Consider the Lagrangian model of the scotch yoke mechanism driven by an electric motor [69], cf. Fig.3.7:

$$\left(\frac{J}{L^2 \sin^2 q}+m\right)\ddot{x}(t)-\frac{J\cos q}{L^3 \sin^4 q}\dot{x}^2(t)+\left(\frac{C_q}{L^2 \sin^2 q}+C_x\right)\dot{x}(t)=\frac{u(t)}{L\sin q}, \quad (3.60)$$

with q the rod angle, $L = 0.5, m = 5, J = 8.75 \times 10^{-3}, C_q = 1, C_x = 6, x = L(1 - \cos q)$, and $u(t)$ the system input.

Defining $\xi = \left[x \ \dot{x}\right]^T$ as the model state, a nonlinear descriptor form (3.57) of model (3.60) can be obtained by simply rewriting it as follows:

$$\begin{bmatrix} 1 & 0 \\ \dfrac{c_q}{L^2 \sin^2 q} + C_x & \dfrac{J}{L^2 \sin^2 q} + m \end{bmatrix} \dot{\xi}(t) = \begin{bmatrix} 0 & 1 \\ 0 & \dfrac{J\cos q}{L^3 \sin^4 q}\dot{x} \end{bmatrix} \xi(t) + \begin{bmatrix} 0 \\ \dfrac{1}{L\sin q} \end{bmatrix} u(t). \quad (3.61)$$

Note that there is one missing model equation $x = L(1 - \cos q)$ or, equivalently, $q = \arccos(1 - L^{-1}x)$ should have been replaced above. This is intentional, as its effect will be modelled within some convex bounds in order to get a descriptor TS model representation.

Three nonlinearities are considered: $z_1 = 1/(\sin^2 q)$, $z_2 = (\cos q)\dot{x}/(\sin^4 q)$, and $z_3 = 1/(\sin q)$, whose bounds can be inferred from bound on x, $\dot{x}$, q, and $\dot{q}$, thus inducing the compact set

$$\mathcal{C} = \{z : z_i^0 \leq z_i \leq z_i^1, i \in 1, 2, 3\},$$

with z_i^0, z_i^1, $i \in \{1, 2, 3\}$ being the minimum and the maximum of nonlinearities z_1, z_2, and z_3, respectively.

Thus, the following pairs of weighting functions arise:

$$w_0^1(z_1) = \frac{z_1 - z_1^0}{z_1^1 - z_1^0}, \quad w_1^1(z_1) = 1 - w_0^1(z_1),$$
$$w_0^2(z_2) = \frac{z_2 - z_2^0}{z_2^1 - z_2^0}, \quad w_1^2(z_2) = 1 - w_0^2(z_2),$$
$$w_0^3(z_3) = \frac{z_3 - z_3^0}{z_3^1 - z_3^0}, \quad w_1^3(z_3) = 1 - w_0^3(z_3).$$

To obtain a TS descriptor model of the form (3.59), two MFs are defined for the left-hand side based on w_0^1 and w_1^1; four others for the right-hand side model based on w_0^i and w_1^i, $i = 2, 3$, i.e. $v_1(z_1) = w_0^1(z_1)$, $v_2(z_1) = w_1^1(z_1)$, $h_1(z_2, z_3) = w_0^2(z_2)w_0^3(z_3)$, $h_2(z_2, z_3) = w_0^2(z_2)w_1^3(z_3)$, $h_3(z_2, z_3) = w_1^2(z_2)w_0^3(z_3)$, and $h_4(z_2, z_3) = w_1^2(z_2)w_1^3(z_3)$. At last, (3.61) is equivalent to

$$\sum_{k=1}^{2} v_k(z)E_k\dot{\xi}(t) = \sum_{i=1}^{4} h_i(z)(A_i\xi(t) + B_iu(t)), \tag{3.62}$$

with $z = [z_1\, z_2\, z_3]^T$,

$$E_1 = \begin{bmatrix} 1 & 0 \\ \frac{C_q}{L^2}z_1^1 + C_x & \frac{J}{L^2}z_1^1 + m \end{bmatrix}, \quad E_2 = \begin{bmatrix} 1 & 0 \\ \frac{C_q}{L^2}z_1^0 + C_x & \frac{J}{L^2}z_1^0 + m \end{bmatrix},$$

$$A_1 = A_2 = \begin{bmatrix} 0 & 1 \\ 0 & \frac{J}{L^3}z_2^1 \end{bmatrix}, \quad A_3 = A_4 = \begin{bmatrix} 0 & 1 \\ 0 & \frac{J}{L^3}z_2^0 \end{bmatrix},$$

$$B_1 = B_3 = \begin{bmatrix} 0 \\ \frac{1}{L}z_3^1 \end{bmatrix}, \quad B_2 = B_4 = \begin{bmatrix} 0 \\ \frac{1}{L}z_3^0 \end{bmatrix}.$$

□

Example 3.13 Consider the following inverted pendulum on a cart in the nonlinear descriptor form (3.57) (cf. Fig.3.8):

$$\begin{aligned} \dot{x}_1(t) &= x_2(t), \\ e(x_1)\dot{x}_2(t) &= f(x_1, x_2) + g(x_1)u(t), \end{aligned} \tag{3.63}$$

where $e(x_1) = \left(ml\left(1/3 + \sin^2 x_1\right) + 4/3Ml\right)$; $f(x_1, x_2) = (m + M)\, g \sin x_1 - mlx_2^2(t) \sin x_1 \cos x_1$; $g(x_1) = -\cos x_1$; $M = 1$ and $m = 0.1$ are the masses of the car and the rod, respectively; $l = 0.5$ is the length of the rod; $g = 9.8$ is the gravity

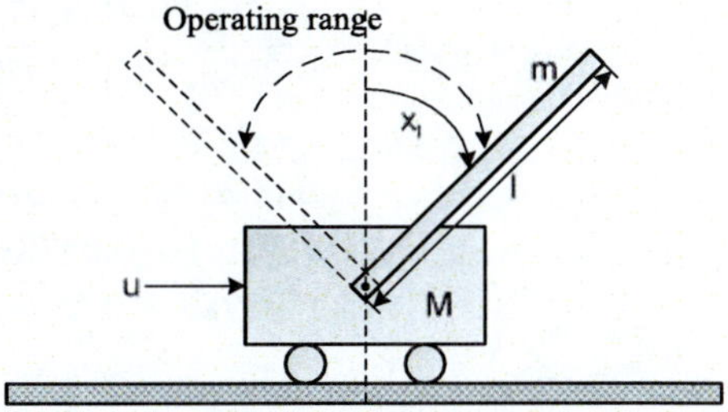

Fig. 3.8 Inverted pendulum

constant; x_1 is the angle from the vertical axis to the pendulum rod; and x_2 is the pendulum angular velocity. It is assumed that $|x_1| \le 22\pi/45$ and $|x_2| \le 0.1$. We pursue a polynomial descriptor model of (3.63).

In order to rewrite the three non-polynomial terms in (3.63), $\sin x_1$, $\cos x_1$, and $\sin^2 x_1$, as convex sums of fifth-degree polynomials ($N = 5$), let us define $z_1 = x_1$, $z_2 = x_1$, and $z_3 = x_1$, along with $\xi_1(z_1) = \sin z_1$, $\xi_2(z_2) = \cos z_2$, and $\xi_3(z_3) = \sin^2 z_3$. The following expressions are defined from the Taylor series expansion of the non-polynomial terms:

$$\xi_1^5(z_1) = z_1 - \frac{z_1^3}{6} = x_1 - \frac{x_1^3}{6},$$
$$\xi_2^5(z_2) = 1 - \frac{z_2^2}{2} + \frac{z_2^4}{24} = 1 - \frac{x_1^2}{2} + \frac{x_1^4}{24},$$
$$\xi_3^5(z_3) = z_3^2 - \frac{z_3^4}{3} = x_1^2 - \frac{x_1^4}{3}.$$

Based on these terms, the following expressions and their bounds are found:

$$T_1^5(z_1) = \frac{\sin z_1 - z_1 + \frac{z_1^3}{3!}}{z_1^5} = \frac{\sin x_1 - x_1 + \frac{x_1^3}{3!}}{x_1^5},$$
$$T_2^5(z_2) = \frac{\cos z_2 - 1 + \frac{z_2^2}{2!} - \frac{z_2^4}{4!}}{z_2^5} = \frac{\cos x_1 - 1 + \frac{x_1^2}{2!} - \frac{x_1^4}{4!}}{x_1^5},$$
$$T_3^5(z_3) = \frac{\sin^2 z_3 - z_3^2 + \frac{z_3^4}{3}}{z_3^5} = \frac{\sin^2 x_1 - x_1^2 + \frac{x_1^4}{3}}{x_1^5},$$

$$0.00788 = \psi_1^1 \le T_1^5(z_1) \le \psi_1^2 = 0.008333,$$
$$-0.002046 = \psi_2^1 \le T_2^5(z_2) \le \psi_2^2 = 0.002046,$$
$$-0.05788 = \psi_3^1 \le T_3^5(z_3) \le \psi_3^2 = 0.05788.$$

The corresponding weights are, thus, defined as follows:

$$w_0^1(z_1) = \frac{T_1^5(z_1) - 0.00788}{0.008333 - 0.00788}, \quad w_1^1(z_1) = 1 - w_0^1(z_1),$$
$$w_0^2(z_2) = \frac{T_2^5(z_2) + 0.002046}{0.002046 + 0.002046}, \quad w_1^2(z_2) = 1 - w_0^2(z_2),$$
$$w_0^3(z_3) = \frac{T_3^5(z_3) + 0.05788}{0.05788 + 0.05788}, \quad w_1^3(z_3) = 1 - w_0^3(z_3),$$

from which two sets of MFs can be formed

$$v_1(\boldsymbol{\xi}) = w_0^3(z_3),\ v_2(\boldsymbol{\xi}) = w_1^3(z_3),$$
$$h_1(\boldsymbol{\xi}) = w_0^1(z_1)w_0^2(z_2),\ h_2(\boldsymbol{\xi}) = w_0^1(z_1)w_1^2(z_2),$$
$$h_3(\boldsymbol{\xi}) = w_1^1(z_1)w_0^2(z_2),\ h_4(\boldsymbol{\xi}) = w_1^1(z_1)w_1^2(z_2).$$

The corresponding polynomials are

$$p_0^1(z_1) = z_1 - \frac{z_1^3}{3!} + 0.008333z_1^5, \quad p_1^1(z_1) = z_1 - \frac{z_1^3}{3!} + 0.007788z_1^5,$$

$$p_0^2(z_2) = 1 - \frac{z_2^2}{2!} + \frac{z_2^4}{4!} + 0.002046z_2^5, \quad p_1^2(z_2) = 1 - \frac{z_2^2}{2!} + \frac{z_2^4}{4!} - 0.002046z_2^5,$$

$$p_0^3(z_3) = \frac{2z_3^2}{2!} - \frac{8z_3^4}{4!} + 0.05788z_3^5, \quad p_1^3(z_3) = \frac{2z_3^2}{2!} - \frac{8z_3^4}{4!} - 0.05788z_3^5.$$

Therefore, under the previous definitions, an exact fifth-degree polynomial convex rewriting of (3.63) is given by

$$\sum_{i=1}^{2} v_i(\boldsymbol{\xi})E_i\dot{\boldsymbol{x}}(t) = \sum_{i=1}^{4} h_i(\boldsymbol{\xi})\widehat{\boldsymbol{f}}_i, \tag{3.64}$$

with

$$E_1 = \begin{bmatrix} 1 & 0 \\ 0 & ml\left(\frac{1}{3} + p_0^3(z_3)\right) + \frac{4}{3}lM \end{bmatrix},$$
$$E_2 = \begin{bmatrix} 1 & 0 \\ 0 & ml\left(\frac{1}{3} + p_1^3(z_3)\right) + \frac{4}{3}lM \end{bmatrix},$$

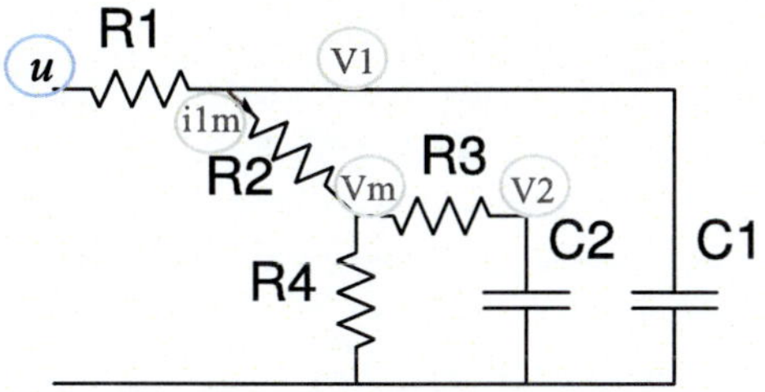

Fig. 3.9 Electrical circuit

$$
\widehat{f}_1 = \begin{bmatrix} x_2 \\ (m+M)gp_0^1 - mlx_2^2 p_0^1(z_1)p_0^2(z_2) - p_0^2(z_2)u(t) \end{bmatrix},
$$
$$
\widehat{f}_2 = \begin{bmatrix} x_2 \\ (m+M)gp_0^1 - mlx_2^2 p_0^1(z_1)p_1^2(z_2) - p_1^2(z_2)u(t) \end{bmatrix},
$$
$$
\widehat{f}_3 = \begin{bmatrix} x_2 \\ (m+M)gp_1^1 - mlx_2^2 p_1^1(z_1)p_0^2(z_2) - p_0^2(z_2)u(t) \end{bmatrix},
$$
$$
\widehat{f}_4 = \begin{bmatrix} x_2 \\ (m+M)gp_1^1 - mlx_2^2 p_1^1(z_1)p_1^2(z_2) - p_1^2(z_2)u(t) \end{bmatrix}.
$$

As expected, the components of E_i, $i \in 1, 2$, and $\widehat{\boldsymbol{f}}_j$, $j \in \{1, 2, 3, 4\}$ are polynomials. □

If $E(\boldsymbol{x}, \boldsymbol{u}, \boldsymbol{d}, \boldsymbol{\theta}, t)$ has constant, but not full rank, the descriptor amounts to a class of singular systems, whose LTI case is discussed in [53, 70] and references therein. Even in the LTI case, such systems can be not only unstable, but even improper (yielding non-causal predictors in the discrete-time case, or impulsive solutions in the continuous-time one). If $E(\boldsymbol{x}, \boldsymbol{u}, \boldsymbol{d}, \boldsymbol{\theta}, t)$ changes rank depending on the values of its arguments, then we have a system whose state-space order can change with time. These systems will not be considered.

Descriptor systems with non-invertible E naturally arise from plainly juxtaposing differential equations with algebraic ones. In fact, this is the historical root of the denomination, as the state vector is changed to a descriptor vector whose variables naturally describe the model dynamics [53]. In order to briefly illustrate singular systems, their descriptor forms, as well as concepts such as differentiability index and the Pantelides algorithm, the next two linear examples are considered.

Example 3.14 Consider the electrical circuit in Fig. 3.9. The involved descriptor variables are voltages at capacitors $C1$ and $C2$ (V_1 and V_2); voltage at the node joining $R2$, $R3$, and $R4$ (denoted as V_m); and current through $R2$ (denoted as i_{1m}). Apart from those variables, we have the system input voltage, denoted as u.

Choosing a descriptor vector $\boldsymbol{x} = [V_1\ V_2\ i_{1m}\ V_m]^T$, we can write the model equations as follows:

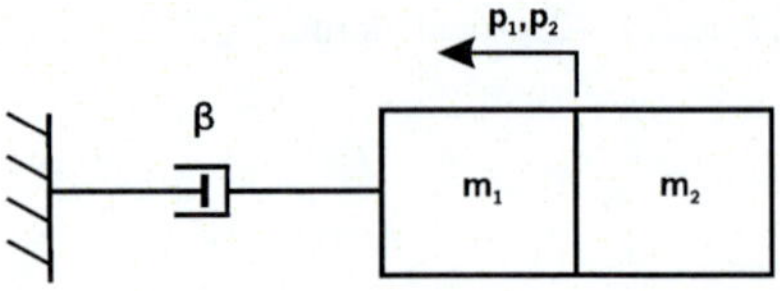

Fig. 3.10 Two-mass-damper system scheme

$$\begin{bmatrix} C_1 & 0 & 0 & 0 \\ 0 & C_2 & 0 & 0 \\ 0 & 0 & 0 & 0 \\ 0 & 0 & 0 & 0 \end{bmatrix} \dot{\boldsymbol{x}}(t) = \begin{bmatrix} -\frac{1}{R1} & 0 & -1 & 0 \\ 0 & -\frac{1}{R3} & 0 & \frac{1}{R3} \\ 0 & -\frac{1}{R3} & -1 & \frac{1}{R4} + \frac{1}{R3} \\ 1 & 0 & R2 & -1 \end{bmatrix} \boldsymbol{x}(t) + \begin{bmatrix} \frac{1}{R1} \\ 0 \\ 0 \\ 0 \end{bmatrix} u(t). \quad (3.65)$$

Note that at the left-hand side of the equation above, the matrix operations actually result in $\left[C_1 \dot{V}_1 \; C_2 \dot{V}_2 \; 0 \; 0\right]^T$, so the last two equations are actually algebraic ones as the derivatives of i_{1m} and V_m are multiplied by zero: any dummy variable could appear instead of the last two derivatives, but, by convention, the derivatives of the descriptor vector are used when writing (3.57). The system in the example is linear, but a TS descriptor system can be considered if uncertain/nonlinear/time-varying resistors or capacitors were considered instead. □

Example 3.15 Consider a two-mass-damper system as the one shown in Fig. 3.10 (units omitted for convenience), where the masses $m_1 = m_2 = 1$ are linked and the damping coefficient is $\beta = 1$.

The linkage condition can be expressed as the constraint $p_1 - p_2 = 0$, where p_1 and p_2 stand for the positions of masses m_1 and m_2, respectively. Therefore, the equations of motion can be obtained as in [71]:

$$\frac{d}{dt}\left(\frac{\partial L}{\partial \dot{p}_i}\right) - \frac{\partial L}{\partial p_i} = -\frac{\partial R}{\partial \dot{p}_i} - \lambda \frac{\partial f}{\partial p_i}, \; i = 1, 2,$$

where $L = (m_1 \dot{p}_1^2)/2 + (m_2 \dot{p}_2^2)/2$ is the Lagrangian, $R = (\beta \dot{p}_1^2)/2$ the Rayleigh dissipation function, $f(p_i) = p_1 - p_2$, and λ is the Lagrange multiplier used to include the constraint [72]. Computing this expression, we obtain

$$\ddot{p}_1(t) = -\dot{p}_1(t) - \lambda, \quad \ddot{p}_2(t) = \lambda.$$

If we define $v_i = \dot{p}_i$, $i \in \{1, 2\}$, (velocities), and the force constraint as $F = \lambda(\partial f(p_i)/p_1) = \lambda$, then the system

$$\begin{aligned} \dot{v}_1(t) &= -v_1(t) - F(t), \\ \dot{v}_2(t) &= F(t) \end{aligned} \quad (3.66)$$

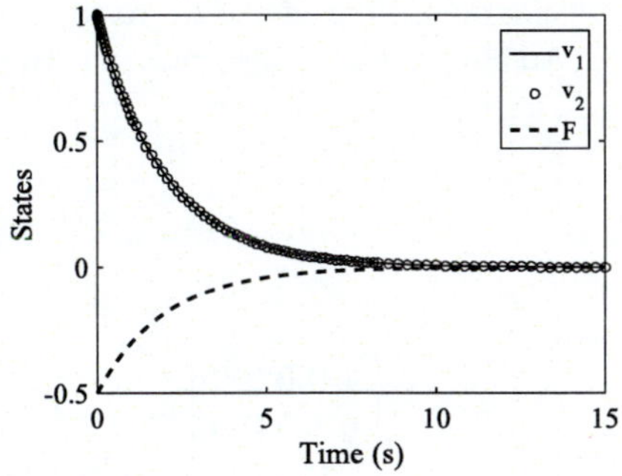

Fig. 3.11 Time evolution of the states of (3.67)

arises, subject to restriction $v_1 - v_2 = 0$, which reflects the fact that the masses are coupled.

If written in a matrix form (and taking F as a third descriptor variable), we have

$$\begin{bmatrix} 1 & 0 & 0 \\ 0 & 1 & 0 \\ 0 & 0 & 0 \end{bmatrix} \begin{bmatrix} \dot{v}_1 \\ \dot{v}_2 \\ \dot{F} \end{bmatrix} = \begin{bmatrix} -1 & 0 & -1 \\ 0 & 0 & 1 \\ 1 & -1 & 0 \end{bmatrix} \begin{bmatrix} v_1 \\ v_2 \\ F \end{bmatrix}. \tag{3.67}$$

Note that this is a singular descriptor system of the form:

$$E\dot{\boldsymbol{x}}(t) = A\boldsymbol{x}(t), \tag{3.68}$$

with

$$E = \begin{bmatrix} 1 & 0 & 0 \\ 0 & 1 & 0 \\ 0 & 0 & 0 \end{bmatrix}, A = \begin{bmatrix} -1 & 0 & -1 \\ 0 & 0 & 1 \\ 1 & -1 & 0 \end{bmatrix}, \boldsymbol{x} = \begin{bmatrix} v_1 \\ v_2 \\ F \end{bmatrix}.$$

Descriptor system (3.67) has index 2 as follows from deriving the algebraic restriction until the time derivative of the third state F appears:

$$v_1 - v_2 = 0 \Leftrightarrow \dot{v}_1 - \dot{v}_2 = 0 \Leftrightarrow -v_1 - F - F = 0, \tag{3.69}$$

$$\Leftrightarrow -\dot{v}_1 - 2\dot{F} = 0 \Leftrightarrow v_1 + F - 2\dot{F} = 0 \Leftrightarrow \dot{F} = \frac{v_1 + F}{2}. \tag{3.70}$$

For simulation purposes, equations in (3.66) should include a third one, i.e. (3.70), with appropriate initial conditions as to hold the original algebraic restriction $v_1 - v_2 = 0$ as well as the derived one in (3.69); for instance, $\boldsymbol{x}(0) = \begin{bmatrix} 1 & 1 & -0.5 \end{bmatrix}^T$. Simulations carried out from this initial condition using the `DAE Toolbox` of MATLAB®are shown in Fig. 3.11. It is apparent that the system is stable.

Once the system (3.68) is in this form, the Pantelides algorithm [55] might be translated into obtaining as many dynamical equations as states in it and as many algebraic restrictions as derived during the process: in the example under consideration, the algebraic restriction (3.69) as well as the dynamical equation (3.70) have yet to be incorporated. To preserve the square form of the system matrices, spurious states (z_1 and z_2) may be added without loss of generality.

For the example above, this is

$$\bar{E}\dot{\bar{x}} = \bar{A}\bar{x}, \tag{3.71}$$

with

$$\bar{E} = \begin{bmatrix} 1&0&0&0&0 \\ 0&1&0&0&0 \\ 0&0&1&0&0 \\ 0&0&0&0&0 \\ 0&0&0&0&0 \end{bmatrix}, \; \bar{A} = \begin{bmatrix} -1&0&-1&0&0 \\ 0&0&1&0&0 \\ 1/2&0&1/2&0&0 \\ 1&-1&0&0&0 \\ -1&0&-2&0&0 \end{bmatrix}, \; \bar{x} = \begin{bmatrix} v_1 \\ v_2 \\ F \\ z_1 \\ z_2 \end{bmatrix}. \tag{3.72}$$

If algebraic restrictions and dynamical equations are consistent, i.e. they have solutions[1], the state space $\mathbb{R}^n$, where n corresponds to the number of dynamical equations, will be cut by m algebraic restrictions [73]. In this example, this translates into any trajectory of the system (3.71) lying in the kernel of

$$\begin{bmatrix} 1 & -1 & 0 \\ -1 & 0 & -2 \end{bmatrix},$$

with $m = 2$, $n = 3$, i.e. in a reduced space of dimension 1. This is consistent with the fact that a direct substitution of the two algebraic restrictions in *any* of the dynamical equations will reduce the dynamical system to a single state, for instance, $\dot{v}_1 = -v_1/2$. □

3.5 The Linear Fractional Transformation Structure

Closely related to the just described classes of convex models, but independently developed, linear fractional transformations (LFTs) have appeared in the work of [74] to represent linear parameter-varying (LPV) plants; this framework has been more convenient than small-gain approaches for controller synthesis when the parameters θ vary somehow slowly in time.

As in the TS framework and the closely related descriptor forms, LFT representations lead to LMI design conditions when the controller dependence on the measured parameters is linear fractional. In [75], the LFT approach has been employed along with gridding techniques (dividing the state space) to solve otherwise complex output feedback or multi-objective control problems. LFTs also serve to synthesise

[1] Existence and uniqueness of solutions for systems of the form $E\dot{x}(t) = Ax(t)$, with A and E square matrices of the same size, are guaranteed if $\det(sE - A) \neq 0$ for some s; such systems are called *regular* [52]. Since this condition is only sufficient, the problem remains open for systems of the form (3.71).

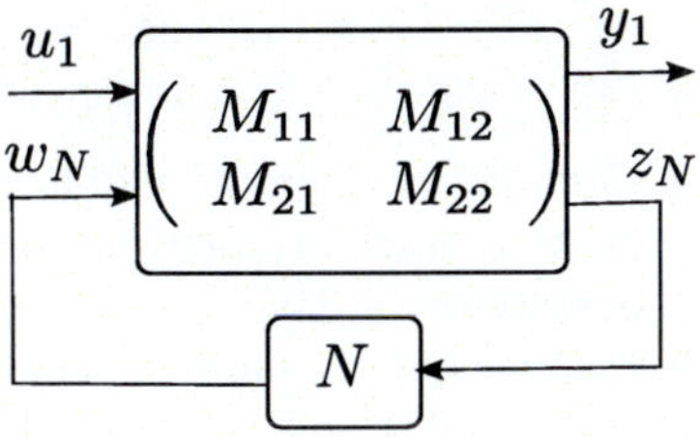

Fig. 3.12 Interpretation of an LFT as a closed-loop block diagram

LPV controllers via parameter-dependent LMIs in [76] or the parameter-dependent Lyapunov functions for gain scheduling in [77]; an old review can be found in [78].

Definition 3.13 Given the suitably partitioned matrix M below and a matrix N of compatible size, the expressions on the right

$$M = \begin{bmatrix} M_{11} & M_{12} \\ M_{21} & M_{22} \end{bmatrix}, \begin{array}{l} F_l(M, N) := M_{11} + M_{12}N(I - M_{22}N)^{-1}M_{21} \\ F_u(M, N) := M_{22} + M_{21}N(I - M_{11}N)^{-1}M_{12} \end{array} \tag{3.73}$$

are denoted as *lower and upper linear fractional transformations*, respectively.

In some literature, the star notation below is used:

$$M \star N := F_l(M, N), \qquad N \star M := F_u(M, N). \tag{3.74}$$

The above-defined LFTs represent rational matrix expressions. Obviously, the number of rows and columns of N must be smaller than those of M for the above expressions to have sense.

LFT representations were introduced in optimal $\mathcal{H}_2$ and $\mathcal{H}_\infty$ control [79], using suitably partitioned $M(s)$ and $N(s)$ to represent linear dynamical system taking the roles of plant models and controllers, respectively. Indeed, the motivation when introducing the above formulae is the fact that they are an abstract way of representing a very common feedback block diagram, see Fig. 3.12, whose equations are

$$\boldsymbol{y}_1 = M_{11}\boldsymbol{u}_1 + M_{12}\boldsymbol{w}_N, \tag{3.75}$$

$$\boldsymbol{z}_N = M_{21}\boldsymbol{u}_1 + M_{22}\boldsymbol{w}_N, \tag{3.76}$$

$$\boldsymbol{w}_N = N\boldsymbol{z}_N. \tag{3.77}$$

Indeed, if we solve for $\boldsymbol{y}_1$, we will obtain $\boldsymbol{y}_1 = F_l(M, N)\boldsymbol{u}_1$ (explicit dependence on s of the sub-matrices omitted; details left to the reader).

So, from an abstract modelling point of view (see next section), LFTs are the form in which solutions of linear systems of equations end up when eliminating certain variables or, equivalently, in a control interpretation, the closed-loop expressions from the block diagram in the above-referred figure once signals $\boldsymbol{z}_N$ and $\boldsymbol{w}_N$ have been eliminated.

In particular, in the examples below, and in most robust control developments, N in (3.77) will usually denote two different entities, either

- *Controllers*, where z_N will be the information available to the controller and $\boldsymbol{w}_N$ the control action (controller output), $\boldsymbol{u}_1$ will denote disturbances and $\boldsymbol{y}_1$ controlled outputs, or
- *Unknown*, *time-varying coefficients*, or *nonlinear* elements of a model; then, the model will be split into (*a*) an LTI part, defined by (3.75) and (3.76), and (*b*) a nonlinear time-varying one (3.77). Usually, such nonlinear time-varying or unknown elements are represented with Greek symbol Δ in literature, instead of N. The expression $\boldsymbol{y} = F_l(M, \Delta)\boldsymbol{u}$ would model an uncertain relationship between $\boldsymbol{y}$ and $\boldsymbol{u}$ that can be expressed as a system matrix which is rationally dependent on the uncertainty Δ.

If Δ is a nonlinear or time-varying block, then (3.73) must be understood in a block-diagram system interconnection sense, as in Fig. 3.12, and not as linear matrix operations.

3.5.1 LFT in Physical Modelling

Let us discuss how the above LFT framework naturally appears when manipulating the first-principle equations of a physical system trying to solve for some variables.

In general, when modelling a linear system by physical first-principle equations, once a suitable equilibrium point has been chosen as the origin of coordinates, we will have a set of variables and equations that relate them. The variables, representing signals that change with time, can be classified as follows:

- A set of n *state* variables, denoted as $\boldsymbol{x} = [x_1\ x_2\ \cdots\ x_n]^T$ whose derivatives $\dot{\boldsymbol{x}}$ also appear in the model equations.
- A set of q other variables, denoted as $\boldsymbol{z} = [z_1\ z_2\ \cdots\ z_q]^T$, appearing in the physical laws, or being outputs of interest to be computed/simulated/controlled. The derivatives of this second group of variables do *not* appear in the first-principle equations. These variables should not be confused with the premise vector $\boldsymbol{z}$ in the previous sections.
- A total of m input variables, denoted by $\boldsymbol{u} = [u_1\ u_2\ \cdots\ u_m]^T$.

A generic nonlinear differential-algebraic model is in the form $\boldsymbol{f}(\dot{\boldsymbol{x}}, \boldsymbol{x}, \boldsymbol{z}, \boldsymbol{u}) = 0$, with $\boldsymbol{f} : \mathbb{R}^n \times \mathbb{R}^n \times \mathbb{R}^q \times \mathbb{R}^m \times \mapsto \mathbb{R}^e$ holding the usual assumptions on the locally Lipschitz characteristics in a compact set of interest, e being the number of equations. If we restrict ourselves to a linear case, for the linear model to be correctly set up, a total of $e = n + q$ (linearly independent) equations must be available, in the form

$$T_1 \begin{bmatrix} \dot{\boldsymbol{x}}(t) \\ \boldsymbol{z}(t) \end{bmatrix} - T_2 \boldsymbol{x}(t) - T_3 \boldsymbol{u}(t) = 0, \tag{3.78}$$

for some matrices of adequate size T_1, T_2, and T_3. Let us assume matrix T_1 is invertible[2]. In this case, variables $\dot{\boldsymbol{x}}$ and z can be expressed as a linear function of variables $\boldsymbol{x}$ and $\boldsymbol{u}$, yielding the standard state-space representation

$$\begin{bmatrix} \dot{\boldsymbol{x}}(t) \\ z(t) \end{bmatrix} = T_1^{-1} \begin{bmatrix} T_2 & T_3 \end{bmatrix} \begin{bmatrix} \boldsymbol{x}(t) \\ \boldsymbol{u}(t) \end{bmatrix} = \begin{bmatrix} A & B \\ C & D \end{bmatrix} \begin{bmatrix} \boldsymbol{x}(t) \\ \boldsymbol{u}(t) \end{bmatrix}, \tag{3.79}$$

where matrices A, B, C, and D are suitably defined from the corresponding rows and columns of $T_1^{-1}T_2$ and $T_1^{-1}T_3$.

Let us now assume that we have s uncertain coefficients or nonlinearities $\delta_i(\boldsymbol{x}, t)$, $i \in \{1, 2, \dots, s\}$, in the model, let $\boldsymbol{\delta}(\boldsymbol{x}, t) := [\delta_1(\boldsymbol{x}, t) \; \delta_2(\boldsymbol{x}, t) \; \cdots \; \delta_s(\boldsymbol{x}, t)]^T$. For brevity, the explicit state and/or time dependence in $\boldsymbol{\delta}$ will be omitted in the sequel when convenient. This notation should not be confused with the operator distinguishing continuous- and discrete-time systems in the previous sections.

Let us assume that the effect of $\boldsymbol{\delta}$ enters in *multiplicative* form, i.e. it can be expressed (see examples below) as

$$T_1 \begin{bmatrix} \dot{\boldsymbol{x}}(t) \\ z(t) \end{bmatrix} - T_2\boldsymbol{x}(t) - T_3\boldsymbol{u}(t) - \eta(\boldsymbol{\delta}, \boldsymbol{x}, z, \boldsymbol{u}) = 0, \tag{3.80}$$

where $\eta(\boldsymbol{\delta}, \boldsymbol{x}, z, \boldsymbol{u})$ is *bilinear* in $\boldsymbol{\delta}$ and $(\boldsymbol{x}, \boldsymbol{u}, z)$, in particular, in the form:

$$\eta(\boldsymbol{\delta}, \boldsymbol{x}, z, \boldsymbol{u}) = T_4\Delta(H_1\boldsymbol{x} + H_2 z + H_3\boldsymbol{u}) = T_4 \cdot \Delta \cdot z_\delta, \tag{3.81}$$

where $\Delta = \text{block-diag}(\boldsymbol{\delta})$ is a block-diagonal matrix of the individual δ_is, and T_4, H_1, H_2, H_3 are known matrices of adequate sizes. The variable z_δ denotes a *fictitious* output:

$$z_\delta = H_1\boldsymbol{x} + H_2 z + H_3\boldsymbol{u} \tag{3.82}$$

to be later used as input to the uncertainty Δ; it is fictitious because it might not exist in a real system, only in the approximation of it given by our model and the uncertainty block. The bilinearity assumption is required so that the equilibrium point $\boldsymbol{x} = 0$, $\boldsymbol{u} = 0$, $z = 0$ does not change when Δ changes.

The above representation arises from common situations in practice. For instance, we might have the following:

- *Nonlinearity*:

$$\eta_1 = (\sin x_1) \cdot x_2, \quad \delta(x, t) = \sin x_1, \quad \eta_1 = \delta \cdot x_2, \quad w_\delta = x_2;$$

[2] If T_1 were not full rank, additional variable elimination and algebraic manipulations involving differentiation would be needed in order to set up a minimal representation (3.78) with invertible T_1; these are the so-called high-index differential-algebraic equations. Alternatively, a descriptor system framework might be pursued, see Sect. 3.4; in fact, all LFT representations admit a descriptor form, see Sect. 3.5.3.

- *Sector nonlinearity*: an additive term $\sin x_1$ may be expressed as $\sin(x_1)/x_1 \cdot x_1$, thus setting

$$\eta_2 = \sin x_1, \quad \delta(x,t) = \sin x_1/x_1, \ \eta_2 = \delta \cdot x_1, \, w_\delta = x_1;$$

- *Parameter uncertainty*: an unknown constant parameter, say δ_3:

$$\eta_3 = \delta_3 \cdot (x_3 + z_1), \ 1.5 < \delta_3 < 2, \ w_\delta = x_3 + z_1.$$

Then, under the uncertainty structure (3.81), the model equations (3.80) can be rewritten, for invertible T_1, as follows:

$$\begin{bmatrix} \dot{\boldsymbol{x}}(t) \\ \boldsymbol{z}(t) \end{bmatrix} = T_1^{-1} T_2 \boldsymbol{x}(t) + T_1^{-1} T_3 \boldsymbol{u}(t) + T_1^{-1} T_4 \boldsymbol{w}_\delta, \tag{3.83}$$

$$\boldsymbol{w}_\delta = \Delta \boldsymbol{z}_\delta, \tag{3.84}$$

$$\boldsymbol{z}_\delta = H_1 \boldsymbol{x} + H_2 \boldsymbol{z} + H_3 \boldsymbol{u} = (H_1 + Q)\boldsymbol{x} + (H_3 + R)\boldsymbol{u} + S\boldsymbol{w}_\delta, \tag{3.85}$$

where matrices Q, R, and S are obtained by multiplying by H_2 the last q rows of the matrices $T_1^{-1}T_2$, $T_1^{-1}T_3$, and $T_1^{-1}T_4$, respectively, which appear in (3.83) when solving for $\boldsymbol{z}$.

As done in (3.79), suitably renaming the matrices according to standard conventions in literature (i.e. n first rows of $T_1^{-1}T_2$ denoted as A, etc.), we can obtain a representation as defined below:

Definition 3.14 A *linear fractional representation* (LFT) of an uncertain/nonlinear state-space system is a set of first-order ODEs and algebraic equations

$$\dot{\boldsymbol{x}}(t) = A\boldsymbol{x}(t) + B_\delta \boldsymbol{w}_\delta + B\boldsymbol{u}(t), \tag{3.86}$$

$$\boldsymbol{z}_\delta = C_\delta \boldsymbol{x} + D_{\delta u}\boldsymbol{u} + D_{\delta\delta}\boldsymbol{w}_\delta, \tag{3.87}$$

$$\boldsymbol{z}_1 = C_1 \boldsymbol{x} + D_{1\delta}\boldsymbol{w}_\delta + D\boldsymbol{u} \tag{3.88}$$

jointly with a last equation

$$\boldsymbol{w}_\delta = \Delta \boldsymbol{z}_\delta, \tag{3.89}$$

lumping all uncertainty/nonlinearity/time variation or even unmodelled dynamics. Only bounds and maybe some block-diagonal structure are known about Δ.

Indeed, defining the linear transfer function matrix

$$G(s) = \begin{bmatrix} C_1 \\ C_\delta \end{bmatrix} (sI - A)^{-1} \begin{bmatrix} B & B_\delta \end{bmatrix} + \begin{bmatrix} D & D_{1\delta} \\ D_{\delta u} & D_{\delta\delta} \end{bmatrix},$$

we can express (3.86)–(3.88) as

$$\begin{bmatrix} z_1 \\ z_\delta \end{bmatrix} = \begin{bmatrix} G_{11}(s) & G_{12}(s) \\ G_{21}(s) & G_{22}(s) \end{bmatrix} \begin{bmatrix} \boldsymbol{u} \\ \boldsymbol{w}_\delta \end{bmatrix}, \tag{3.90}$$

$$\boldsymbol{w}_\delta = \Delta z_\delta. \tag{3.91}$$

As in (3.75), eliminating $\boldsymbol{w}_\delta$ and z_δ results in

$$z_1 = F_l(G(s), \Delta), \tag{3.92}$$

where the interpretation of F_l, as previously discussed, is in a system (operator) interconnection framework, as Δ may be nonlinear or time-varying and, hence, actual operations with the Laplace transform would not be valid. Nevertheless, the key idea in the LFT approach to control is the fact that the Laplace transform, frequency response, LMIs, etc. do indeed hold on $G(s)$ so the well-developed linear apparatus can be exploited, if so needed, with the linear subsystem G.

Remark 3.3 As later tools will use LTI-based results for analysis or synthesis over (3.86)–(3.88), a translated/scaled representation of Δ centred around the origin and block-diagonal is advised, in order to bound it by the smallest possible ball around the origin into some subspaces. For instance, if some uncertain coefficient verifies $5 \le \delta \le 7$, we could add the mean value 6 to the known (linear) part and express $\delta = 6 + \tilde{\delta}$, where $-1 \le \tilde{\delta} \le 1$. This centring/scaling is assumed in most literature references, pursuing control design with δ inside a suitable unit ball.

Relationship to TS Modelling

Note that in an LFT representation (3.73), if M_{22} is equal to zero, then the resulting expressions of lower LFT $F_l(M, N)$ (in the case $M_{11} = 0$, this would apply to $F_u(M, N)$) are *linear* in the second argument N. Hence, if N can be expressed as a convex sum $N = \sum_{i=1}^r h_i N_i$, where $\sum_{i=1}^r h_i = 1, 0 \le h_i \le 1, i \in \{1, 2, \dots, r\}$, the LFTs will, too, be amenable to such expression. For instance,

$$F_l(M, N) = M_{11} + M_{12} \left(\sum_{i=1}^r h_i N_i \right) M_{21}. \tag{3.93}$$

Thus, the expressive modelling power of LFT in such a case is the same as the TS approach (particular case). In fact, in Sect. 4.7.4, it is shown that any TS model admits an equivalent trivial LFT representation.

Example 3.16 Consider the electrical circuit in Fig. 3.13 with $R1 = 8$, $R3 = 16$, $R4 = 32$, $C1 = 1 \times 10^{-3}$, and $C2 = 2 \times 10^{-3}$, and consider that there are some increments `incR2` of resistance $R2$ to be modelled, so we set up $R2 =$ $R2\_0+$`incR2`, with $R2\_0 = 4$ (all units omitted).

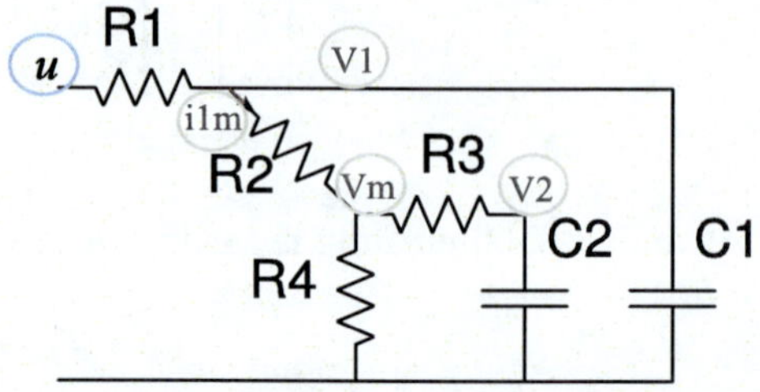

Fig. 3.13 Electrical Circuit with uncertain/time-varying $R2$

In this example, the developments will be carried out using MATLAB®symbolic toolbox in order to symbolically solve systems of linear equations. Basically, the syntax of the `solve` command to be used is

```
Solution=solve(Eq1,Eq2,Eq3,..., Unknown1,Unknown2,...).
```

First, the constant values for all components except $R2$, and the nominal value for $R2$, i.e. $R2\_0$, will be defined, and eight different symbolic variables in the model equations will be declared:

```
C1=1e-3;C2=2e-3; R1=8;R3=16; R4=32; R2_0=4;
syms dv1 dv2 V1 V2 incR2 Vm u i1m.
```

Then, the four model equations are introduced, and they are solved for four unknowns:

```
%% standard model
Model1a={C1*dv1==(u-V1)/R1-i1m, ...
   C2*dv2==(Vm-V2)/R3, ...
   i1m==(Vm-V2)/R3+Vm/R4 };
EqResistor2= [ (V1-Vm)==(R2_0+incR2)*i1m ];
Model1={Model1a{:},EqResistor2};
Unknowns={dv1,i1m,Vm,dv2};
SolvedModel=solve(Model1{:},Unknowns{:}).
```

Note that the four unknowns have been chosen to be the state derivatives (`dv1`, `dv2`) and two other intermediate variables (current `i1m` and voltage `Vm`) to be later discarded. Indeed, the remaining four symbolic variables are the states $V1$ and $V2$, the input u, and the uncertain variation in resistance, `incR2`. The resulting solved for the state derivatives is

```
>> collect(SolvedModel.dv1)
125*(16*V2 - 68*V1  - 3*V1*incR2 )/(3*incR2 + 44)+125*u
>> SolvedModel.dv2
-125*(36*V2 - 32*V1 + V2*incR2)/(4*(3*incR2 + 44)),
```

which can, of course, be readily expressed as a standard state equation:

```
A=jacobian([SolvedModel.dv1;SolvedModel.dv2],[V1,V2])
B=jacobian([SolvedModel.dv1;SolvedModel.dv2],[u]).
```

Denoting

$$l_1(\mathrm{inc}R2) = \frac{1}{3\mathrm{inc}R2 + 44}, \quad l_2(\mathrm{inc}R2) = \frac{\mathrm{inc}R2}{3\mathrm{inc}R2 + 44}, \tag{3.94}$$

the result is (omitting arguments for l_1 and l_2)

$$\dot{\boldsymbol{x}}(t) = \begin{bmatrix} -125(3l_2 + 68l_1) & 2000l_1 \\ 1000l_1 & -125(l_2 + 36l_1) \end{bmatrix} \boldsymbol{x}(t) + \begin{bmatrix} 125 \\ 0 \end{bmatrix} \boldsymbol{u}(t). \tag{3.95}$$

TS modelling: If we wished to set up a TS model, we would need to bound the two functions l_1 and l_2 of the uncertain parameter `incR2` appearing in the A matrix above and write a 4-vertex model. For instance, if $|$`incR2`$| \le 2$, we can bound

$$0.02 \le l_1(\mathrm{inc}R2) \le 0.0263, \quad -0.0526 \le l_1(\mathrm{inc}R2) \le 0.04,$$

and four extreme matrices can be set up with the above bounds:

$$\dot{\boldsymbol{x}}(t) = \sum_{i=1}^{2} \sum_{j=1}^{2} w_i^1(\mathrm{inc}R2) \cdot w_j^2(\mathrm{inc}R2) \cdot A_{ij}\boldsymbol{x}(t) + B\boldsymbol{u}(t) \tag{3.96}$$

with

$$A_{11} = \begin{bmatrix} -203.9474 & 52.6316 \\ 26.3158 & -111.8421 \end{bmatrix}, \quad A_{12} = \begin{bmatrix} -238.6842 & 52.6316 \\ 26.3158 & -123.4211 \end{bmatrix}, \tag{3.97}$$

$$A_{21} = \begin{bmatrix} -150.2632 & 40.0000 \\ 20.0000 & -83.4211 \end{bmatrix}, \quad A_{22} = \begin{bmatrix} -185 & 40 \\ 20 & -95 \end{bmatrix}. \tag{3.98}$$

The details are left to the reader.

LFT representation: We can obtain an LFT representation of the same circuit, by trivially rewriting the resistor equation as $V1 - Vm = (R2\_0)i1m + w_\delta$, introducing the uncertain input w_δ as $w_\delta = (\mathtt{incR2})i1m$. Of course, if we add one equation and one unknown, the result is an equivalent way of writing the model. Indeed, executing

```
 %% introducing changes in EqResistor2 to incorporate LFT
syms w_delta      %extra symbolic variable
EqResistor2_LFT= ( (V1-Vm)==R2_0*i1m+w_delta   );
Eq_Uncertainty = ( w_delta==incR2*i1m );
Model2={Model1a{:},EqResistor2_LFT};
Model3={Model2{:}, Eq_Uncertainty};
Unknowns3={Unknowns{:},w_delta};
SolvedModel3=solve(Model3{:}, Unknowns3{:}).
```

`Model3` has 5 equations, so solving for 5 unknowns, the expressions for `dv1` and `dv2` in `SolvedModel3` are identical to those in `SolvedModel` above.

In order to get an LFT expression, the code to be used is

```
SolvedModelLFT=solve(Model2{:},Unknowns{:}),
```

where `Model2` has four equations (same as `Model1`) because `Eq_Uncertainty` is not into it. So, we solve for the same four unknowns as in `Model1`. The remaining symbolic variables in `SolvedModelLFT` are the states $V1$ and $V2$ and the inputs (both the true one u and the artificial one w_δ). As `Eq_Uncertainty` has not been inserted in `Model2`, `incR2` does not appear in the expressions for $dv1$ and $dv2$. Hence, using

```
A=jacobian([SolvedModelLFT.dv1;SolvedModelLFT.dv2],[V1,V2]),
B=jacobian([SolvedModelLFT.dv1;SolvedModelLFT.dv2],[u,w_delta]),
Cdelta=jacobian(SolvedModelLFT.i1m,[V1,V2]),
Ddelta=jacobian(SolvedModelLFT.i1m,[u,w_delta])
```

and separately writing `Eq_Uncertainty` as (3.100) (which was not taken into account in `SolvedModelLFT`) give us the full model, of course, equivalent to `SolvedModel` , but written as the LFT interconnection of the linear system (3.99) and the uncertain one (3.100) below:

$$\begin{aligned} \dot{x}(t) &= \begin{bmatrix} -193.1818 & 45.4545 \\ 22.7273 & -25.5682 \end{bmatrix} x(t) + \begin{bmatrix} 125 & 68.1818 \\ 0 & -22.7273 \end{bmatrix} \begin{bmatrix} u(t) \\ w_\delta \end{bmatrix}, \\ i1m &= \begin{bmatrix} 0.0682 & -0.0455 \end{bmatrix} x(t) + \begin{bmatrix} 0 & -0.0682 \end{bmatrix} \begin{bmatrix} u(t) \\ w_\delta \end{bmatrix}, \end{aligned} \tag{3.99}$$

$$w_\delta = (\text{inc}R2)i1m. \tag{3.100}$$

In this case, a 2-vertex LFT model suffices to model the uncertain resistor, as the only function of the uncertain parameter appearing in the equations is `incR2` itself: the system description is in between $w_\delta = (-2)i1m$ and $w_\delta = (+2)i1m$.

The TS model required 2×2 vertices because two different nonlinear functions of `incR2` appeared in the state-space representation. Of course, regarding modelling, the models are *equivalent* but TS stability results might introduce conservativeness, because of the so-called shape-independence issues (see Sect. 3.2), as LMIs for TS models will assume that $w^1_{(\cdot)}$ and $w^2_{(\cdot)}$ in (3.96) are unrelated while they actually are two rational functions of the same uncertain parameter.

Note that in this modelling example, the source of variation of $R2$ is irrelevant; it might be either

- an *unknown-but-constant* parameter, such as tolerance errors in the actual resistance value;
- a variation of the resistance with time or exogenous variables (temperature, humidity);

- a nonlinearity (variation of the resistance as a function of the current that traverses it); or
- unmodelled nonlinear dynamics (variation of the resistance with temperature, which itself depends on the past current waveform).

Later stability and control design will only consider bounds on such variations and, depending on the availability of sensors, instant knowledge (or lack thereof) of the resistance value. Of course, robust/gain-scheduled control design techniques can have extra information about `incR2`, such as time variation bounds and whether it is dynamic or static, that can be used to avoid conservatism; details on these issues will be revisited in Sect. 5.6.3. □

3.5.2 Structure of the Uncertainty/Nonlinearity Block Δ

When doing physical modelling, the resulting structure of Δ is a block-diagonal matrix with repeated diagonal elements (repeated as many times as the uncertainty in a parameter or the nonlinearity appears in the physical equations).

Also, polynomial-fraction dependence can be expressed as a repeated-block-diagonal structure.

Example 3.17 Consider the case in which a nonlinearity/uncertainty δ_1 appears in some model equations as $\boldsymbol{y}_1 = G(\delta_1) \cdot \boldsymbol{\eta}$, where $G(\delta_1)$ is a quotient of polynomials in δ_1 , and $\boldsymbol{\eta}$ is an arbitrary system variable.

Then, we can express the dependence as the LFT interconnection:

$$\begin{bmatrix} \boldsymbol{z}_\delta \\ \boldsymbol{y}_1 \end{bmatrix} = \begin{bmatrix} A^{-1} & -A^{-1}B \\ CA^{-1} & -CA^{-1}B + D \end{bmatrix} \begin{bmatrix} \boldsymbol{w}_\delta \\ \boldsymbol{\eta} \end{bmatrix}, \qquad \boldsymbol{w}_\delta = \delta_1 I \boldsymbol{z}_\delta \tag{3.101}$$

for suitable matrices A, B, C, and D.

Indeed, replacing the right equation into the left one, we get $\boldsymbol{z}_\delta = A^{-1}\delta I \boldsymbol{z}_\delta - A^{-1}B\boldsymbol{\eta}$. Then, multiplying by A, it results in $A\boldsymbol{z}_\delta = \delta_1 I \boldsymbol{z}_\delta - B\boldsymbol{\eta}$, so $\boldsymbol{z}_\delta = (\delta_1 I - A)^{-1}B\boldsymbol{\eta}$. After straightforward manipulations, we can get $\boldsymbol{y}_1 = C\boldsymbol{z}_\delta + D\boldsymbol{\eta} = (C(\delta_1 I - A)^{-1}B + D)\boldsymbol{\eta}$. Note the close relationship with state-space realisations. For the above to hold, we need A invertible, i.e. $G(0)$ bounded, and selecting matrices A, B, C, and D such that their transfer function is $G(\delta_1)$. Note that, of course, δ_1 here has no relationship to the Laplace transform.

The advantage of the above representation is that, if δ_1 appears elsewhere as a new rational expression multiplied by other variables, we can express the overall setup with δ_1 repeated more times so the multipliers in later sections will be able to exploit the fact that δ_1 is actually the same in *all* expressions; basically, the idea amounts to considering a matrix $G(\delta_1)$ and auxiliary minimal state-space realisation matrices of it. □

Example 3.18 In [64], it is shown how a rotational–translational actuator model

$$\dot{x}_1 = x_2, \quad \dot{x}_2 = x_4, \tag{3.102}$$

$$\dot{x}_3 = \frac{-x_1 + \epsilon x_4^2 \sin x_3 - \epsilon \cos x_3 u + w}{1 - \epsilon^2 \cos^2 x_3}, \tag{3.103}$$

$$\dot{x}_4 = \frac{u + \epsilon \cos x_3 (x_1 - w) + \epsilon x_4^2 \sin x_3}{1 - \epsilon^2 \cos^2 x_3} \tag{3.104}$$

can be expressed as a 4-vertex LFT system from the 2-vertex bounds of $\delta_1 = \epsilon \cos x_3$ and $\delta_2 = x_4 \sin x_3$. Indeed, the model can be written as follows:

$$\dot{x}_1 = x_2, \quad \dot{x}_2 = x_4, \tag{3.105}$$

$$\dot{x}_3 = \frac{-x_1 + \epsilon \delta_2 x_4 - \delta_1 u + w}{1 - \delta_1^2}, \tag{3.106}$$

$$\dot{x}_4 = \frac{u + \delta_1 (x_1 - w) + \epsilon x_4 \delta_2}{1 - \delta_1^2} \tag{3.107}$$

so all elements in the state-space matrices are rational expressions of uncertain elements δ_1 and δ_2 (see the cited reference for details).

A TS model can only do a similar bounding with linear expressions of uncertain elements. For instance, the model could be rewritten by introducing

$$\delta_1 = \frac{1}{1 - \epsilon^2 \cos^2 x_3}, \quad \delta_2 = \frac{-\epsilon \cos x_3}{1 - \epsilon^2 \cos^2 x_3}, \quad \delta_3 = \frac{x_4 \sin x_3}{1 - \epsilon^2 \cos^2 x_3} \tag{3.108}$$

resulting in

$$\dot{x}_1 = x_2, \quad \dot{x}_2 = x_4, \tag{3.109}$$

$$\dot{x}_3 = -\delta_1 x_1 + \epsilon \delta_3 x_4 - \delta_2 u + \delta_1 w, \tag{3.110}$$

$$\dot{x}_4 = \delta_1 u + \delta_2 (x_1 - w) + \epsilon x_4 \delta_3, \tag{3.111}$$

so, upper and lower bounds of δ_1, δ_2, and δ_3 would yield a TS representation with 2^3 vertices. This would be an 8-vertex model, which, apart from having more vertices, may increase the conservativeness by thinking that the three nonlinearities are totally unrelated (instead of having only two of them as in the LFT case). □

As shown in the above example, shape-independent sector nonlinearity TS results cannot exploit, in general, that the argument of two distinct *rational* expressions is the same (it can only do so in the case of *linear* expressions), so LFT is a more powerful modelling approach.

3.5.3 Relationship with Descriptor Models

As descriptors are another way of expressing systems of differential-algebraic equations (DAEs), it comes at no surprise that any dynamic system with LFT representation can be expressed in descriptor form (3.57). Indeed, consider the LFT interconnection of systems (3.112) and (3.113) below:

$$\begin{aligned} \dot{\boldsymbol{x}}(t) &= A\boldsymbol{x}(t) + B_\delta \boldsymbol{u}_\delta, \\ z_\delta &= C_\delta \boldsymbol{x} + D_{\delta\delta}\boldsymbol{u}_\delta, \end{aligned} \tag{3.112}$$

$$\boldsymbol{u}_\delta = \Delta z_\delta. \tag{3.113}$$

Then, the following descriptor representation with descriptor vector $\psi = [\boldsymbol{x}^T \ \boldsymbol{u}_\delta^T]^T$ models the same system:

$$\begin{bmatrix} I & 0 \\ 0 & 0 \end{bmatrix} \dot{\psi}(t) = \begin{bmatrix} A & B_\delta \\ \Delta C_\delta & \Delta D_{\delta\delta} - I \end{bmatrix} \psi(t) \tag{3.114}$$

as, indeed, the second equation amounts to $\boldsymbol{u}_\delta = \Delta C_\delta \boldsymbol{x} + \Delta D_{\delta\delta}\boldsymbol{u}_\delta$. Thus, any LFT with polytopic Δ admits a polytopic descriptor representation.

The converse is also true; a descriptor model

$$\begin{bmatrix} I & 0 \\ 0 & 0 \end{bmatrix} \dot{\psi}(t) = \begin{bmatrix} A(\Delta) & B(\Delta) \\ C(\Delta) & D(\Delta) \end{bmatrix} \psi(t) \tag{3.115}$$

can be written (assuming that $D(\Delta)$ is invertible for all Δ in a particular prescribed set) as the standard state-space $\dot{\boldsymbol{x}}(t) = (A - BD^{-1}C)\boldsymbol{x}(t)$, and if matrices A, B, C, and D are rational in the scheduling parameters Δ, the state-space equation admits an LFT representation (details omitted for brevity).

Note, however, that descriptor models (DAE) are a more general representation, in the sense that DAEs with differentiability index larger or equal to 2 can be represented in descriptor form with non-invertible $D(\Delta)$ but in such a case, the LFT representation would not be well-posed. Such systems would be amenable to an LFT representation only after differentiability-index reduction / the Pantelides steps are taken on the original nonlinear DAEs (which, basically, amounts to appending the derivative of some model equations to such model) and sector nonlinearity bounds are subsequently computed.

3.5.4 Uncertain Models

The main drawback of the sector nonlinearity approach is that the resulting model has $r = 2^p$ vertices, where p is the number of nonlinearities. In high-order, complex models, this may result in an unsurmountable computational burden. For instance, a fourth-order model with 16 interval-valued coefficients in its state matrix would have 65536 vertices. This fact requires to make some simplifications or to bound some of the variations as uncertainty.

Parametric Uncertainty as Unmeasurable Premises

In principle, the convex (TS) description of a system does not depend on the premise variables being known or not. The only relevant feature of a parameter is knowing if it is varying in time or not. If it is, the possibly nonlinear expressions in which it appears in the model can be upper and lower bounded and the varying parameter made part of the premise vector.

Example 3.19 For an uncertain linear system of the form

$$\dot{x}(t) = \begin{bmatrix} \theta & 1 \\ 0 & -2 \end{bmatrix} x(t),$$

it is usually understood that parameters such as θ are fixed, but unknown. Such systems can be analysed via characteristic polynomial approaches [25] and/or the Lyapunov-based stability analysis via a TS embedding [7]. Yet, the latter tool may guarantee the stability of this system even if θ is time-varying or state-dependent as long as it remains bounded by the same values. In each case, the corresponding polytope will have the same vertices, though the membership functions will retain the "constant, but unknown" or the time-varying characteristics. But, of course, the Kharitonov or structured-singular-value or parameter-dependent Lyapunov functions will yield better results if we do know that θ is a constant-but-unknown uncertain parameter. □

Then, when actually solving a control problem, if the parameter is unknown, the relevant membership functions will be unknown. However, as discussed above, if such an approach is pursued, the number of vertices may get too high. An alternative option is (conservatively) "burying" the uncertain parameter in a ball of uncertainty, discussed below.

Parametric Uncertainty as Norm-Bounded Uncertain Matrix

For simplicity in the presentation, let us consider the linear TS case instead of rational/descriptor options. In general, let us assume that in the model, a linear function of the state $E_{1\times n}x$ is multiplied by a nonlinear function $\eta(z)$ of some variables which can be bounded in an interval with centre p_0 and radius p_1, i.e. $\eta(z) \in [p_0 - \delta p_1, p_0 + \delta p_1]$, $|\delta| \leq 1$.

If, later, $\eta(z)$ appears linearly in the model matrices, i.e. the expression $Q = H\eta(z)E$ is present in the model, then Q can be expressed as $Q = Hp_0E + H\delta p_1 E$.

In the general case, the uncertain element will appear in several positions of the system matrix, so we will have $A = A_0 + \delta A_1$ where A_1 is a matrix of a particular rank, say ρ, so that a (obviously, non-unique) decomposition $A_1 = U_1 \cdot V_1$ can be easily made (for instance, using svd) with U_1 being a matrix of dimension $n \times \rho$ and V_1 having dimensions $\rho \times n$.

In this case, the uncertain matrix can be expressed as $A = A_0 + U_1 \cdot \delta I_{\rho\times\rho} \cdot V_1$, $|\delta| \leq 1$.

If there are more parameters to be lumped into uncertainty, entering as $A = A_0 + \sum_{i=1}^{s} \delta_i A_i$, then the overall description can be written as

$$A = A_0 + \begin{bmatrix} U_1 \; U_2 \; \cdots \; U_s \end{bmatrix} \Delta \begin{bmatrix} V_1 \\ V_2 \\ \vdots \\ V_s \end{bmatrix} \tag{3.116}$$

Δ being a block-diagonal matrix with each element repeated as many times as its rank, whose elements range in the interval $[-1, 1]$. It is easy to see that, given an arbitrary vector τ compatible with the size of Δ, the inequality $\|\Delta\tau\| \leq \|\tau\|$ holds, so $\bar{\sigma}(\Delta) \leq 1$.

Most of the control results in this book will prove stability or stabilisation for all matrices in the form $A + H\Delta E$, $\Delta^T\Delta \leq I$ which, indeed, include the previous description of the uncertain parameters. However, only using $\Delta^T\Delta \leq I$ is conservative, due to two reasons: first, there are many more non-diagonal matrices Δ fulfilling the norm bound; second, the "repeated parameter" block-diagonal structure is not taken into account; removing the conservatism would require significant complications and solving non-convex BMI problems.

Example 3.20 Consider the system

$$\dot{\boldsymbol{x}}(t) = \begin{bmatrix} 1 & 2+p_1 & 3 & 4 \\ 5+\sin x_1 & 6 & 7+4p_1 & 8 \\ 9-p_2 & 10 & 11 & 12 \\ 13+5p_2 & 14 & 15+p_2 & 16 \end{bmatrix} \boldsymbol{x}(t),$$

with $\boldsymbol{x} = [x_1\, x_2\, x_3\, x_4]^T$, p_1 and p_2 being uncertain terms.

Clearly, $\sin x_1$ can be turned into premises, but p_1 and p_2 can be treated as lying in a ball Δ (maybe a square, if separate intervals are known for each uncertain parameter). Indeed, we can express the system as follows:

$$\dot{x}(t) = \begin{bmatrix} 1 & 2 & 3 & 4 \\ 5 & 6 & 7 & 8 \\ 9 & 10 & 11 & 12 \\ 13 & 14 & 15 & 16 \end{bmatrix} x(t) + \begin{bmatrix} 0 \\ 1 \\ 0 \\ 0 \end{bmatrix} \sin(x_1) \begin{bmatrix} 1 & 0 & 0 & 0 \end{bmatrix} + \begin{bmatrix} 0 & 1 \\ 2 & 0 \\ 0 & 0 \\ 0 & 0 \end{bmatrix} p_1 I_{2\times 2} \begin{bmatrix} 0 & 0 & 2 & 0 \\ 0 & 1 & 0 & 0 \end{bmatrix}$$
$$+ \begin{bmatrix} 0 & 0 \\ 0 & 0 \\ 0.431 & -0.431 \\ -2.238 & -0.083 \end{bmatrix} p_2 I_{2\times 2} \begin{bmatrix} -2.234 & 0 & -0.431 & 0 \\ 0.083 & 0 & -0.431 & 0 \end{bmatrix}. \tag{3.117}$$

Thus, depending on the computational resources or the wished performance level, some of the uncertainty can be (conservatively) lumped in a ball, such as assuming $|p_1| < 1$ and $|p_2| < 1$, hence $\|\Delta\| < 1$:

$$\dot{x}(t) = \begin{bmatrix} 1 & 2 & 3 & 4 \\ 5 & 6 & 7 & 8 \\ 9 & 10 & 11 & 12 \\ 13 & 14 & 15 & 16 \end{bmatrix} x(t) + \begin{bmatrix} 0 \\ 1 \\ 0 \\ 0 \end{bmatrix} \sin(x_1) \begin{bmatrix} 1 & 0 & 0 & 0 \end{bmatrix}$$
$$+ \begin{bmatrix} 0 & 1 & 0 & 0 \\ 2 & 0 & 0 & 0 \\ 0 & 0 & 0.431 & -0.431 \\ 0 & 0 & -2.238 & -0.083 \end{bmatrix} \Delta \begin{bmatrix} 0 & 0 & 2 & 0 \\ 0 & 1 & 0 & 0 \\ -2.234 & 0 & -0.431 & 0 \\ 0.083 & 0 & -0.431 & 0 \end{bmatrix}.$$

□

Note that if the expression is more complex and the matrices are dependent on premise variables or polynomials, the size of the matrix Δ and the number of times the parameters are repeated would be larger (and, hence, more conservative) than a carefully studied minimal representation which, however, might be harder to obtain.

Principal Component Vertex Reduction

Though in this book a special emphasis has been put on exactly rewriting nonlinear models so that they can be treated as polytopes, it must be clear by now that the number of nonlinearities or uncertainties may cause the polytopic representation to explode in the number of vertexes. A variety of techniques for vertex reduction can be found in the literature, some of them based on principal component analysis (PCA) to create approximations of LPV models that hold a tradeoff between the number of parameters and the model accuracy [34]; PCA has been applied in [80].

PCA consists in sampling the parameter space from a dense-enough grid and gathering the information in a matrix Θ decomposed via singular-value decomposition (SVD) or, equivalently, via the eigenvalues of the covariance matrix. This decomposition helps the designer to detect which singular values should be neglected in order to find a mapping $\phi(t) = q(\rho(t))$ with a less number of entries than the original

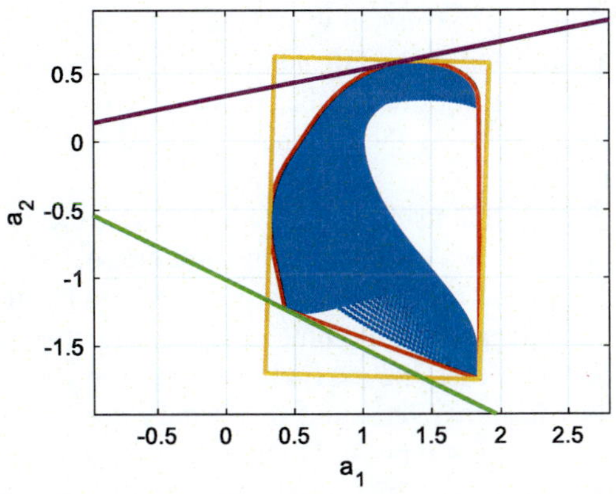

Fig. 3.14 Vertex reduction in general nonlinear models

parameter set. The quality of the approximation against complexity can be measured for each choice and the reduced parameter vector $\phi(t)$ can be employed to approximate the original (larger) LPV model. The interested reader is referred to [29, 81].

Example 3.21 Let us consider a parameter-varying expression

$$f(x, z) = a_1(z) \cdot x_1 + a_2(z) \cdot x_2,$$

where $z = (z_1, z_2)$ may be any exogenous variable (LPV) or part of the state (TS) premise variable known to lie in $[-1, 1] \times [-1, 1]$. The expressions for the uncertain model are

$$a_1(z) = \cos(z_1 + z_2) + \sin(z_1^2), \quad a_2(z) = \sin(z_1 - 0.5z_2^3) - 0.5z_2^2 - 0.25z_1^2.$$

An standard TS approach might consider $\psi_1 = \cos(z_1 + z_2)$, $\psi_2 = \sin(z_1^2)$, $\psi_3 = \sin(z_1 - 0.5z_2^3)$, $\psi_4 = z_2^2$, and $\psi_5 = z_1^2$, which would entail 2^5 vertices.

Alternatively, a dense grid in the square where z takes values may produce a cloud of points (blue dots in Fig. 3.14).

The least-conservative convex modelling would entail choosing the convex hull of the cloud of points (red line), with a theoretically infinite number of vertices (rational approximations of the nonlinearities might allow for LFT versions of the example, omitted for brevity).

If neither the convex hull (well, an approximation to it with, say, 200 vertices) nor the 2^5 vertex sector nonlinearity bounds are considered viable, informally, the directions of maximum and minimum quadratic variation would be aligned with the eigenvectors of the covariance matrix of the cloud of points. Linear programming easily computes the bounding lines in these directions (yellow rectangle in Fig. 3.14), and, in fact, in any arbitrarily chosen ones (green and violet lines). Although there is a curse-of-dimensionality in the grid-generation step, covariance and linear-programming computations do scale efficiently to higher dimensions.

A 4-vertex model would naturally arise from the covariance-based rectangle, and a judicious choice of additional lines would reduce conservatism at the expense of increasing the number of vertices until the actual convex hull was reached; for instance, using the rectangle plus red and violet lines would produce a 6-vertex TS model.

Note that using these methodologies the number of vertices does not need to be a power of two once additional oblique cuts are considered. Note also that the vertices originated by the vertices of the five nonlinearities $\psi_1,\ldots, \psi_5$ are not necessarily tight (tangent to the blue dots) as, say, the sum of the maxima of ψ_1 and ψ_2 may be larger than the maximum of the sum $A_1(z) = \psi_1 + \psi_2$: sector nonlinearity may significantly overbound an arbitrary nonlinear model.

As a last remark, if the resulting polyhedron was significantly narrow in any dimension, it might be collapsed (onto a line in this case) and the remaining parameter variation is considered to be uncertain. This is the motivation to include them in the uncertain/approximate model section: in high-order systems, the covariance-based box will have up to 2^p vertices, p being the number of model coefficients where nonlinearities do appear; for instance, a sixth-order model $\dot{x} = A(x)x$ full of nonlinearities would have 36 elements in the A matrix which, in a worst-case situation, would be bounded by a box with 2^{36} vertices. This can quickly become intractable so, indeed, the only way to go seems to be neglecting the smallest covariance matrix eigenvalues. Of course, the same intractability and need for approximations may likely happen if a lot of different nonlinearities appear in $A(x)$ and sector nonlinearity /LFT options are pursued.

Non-Parametric (Dynamic) Uncertainty

In general, dynamic uncertainty can also be conceived as above. The only difference is that Δ would be composed of dynamic systems. A general setting, thus, will feed a set of system outputs to a block of uncertain elements which will be later fed to the system through some output channels.

In Fig. 3.15, system M will be described by a convex structure, and the stability analysis or stabilisation may consider the nature of Δ in different ways: either as a full time-varying dynamical system (fully unstructured uncertainty) or as a block-diagonal matrix comprised of dynamical systems or static (time-varying) elements.

As discussed before, when comparing fixed-but-unknown parameters against time- or state-dependent ones, LFT models would imply a known $\Delta(x)$ or $\Delta(t)$, usable for control design, whereas uncertainty would imply an unknown Δ whose norm is bounded by 1.

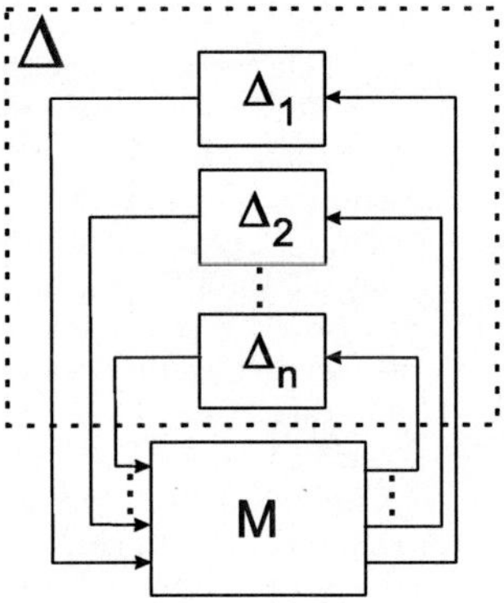

Fig. 3.15 The linear fractional transformation view of uncertainty

3.6 Summary and Conclusions

Once a convex model of a nonlinear system is available (TS or polynomial, descriptor or ordinary, approximate or exact), we can proceed to analyse its stability properties and perform control or observation via the direct Lyapunov method: the convenience of the convex structure will be appreciated then. This is the matter of the following chapters.

References

1. Kailath, T.: Linear Systems. Prentice Hall, Upper Saddle River, NJ (1980)
2. Franklin, G.F., Powell, J.D., Workman, M.L.: Digital Control of Dynamic Systems. Addison-Wesley, Boston, MA (1990)
3. Åström, K.J., Wittenmark, B.: Computer Controlled Systems. Prentice-Hall, Upper Saddle River, NJ (1990)
4. Boyd, S., El Ghaoui, L., Féron, E., Balakrishnan, V.: Linear Matrix Inequalities in System and Control Theory. Studies in Applied Mathematics, Society for Industrial and Applied Mathematics, Philadelphia, PA (1994)
5. Takagi, T., Sugeno, M.: Fuzzy identification of systems and its applications to modeling and control. IEEE Trans. Syst. Man Cybern. **15**(1), 116–132 (1985)
6. Passino, K., Yurkovich, S.: Fuzzy Control. Addisson-Wesley Longman, Menlo Park, CA (1998)
7. Tanaka, K., Wang, H.O.: Fuzzy Control System Design and Analysis: A Linear Matrix Inequality Approach. Wiley, New York (2001)
8. Tanaka, K., Wang, H.O.: Fuzzy Control Systems Design and Analysis. Wiley, New York (2001)
9. Sala, A.: Generalising quasi-lpv and cdi models to quasi-convex difference inclusions. IFAC-PapersOnLine **50**(1), 7560–7565 (2017)
10. Driankov, D., Hellendoorn, H., Reinfrank, M.: An Introduction to Fuzzy Control. Springer, New York (1993)
11. Abonyi, J., Babuška, R., Szeifert, F.: Modified Gath-Geva fuzzy clustering for identification of Takagi–Sugeno fuzzy models. IEEE Trans. Syst. Man Cybern. B Cybern. **32**(5), 612–621 (2002)

12. Babuška, R., van der Veen, P., Kaymak, U.: Improved covariance estimation for Gustafson-Kessel clustering. In: Proceedings of the IEEE International Conference on Fuzzy Systems, vol. 2. IEEE, New York, pp. 1081–1085 (2002)
13. Johansen, T., Babuška, R.: Multiobjective identification of Takagi–Sugeno fuzzy models. IEEE Trans. Fuzzy Syst. **11**(6), 847–860 (2003)
14. Kukolj, D., Levi, E.: Identification of complex systems based on neural and Takagi–Sugeno fuzzy model. IEEE Trans. Syst. Man Cybern. B Cybern. **34**(1), 272–282 (2004)
15. Kaymak, U., van den Berg, J.: On constructing probabilistic fuzzy classifiers from weighted fuzzy clustering. In: Proceedings of the International Conference on Fuzzy Systems, vol. 1, pp. 395–400 (2004)
16. Angelov, P., Filev, D.: An approach to online identification of Takagi–Sugeno fuzzy models. IEEE Trans. Syst. Man Cybern. B Cybern. **34**(1), 484–498 (2004)
17. Ohtake, H., Tanaka, K., Wang, H.: Fuzzy modeling via sector nonlinearity concept. In: Proceedings of the Joint 9th IFSA World Congress and 20th NAFIPS International Conference, vol. 1, pp. 127–132 (2001)
18. Wang, L.-X., Mendel, J.: Fuzzy basis functions, universal approximation, and orthogonal least-squares learning. IEEE Trans. Neural Networks **3**(5), 807–814 (1992)
19. Kosko, B.: Fuzzy systems as universal approximators. IEEE Trans. Comput. **43**(11), 1329–1333 (1994)
20. Ying, H.: Sufficient conditions on general fuzzy systems as function approximators. Automatica **30**(3), 521–525 (1994)
21. Fantuzzi, C., Rovatti, R.: On the approximation capabilities of the homogeneous Takagi–Sugeno model. In: Proceedings of the Fifth IEEE International Conference on Fuzzy Systems, pp. 1067–1072. IEEE, New York (1996)
22. Johansen, T.A., Shorten, R., Murray-Smith, R.: On the interpretation and identification of dynamic Takagi–Sugeno fuzzy models. IEEE Trans. Fuzzy Syst. **8**(3), 297–313 (2000)
23. Kiriakidis, K.: Nonlinear modeling by interpolation between linear dynamics and its application in control. J. Dyn. Syst. Meas. Contr. **129**(6), 813–824 (2007)
24. Taniguchi, T., Tanaka, K., Ohtake, H., Wang, H.: Model construction, rule reduction, and robust compensation for generalized form of Takagi–Sugeno fuzzy systems. IEEE Trans. Fuzzy Syst. **9**(4), 525–538 (2001)
25. Sanchez, M., Bernal, M.: A convex approach for reducing conservativeness of Kharitonovs-based robustness analysis. In: Proceedings of the 20th IFAC World Congress, pp. 855–860 (2017)
26. Ariño, C., Sala, A.: Relaxed LMI conditions for closed-loop fuzzy systems with tensor-product structure. Eng. Appl. Artif. Intell. **20**(8), 1036–1046 (2007)
27. Guerra, T., Guerra, J., Bernal, M.: LMI-based dynamic control of non-affine nonlinear systems via Takagi–Sugeno models. In: 2018 IEEE International Conference on Fuzzy Systems (FUZZ-IEEE), pp. 1–6. IEEE, New York (2018)
28. Robles, R., Sala, A., Bernal, M., Gonzalez, T.: Choosing a Takagi–Sugeno model for improved performance. In: Proceedings of the 2015 IEEE International Conference on Fuzzy Systems, pp. 1–6. IEEE, New York (2015)
29. Robles, R., Sala, A., Bernal, M., González, T.: Subspace-based Takagi–Sugeno modeling for improved LMI performance. IEEE Trans. Fuzzy Syst. **25**(4), 754–767 (2017)
30. Robles, R., Sala, A., Bernal, M., Gonzalez, T.: Optimal-performance Takagi–Sugeno models via the LMI null space. IFAC-PapersOnLine **49**(5), 13–18 (2016)
31. Khalil, H.K.: Nonlinear Systems. Prentice-Hall, Upper Saddle River, NJ (2002)
32. Johansson, M., Rantzer, A., Arzen, K.: Piecewise quadratic stability of fuzzy systems. IEEE Trans. Fuzzy Syst. **7**(6), 713–722 (1999)
33. Feng, M., Harris, C.J.: Piecewise Lyapunov stability conditions of fuzzy systems. IEEE Trans. Syst. Man Cybern. B Cybern. **31**(2), 259–262 (2001)
34. Kwiatkowski, A., Werner, H.: PCA-based parameter set mappings for LPV models with fewer parameters and less overbounding. IEEE Trans. Control Syst. Technol. **16**(4), 781–788 (2008)

35. Loke, M., Barker, R.: Rapid least-squares inversion of apparent resistivity pseudosections by a quasi-newton method 1. Geophys. Prospect. **44**(1), 131–152 (1996)
36. Moré, J.J.: The Levenberg-Marquardt algorithm: implementation and theory. In: Numerical Analysis, pp. 105–116. Springer, Berlin (1978)
37. Godefroid, P., Klarlund, N., Sen, K.: Dart: directed automated random testing. ACM Sigplan Notices **40**(6), 213–223 (2005)
38. Krstic, M., Kanellakopoulos, I., Kokotovic, P.V., et al.: Nonlinear and adaptive control design, vol. 222. Wiley, New York (1995)
39. Polycarpou, M.: Stable adaptive neural control scheme for nonlinear systems. IEEE Trans. Autom. Control **41**(3), 447–451 (1996)
40. Han, H., Su, C.-Y., Stepanenko, Y.: Adaptive control of a class of nonlinear systems with nonlinearly parameterized fuzzy approximators. IEEE Trans. Fuzzy Syst. **9**(2), 315–323 (2001)
41. Rantzer, A., Johansson, M.: Piecewise linear quadratic optimal control. IEEE Trans. Autom. Control **45**(4), 629–637 (2000)
42. Gonzalez, T., Sala, A., Bernal, M., Robles, R.: Invariant sets of nonlinear models via piecewise affine Takagi–Sugeno model. In: Proceedings of the 2015 IEEE International Conference on Fuzzy Systems. IEEE, New York (2015)
43. Gonzalez, T., Bernal, M.: Progressively better estimates of the domain of attraction for nonlinear systems via piecewise Takagi–Sugeno models: Stability and stabilization issues. Fuzzy Sets Syst. **297**, 73–95 (2016)
44. Gonzalez, T., Sala, A., Bernal, M., Aguiar, B.: Piecewise-Takagi–Sugeno asymptotically exact estimation of the domain of attraction of nonlinear systems. J. Franklin Inst. **354**(3), 1514–1541 (2017)
45. Gonzalez, T., Bernal, M., Marquez, R.: Stability analysis of nonlinear models via exact piecewise Takagi–Sugeno models. In: 19th World Congress of the International Federation of Automatic Control (IFAC) (2014)
46. Hedlund, S., Johansson, M.: A Toolbox for Computational Analysis of Piecewise Quadratic Linear Systems. Department of Automatic Control, Lund, Sweden (1999)
47. Sala, A., Ariño, C.: Polynomial fuzzy models for nonlinear control: A Taylor series approach. IEEE Trans. Fuzzy Syst. **17**(6), 1284–1295 (2009)
48. Tanaka, K., Yoshida, H., Ohtake, H., Wang, H.: A sum of squares approach to modeling and control of nonlinear dynamical systems with polynomial fuzzy systems. IEEE Trans. Fuzzy Syst. **17**(4), 911–922 (2009)
49. Prajna, S., Papachristodoulou, A., Seiler, P., Parrilo, P.A.: SOSTOOLS: Sum of Squares Optimization Toolbox for MATLAB. California Inst. Technol, Pasadena, CA (2004)
50. Campbell, S.: Singular Systems of Differential Equations II. Research Notes in Mathematics. Pitman Publishing, London (1982)
51. Louis, F.L.: A survey of linear singular systems. J. Circuits, Systems Signal Process. **5**(1), 3–36 (1986)
52. Dai, L.: Singular control systems. Lecture Notes in Control and Information Sciences, Springer, Berlin (1989)
53. Luenberger, D.G.: Dynamic equations in descriptor form. IEEE Trans. Autom. Control **22**(3), 312–321 (1977)
54. Luenberger, D.G.: Non-linear descriptor systems. J. Econ. Dyn. Control **1**(3), 219–242 (1979)
55. Pantelides, C.C.: The consistent initialization of differential-algebraic systems. SIAM J. Sci. Stat. Comput. **9**(2), 213–231 (1988)
56. Yang, C., Sun, J., Zhang, Q., Ma, X.: Lyapunov stability and strong passivity analysis for nonlinear descriptor systems. IEEE Trans. Circuits Syst. I: Regular Papers **60**(4), 1003–1012 (2013)
57. Kumar, A., Daoutidis, P.: Control of nonlinear differential algebraic equation systems: an overview. In: Nonlinear Model Based Process Control, pp. 311–344. Springer, Berlin (1998)
58. Aliyu, M., Boukas, E.-K.: H ∞ filtering for nonlinear singular systems. IEEE Trans. Circuits Syst. I: Regular Papers **59**(10), 2395–2404 (2012)

59. Wu, L., Shi, P., Gao, H.: State estimation and sliding-mode control of Markovian jump singular systems. IEEE Trans. Autom. Control **55**(5), 1213–1219 (2010)
60. Li, F., Du, C., Yang, C., Gui, W.: Passivity-based asynchronous sliding mode control for delayed singular Markovian jump systems. IEEE Trans. Autom. Control **63**(8), 2715–2721 (2017)
61. Duan, G.R.: Analysis and Design of Descriptor Linear Systems. Springer, New York (2010)
62. Xu, S., Lam, J.: Robust Control and Filtering of Singular Systems. Springer, Berlin (2006)
63. Nedialkov, N., Pryce, J., Tan, G.: Algorithm 948 DAESA: A MATLAB tool for structural analysis of differential-algebraic equations software. ACM Trans. Math. Software (TOMS) **41**(2), 12 (2015)
64. Tuan, H.D., Apkarian, P., Narikiyo, T., Kanota, M.: New fuzzy control model and dynamic output feedback parallel distributed compensation. IEEE Trans. Fuzzy Syst. **12**(1), 13–21 (2004)
65. Estrada-Manzo, V., Lendek, Z., Guerra, T.M., Pudlo, P.: Controller design for discrete-time descriptor models: a systematic LMI approach. IEEE Trans. Fuzzy Syst. **23**(5), 1608–1621 (2015)
66. Lewis, F., Dawson, D., Abdallah, C.: Robot Manipulator Control: Theory and Practice. Marcel Dekker Inc, New York (2004)
67. Guelton, K., Delprat, S., Guerra, T.-M.: An alternative to inverse dynamics joint torques estimation in human stance based on a Takagi–Sugeno unknown-inputs observer in the descriptor form. Control. Eng. Pract. **16**(12), 1414–1426 (2008)
68. Schulte, H., Guelton, K.: Descriptor modelling towards control of a two-link pneumatic robot manipulator: a T–S multimodel approach. Nonlinear Anal. Hybrid Syst **3**(2), 124–132 (2009)
69. Vermeiren, L., Dequidt, A., Afroun, M., Guerra, T.M.: Motion control of planar parallel robot using the fuzzy descriptor system approach. ISA Trans. **51**, 596–608 (2012)
70. Verghese, G.C., Levy, B.C., Kailath, T.: A generalized state-space for singular systems. IEEE Trans. Autom. Control **26**(4), 811–831 (1981)
71. Goldstein, H., Poole, C.P., Safko, J.L.: Classical Mechanics: Pearson New International Edition. Pearson Higher Ed, London (2014)
72. Kibble, T., Berkshire, F.H.: Classical Mechanics. Imperial College Press, London (2004)
73. Arceo, J.C., Sánchez, M., Estrada-Manzo, V., Bernal, M.: Convex stability analysis of nonlinear singular systems via linear matrix inequalities. IEEE Trans. Autom. Control **64**(4), 1740–1745 (2018)
74. Apkarian, P., Gahinet, P.: A convex characterization of gain-scheduled h_∞ controllers. IEEE Trans. Autom. Control **40**(5), 853–864 (1995)
75. Apkarian, P., Adams, R.J.: Advanced gain-scheduling techniques for uncertain systems. IEEE Trans. Control Syst. Technol. **6**(1), 21–32 (1998)
76. Scherer, C.W.: LPV control and full block multipliers. Automatica **37**(3), 361–375 (2001)
77. Wu, F., Dong, K.: Gain-scheduling control of LFT systems using parameter-dependent Lyapunov functions. Automatica **42**(1), 39–50 (2006)
78. Rugh, W.J., Shamma, J.S.: Research on gain scheduling. Automatica **36**(10), 1401–1425 (2000)
79. Zhou, K., Doyle, J.C., Glover, K., et al.: Robust and Optimal Control, vol. 272. Prentice Hall, Upper Saddle River, NJ (1996)
80. Gharibzahedi, S.M.T., Mousavi, S.M., Khodaiyan, F., Hamedi, M.: Optimization and characterization of walnut beverage emulsions in relation to their composition and structure. Int. J. Biol. Macromol. **50**(2), 376–384 (2012)
81. Robles, R., Sala, A., Bernal, M.: Performance-oriented quasi-LPV modeling of nonlinear systems. Int. J. Robust Nonlinear Control **29**(5), 1230–1248 (2019)

Chapter 4
Stability Analysis

4.1 Introduction

This chapter reviews various results concerning the stability analysis of nonlinear models with convex structures such as *Takagi–Sugeno* (TS) or polynomial, in standard or descriptor form; closely related results for *linear fractional transformation* (LFT) models are also visited. We begin by focussing on the application of direct Lyapunov method to derive stability conditions for TS models; if the reader is not familiar with Lyapunov stability, reading Appendix B is strongly advised. In the literature, for the sake of simplicity and to write the problems in the form of *linear matrix inequalities* (LMIs), mainly a quadratic Lyapunov function is considered, thus reducing the notion of stability to that of quadratic stability. This reduction encompasses another drawback: it disregards the TS nature of the model by subsuming it into the larger family of *linear parameter-varying* (LPV) systems. For these reasons, we also present results that go beyond the quadratic framework.

Conditions for analysis and design are expressed as LMIs because there exist efficient algorithms for solving them. Though LMIs as well as LMI-based control are well-established scientific fields, these constraints do not appear naturally in control problems. Therefore, one of the goals when encountering control problems is to recast them as LMI expressions by making use of their properties. A brief overview of these properties is found in the Appendix A; an overview of common LMI problems as well as solvers is given in Appendix C. LMI conditions emerge once the membership functions (MFs) in the expressions involving TS models (or any other model with convex structure) are dropped off. These MFs are involved nested convex sums whose positivity has to be determined by solving a multiple-sum copositivity problem, for which several solutions have been given; these solutions are known as sum relaxations. An overview of these schemes is given in Appendix D.

Stability analysis of equilibrium points of nonlinear system (3.8) is investigated under some assumptions: the system is unforced, meaning that the input and disturbance signals are identically zero, i.e. $\boldsymbol{u}(t) \equiv 0$, $\boldsymbol{d}(t) \equiv 0$, $\forall t$; it is assumed that $\boldsymbol{x} = 0$ is an equilibrium point which remains so for any variation of θ in a compact

M. Bernal et al., *Analysis and Synthesis of Nonlinear Control Systems*, Studies in Systems, Decision and Control 408, https://doi.org/10.1007/978-3-030-90773-0_4

set (any other equilibrium point can be shifted to the origin by an adequate change of coordinates). Therefore, before any convex model is adopted, nonlinear systems under consideration in this chapter have the form

$$E(\boldsymbol{x}, \boldsymbol{\theta}, t)\delta \boldsymbol{x}(t) = \boldsymbol{f}(\boldsymbol{x}, \boldsymbol{\theta}, t), \tag{4.1}$$

under the already presented definitions and assumptions on vector field $\boldsymbol{f}(\cdot)$, matrix function $E(\cdot)$, state $\boldsymbol{x}$, parametric uncertainty $\boldsymbol{\theta}$, and time t. Naturally, $E(\boldsymbol{x}, \boldsymbol{\theta}, t) = I$ yields ordinary state space representation; otherwise it leads to a descriptor model.

The following general form for Lyapunov function candidates will be used whenever TS models (both ordinary or descriptor) are involved:

$$V(\boldsymbol{x}) = \boldsymbol{x}^T \boldsymbol{P}(\cdot)\boldsymbol{x}, \tag{4.2}$$

where notation $(\cdot)$ implies a dependence on arguments such as $z(t+1)$, $z(t)$, $z(t-1)$, or different combinations of these variables, that will be described further on, according to the problem solved. Of course, if $\boldsymbol{P}(\cdot) = P$, the quadratic case is recovered.

Notation

In this chapter, definitions concerning positive (negative) definiteness of matrices are crucial; they are given in Appendix A, but recalled briefly here for convenience. For a given matrix, expression A holding the property $A = A^T \in \mathbb{R}^{n \times n}$ (symmetry), $A > 0$ ($A < 0$) stands for positive (negative) definiteness, meaning that $\boldsymbol{x}^T A\boldsymbol{x} > 0$ ($\boldsymbol{x}^T A\boldsymbol{x} < 0$) $\forall \boldsymbol{x} \neq 0$; as it is well known, this translates into the equivalent conditions $\text{Re}(\sigma(A)) > 0$ ($\text{Re}(\sigma(A)) < 0$), where $\sigma(A)$ denotes the spectrum of A. Whenever an expression is written as $A > 0$ or $A < 0$, it is assumed that it is symmetric, even if the explicit symmetry condition is omitted. Let $A,\ B \in \mathbb{R}^{n \times n}$ be two symmetric matrices. Then $A > B$ is equivalent to $A - B > 0$. Similar definitions hold for $A < B$ or for semidefinite expressions with symbols "$\geq$" and "$\leq$".

A star $(*)$ as a block matrix entry indicates the transposed of the quantity in the symmetric position, e.g.

$$\begin{bmatrix} A_1 & (*) \\ A_2 & A_3 \end{bmatrix} := \begin{bmatrix} A_1 & A_2^T \\ A_2 & A_3 \end{bmatrix}.$$

For inline expressions, $A^T P(*) := A^T P A$, and $A + (*) := A + A^T$.

Nested convex sums of matrix expressions appear in some stability approaches as well as in most design techniques based on convex models. As handling them might be cumbersome, some notation is adopted based on the simplest one in (3.16), namely:

1. Nested convex sums, e.g.

$$\Upsilon_{hh} = \sum_{i=1}^{r}\sum_{j=1}^{r} h_i(z)h_j(z)\Upsilon_{ij},$$

for 2 sums, where $h_i(z)$ are MFs and Υ_{ij} are matrix expressions.

2. Inverse of a convex sum, e.g.

$$\Upsilon_h^{-1} = \left(\sum_{i=1}^{r} h_i(z)\Upsilon_i\right)^{-1},$$

where $h_i(z)$ are MFs and Υ_i are matrix expressions.

3. Nested convex sums involving different sets of MFs, e.g.

$$\begin{aligned}\Upsilon_{hh\eta} &\equiv \sum_{i=1}^{r_1} h_i(z)\left(\sum_{j=1}^{r_1} h_j(z)\left(\sum_{k=1}^{r_2}\eta_k(\boldsymbol{w})\Upsilon_{ijk}\right)\right)\\ &= \sum_{i=1}^{r_1}\sum_{j=1}^{r_1}\sum_{k=1}^{r_2} h_i(z)h_j(z)\eta_k(\boldsymbol{w})\Upsilon_{ijk},\end{aligned}$$

where MFs $h_i(z)$ hold for the first two sums and $\eta_i(\boldsymbol{w})$ for the third one; z and $\boldsymbol{w}$ being scheduling variables; Υ_{ijk} are matrix expressions.

4. Convex sums at different time instances, e.g.

$$\Upsilon_{h(t+1)} \equiv \sum_{i=1}^{r} h_i(z(t+1))\Upsilon_i,$$

$$\Upsilon_{h(t+\alpha)} \equiv \sum_{i=1}^{r} h_i(z(t+\alpha))\Upsilon_i,$$

with $\alpha \in \mathbb{Z}$ standing for time advance or delay; $h_i(z)$ being MFs and Υ_i being matrix expressions.

Similar notation can be adopted for weight-dependent expressions (3.18) employed in tensor-product (TP) models.

In the literature, notation for convex sums vary wildly, sometimes to stress an aspect over another, e.g. the dependency of the convex sum on the scheduling variable (using Υ_z instead of Υ_h), the ultimate dependency of the scheduling variable on nonlinear expressions such as the state (Υ_x instead of Υ_z), omitting time dependence for brevity (Υ_{h+1}, instead of $\Upsilon_{h(t+1)}$), or including every detail of such dependency ($\Upsilon_{h(z(t+1))}$, instead of $\Upsilon_{h(t+1)}$). Unless otherwise stated, in the sequel, for the sake of uniformity, we will use the notation defined above.

4.2 Quadratic Stability of TS Models

In this section, we present LMI-based stability analysis of unforced TS models, namely:

$$\delta x(t) = \sum_{i=1}^{r} h_i(z) A_i x(t) = A_h x(t), \tag{4.3}$$

where, as usual, $\delta x(t) = \dot{x}(t)$, $t \in \mathbb{R}$ for continuous time; $\delta x(t) = x(t+1)$, $t \in \mathbb{Z}$ for discrete time; $x \in \mathbb{R}^n$ and $z \in \mathbb{R}^p$ stand for the state and scheduling vectors, respectively; $h_i(z)$ are membership functions (MFs) holding the convex sum property

$$\sum_{i=1}^{r} h_i(z) = 1, \; 0 \leq h_i(z) \leq 1,$$

for every z in a compact set $\mathcal{C} \subseteq \mathbb{R}^p$, and $A_i \in \mathbb{R}^{n \times n}$, $i \in \{1, 2, \ldots, r\}$ are constant matrices.

We are no longer concerned about the source of the convex representation, but the reader should keep in mind that it might be the result of the sector nonlinearity approach allowing *rewriting* an autonomous nonlinear system such as (4.1) with $E(x, \theta, t) = I$, by subsuming its nonlinearities in a scheduling vector z, which may depend on x, θ, or t. The method guarantees z belongs to a compact set $\mathcal{C} \subseteq \mathbb{R}^p$, which we call "induced" as it might be the consequence of boundedness of expressions depending on states, parameters, or time.

Unless otherwise stated, whenever a convex model is considered (TS, descriptor, polynomial, or LFT), the following assumptions will be made:

Assumption 4.1 Given a convex model, its origin $x = 0$ is an equilibrium point $\forall z \in \mathcal{C}$.

Assumption 4.2 Given a convex model, there exists a compact set $\mathcal{X}$ of the state space containing the origin $x = 0$ such that $\forall x \in \mathcal{X}$ then $z \in \mathcal{C}$.

Stability of the origin $x = 0$ (which, by Assumption 4.1 is an equilibrium point) is investigated using the direct Lyapunov method (see Appendix B). The Lyapunov function candidate commonly used is the quadratic one,

$$V(x) = x^T P x \tag{4.4}$$

with $P = P^T > 0$, which corresponds to (4.2) with $\boldsymbol{P}(\cdot) = P$. If proven a valid Lyapunov function based on the MFs $h_i(z)$ holding the convex sum property, it will directly follow that any trajectory $\boldsymbol{x}(t)$ contained in the outermost Lyapunov level within $\mathcal{X}$ will go asymptotically to $\boldsymbol{x} = 0$, since by Assumption (4.2) $\mathcal{X}$ contains the origin $\boldsymbol{x} = 0$ and is in concord with $z \in \mathcal{C}$, where convexity of $h_i(z)$ is guaranteed.

Since TS model (4.3) depends on z, which, in turn, may depend on time t, state $\boldsymbol{x}$, and/or parameters $\boldsymbol{\theta}$, asymptotic stability of $\boldsymbol{x} = 0$ can be refined as follows:

1. Should the scheduling vector z depend exclusively on the state $\boldsymbol{x}$ or fixed-but-unknown bounded parameters $\boldsymbol{\theta}$ (assumed not to change the equilibrium point), the TS model (4.3) is time invariant and asymptotic stability in the sense of Definition B.1 is investigated via Theorems B.1 and B.5 depending on whether the system is continuous or discrete in time. If the Lyapunov function candidate (4.4) is employed, we have $V(0) = 0$, $V(\boldsymbol{x}) > 0$, $\forall \boldsymbol{x} \neq 0$ in Theorems B.1 and B.5.
2. Should the scheduling vector z include time-dependent variables t or $\boldsymbol{\theta}$ (assumed not to change the equilibrium point), the TS model (4.3) is time variant and uniform asymptotic stability in the sense of Definition B.8 is investigated via Theorems B.3 and B.7 depending on whether the system is continuous or discrete in time. If the Lyapunov function candidate (4.4) is employed, clearly $W_1(\boldsymbol{x}) = W_2(\boldsymbol{x}) = V(\boldsymbol{x})$ and $\partial V/\partial t = 0$ in Theorem B.3 and $\alpha_1(\|\boldsymbol{x}\|) = \lambda_{\min}(P)\|\boldsymbol{x}\|^2$ in Theorem B.7.

For simplicity, results in the sequel will pay no attention to the possibly time-variant nature of TS model (4.3) as proofs can easily be adapted as suggested above. Nevertheless, the reader should keep in mind that in this case uniform asymptotic stability follows. Whenever the results are restricted to time-invariant models, it is explicitly stated.

Evidently, since TS system (4.3) is a nonlinear one, using a quadratic Lyapunov function candidate (4.4) to investigate the stability of $\boldsymbol{x} = 0$ can only yield sufficient conditions. This means that, if stability conditions derived from using the Lyapunov function candidate (4.4) fail, nothing can be directly said about stability or instability of the equilibrium point under analysis.

In LPV or LTV contexts, the existence of a quadratic Lyapunov function proving asymptotic stability of the origin $\boldsymbol{x} = 0$ is referred to as *quadratic stability* [1, 2]. We extend this definition to the TS models under consideration

Definition 4.1 The origin $\boldsymbol{x} = 0$ of the TS model (4.3) is *quadratically stable* if there exists a quadratic Lyapunov function (4.4) establishing its (uniform) asymptotic stability.

The aforementioned sufficiency of results can be reformulated as the fact that quadratic stability implies asymptotic stability of the origin, but not conversely.

4.2.1 The Continuous-Time Case

Let us consider first the TS model (4.3) with $\delta x(t) = \dot{x}(t)$, i.e. the continuous-time case. The results are based on Lyapunov Theorems B.1 and B.3 in Appendix B, depending on whether the system is time invariant or variant. Thus, the following generalisation of a well-known result in [3] is obtained

Theorem 4.1 *The origin $\boldsymbol{x} = 0$ of the unforced TS model (4.3) with $\delta \boldsymbol{x}(t) = \dot{\boldsymbol{x}}(t)$, $t \in \mathbb{R}$, is asymptotically stable if there exists a matrix $P = P^T > 0$ such that the following LMIs hold:*

$$PA_i + A_i^T P < 0, \; i \in \{1, 2, \ldots, r\}. \tag{4.5}$$

In this case, any trajectory $\boldsymbol{x}(t)$ starting within the outermost Lyapunov level $V(\boldsymbol{x}) = \boldsymbol{x}^T P \boldsymbol{x} \le k$, $k > 0$, in a compact set of the state space $\mathcal{X}$ guaranteeing $z \in \mathcal{C} \subseteq \mathbb{R}^p$ goes asymptotically to $\boldsymbol{x} = 0$.

Proof Since $P = P^T > 0$, $V(\boldsymbol{x}) = \boldsymbol{x}^T P \boldsymbol{x}$ is a valid Lyapunov function candidate; its time derivative along the trajectories of the unforced model (4.3) is

$$\begin{aligned} \dot{V}(t) &= \dot{\boldsymbol{x}}^T(t) P \boldsymbol{x}(t) + \boldsymbol{x}^T(t) P \dot{\boldsymbol{x}}(t) = (A_h \boldsymbol{x}(t))^T P \boldsymbol{x}(t) + \boldsymbol{x}^T(t) P (A_h \boldsymbol{x}(t)) \\ &= \left(\sum_{i=1}^r h_i(z) A_i \boldsymbol{x}(t)\right)^T P \boldsymbol{x}(t) + \boldsymbol{x}^T(t) P \left(\sum_{i=1}^r h_i(z) A_i \boldsymbol{x}(t)\right) \\ &= \sum_{i=1}^r h_i(z) \boldsymbol{x}^T(t) \left(A_i^T P + P A_i\right) \boldsymbol{x}(t), \end{aligned} \tag{4.6}$$

where the last equality made use of the property $\sum_{i=1}^r h_i(z) = 1$. Since $h_i(z) \ge 0$, $i \in \{1, 2, \ldots, r\}$, $\forall z \in \mathcal{C}$, not all simultaneously 0, it is clear that LMIs (4.5) imply $\forall \boldsymbol{x} \neq 0$ then $\dot{V} < 0$, which means $V(\boldsymbol{x})$ is a valid Lyapunov function establishing asymptotic stability of $\boldsymbol{x} = 0$ (see Theorem B.1). By assumption 4.2, there exists a compact set $\mathcal{X}$ of the state space with $0 \in \mathcal{X}$, guaranteeing $z \in \mathcal{C}$, which means that any trajectory $\boldsymbol{x}(t)$ in the largest Lyapunov level within $\mathcal{X}$ goes asymptotically to $\boldsymbol{x} = 0$. □

If the convex sum property holds everywhere, i.e. if $z \in \mathcal{C}$ irrespective of $\boldsymbol{x}$, the results above are global, i.e. $\mathcal{X} = \mathbb{R}^n$ [4]. This is the case with most TS fuzzy systems whose MFs are usually normalised as to add up to 1 everywhere, regardless of the state value. Nevertheless, whenever the sector nonlinearity approach is employed, a compact set $\mathcal{C} \subseteq \mathbb{R}^p$ is possibly induced by the bounds on nonlinearities and parameters, restricting the results to the aforementioned outermost Lyapunov level.

Example 4.1 Consider the continuous-time nonlinear system in (3.7), for example 3.1, reproduced below for convenience:

$$\dot{x}(t) = \begin{bmatrix} -2 + \sin x_1 & 1 \\ 0 & -3 - \cos^2 x_2 \end{bmatrix} x(t),$$

where $x = [x_1 \; x_2]^T$ is the state vector.

Via the sector nonlinearity approach, the previous system can be rewritten as a TS model by choosing $z_1 = \sin x_1 \in [-1, 1]$ and $z_2 = \cos^2 x_2 \in [0, 1]$ to define $w_0^1(z_1) = 0.5(1 - z_1)$, $w_0^2(z_2) = z_2$, $w_1^1(z_1) = 1 - w_0^1(z_1)$, $w_1^2(z_2) = 1 - w_0^2(z_2)$ as to obtain:

$$\dot{x}(t) = \sum_{i=1}^{4} h_i(z) A_i x(t), \tag{4.7}$$

with $h_1(z) = w_0^1(z_1)w_0^2(z_2)$, $h_2(z) = w_0^1(z_1)w_1^2(z_2)$, $h_3(z) = w_1^1(z_1)w_0^2(z_2)$, $h_4(z) = w_1^1(z_1)w_1^2(z_2)$, and

$$A_1 = \begin{bmatrix} -3 & 1 \\ 0 & -3 \end{bmatrix}, \quad A_2 = \begin{bmatrix} -3 & 1 \\ 0 & -4 \end{bmatrix}, \quad A_3 = \begin{bmatrix} -1 & 1 \\ 0 & -3 \end{bmatrix}, \quad A_4 = \begin{bmatrix} -1 & 1 \\ 0 & -4 \end{bmatrix}.$$

According to Theorem 4.1, the origin $x = 0$ of this TS model is asymptotically stable, if there exists $P = P^T > 0$ such that

$$\begin{aligned} A_1^T P + P A_1 &< 0, & A_2^T P + P A_2 &< 0, \\ A_3^T P + P A_3 &< 0, & A_4^T P + P A_4 &< 0. \end{aligned}$$

Solving the LMIs above with the MATLAB®LMI Toolbox, we obtain a feasible solution

$$P = \begin{bmatrix} 0.3327 & 0.0580 \\ 0.0580 & 0.2458 \end{bmatrix} > 0,$$

which means the equilibrium point $x = 0$ both of the original nonlinear system and the TS model (they are the same) is asymptotically stable.

Since MFs $h_i(z)$, $i \in \{1, 2, 3, 4\}$ hold the convex sum property for every value of the states, the origin $x = 0$ is also globally asymptotically stable. □

As follows from the development in (4.6), conditions in Theorem 4.1 are only sufficient, not only because a quadratic Lyapunov function is used, but also because MFs are removed from the convex sum, a process which is known as sum relaxation scheme (see Appendix D). Moreover, once the LMI conditions in (4.5) hold, the stability of *any* convex model with vertex matrices $A_1, A_2, \ldots, A_r$ is established, regardless of its LPV, LTV or TS nature. This lack of specificity of the quadratic approach is schematically illustrated by the following implications (see Fig. 4.1):

$$\begin{matrix} PA_i + (*) < 0 \\ i = 1, 2, \ldots, r \end{matrix} \underset{\text{sum relaxation}}{\Longrightarrow} PA_h + (*) < 0 \underset{\text{Lyapunov function}}{\Longrightarrow} \begin{cases} \dot{x} = A_h x \\ \text{LPV-stable } z(\theta) \\ \text{LTV-stable } z(t) \\ \text{TS-stable } \; z(x, \theta, t). \end{cases}$$

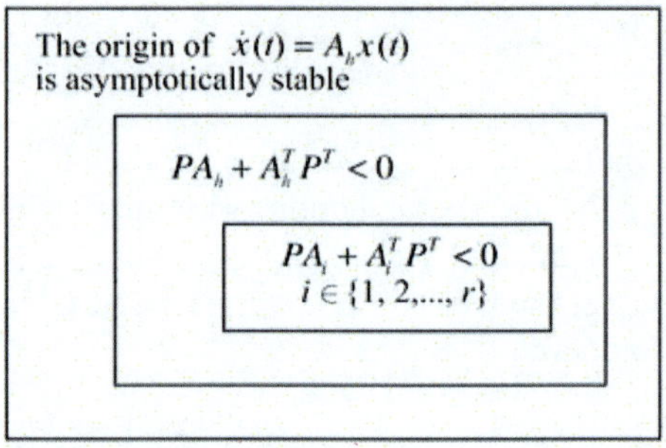

Fig. 4.1 Sufficiency of quadratic stability of TS models

The next example illustrates this discussion.

Example 4.2 Consider the LPV, LTV, and nonlinear models in (3.5), (3.6), and (3.7), respectively, taken from Example 3.1 and reproduced here for convenience:

LPV model:

$$\dot{\boldsymbol{x}}(t) = \begin{bmatrix} -2 + \theta_1 & 1 \\ 0 & -3 - \theta_2 \end{bmatrix} \boldsymbol{x}(t)$$

with $\theta_1 \in [-1, 1]$ and $\theta_2 \in [0, 1]$.

LTV model:

$$\dot{\boldsymbol{x}}(t) = \begin{bmatrix} -2 + \sin t & 1 \\ 0 & -3 - \cos^2 t \end{bmatrix} \boldsymbol{x}(t).$$

Nonlinear model:

$$\dot{\boldsymbol{x}}(t) = \begin{bmatrix} -2 + \sin x_1 & 1 \\ 0 & -3 - \cos^2 x_2 \end{bmatrix} \boldsymbol{x}(t).$$

Note that these models have the same convex representation as the example in (4.7), except for the MFs whose premise vector $\boldsymbol{z}$ may depend on $\boldsymbol{\theta}$ (LPV) or t (LTV), instead of $\boldsymbol{x}$. This means that the same quadratic stability analysis can be applied to all of them to draw the same conclusions, despite their different dependence on parameters, time, or state. □

Naturally, Theorem 4.1 can also be used to investigate asymptotic stability of equilibria of more complex nonlinear systems in an adequately defined compact, no matter if time-variant or parameter-varying. This is illustrated in the next example:

Example 4.3 Consider the following uncertain time-varying nonlinear system:

$$\dot{\boldsymbol{x}}(t) = \begin{bmatrix} -\cos^2 t & \theta(t) \\ -1 + x_1^2/3 & -1 \end{bmatrix} \boldsymbol{x}(t),$$

where $\boldsymbol{x}(t) = \begin{bmatrix} x_1 & x_2 \end{bmatrix}^T$ is the state vector and $\theta(t) \in [0.37, 2.2]$ a bounded time-varying parameter.

We are interested in investigating the stability of the origin via convex modelling, for which we employ the sector nonlinearity approach; to this end, consider the non-constant terms in the matrix above, namely: $z_1 = \theta$, $z_2 = -1 + x_1^2/3$, and $z_3 = \cos^2 t$; the first signal is bounded by assumption, the last one is naturally bounded, the second one requires defining a compact set of the state space that guarantees boundedness of z_2, e.g. $\mathcal{X} = \{\boldsymbol{x} : x_1 \in [-1.5, 1.5]\}$, which guarantees $z_2 \in [-1, -0.25]$, i.e, z to lie in $\mathcal{Z} = [0.37, 2.2] \times [-1, -0.25] \times [0, 1]$ as well as $0 \in \mathcal{X}$.

Thus, defining $w_0^1(z_1) = 1.2022 - 0.5464z_1$, $w_0^2(z_2) = -(4z_2 + 1)/3$, $w_0^3(z_3) = 1 - z_3$, $w_1^1(z_1) = 1 - w_0^1(z_1)$, $w_1^2(z_2) = 1 - w_0^2(z_2)$, and $w_1^3(z_3) = 1 - w_0^3(z_3)$, the system can be rewritten as

$$\dot{x}(t) = \sum_{i=1}^{8} h_i(z)A_i x(t), \tag{4.8}$$

where $h_i = w_{i_1}^1(z_1)w_{i_2}^2(z_2)w_{i_3}^3(z_3)$, (i_1, i_2, i_3) the 3-bit representation of $i - 1$, with the vertex matrices

$$A_1 = \begin{bmatrix} 0 & 0.37 \\ -1 & -1 \end{bmatrix}, A_2 = \begin{bmatrix} -1 & 0.37 \\ -1 & -1 \end{bmatrix}, A_3 = \begin{bmatrix} 0 & 0.37 \\ -0.25 & -1 \end{bmatrix}, A_4 = \begin{bmatrix} -1 & 0.37 \\ -0.25 & -1 \end{bmatrix},$$
$$A_5 = \begin{bmatrix} 0 & 2.2 \\ -1 & -1 \end{bmatrix}, A_6 = \begin{bmatrix} -1 & 2.2 \\ -1 & -1 \end{bmatrix}, A_7 = \begin{bmatrix} 0 & 2.2 \\ -0.25 & -1 \end{bmatrix}, A_8 = \begin{bmatrix} -1 & 2.2 \\ -0.25 & -1 \end{bmatrix}.$$

Conditions in Theorem 4.1 are feasible with

$$P = \begin{bmatrix} 1.0426 & 0.2255 \\ 0.2255 & 0.6536 \end{bmatrix} > 0,$$

which means the equilibrium point $\boldsymbol{x} = 0$ is asymptotically stable. The outermost Lyapunov level as well as trajectories of the system are shown in Fig. 4.2a, where $\theta(t)$ has been chosen as a random signal within the given interval, shown in Fig. 4.2b.

Theorem 4.1 also expresses that it is not enough to have all the vertices globally asymptotically stable to ensure the stability of the TS model. The reason for this is that the domain of Hurwitz matrices—those with eigenvalues with strictly negative real part—is non-convex. To illustrate this, consider the following example:

Example 4.4 Consider the matrices

$$A_1 = \begin{bmatrix} -1 & 6 \\ 0 & -2 \end{bmatrix}, \quad A_2 = \begin{bmatrix} -1 & 0 \\ 6 & -2 \end{bmatrix},$$

which are Hurwitz with eigenvalues at -1 and -2. Consider now the convex combination

$$A = 0.5 \times A_1 + 0.5 \times A_2 = \begin{bmatrix} -1 & 3 \\ 3 & -2 \end{bmatrix},$$

whose eigenvalues are -4.5414 and 1.5414, thus being non-Hurwitz. □

A test of non-existence of a common matrix $P = P^T$ has been given by [5]:

Theorem 4.2 *If there exist matrices $R_i = R_i^T > 0$, $i \in \{1, 2, \ldots, r\}$, such that*

$$\sum_{i=1}^{r} \left(A_i^T R_i + R_i A_i \right) > 0, \tag{4.9}$$

then there is no matrix $P = P^T > 0$ such that conditions (4.5) hold.

Proof The proof uses the facts (i) $\text{tr}(ABC) = \text{tr}(BCA)$, and (ii) $\text{tr}(AB) > 0$ if $A > 0$ and $B > 0$. Considering a matrix $P > 0$, we have, by (ii):

$$0 < \text{tr} \left(P \sum_{i=1}^{r} (A_i^T R_i + R_i A_i) \right).$$

Then, by (i):

$$0 < \text{tr} \left(P \sum_{i=1}^{r} (A_i^T R_i + R_i A_i) \right) = \sum_{i=1}^{r} \text{tr} \left(R_i A_i^T P + R_i P A_i \right)$$

and, hence, there exist at least one i where

$$0 < \text{tr} \left(R_i \left(A_i^T P + P A_i \right) \right),$$

so $A_i^T P + P A_i$ cannot be negative definite, because by (ii) $\text{tr}(R_i(-A_i^T P - P A_i)) > 0$, so we reach a contradiction, thus concluding the proof. □

Example 4.5 With the two matrices A_1 and A_2 defined in Example 4.4, a result of the LMI problem (4.9) is

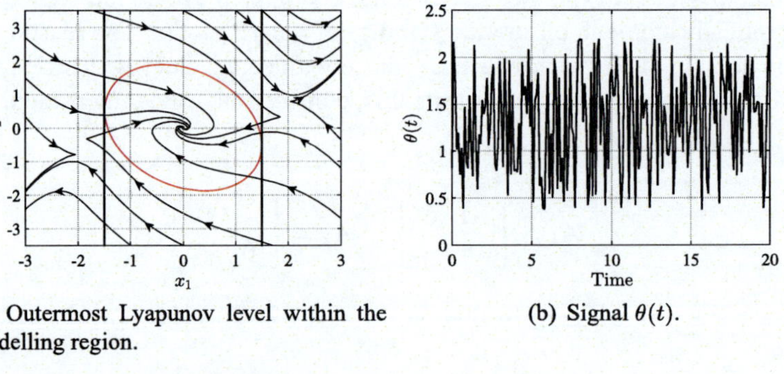

(a) Outermost Lyapunov level within the modelling region.

(b) Signal $\theta(t)$.

Fig. 4.2 Simulation results in Example 4.3

$$R_1 = \begin{bmatrix} 4 & 7 \\ 7 & 15 \end{bmatrix}, \; R_2 = \begin{bmatrix} 17 & 4 \\ 4 & 2 \end{bmatrix} \Rightarrow \sum_{i=1}^{2} (A_i^T R_i + R_i A_i) = \begin{bmatrix} 6 & 3 \\ 3 & 16 \end{bmatrix} > 0.$$

Therefore, there is no matrix $P = P^T > 0$ such that conditions (4.5) hold. □

Meeting conditions in Theorem 4.1 implies proving that $\boldsymbol{x} = 0$ is quadratically stable, as stated in Definition 4.1. As it turns out, it also ensures $\boldsymbol{x} = 0$ is an exponentially stable equilibrium point. To see this, consider (4.6). If $\dot{V}(t) < 0$ then there always exists a sufficiently small $\varepsilon > 0$ such that

$$\sum_{i=1}^{r} h_i(z) \left(A_i^T P + P A_i \right) + 2\varepsilon P < 0. \tag{4.10}$$

Thus, for all $t \in \mathbb{R}$ and $h_i(z) \geq 0, i \in \{1, 2, \ldots, r\}$,

$$\dot{V} + 2\varepsilon V = \boldsymbol{x}^T \left(\sum_{i=1}^{r} h_i(z) \left(A_i^T P + P A_i \right) + 2\varepsilon P \right) \boldsymbol{x} < 0, \tag{4.11}$$

and integrating the left-hand-side of (4.11) over $[t_0, \; t]$ yields that $V(\boldsymbol{x})$ has an exponential decay

$$V\left(\boldsymbol{x}(t)\right) \leq V\left(\boldsymbol{x}\left(t_0\right)\right) e^{-2\varepsilon(t-t_0)}.$$

Since for a positive definite matrix P, it holds that $\lambda_{\min}(P) \|\boldsymbol{x}\|^2 \leq \boldsymbol{x}^T P \boldsymbol{x} \leq \lambda_{\max}(P) \|\boldsymbol{x}\|^2$, we have

$$\lambda_{\min}(P) \|\boldsymbol{x}\|^2 \leq V\left(\boldsymbol{x}(t)\right) \leq V\left(\boldsymbol{x}\left(t_0\right)\right) e^{-2\varepsilon(t-t_0)} \leq \lambda_{\max}(P) \|\boldsymbol{x}_0\|^2 e^{-2\varepsilon(t-t_0)},$$

which means that $\|\boldsymbol{x}(t)\|$ has a decay rate of ε as follows from

$$\|\boldsymbol{x}(t)\|^2 \leq \frac{\lambda_{\max}(P)}{\lambda_{\min}(P)} \|\boldsymbol{x}_0\|^2 e^{-2\varepsilon(t-t_0)}$$

by taking square root in both sides of the inequality (see Definition B.3).

Thus, $\lim\limits_{t \to \infty} \boldsymbol{x}(t) = 0$ with exponential decay rate ε irrespective of the initial condition $\boldsymbol{x}_0(t_0)$, and for all $h_i(z) \geq 0, i \in \{1, 2, \ldots, r\}$. Clearly, this fulfils the hypotheses of Theorem B.2 in Appendix B with $a = 2$, $k_1 = \lambda_{\min}(P)$, $k_2 = \lambda_{\max}(P)$, and $k_3 = 2\varepsilon\lambda_{\min}(P)$.

From the previous development, it is easy to see that the solutions of a TS model hold a given decay rate $\alpha > 0$ if the following holds:

Theorem 4.3 *The origin $\boldsymbol{x} = 0$ of the unforced TS model (4.3) with $\delta\boldsymbol{x}(t) = \dot{\boldsymbol{x}}$, $t \in \mathbb{R}$, is exponentially stable with decay rate of $\alpha > 0$, if there exists a matrix $P = P^T > 0$ such that the following LMIs hold:*

$$A_i^T P + PA_i + 2\alpha P < 0, \ i \in \{1, 2, \ldots, r\}. \tag{4.12}$$

Proof With the Lyapunov function candidate $V(\boldsymbol{x}) = \boldsymbol{x}^T P \boldsymbol{x}$ and conditions (4.12), convexity of MFs $h_i(z)$ in the compact set $\mathcal{C}$ (on which MFs of the TS model (4.3) are supposed to hold the convex sum property), allows writing

$$\sum_{i=1}^{r} h_i(z)\left(A_i^T P + PA_i\right) + 2\alpha P < 0, \ i \in \{1, 2, \ldots, r\},$$

from which the proof follows the same path as the previous development replacing ε by α from (4.10) on. □

The maximum decay rate $\alpha > 0$ holding the conditions of the previous theorem can be found via a generalised eigenvalue problem (GEVP) (see Appendix C), which can be solved in most LMI toolboxes such as the MATLAB®LMI Toolbox [4, 6].

4.2.2 The Discrete-Time Case

Let us now consider the discrete-time case, i.e. TS models of the form (4.3) with $\delta \boldsymbol{x}(t) = \boldsymbol{x}(t+1)$. Using the quadratic Lyapunov function (4.4), stability analysis is performed by taking the forward difference in the Lyapunov function instead of the time derivative. By means of the Lyapunov Theorem B.5 in Appendix B, the following generalisation of the result for TS fuzzy models in [7] can be established:

Theorem 4.4 *The origin $\boldsymbol{x} = 0$ of the unforced TS model (4.3) with $\delta \boldsymbol{x}(t) = x(t+1)$, $t \in \mathbb{Z}$, is asymptotically stable if there exists a matrix $P = P^T > 0$ such that the following LMIs hold:*

$$\begin{bmatrix} P & A_i^T P \\ PA_i & P \end{bmatrix} > 0, \ i \in \{1, 2, \ldots, r\}. \tag{4.13}$$

In this case, any trajectory $\boldsymbol{x}(t)$ starting within the outermost Lyapunov level $V(\boldsymbol{x}) = \boldsymbol{x}^T P \boldsymbol{x} \le k$, $k \in \mathbb{R}$, in a compact set $\mathcal{X}$ of the state-space guaranteeing $z \in \mathcal{C} \subseteq \mathbb{R}^p$ goes asymptotically to $\boldsymbol{x} = 0$.

Proof Clearly, $P = P^T > 0$ makes $V(\boldsymbol{x}) = \boldsymbol{x}^T P \boldsymbol{x}$ a valid Lyapunov function candidate. The forward difference in the Lyapunov function along the trajectories of the system is calculated as follows:

$$\begin{aligned} \Delta V &= V(\boldsymbol{x}(t+1)) - V(\boldsymbol{x}(t)) \\ &= \boldsymbol{x}^T(t) A_h^T P A_h \boldsymbol{x}(t) - \boldsymbol{x}^T(t) P \boldsymbol{x}(t) \\ &= \boldsymbol{x}^T(t)\left(A_h^T P A_h - P\right)\boldsymbol{x}(t) \\ &= \boldsymbol{x}^T(t)\left(A_h^T P P^{-1} P A_h - P\right)\boldsymbol{x}(t). \end{aligned}$$

By Schur complement (see Lemma A.1 in Appendix A), it is clear that $\Delta V < 0$ if

$$\begin{bmatrix} P & A_h^T P \\ P A_h & P \end{bmatrix} = \begin{bmatrix} P & \left(\sum_{i=1}^{r} h_i(z) A_i\right)^T P \\ P\left(\sum_{i=1}^{r} h_i(z) A_i\right) & P \end{bmatrix}$$
$$= \sum_{i=1}^{r} h_i(z) \begin{bmatrix} P & A_i^T P \\ P A_i & P \end{bmatrix} > 0,$$

where the last equality made use of the fact that the MFs $h_i(z)$ add up to 1. Since $h_i(z) \geq 0, i \in \{1, 2, \ldots, r\}$, as long as $z \in \mathcal{C}$, LMI conditions (4.13) guarantee the inequality above. As a consequence, $V(x)$ is a valid Lyapunov function and $x = 0$ is proven to be asymptotically stable.

Since the derivation above is valid only for the compact set $\mathcal{C}$ where the MFs $h_i(z), i \in \{1, 2, \ldots, r\}$, hold the convex sum property, and $x \in \mathcal{X}$ guarantees $z \in \mathcal{C}$, it follows that any trajectory within the outermost Lyapunov level in $\mathcal{X}$ goes asymptotically to $x = 0$, thus concluding the proof. □

Example 4.6 Consider a discrete-time TS model with two vertex models described by

$$A_1 = \begin{bmatrix} -0.43 & 0.12 \\ -1.66 & 0.3 \end{bmatrix}, \quad A_2 = \begin{bmatrix} 0.32 & -0.18 \\ 0.17 & 0.72 \end{bmatrix}.$$

According to Theorem 4.4, this TS model is asymptotically stable, if there exists $P = P^T > 0$ so that

$$\begin{bmatrix} P & A_1^T P \\ P A_1 & P \end{bmatrix} > 0, \quad \begin{bmatrix} P & A_2^T P \\ P A_2 & P \end{bmatrix} > 0.$$

Solving the LMIs above, we obtain

$$P = \begin{bmatrix} 1.53 & -0.18 \\ -0.18 & 0.46 \end{bmatrix},$$

and, consequently, stability of the origin $x = 0$ of the corresponding TS model is proven.

Notice that no system has been given from which this TS model is obtained; this should be interpreted as if the nature of the MFs $h_i(z)$ is ignored, except for the crucial fact that "somehow" they satisfy the convex sum property. This is the reason TS models and LPV systems have been considered by many as "equivalent". Yet, as we have already seen in other examples and will keep seeing throughout the book, this equivalence does not apply to TS models arising from exactly rewriting of nonlinear systems via the sector nonlinearity approach; in this case, TS models can be considered as quasi-LPV systems. □

The next example considers a system with multiple equilibria which requires modelling on a compact set of the state space including the origin.

Example 4.7 Consider the following discrete-time nonlinear system

$$\boldsymbol{x}(t+1) = \begin{bmatrix} -0.8 + 0.72\tanh t & -0.71 \\ -0.52 + \theta(t) & 0.5 + 0.2x_2^2 \end{bmatrix} \boldsymbol{x}(t),$$

where $\boldsymbol{x}(t) = \begin{bmatrix} x_1 & x_2 \end{bmatrix}^T$ is the state vector and $\theta(t) \in [0.2, 0.5]$ a bounded time-varying parameter.

This system can be rewritten as a TS model considering the modelling region $\mathcal{X} = \{\boldsymbol{x} : -0.2 \leq x_2 \leq 0.2\}$ (as to include the origin) and choosing $z_1 = \tanh t \in [0, 1]$, $z_2 = \theta(t) \in [0.2, 0.5]$, and $z_3 = x_2^2 \in [0, 0.04]$. The vertex matrices are

$$A_1 = \begin{bmatrix} -0.8 & -0.71 \\ -0.32 & 0.5 \end{bmatrix}, A_2 = \begin{bmatrix} -0.8 & -0.71 \\ -0.32 & 0.508 \end{bmatrix}, A_3 = \begin{bmatrix} -0.8 & -0.71 \\ -0.02 & 0.5 \end{bmatrix},$$
$$A_4 = \begin{bmatrix} -0.8 & -0.71 \\ -0.02 & 0.508 \end{bmatrix}, A_5 = \begin{bmatrix} -0.08 & -0.71 \\ -0.32 & 0.5 \end{bmatrix}, A_6 = \begin{bmatrix} -0.08 & -0.71 \\ -0.32 & 0.508 \end{bmatrix},$$
$$A_7 = \begin{bmatrix} -0.08 & -0.71 \\ -0.02 & 0.5 \end{bmatrix}, A_8 = \begin{bmatrix} -0.08 & -0.71 \\ -0.02 & 0.508 \end{bmatrix}.$$

Theorem 4.4 renders a feasible solution with

$$P = \begin{bmatrix} 0.6636 & 0.1443 \\ 0.1443 & 0.9285 \end{bmatrix},$$

which proves the origin $\boldsymbol{x} = 0$ is asymptotically stable. The outermost Lyapunov level as well as trajectories of the system are shown in Fig. 4.3(a), where $\theta(t)$ is a random signal within the prescribed bounds, shown in Fig. 4.3(b).

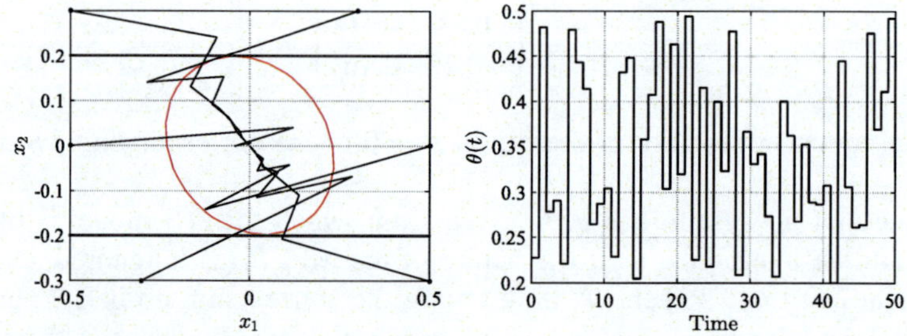

(a) Outermost Lyapunov level within the modelling region.

(b) Signal $\theta(t)$.

Fig. 4.3 Simulation results in Example 4.7

A simpler proof can be obtained using Finsler's Lemma A.3 (see Appendix A), which has been used in several state-of-the-art results in the discrete-time case. Consider again the quadratic Lyapunov function (4.4). The forward difference in the Lyapunov function along the trajectories of the system is:

$$\begin{aligned} \Delta V &= \boldsymbol{x}^T(t+1)P\boldsymbol{x}(t+1) - \boldsymbol{x}^T(t)P\boldsymbol{x}(t) \\ &= \begin{bmatrix} \boldsymbol{x}(t) \\ \boldsymbol{x}(t+1) \end{bmatrix}^T \begin{bmatrix} -P & 0 \\ 0 & P \end{bmatrix} \begin{bmatrix} \boldsymbol{x}(t) \\ \boldsymbol{x}(t+1) \end{bmatrix}. \end{aligned}$$

At the same time, the dynamics of the discrete-time system can be written as

$$[A_h \ -I] \begin{bmatrix} \boldsymbol{x}(t) \\ \boldsymbol{x}(t+1) \end{bmatrix} = 0.$$

Using Finsler's Lemma A.3 and choosing $M = \begin{bmatrix} 0 \\ H \end{bmatrix}$, with H being a free matrix, one obtains the conditions:

$$\begin{bmatrix} -P & 0 \\ 0 & P \end{bmatrix} + \begin{bmatrix} 0 \\ H \end{bmatrix} [A_h \ -I] + \left(\begin{bmatrix} 0 \\ H \end{bmatrix} [A_h \ -I] \right)^T < 0,$$

leading to

$$\begin{bmatrix} -P & A_h^T H^T \\ H A_h & P - H - H^T \end{bmatrix} < 0.$$

A sufficient LMI condition can be formulated as

$$\begin{bmatrix} -P & A_i^T H^T \\ H A_i & P - H - H^T \end{bmatrix} < 0, \qquad i \in \{1, 2, \ldots, r\}.$$

This result is not worse than Theorem 4.4. To see this, lets us choose $H = H^T = P$. With this choice, we have

$$\begin{bmatrix} -P & A_h^T P \\ P A_h & -P \end{bmatrix} < 0,$$

which, writing the corresponding LMI conditions, is equivalent to (4.13).

Similar to the continuous-time case, quadratic stability of discrete-time TS systems ensure exponential decay of the state values. To verify whether this decay rate is more than a certain value or finding an upper bound on the decay rate, one can use the following result [7]:

Theorem 4.5 *The origin $\boldsymbol{x} = 0$ of the unforced TS model (4.3) with $\delta \boldsymbol{x}(t) = x(t+1)$, $t \in \mathbb{Z}$, is exponentially stable with decay rate of $\alpha \in [0, 1]$, if there exists $\beta \in [0, 1]$ and $P = P^T > 0$ such that the following LMIs hold:*

$$\begin{bmatrix} \beta P & A_i^T P \\ P A_i & P \end{bmatrix} > 0,\ i \in \{1, 2, \dots, r\}. \tag{4.14}$$

The decay rate is thus $\alpha = \beta^{1/2}$.

Proof Since $P > 0$, it is clear that the quadratic nature of the Lyapunov function candidate $V(\boldsymbol{x}) = \boldsymbol{x}^T P \boldsymbol{x}$ guarantees

$$\lambda_{\min}(P)\|\boldsymbol{x}\|^2 \leq V(\boldsymbol{x}) \leq \lambda_{\max}(P)\|\boldsymbol{x}\|^2.$$

Now, by Schur complement (Lemma A.1 in Appendix A), conditions (4.14) imply that

$$\beta P - A_i^T P A_i > 0,\ i \in \{1, 2, \dots, r\},$$

which after pre- and post-multiplying by $\boldsymbol{x}^T \neq 0$ and $\boldsymbol{x} \neq 0$, as well as an algebraic rearrangement involving MFs $h_i(z)$, $i \in \{1, 2, \dots, r\}$, is equivalent to:

$$\boldsymbol{x}^T \left(\sum_{i=1}^{r} h_i(z) A_i^T\right) P \left(\sum_{i=1}^{r} h_i(z) A_i\right) \boldsymbol{x} - \beta \boldsymbol{x}^T P \boldsymbol{x} = V(\boldsymbol{x}(t+1)) - \beta V(\boldsymbol{x}(t)),$$

which in turn is equivalent to:

$$V(\boldsymbol{x}(t+1)) - V(\boldsymbol{x}(t)) < (\beta - 1)V(\boldsymbol{x}(t)) \leq -(1-\beta)\lambda_{\min}(P)\|\boldsymbol{x}\|^2.$$

Invoking Theorem B.6 in Appendix B, the last inequality guarantees that the origin is exponentially stable with $k_1 = \lambda_{\min}(P)$, $k_2 = \lambda_{\max}(P)$, and $k_3 = (1-\beta)\lambda_{\min}(P)$ for the outermost Lyapunov level within the induced compact set $\mathcal{C}$ (on which the MFs of the TS model (4.3) holds the convex sum property).

As for the exact value of the decay rate, notice that after t steps from the initial condition $\boldsymbol{x}(t_0)$ we have:

$$\lambda_{\min}(P)\|\boldsymbol{x}\|^2 \leq V(\boldsymbol{x}) \leq \beta^{t-t_0} V(\boldsymbol{x}(t_0)) \leq \lambda_{\max}(P)\|\boldsymbol{x}(t_0)\|^2 \beta^{t-t_0},$$

which implies

$$\|x(t)\|^2 \leq \frac{\lambda_{\max}(P)}{\lambda_{\min}(P)} \|\boldsymbol{x}(t_0)\|^2 \beta^{t-t_0}.$$

Taking the square root of both sides of the last inequality and invoking Definition B.12, a decay rate of $\alpha = \beta^{1/2}$ is thus proven. □

Note that, in contrast with the continuous-time case, a "maximum decay" of solutions for a discrete-time system means a minimum value of α; finding such a minimum is a generalised eigenvalue problem (GEVP) (see Appendix C), which can be solved iteratively [4].

4.2.3 Shape-Dependent Quadratic Stability

As previously mentioned, the conditions presented in the previous section are only sufficient. They rely on the convex sum property of the MFs $h_i(z)$ without involving them in the analysis and reduce the stability issue to quadratic stability. Nevertheless, a system may be proven stable if more information on the MFs is taken into account, as illustrated by the following example [8]:

Example 4.8 Consider the nonlinear model

$$\dot{x}(t) = \begin{bmatrix} -5 - 10\cos^2 x_1 & 30 - 20|\cos x_1| \\ 4(1 - |\cos x_1|)\cos^2 x_1 & -5-10(1-\cos^2 x_1) \end{bmatrix} x(t), \tag{4.15}$$

with $x = [x_1\, x_2]^T$.

By sector nonlinearity approach, two nonlinearities can be identified: $z_1 = |\cos x_1| \in [0, 1]$, $z_2 = \cos^2 x_1 \in [0, 1]$, based on which the following equivalent TS model can be obtained:

$$\dot{x} = \sum_{i=1}^{4} h_i(z) A_i x, \tag{4.16}$$

with $w_0^1(z_1) = z_1$, $w_1^1(z_1) = 1 - w_0^1(z_1)$, $w_0^2(z_2) = z_2$, $w_1^2(z_2) = 1 - w_0^2(z_2)$, $h_1(z) = w_0^1(z_1)w_0^2(z_2)$, $h_2(z) = w_1^1(z_1)w_0^2(z_2)$, $h_3(z) = w_0^1(z_1)w_1^2(z_2)$, $h_4(z) = w_1^1(z_1)w_1^2(z_2)$, and

$$A_1 = \begin{bmatrix} -15 & 10 \\ 0 & -5 \end{bmatrix},\; A_2 = \begin{bmatrix} -15 & 30 \\ 4 & -5 \end{bmatrix},\; A_3 = \begin{bmatrix} -5 & 10 \\ 0 & -15 \end{bmatrix},\; A_4 = \begin{bmatrix} -5 & 30 \\ 0 & -15 \end{bmatrix}.$$

For this model, there is no P such that LMI conditions (4.5) are feasible. Nevertheless, the origin $x = 0$ of (4.16) is asymptotically stable, as can be seen by rewriting the stability condition $PA_h + A_h^T P < 0$ with $L_i = PA_i + A_i^T P$ and taking into account that $(1 - \cos^2 x_1)|\cos x_1| \geq (1 - |\cos x_1|)\cos^2 x_1$, i.e.

$$\begin{aligned} PA_h + A_h^T P &= \cos^2 x_1 |\cos x_1| L_1 + \cos^2 x_1 (1 - |\cos x_1|) L_2 \\ &+ (1 - \cos^2 x_1)|\cos x_1| L_3 + (1 - \cos^2 x_1)(1 - |\cos x_1|) L_4 \\ &< \cos^2 x_1 |\cos x_1| L_1 + \cos^2 x_1 (1 - |\cos x_1|)(L_2 + L_3) \\ &+ (1 - \cos^2 x_1)(1 - |\cos x_1|) L_4 < 0 \end{aligned}$$

if $L_1 < 0$, $L_3 < 0$, $L_4 < 0$, and $L_2 + L_3 < 0$. These conditions are actually LMIs; with

$$P = \begin{bmatrix} 1 & -2 \\ -2 & 20 \end{bmatrix}.$$

they yield a feasible solution, thus proving the asymptotic stability of the origin of (4.16). □

The example above does not rely exclusively on the convex sum property of the MFs, but on additional relationships among them, i.e. on their *shape*. Originally, the MFs associated with a TS model were nothing else than (possibly) exogenous signals holding the convex sum property, which corresponds to the original fuzzy interpretation. Nevertheless, when exact TS modelling is employed for nonlinear, time-varying, or parameter-dependent systems, the actual shape of the MFs is known and can be used in a variety of ways:

1. affine constraints on memberships, say $h_1(\cdot) - 0.2h_3(\cdot) - 0.2 \geq 0$ [9–11];
2. polynomial constraints on memberships, say $h_1^2(\cdot) - h_2(\cdot)h_3(\cdot) \geq 0$, [12, 13];
3. tensor-product structures such as those employed in the use of derivative bounds [14, 15].

Some of the forthcoming results (derivative bounds, line-integral in Sect. 4.4) can also be considered to be shape-dependent, as they require the memberships to satisfy additional assumptions to that of a mere convex sum property (time derivative bounds, partial derivative bounds, number and structure of arguments, etc.).

4.3 Piecewise Lyapunov Functions

The previous examples show that although quadratic stability cannot be proven, introducing some knowledge into the Lyapunov function can eliminate some drawbacks. For details on how to introduce information from the MFs into the *quadratic* stability analysis as in Example 4.8, the interested reader is referred to [8, 16, 17]. As for how to use the MFs to induce a state-space partition according to the scheduling variables as suggested in Example 4.9, two approaches are considered: the first one is the use of piecewise quadratic [18–21] Lyapunov functions (PWLF); the second one is the use of a fuzzy Lyapunov function (FLF) proposed by [22, 23], which was later generalised and referred to as *parameter-dependent Lyapunov function* (PDLF). The use of these non-quadratic Lyapunov functions for stability analysis of TS models is the matter of the next subsections, both for continuous and discrete-time cases. Other piecewise options are considered in [24–26] for stability analysis.

4.3.1 The Continuous-Time Case

In [19], a MF-induced polyhedral partition of the state space was used to define a PWLF for affine continuous-time TS models, i.e. models of the form

$$\dot{\boldsymbol{x}}(t) = \sum_{l=1}^{r} h_l(z(\boldsymbol{x}))\,(A_l\boldsymbol{x}(t) + \boldsymbol{a}_l)\,, \tag{4.17}$$

where $A_l \in \mathbb{R}^{n\times n}$ and $\boldsymbol{a}_l \in \mathbb{R}^{n\times 1}$ are vertex matrices and affine vector terms, respectively, and MFs $h_l(\boldsymbol{x})$, $l \in \{1, 2, \ldots, r\}$ depend on the premise vector $z(\cdot)$, which, in turn, is assumed to depend linearly on the system state $\boldsymbol{x}$, i.e. $z(\boldsymbol{x}) = C\boldsymbol{x}$, with $C \in \mathbb{R}^{p\times n}$. Importantly, MFs are assumed to hold the convex sum property in the whole state space; therefore, with few exceptions, these models can only come from approximations of nonlinear ones or from fuzzy systems, not from sector nonlinearity approach.

Regions where $h_l(\boldsymbol{x}) = 1$ for some l are called *operating regimes* since only the l-th subsystem $\dot{\boldsymbol{x}}(t) = A_l\boldsymbol{x}(t) + \boldsymbol{a}_l$ is active on them. Otherwise, in between operating regimes, regions will be called *interpolation regimes*. Both these regions have also a geometrical interpretation, provided that the premise vector $z(\boldsymbol{x})$ depends linearly on the states $\boldsymbol{x}$: they form a polyhedral collection $\{X_i\}_{i\in I} \subseteq \mathbb{R}^n$, where I is the set of cell indices.

For each cell X_i, a set $K(i)$ will be defined as the set of indices of the system matrices used in the interpolation within that cell. Naturally, for operating regimes, $K(i)$ contains only a single element. Since this approach investigates exponential stability of the origin, I is divided into two sets: I_0 gathering the indices for cells that contain the origin and I_1 having the indices for cells that do not contain the origin.

Defining

$$\overline{A}_k = \begin{bmatrix} A_k & \boldsymbol{a}_k \\ 0 & 0 \end{bmatrix}, \quad \overline{\boldsymbol{x}} = \begin{bmatrix} \boldsymbol{x} \\ 1 \end{bmatrix} \tag{4.18}$$

where it is assumed that $\boldsymbol{a}_k = 0$ for all $k \in K(i)$ with $i \in I_0$, the system (4.17) can be rewritten as

$$\dot{\overline{\boldsymbol{x}}}(t) = \sum_{k\in K(i)} h_k(z(\boldsymbol{x}))\overline{A}_k\overline{\boldsymbol{x}}(t), \quad \boldsymbol{x} \in X_i.$$

Traditionally, this approach partitions the state space according to the activation of the linear models, allowing the Lyapunov function to change from one region to another. To see this, consider a Lyapunov function candidate (4.2) with the following definition:

$$V(\boldsymbol{x}) = \overline{\boldsymbol{x}}^T \boldsymbol{P}(\boldsymbol{x})\overline{\boldsymbol{x}}, \tag{4.19}$$

where

$$\boldsymbol{P}(\boldsymbol{x}) = \begin{cases} \text{block-diag}(P_i, 0), & \boldsymbol{x} \in X_i,\ i \in I_0 \\ \overline{P}_i, & \boldsymbol{x} \in X_i,\ i \in I_1, \end{cases}$$

where $P_i \in \mathbb{R}^{n\times n} : P_i = P_i^T > 0, i \in I_0, \overline{P}_i \in \mathbb{R}^{(n+1)\times(n+1)} : \overline{P}_i = \overline{P}_i^T > 0, i \in I_1$.

The above partition is natural for those TS models that do not have all their linear models activated at once. Nevertheless, this assumption does not hold for TS models built using the sector nonlinearity approach, but it holds, e.g. for TS models obtained by dynamic linearisation or substitution (see Chapter 3).

In order to guarantee continuity of the PWLF across the borders between regimes, this function is parameterised by matrices $\overline{F}_i = [F_i \ \ f_i]$, $i \in I$, with $F_i \in \mathbb{R}^{n \times n}$, $f_i \in \mathbb{R}^{n \times 1}$, $f_i = 0$ for $i \in I_0$, such that:

$$\overline{F}_i \begin{bmatrix} x \\ 1 \end{bmatrix} = \overline{F}_j \begin{bmatrix} x \\ 1 \end{bmatrix}, \quad x \in \{X_i \cap X_j\}, \ i, j \in I. \tag{4.20}$$

A systematic procedure for constructing these matrices is given by [19]. Then, P_i and $\overline{P}_i$ in (4.19) are parameterised as follows:

$$\begin{aligned} P_i &= F_i^T T F_i, \quad i \in I_0, \\ \overline{P}_i &= \overline{F}_i^T T \overline{F}_i, \quad i \in I_1, \end{aligned} \tag{4.21}$$

with T being a symmetric matrix of appropriate dimensions which collects the free parameters of the Lyapunov function. Note that this arrangement is a compromise between continuity and LMI formulation of results, which is hereby possible.

Since matrices P_i or $\overline{P}_i$ are only used to describe the Lyapunov function in cell X_i, it is possible to use the S-procedure (Property A.2) to reduce conservativeness by constructing matrices $\overline{E}_i = [E_i \ e_i]$ with $E_i \in \mathbb{R}^{n \times n}$, $e_i \in \mathbb{R}^{n \times 1}$, $e_i = 0$ for $i \in I_0$, that satisfy

$$\overline{E}_i \begin{bmatrix} x \\ 1 \end{bmatrix} \succeq 0, \quad x \in X_i, \ i \in I, \tag{4.22}$$

where for every matrix W_i with nonnegative entries $W_i \succeq 0$, condition (4.22) implies that

$$\begin{bmatrix} x \\ 1 \end{bmatrix}^T \overline{E}_i^T W_i \overline{E}_i \begin{bmatrix} x \\ 1 \end{bmatrix} > 0, \quad \forall x \in X_i, \ i \in I.$$

As with $\overline{F}_i$, matrices $\overline{E}_i$ can also be systematically constructed [19]. Then, the following result on piecewise quadratic stability of continuous-time affine TS models can be stated:

Theorem 4.6 *[19] If there exist symmetric matrices T, $U_i \succeq 0$ and $W_{ik} \succeq 0$ such that*

$$\begin{aligned} P_i &= F_i^T T F_i, \quad i \in I_0, \\ \overline{P}_i &= \overline{F}_i^T T \overline{F}_i, \quad i \in I_1 \end{aligned} \tag{4.23}$$

satisfy

$$\begin{aligned} & P_i - E_i^T U_i E_i > 0, \\ & A_k^T P_i + P_i A_k + E_i^T W_{ik} E_i < 0 \end{aligned} \tag{4.24}$$

for $i \in I_0$, $k \in K(i)$, and

$$\overline{P}_i - \overline{E}_i^T U_i \overline{E}_i > 0,$$
$$\overline{A}_k^T \overline{P}_i + \overline{P}_i \overline{A}_k + \overline{E}_i^T W_{ik} \overline{E}_i < 0 \tag{4.25}$$

for $i \in I_1$, $k \in K(i)$, then $x(t)$ tends to zero exponentially for every continuous piecewise differentiable trajectory in $\cup_{i \in I} X_i$ satisfying (4.17).

We do not intend to reproduce the details given in [19], but it is important to notice that LMIs (4.24) and (4.25) guarantee the PWLF candidate (4.19) to be positive everywhere and its time derivative negative in every state-space partition. Once the system trajectory enters the regions containing the origin, the results in Theorem B.4 of Appendix B guarantee exponential convergence to the origin. Auxiliary matrices $\overline{F}_i$ and $\overline{E}_i$ guarantee continuity of the PWLF as well as the inclusion of specific partition-information via an adaptation of the S-procedure in Property A.2.

Example 4.9 Consider the model

$$\dot{x}(t) = \begin{cases} A_1 x(t), & \text{if } x_1 < 0 \\ A_2 x(t), & \text{if } x_1 \geq 0 \end{cases} \text{ with } A_1 = \begin{bmatrix} -10 & -8 \\ -2 & -4 \end{bmatrix}, \; A_2 = \begin{bmatrix} -1 & -2 \\ 10 & -1 \end{bmatrix}, \tag{4.26}$$

where $x = [x_1 \; x_2]^T$.

Ordinary quadratic stability reduces to find $P > 0$ such that

$$A_1^T P + P A_1 < 0, \; A_2^T P + P A_2 < 0.$$

Nevertheless, no such P exists. Indeed, according to Property 4.2, the fact that

$$\sum_{i=1}^{2} (A_i^T R_i + R_i A_i) = \begin{bmatrix} 0.74 & 0.04 \\ 0.04 & 0.53 \end{bmatrix} > 0,$$

with

$$R_1 = \begin{bmatrix} 0.85 & -0.7 \\ -0.7 & 0.67 \end{bmatrix}, \; R_2 = \begin{bmatrix} 2.62 & 1.01 \\ 1.01 & 0.57 \end{bmatrix},$$

proves that no such P exists. Therefore, quadratic stability of (4.26) cannot be established in this way.

Now consider the piecewise Lyapunov function

$$V(x) = \begin{cases} x^T P x, & \text{if } x_1 < 0, \\ x^T \left(P + \eta \begin{bmatrix} 1 \\ 0 \end{bmatrix} \begin{bmatrix} 1 & 0 \end{bmatrix} \right) x, & \text{if } x_1 \geq 0. \end{cases} \tag{4.27}$$

With $C = \begin{bmatrix} 1 & 0 \end{bmatrix}$, the LMI problem for stability consists in finding $P = P^T > 0$ and η such that

$$\begin{aligned}&P+\eta C^T C>0,\\&PA_1+(*)<0,\\&(P+\eta C^T C)A_2+(*)<0.\end{aligned}$$

The solution

$$P=\begin{bmatrix}0.05 & -0.01\\ -0.01 & 0.20\end{bmatrix},\ \eta=0.89$$

was found. In fact,

$$PA_1+A_1^T P=\begin{bmatrix}-0.95 & -0.68\\ -0.68 & -1.5\end{bmatrix},$$

$$(P+\eta C^T C)A_2+A_2^T(P+\eta C^T C)=\begin{bmatrix}-2.04 & 0.16\\ 0.16 & -0.37\end{bmatrix}$$

are both negative definite. Thus, the origin $\boldsymbol{x}=0$ of system (4.26) is asymptotically stable. □

Example 4.10 Consider the following TS model [23]:

$$\dot{\boldsymbol{x}}(t)=\begin{cases}A_h^1\boldsymbol{x}(t)=\sum_{i=1}^2 h_i^1(\boldsymbol{x})A_h^1\boldsymbol{x}(t), & \boldsymbol{x}\in\mathcal{C}_1,\\ A_h^2\boldsymbol{x}(t)=\sum_{i=1}^2 h_i^2(\boldsymbol{x})A_h^2\boldsymbol{x}(t), & \boldsymbol{x}\in\mathcal{C}_2,\end{cases} \tag{4.28}$$

where $h_1^1(x_1)=1-\sin x_1,\ h_2^1(x_1)=1-h_1^1(x_1),\ h_1^2(x_1)=\sin x_1,\ h_2^2(x_1)=1-h_1^2(x_1)$, and

$$A_1^1=\begin{bmatrix}-3.5 & -4\\ 9.5 & -2\end{bmatrix},\ A_2^1=\begin{bmatrix}-5 & -4\\ -1 & -2\end{bmatrix},\ A_1^2=\begin{bmatrix}-2 & -4\\ 20 & -2\end{bmatrix},\ A_2^2=\begin{bmatrix}-3.5 & -4\\ 9.5 & -2\end{bmatrix}.$$

Conditions for quadratic stability fail, though simulations suggest that the system is stable. Consider a partition into two sets containing the origin: $X_1=\{\boldsymbol{x}:x_1\geq 0\}$ and $X_2=\{\boldsymbol{x}:x_1\leq 0\}$.

The following matrices F_k and E_k, $k\in\{1,2\}$ can be constructed via the PWL-TOOL [27]:

$$F_1=\begin{bmatrix}1 & 0\\ 1 & 0\\ 0 & 1\end{bmatrix},\quad F_2=\begin{bmatrix}0 & 0\\ 1 & 0\\ 0 & 1\end{bmatrix},\quad E_1=\begin{bmatrix}1\\ 0\end{bmatrix},\quad E_2=\begin{bmatrix}-1\\ 0\end{bmatrix}.$$

Conditions (4.24) are feasible. The following matrices were obtained for the PWLF (4.19):

$$P_1=\begin{bmatrix}9.7663 & 0.6145\\ 0.6145 & 6.4884\end{bmatrix},\quad P_2=\begin{bmatrix}26.0256 & 1.03\\ 1.03 & 6.4884\end{bmatrix}.$$

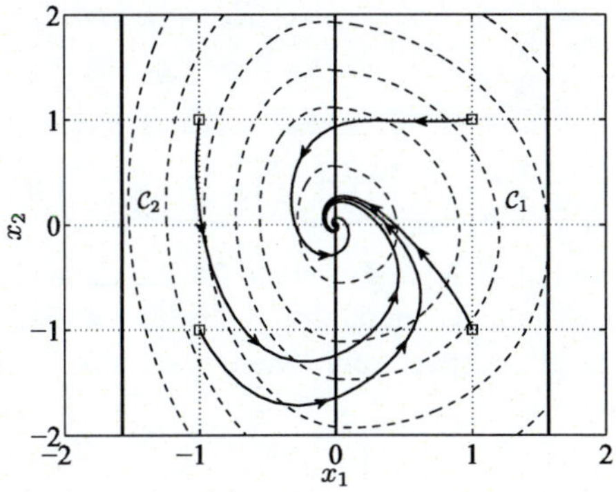

Fig. 4.4 Level surfaces of the computed Lyapunov function

Some level surfaces of the computed PWLF are shown in Fig. 4.4 with dashed lines along with some system trajectories in solid lines. Clearly, the origin $\boldsymbol{x} = 0$ of the system is asymptotically stable. □

In [28, 29], stability analysis of nonlinear systems via piecewise affine Takagi–Sugeno models (PWATS) allowed enhancing the aforementioned conditions by including more geometric considerations than the sole partition cells. The sector nonlinearity approach is generalised to include piecewise considerations, allowing exact rewriting of nonlinear systems instead of approximate or fuzzy models. Via Lagrange multipliers, different instances of the *Positivstellensatz* argumentation [30] are included to remove already-proven regions of the domain of attraction (ellipsoids or faces), easing the continuity of the piecewise Lyapunov function, and iteratively increasing a piecewise-based estimation of the domain of attraction.

4.3.2 The Discrete-Time Case

Results similar to those of the continuous-time case are available in the discrete-time domain; in the remainder of this section, we briefly describe them based on the work of [31]. Following the same notation of the previous section, consider discrete-time affine TS models of the form

$$\boldsymbol{x}(t+1) = \sum_{l=1}^{r} h_l(z(\boldsymbol{x}))\,(A_l\boldsymbol{x}(t) + \boldsymbol{a}_l) = A_h\boldsymbol{x}(t) + \boldsymbol{a}_h, \qquad (4.29)$$

where, as in the previous section, MFs are assumed to hold the convex sum property in the whole state space, which means these models are, in most cases, either fuzzy or approximate.

As before, the state space is partitioned, though in a slightly different way. Define r partitions as follows:

$$\overline{X}_l = X_l \cup \partial X_l, \quad l = 1, 2, \ldots, r, \tag{4.30}$$

where

$$X_l = \{x : h_l(\boldsymbol{x}) > h_i(\boldsymbol{x}),\ i = 1, 2, \ldots, r,\ i \neq l\},$$
$$\partial X_l = \{\boldsymbol{x} : h_l(\boldsymbol{x}) = h_i(\boldsymbol{x}),\ i = 1, 2, \ldots, r,\ i \neq l\},$$

and let I, I_0, and I_1 stand for the same as before, i.e. the set of cell indices, the set of cell indices containing the origin, and the set of cell indices that do not contain the origin, respectively. Then, system (4.29) can be rewritten as

$$\boldsymbol{x}(t+1) = (A_l + \Delta A_l(h))\boldsymbol{x}(t) + \boldsymbol{a}_l + \Delta \boldsymbol{a}_l(h), \quad \boldsymbol{x}(t) \in \overline{S}_l,$$

where

$$\Delta A_l(h) = \sum_{i=1, i\neq l}^{r_l} \overline{h}_i \Delta A_{li}, \ \Delta \boldsymbol{a}_l(h) = \sum_{i=1, i\neq l}^{r_l} \overline{h}_i \Delta \boldsymbol{a}_{li},$$
$$\Delta A_{li} = A_i - A_l, \ \Delta \boldsymbol{a}_{li} = \boldsymbol{a}_i - \boldsymbol{a}_l, \ i = 1, \ldots, r_l,$$

and $\overline{h}_1, \ldots, \overline{h}_{r_l}$ are r_l membership functions which are not equal to zero when the l-th subsystem has the biggest influence in the system dynamics.

Using the notation in (4.18), the previous expression can be further simplified as

$$\overline{\boldsymbol{x}}(t+1) = \left(\overline{A}_l + \Delta\overline{A}_l(h)\right)\overline{\boldsymbol{x}}(t), \ \boldsymbol{x}(t) \in \overline{S}_l.$$

As in the continuous-time case, matrices $\overline{E}_l$ are defined as in (4.22) to relax conservativeness by including information from the regions where local models are valid. Moreover, new upper bound matrices E_{lA} and E_{la} will be defined as

$$\left[\Delta\overline{A}_l(h)\right]^T \left[\Delta\overline{A}_l(h)\right] \leq \left[E_{lA}\ E_{la}\right]^T \left[E_{lA}\ E_{la}\right].$$

Systematic ways to calculate these new matrices can be found in [32]. We are now ready to provide the following main result:

Theorem 4.7 *[31] If there exist $\epsilon_l > 0$, $l \in \{1, 2, \cdots, r\}$, a set of symmetric matrices P_l, $l \in I_0$, $\overline{P}_l$, $l \in I_1$, symmetric matrices $U_l \succeq 0$, $W_l \succeq 0$ and $Q_{lj} \succeq 0$ such that the following LMIs are satisfied*

$$P_l - E_l^T U_l E_l > 0, \ l \in I_0;$$

$$\begin{bmatrix} A_l^T P_l A_l - P_l + \epsilon_l^{-1} E_{lA}^T E_{lA} + E_l^T W_l E_l & A_l^T P_l \\ P_l A_l - \left(\epsilon_l^{-1} - P_l\right) \end{bmatrix} < 0, \ l \in I_0;$$

$$\overline{P}_l - \overline{E}_l^T U_l \overline{E}_l > 0, \ l \in I_1;$$

$$\begin{bmatrix} \overline{A}_l^T P_l \overline{A}_l - \overline{P}_l + \epsilon_l^{-1} E_{l\overline{A}}^T E_{l\overline{A}} + \overline{E}_l^T W_l \overline{E}_l & \overline{A}_l^T \overline{P}_l \\ \overline{P}_l \overline{A}_l & -\left(\epsilon_l^{-1} - \overline{P}_l\right) \end{bmatrix} < 0, \ l \in I_1;$$

$$\begin{bmatrix} A_l^T P_j A_l - P_l + \epsilon_l^{-1} E_{lA}^T E_{lA} + E_l^T Q_{lj} E_l & A_l^T P_j \\ P_j A_l & -\left(\epsilon_l^{-1} - P_j\right) \end{bmatrix} < 0, \ l, j \in (\Omega \cap I_0);$$

$$\begin{bmatrix} \overline{A}_l^T P_j \overline{A}_l - \overline{P}_l + \epsilon_l^{-1} E_{l\overline{A}}^T E_{l\overline{A}} + \overline{E}_l^T Q_{lj} \overline{E}_l & \overline{A}_l^T \overline{P}_j \\ \overline{P}_j \overline{A}_l & -\left(\epsilon_l^{-1} - \overline{P}_j\right) \end{bmatrix} < 0, l \in I_1, l, j \in \Omega;$$

$$\begin{bmatrix} A_l^T \hat{P}_j A_l - P_l + \epsilon_l^{-1} E_{lA}^T E_{lA} + E_l^T Q_{lj} E_l & A_l^T \hat{P}_j \\ \hat{P}_j A_l & -\left(\epsilon_l^{-1} - \hat{P}_j\right) \end{bmatrix} < 0, \quad \begin{array}{c} l \in I_0, \\ l, j \in \Omega \end{array}$$

with

$$\overline{P}_j = [I_{n\times 1}\ 0_{n\times 1}]^T P_j [I_{n\times n}\ 0_{n\times 1}], \ j \in I_0,$$
$$\hat{P} = [I_{n\times 1}\ 0_{n\times 1}] \overline{P}_j [I_{n\times n}\ 0_{n\times 1}]^T \ j \in I_1,$$

and the set Ω represents all possible transitions from one region to another, i.e.

$$\Omega = \{l, j : \boldsymbol{x}(t) \in X_l, \ \boldsymbol{x}(t+1) \in X_j, \ j \neq l\},$$

then the discrete-time affine TS model (4.29) is globally asymptotically stable, i.e. $\boldsymbol{x}(t)$ tends to zero exponentially for every continuous piecewise trajectory in the state space.

Although this result mimics that for the continuous-time case, an important difference should be noticed: the continuity of the PWLF is no longer required due to the discrete nature of the system, but for the same reason, extra LMI conditions have to be added to ensure that the PWLF monotonically decreases despite the system "jumping" from one cell to another. Recall that information from the MFs is still incorporated via an adaptation of the S-procedure.

Recall that the global characteristic of most of these results is only valid for TS models with membership functions h_i which hold the convex sum property *everywhere*; should a TS model be valid only within a compact set, the results will hold only for the outermost Lyapunov level within that set.

4.4 Parameter-Dependent Lyapunov Functions

We will now focus on parameter-dependent Lyapunov functions (PDLFs), i.e. the family of all Lyapunov functions that share the MFs $h_i(\cdot)$ of the TS model. Originally proposed in [33, 34] and later considered in the TS context, these functions have been called in a very unspecific way as non-quadratic [22, 35], multiple [23], fuzzy [36], basis-dependent [37], or poly-quadratic [38]. The next subsections summarises the developments as well as the issues for both the continuous- and the discrete-time case.

Parameter-dependent Lyapunov functions have two interpretations:

- As a nonlinear stability tool if $h_i(\cdot)$ are actually functions of the system state; in this way, a Lyapunov function $V(\boldsymbol{x}, h)$ is a shorthand for $V(\boldsymbol{x}) = V(\boldsymbol{x}, h(\boldsymbol{x}))$.
- As a robust stability tool (see Sect. 5.6.3) if the arguments of h_i are unknown, possibly time-varying quantities influencing the dynamics. PDLFs in this context make use of time-derivative bounds of these quantities.

In this section, the first of the above interpretations is the one to be implicitly assumed. However, the second option can be also considered: if a robust stability result holds for unknown parameters $\boldsymbol{\rho}$ up to certain rate ($|\dot{\boldsymbol{\rho}}| \leq \nu$, derivative bound), when the explicit state dependence $\rho(\boldsymbol{x})$ is considered, an immediate consequence is that the largest Lyapunov level set $|(\partial\rho/\partial \boldsymbol{x})\boldsymbol{f}(\boldsymbol{x})| \leq \nu$ is guaranteed to be invariant. Of course, there is an interaction between the derivative bound (the larger it is, the harder it is for conditions guaranteeing the Lyapunov function decrease to be met) and the level set size (which increases with ν): in nonlinear stability settings, ν may be considered as a design parameter balancing conservatism and domain of attraction size.

We will introduce the basic ideas of the approach here, although they will be recalled, based on the above two interpretations, in Sect. 5.6.3.

4.4.1 The Continuous-Time Case: Box LPV Model

Consider a continuous-time system

$$\dot{\boldsymbol{x}}(t) = A(\rho)\boldsymbol{x}(t), \tag{4.31}$$

$\rho(t) \in \mathbb{R}^s$ being a parameter vector, which is, by assumption, known to lie in a region Ω. Considering a Lyapunov function candidate $V(\boldsymbol{x}) = \boldsymbol{x}^T P(\rho)\boldsymbol{x}$, with $P(\rho) > 0$, we would have

$$\dot{V} = \boldsymbol{x}^T \left(P(\rho)A(\rho) + A^T(\rho)P(\rho) + \sum_{i=1}^{s} \dot{\rho}_i \frac{\partial P(\rho)}{\partial \rho_i} \right) \boldsymbol{x}. \tag{4.32}$$

Then, if bounds for the time derivatives of ρ_i were known, for instance $|\dot{\rho}_i| \leq \nu_i$, the system would be stable if the 2^s conditions below hold

$$P(\rho)A(\rho) + A^T(\rho)P(\rho) + \sum_{i=1}^{s} (\pm\nu_i) \frac{\partial P(\rho)}{\partial \rho_i} < 0 \quad \forall \rho \in \Omega. \tag{4.33}$$

However, these conditions are, in general, non-convex in ρ and usually they are solved via parameter-gridding in ρ, as proposed by [33, 39]. Obviously, if all $\nu_i = 0$, the case in consideration would be the so-called *unknown-but-constant parameter* case, which is the common interpretation when robust analysis of systems with parametric uncertainty is considered.

In order to avoid the gridding, in the particular case

$$A(\rho) = A_0 + \rho_1 A_1 + \cdots + \rho_s A_s \tag{4.34}$$

$$P(\rho) = P_0 + \rho_1 P_1 + \cdots + \rho_s P_s \tag{4.35}$$

with ρ ranging in a polytope with vertices $\mathcal{V}$, and $\dot{\rho}$ in a polytope with vertices $\mathcal{T}$, sufficient conditions for expression (4.33) can be written as

$$A^T(w)P(w) + P(w)A(w) + (P(\tau) - P_0) < 0, \tag{4.36}$$

$$P(w) > I, \tag{4.37}$$

$${A_i}^T P_i + P_i A_i \geq 0 \tag{4.38}$$

for all $i = 1, \ldots, s$ and all $w \in \mathcal{V}$, $\tau \in \mathcal{T}$ [34]. The above comes from the fact that

$$\frac{dP}{dt} = \sum_{i=1}^{s} \dot{\rho}_i P_i = P(\dot{\rho}) - P_0, \tag{4.39}$$

and the third condition (4.38) is required for convexity of (4.33) in ρ; see the cited reference for details. Getting sufficient LMI conditions to guarantee (4.36) requires any of the relaxation schemes in Appendix D.

4.4.2 The Continuous-Time Case for Takagi–Sugeno Models

For the continuous-time TS system

$$\dot{x}(t) = \sum_{i=1}^{r} h_i(z) A_i x(t) = A_h x(t), \tag{4.40}$$

the Lyapunov function candidate

$$V(x) = x^T \left(\sum_{i=1}^{r} h_i(z) P_i \right) x = x^T P_h x \tag{4.41}$$

with $P_i = P_i^T > 0$, $i \in \{1, 2, \ldots, r\}$, is usually considered. It clearly corresponds to (4.2) with $\boldsymbol{P}(z) = P_h$, which is a natural way of introducing the nonlinear membership functions of the model into the Lyapunov function.

Unfortunately, the results concerning this approach face the problem of dealing with the time derivative of the MFs. To see this, consider the derivative of (4.41)

$$\dot{V} = \boldsymbol{x}^T \left(A_h^T P_h + P_h A_h + \dot{P}_h \right) \boldsymbol{x}, \tag{4.42}$$

where the shorthand notation

$$\dot{P}_h = \sum_{i=1}^{r} \dot{h}_i(z) P_i$$

has been employed.

Let us now assume that premise variables z are actually a function of the state, instead of possibly exogenous inputs. Then, we may write

$$\dot{h}_i = \left(\frac{\partial h_i}{\partial z} \right)^T \dot{z} = \left(\frac{\partial h_i}{\partial z} \right)^T \left(\frac{\partial z}{\partial \boldsymbol{x}} \right)^T \dot{\boldsymbol{x}}. \tag{4.43}$$

Therefore, $\dot{P}_h$ depends on the time derivative of the state $\dot{\boldsymbol{x}}$ and the best option so far is to consider *local* stability/stabilisation as in [14, 40][1]. The next theorem gives an existence condition for local stability, based on the constant-but-unknown LPV conditions referred to earlier.

Indeed, the explicit nonlinearity structure and the fact that the origin is an equilibrium point allows asserting that $\dot{h}_i$ tends to zero as $\boldsymbol{x}$ tends to zero in (4.43). Continuity of both h, z and partial derivatives of z is needed.

Theorem 4.8 (Local stability) *The origin $\boldsymbol{x} = 0$ of the continuous-time TS model (4.40) is asymptotically stable in a domain $\mathcal{D}$ including the origin, if there exist matrices of appropriate dimension $P_i = P_i^T > 0$, $i \in \{1, 2, \ldots, r\}$ such that the following holds*

$$A_h^T P_h + P_h A_h < 0. \tag{4.44}$$

Proof The non-quadratic Lyapunov function candidate (4.41) satisfies $V(0) = 0$, $V(\boldsymbol{x}) \geq 0$; its time-derivative (4.42) satisfies $\dot{V}(0) = 0$. From (4.44), it is implied that there exists a sufficiently small $\lambda > 0$ such that $A_h^T P_h + P_h A_h + \lambda I < 0$, which can be used to define $\mathcal{D} = \{\boldsymbol{x} : \boldsymbol{x} \in \mathcal{C}, \|\dot{P}_h\| < \lambda\}$. The origin belongs to this domain D since from (4.43) we know that

[1] We are considering here the common case of z depending on the state $\boldsymbol{x}$, but the reader should keep in mind that more complex situations may arise since $\dot{z}$ is a priori unknown and may depend on states or exogenous signals.

$$\dot{P}_h = \sum_{i=1}^{r} \left(\left(\frac{\partial h_i}{\partial z} \right)^T \left(\frac{\partial z}{\partial x} \right)^T A_h x \right) P_i$$

depends on the state vector $x(t)$. Since $V(x) > 0$ and $\dot{V}(x) < 0$ in $\mathcal{D} - \{0\}$, the equilibrium point $x = 0$ is locally asymptotically stable since conditions of Theorem B.1 are satisfied, thus concluding the proof. □

Example 4.11 Consider a continuous-time TS system of the form (4.40) with two vertex models

$$A_1 = \begin{bmatrix} -0.23 & 0.30 \\ -2 & -0.75 \end{bmatrix}, \quad A_2 = \begin{bmatrix} -2 & 10.4 \\ 1.3 & -14.3 \end{bmatrix}.$$

Quadratic stability of this system cannot be proven, as the LMIs from Theorem 4.1 are infeasible. However, we can prove local asymptotic stability. Using the relaxation scheme in Lemma D.1 with the conditions of Theorem 4.8, i.e. solving the LMIs

$$\begin{aligned} &P_i > 0, \\ &2P_i A_i + P_i A_j + P_j A_i + (*) < 0, \end{aligned}$$

$i,\ j = 1,\ 2$, we obtain

$$P_1 = \begin{bmatrix} 0.17 & 0.04 \\ 0.04 & 0.06 \end{bmatrix}, \quad P_2 = \begin{bmatrix} 0.2 & 0.13 \\ 0.13 & 0.11 \end{bmatrix},$$

and therefore local asymptotic stability is proven. □

Whereas Theorem 4.8 shows the interest of reducing the problem from global to local conditions, the last step is to proceed with $\dot{P}_h$. In (4.43), $\partial h_i / \partial z$ can be easily derived and thus bounded, but $\dot{z}$ is a priori unknown and may depend on states or exogenous signals. There are two schools: the first one uses a priori bounds on $\dot{z}$ or $\dot{h}_i$ [23, 41], the second one uses known bounds on $\partial h_i / \partial z$ and the local sector of the state [40]. The next results are shown without any sum relaxation scheme; once any of them is applied (for instance, Lemmas D.1 or D.2), LMI conditions arise.

First Approach: A Priori Time-Derivative Bounds

In general, a priori bounds $|\dot{h}_i(z)| \le \phi_i$ are used. Philosophically, this implies considering the nonlinear quasi-LPV (TS) system as an LPV one, disregarding the relationship between parameters and state. Consider the following result which generalises that in [23]:

Theorem 4.9 *Assume that $|\dot{h}_i(z)| \le \phi_i$, $i \in \{1, 2, \ldots, r\}$. The origin $x = 0$ of the TS model (4.40) is asymptotically stable if there exists matrices $P_i = P_i^T > 0$, $i \in \{1,\ 2,\ \ldots,\ r\}$, such that the following inequality hold:*

$$A_h^T P_h + P_h A_h + \sum_{i=1}^{r} \phi_i P_i < 0. \tag{4.45}$$

Any trajectory $\boldsymbol{x}(t)$ starting within the outermost Lyapunov level $V(\boldsymbol{x}) = \boldsymbol{x}^T P_h \boldsymbol{x} \leq k$, $k > 0$, in a compact set of the state space $\mathcal{X}$ guaranteeing $z \in \mathcal{C} \subseteq \mathbb{R}^p$ and $|\dot{h}_i(z)| \leq \phi_i$ goes asymptotically to $\boldsymbol{x} = 0$.

Proof Given $P_i = P_i^T > 0$, $V(\boldsymbol{x}) = \boldsymbol{x}^T P_h \boldsymbol{x}$ is a valid Lyapunov function candidate; its time derivative includes the term:

$$\dot{P}_h = \sum_{i=1}^{r} \dot{h}_i(z) P_i \leq \sum_{i=1}^{r} |\dot{h}_i(z)| P_i \leq \sum_{i=1}^{r} \phi_i P_i.$$

And from (4.41)–(4.42), we see that $\dot{V} < 0, \forall \boldsymbol{x} \neq 0$ if

$$A_h^T P_h + P_h A_h + \dot{P}_h \leq A_h^T P_h + P_h A_h + \sum_{i=1}^{r} \phi_i P_i < 0,$$

which is guaranteed by inequality (4.45), thus proving $V(\boldsymbol{x})$ is a valid Lyapunov function establishing asymptotic stability of $\boldsymbol{x} = 0$. Trajectories within the outermost Lyapunov level in a compact set of the state space $\mathcal{X}$ holding assumption 4.2 (which implies $z \in \mathcal{C}$, which in turn ensures the MFs hold the convex sum property) are guaranteed to go to $\boldsymbol{x} = 0$ as long as this compact also guarantees $|\dot{h}_i(z)| \leq \phi_i$, $i \in \{1, 2, \ldots, r\}$. □

Several refinements exist. Introducing a slack variable X and noticing that

$$\sum_{i=1}^{r} \dot{h}_i(z) X = 0$$

gives conditions in [41] with $X = X^T$, $P_i = P_i^T > 0$:

$$P_h + X \geq 0, \tag{4.46}$$

$$A_h^T P_h + P_h A_h + \sum_{i=1}^{r} \phi_i (P_i + X) < 0. \tag{4.47}$$

Notice that these conditions mimic those in (4.36)–(4.38) [34]. Moreover, condition (4.47) is quadratic in h, due to the product of the summations in A_h and P_h. Hence, evaluating them at the vertices $h_i = 1$, $h_j = 0$ for $j \neq i$ is not enough to prove that (4.47) holds for all intermediate values, as it was in the quadratic case. In the same way, as condition (4.38) was needed in the box case, some sufficient conditions are needed here. From the fact that

$$A_h^T P_h + P_h A_h = \sum_{i=1}^{r} (h_i^2 (A_i^T P_i + P_i A_i) + \sum_{j>i}^{r} h_i h_j (A_i^T P_j + P_i A_j)),$$

the following sufficient conditions guaranteeing the above inequality are obtained:

$$A_i^T P_j + P_j A_i + A_j^T P_i + P_i A_j + \sum_{k=1}^{r} \phi_k (P_k + X) < 0, \quad 1 \le i \le r, \ i < j \le r, \tag{4.48}$$

$$A_i^T P_i + P_i A_i + \sum_{k=1}^{r} \phi_k (P_k + X) < 0, \quad 1 \le i \le r. \tag{4.49}$$

However, less conservative conditions can be conceived for such double summations: in fact, Appendix D is entirely devoted to such issues. Note also, as remarked in Chap. 3, that h_i have, usually, tensor-product structure so further conservatism can be eliminated if using the ideas outlined in Sect. D.4.

The main drawback of this approach is the necessity to verify a posteriori that the future trajectory does not escape from the boundaries. Clearly, by the chain rule, $\dot{h}_i(z) = (\partial h / \partial \boldsymbol{x}) \dot{\boldsymbol{x}}$, which means conditions $\dot{h}_i(z) \le \phi_i$, $i \in \{1, 2, \ldots, r\}$, translate into an intersection of regions in the state-space containing the origin; how large this region depends directly on the system dynamics for specific initial conditions.

Second Approach: Known Partial-Derivative Bounds

In this section, the tensor-product representation (see Sect. 2.3.2) will be exploited along with specific knowledge of the dependency of MFs h_i on the weights w_i^j, which allows setting up bounds on the partial derivatives of w_i^j. To see this, consider a 2nd order TS model based on weights

$$w_0^1(x_1), \ w_0^2(x_2), \ w_1^1(x_1) = 1 - w_0^1(x_1), \ w_1^2(x_2) = 1 - w_0^2(x_2),$$

where the premise vector z has been directly substituted by $\boldsymbol{x}$ to ease the explanation; membership functions can thus be defined:

$$h_1(\boldsymbol{x}) = w_0^1(x_1) w_0^2(x_2), \ h_2(\boldsymbol{x}) = w_0^1(x_1) w_1^2(x_2),$$
$$h_3(\boldsymbol{x}) = w_1^1(x_1) w_0^2(x_2), \ h_4(\boldsymbol{x}) = w_1^1(x_1) w_1^2(x_2).$$

From the expression

$$P_h = \sum_{i=1}^{4} h_i(\boldsymbol{x}) P_i,$$

it follows that, omitting arguments when convenient,

$$\begin{aligned}\dot{P}_h =& \sum_{i=1}^{4}\sum_{k=1}^{2}\frac{\partial h_i}{\partial x_k}\dot{x}_k P_i = \sum_{i=1}^{4}\sum_{k=1}^{2}\frac{\partial}{\partial x_k}\left(\prod_{j=1}^{2} w_{i_j}^j\right)\dot{x}_k P_i \\ =& \left(\frac{\partial w_0^1}{\partial x_1}w_0^2\dot{x}_1 + \frac{\partial w_0^2}{\partial x_2}w_0^1\dot{x}_2\right)P_1 + \left(\frac{\partial w_0^1}{\partial x_1}w_1^2\dot{x}_1 + \frac{\partial w_1^2}{\partial x_2}w_0^1\dot{x}_2\right)P_2 \\ &+ \left(\frac{\partial w_1^1}{\partial x_1}w_0^2\dot{x}_1 + \frac{\partial w_0^2}{\partial x_2}w_1^1\dot{x}_2\right)P_3 + \left(\frac{\partial w_1^1}{\partial x_1}w_1^2\dot{x}_1 + \frac{\partial w_1^2}{\partial x_2}w_1^1\dot{x}_2\right)P_4 \\ =& \frac{\partial w_0^1}{\partial x_1}w_0^2\dot{x}_1(P_1 - P_3) + \frac{\partial w_0^2}{\partial x_2}w_0^1\dot{x}_2(P_1 - P_2) \\ &+ \frac{\partial w_0^1}{\partial x_1}w_1^2\dot{x}_1(P_2 - P_4) + \frac{\partial w_0^2}{\partial x_2}w_1^1\dot{x}_2(P_3 - P_4),\end{aligned}$$

where the fact that $\partial w_0^k/\partial x_k = -\partial w_1^k/\partial x_k$ has been used.

Generalising this idea, it has been shown in [40] that:

$$\begin{aligned}\dot{P}_h &= \sum_{i=1}^{r}\sum_{k=1}^{p} h_i(z)\frac{\partial w_0^k}{\partial z_k}\left(P_{g_1(i,k)} - P_{g_2(i,k)}\right)\dot{z}_k \\ &= \sum_{k=1}^{p}\frac{\partial w_0^k}{\partial z_k}\left(P_{g_1(h,k)} - P_{g_2(h,k)}\right)\dot{z}_k,\end{aligned} \tag{4.50}$$

where

$$P_{g_l(h,k)} = \sum_{i=1}^{r} h_i(z)P_{g_l(i,k)},\ l \in \{1, 2\},$$

and $g_1(i,k) = \lfloor (i-1)/2^{p+1-k}\rfloor \times 2^{p+1-k} + 1 + (i-1)\text{mod}\, 2^{p-k}$ and $g_2(i,k) = g_1(i,k) + 2^{p-k}$, $\lfloor\cdot\rfloor$ standing for the floor function. An interesting fact in (4.50) is that for the quadratic case, i.e. $P_i = P, i \in \{1, \ldots, r\}$, this term vanishes.

Based on the previous fact, the following theorem can be stated considering the premise vector as a linear combination of the states, i.e. $z(t) = Lx(t)$ with $L \in \mathbb{R}^{p\times n}$; the element at the k-th row and u-th column of LA_h is denoted as $[LA_h]_{ku}$:

Theorem 4.10 *The origin $x = 0$ of the TS model (4.40) is asymptotically stable if there exists matrices $P_i = P_i^T > 0$, $i \in \{1, 2, \ldots, r\}$, such that the following inequality hold*

$$A_h^T P_h + P_h A_h + \sum_{k=1}^{p}\sum_{u=1}^{n}(-1)^{d_{ku}^m}\lambda_{ku}(LA_h)_{ku}\left(P_{g_1(z,k)} - P_{g_2(z,k)}\right) < 0, \tag{4.51}$$

for $m \in \{1, \ldots, 2^{p\times n}\}$ with d_{ku}^m defined from the binary equivalence of $m - 1 = d_{pn}^m + d_{p(n-1)}^m \times 2 + \cdots + d_{11}^m \times 2^{p\times(n-1)}$ and $g_1(i,k)$, $g_2(i,k)$ defined as in (4.50). Any trajectory $x(t)$ starting within the outermost Lyapunov level $V(x) = x^T P_h x \le$

k, $k > 0$, in a compact set $\mathcal{X}$ of the state-space guaranteeing $z \in \mathcal{C} \subseteq \mathbb{R}^p$ and $\|(\partial w_0^k/\partial z_k)x_u\| \leq \lambda_{ku}$, $k \in \{1, 2, \ldots, p\}$, $u \in \{1, 2, \ldots, n\}$, goes asymptotically to $x = 0$.

Proof It has been already established that $V(x) = x^T P_h x$ is a valid Lyapunov function candidate since $P_i = P_i^T > 0, i \in \{1,\ 2,\ \ldots,\ r\}$. From $z(t) = Lx(t)$, it follows that

$$\dot{z}_k = \sum_{u=1}^{n} [LA_h]_{ku} x_u,$$

which can be substituted in (4.50) to give:

$$\dot{P}_h = \sum_{u=1}^{n} \sum_{k=1}^{p} \frac{\partial w_0^k}{\partial z_k} x_u [LA_h]_{ku} \left(P_{g_1(z,k)} - P_{g_2(z,k)} \right).$$

Taking into account

$$\left| \frac{\partial w_0^k}{\partial z_k} x_u \right| \leq \lambda_{ku},\quad k \in \{1, 2, \ldots, p\},\quad u \in \{1, 2, \ldots, n\},$$

the sought non-quadratic stability conditions (4.51) arise as they guarantee $\dot{V}(x) < 0$, $\forall x \neq 0$; hence $V(x)$ is a valid Lyapunov function and the origin $x = 0$ is proven asymptotically stable.

Since assumption 4.2 holds and the partial derivatives above are bounded, it follows that trajectories within the outermost Lyapunov level within $\mathcal{X}$ go to $x = 0$ asymptotically. □

Remark 4.1 Sum relaxations (including the tensor-product considerations, if so wished) in Appendix D must be applied in order to get sufficient LMI conditions, details omitted for brevity. Up to the relaxation scheme, the quadratic case is included by this approach due to the fact that if a quadratic solution holds, then conditions (4.51) are satisfied with $P_i = P$. This is a direct consequence of the left-hand side of (4.51) being simplified to $A_h^T P + PA_h$, which exactly corresponds to the quadratic-case conditions.

Remark 4.2 Inequalities in (4.51) are LMIs since $\lambda_{ku} > 0$ are given. Note that if conditions (4.51) are satisfied, local stability of the origin is ensured, i.e. there exists a set of sufficiently small values $\lambda_{ku} > 0$ for which this assertion is verified. Therefore, if they do not satisfy (4.51), algorithms such as bisection (successively dividing λ_{ku} by a common value $\lambda > 0$) come at hand [42]. The main advantage of this second approach is that, when conditions are satisfied, the trajectories will remain in the estimated domain.

Example 4.12 We resume Example 4.11 to test the two aforementioned approaches for non-quadratic stability analysis: how bad could it be to bound the time derivatives

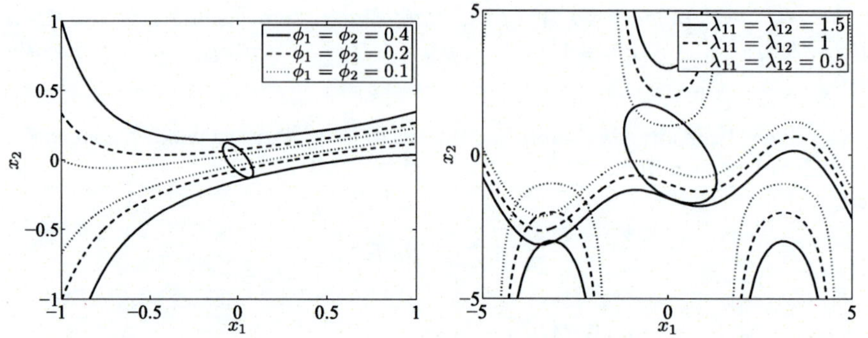

(a) Direct bounds on $\dot{h}_i$. (b) Bounds on $\partial h_i/\partial x$.

Fig. 4.5 Comparing non-quadratic approaches for Example 4.12

of the membership functions? Consider *globally valid* membership functions such as $h_1(x_1) = 0.5(1 - \sin x_1)$ and $h_2(x_1) = 0.5(1 + \sin x_1)$, i.e. they hold the convex sum property everywhere in the state space. Figure 4.5(a) shows different bounds of the form $\dot{h}_i = \phi_i$, i.e. direct bounds on the time derivatives of the membership functions; the borders of the largest Lyapunov levels for the maximum feasible choice of $\phi_1 = \phi_2$ within $\dot{h}_i \leq \phi_i$, have been drawn. Since the whole state space is the modelling area of the system, this approach has significantly contracted the estimation of the domain of attraction. On the other hand, when the second approach is used, i.e. bounds of the sort $\left|(\partial w_0^k/z_k)x_u\right| \leq \lambda_{ku}$, results in Fig. 4.5(b) are obtained: though also limited, they clearly outperform the direct bounds of the first approach. □

Third Approach: Line-Integral LFs

Another way to overcome the disadvantages of the quadratic approach is to consider the following line-integral Lyapunov function [43], whose roots lie in pioneering work around Popov criterion [44]:

$$V(\boldsymbol{x}) = 2\int_{\Gamma(0,\boldsymbol{x})} \boldsymbol{f}(\boldsymbol{\psi}) \cdot d\boldsymbol{\psi}, \tag{4.52}$$

where $\Gamma(0, \boldsymbol{x})$ is a path from the origin to the current state $\boldsymbol{x}$, $\boldsymbol{\psi} \in \mathbb{R}^n$ is a dummy integration vector, $\boldsymbol{f}(\boldsymbol{x}) : \mathbb{R}^n \to \mathbb{R}^n$ is a vector function of the state $\boldsymbol{x}$, $(\cdot)$ denotes the inner product, and $d\boldsymbol{\psi} \in \mathbb{R}^n$ is an infinitesimal displacement vector. In the TS context, a possibility for $\boldsymbol{f}(\boldsymbol{x})$ is

$$\boldsymbol{f}(\boldsymbol{x}) = \boldsymbol{P}(\boldsymbol{x})\boldsymbol{x},$$

along with the convex sum

$$P(x) = \sum_{i=1}^{r} \prod_{j=1}^{p} w_{i_j}^{j}(x_j) P_i = \sum_{i=1}^{r} h_i(z) P_i = P_h,$$

with $P_i = P_i^T > 0, i \in \{1, 2, \ldots, r\}$.

Nevertheless, for (4.52) to have the path-independent property, the following restrictions must hold [43]: $z = x$ as can be clearly deduced from the definition of $P(x)$, each weighting function $w_{i_j}^{j}$ must depend only on x_j (for example, $x_1 x_2$ or $x_1^2 \sin x_2$ cannot be used as premise variables for some weight), and the degrees of freedom for the set of matrices $P_i, i \in \{1, \ldots, r\}$ must be restricted as shown in the next result.

Theorem 4.11 *[43] The origin $x = 0$ of the TS model (4.40) is asymptotically stable if there exist matrices $\bar{P}$, $X \geq 0$, and D_i, $i \in \{1, \ldots, r\}$, satisfying*

$$\begin{aligned} &P_i = \bar{P} + D_i, \quad i \in \{1, 2, \ldots, r\}, \\ &P_i A_i + A_i^T P_i + (s-1)X < 0, \quad i \in \{1, 2, \ldots, r\}, \\ &P_i A_j + A_j^T P_i + P_j A_i + A_i^T P_j - 2X \leq 0, \quad i, j \in \{1, 2, \ldots, r\}, \quad i < j, \end{aligned}$$

where

$$\bar{P} = \begin{bmatrix} 0 & p_{12} & \cdots & p_{1n} \\ p_{12} & 0 & \cdots & p_{2n} \\ \vdots & \vdots & \ddots & \vdots \\ p_{1n} & p_{2n} & \cdots & 0 \end{bmatrix}, \quad D_i = \begin{bmatrix} d_{11}^{\alpha_{i1}} & 0 & \cdots & 0 \\ 0 & d_{22}^{\alpha_{i2}} & \cdots & 0 \\ \vdots & \vdots & \ddots & \vdots \\ 0 & 0 & \cdots & d_{nn}^{\alpha_{in}} \end{bmatrix}.$$

Some generalisations of the conditions above have appeared in the recent decade: one of them allowing convex terms in non-diagonal expressions [45]; another one showing that the path-independent arrangement can be dropped by suitably choosing a fixed expression of the line-integral Lyapunov function which, in addition, can be upgraded to handle the polynomial case with convex polynomial systems [46].

Line-integral-based stabilisation is not considered in this book as it leads to BMI conditions; partial remedies to this limitation can be found in [47, 48], where matrix transformations and Finsler's Lemma (A.3) have led to improvements or extensions to nonlinear systems in the descriptor form.

4.4.3 The Discrete-Time Case for Takagi–Sugeno Models

For the discrete-time case, very interesting results [35, 49] can be found for Lyapunov functions in the form of (4.41). Some of these will be presented in what follows.

Due to its nature, in the discrete-time case, the derivative of the membership functions—and thus one of the main problems when developing stability conditions for continuous-time TS models—does not occur. To see this, consider the discrete-time TS model with $z \in \mathcal{C}$, $\mathcal{C}$ being the compact set where MFs hold the convex sum property:

$$\boldsymbol{x}(t+1) = \sum_{i=1}^{r} h_i(z) A_i \boldsymbol{x}(t) = A_h \boldsymbol{x}(t) \tag{4.53}$$

along with a Lyapunov function candidate (4.2) with $\boldsymbol{P}(\cdot) = P_h$ as defined in (4.41).

Theorem 4.12 *The origin $\boldsymbol{x} = 0$ of the unforced discrete-time TS model (4.53) with $t \in \mathbb{Z}$, is asymptotically stable if there exists matrices $P_i = P_i^T > 0$ such that the following LMIs hold:*

$$\begin{bmatrix} P_i & (*) \\ P_j A_i & P_j \end{bmatrix} > 0, \; i \in \{1, 2, \ldots, r\}, \; j \in \{1, 2, \ldots, r\}. \tag{4.54}$$

Any trajectory $\boldsymbol{x}(t)$ starting within the outermost Lyapunov level $V(\boldsymbol{x}) = \boldsymbol{x}^T P_h \boldsymbol{x} \leq k$, $k \in \mathbb{R}$, in a compact set $\mathcal{X}$ of the state-space guaranteeing $z \in \mathcal{C} \subseteq \mathbb{R}^p$ goes asymptotically to $\boldsymbol{x} = 0$.

Proof Since $P_i = P_i^T > 0$, it is clear that $V(\boldsymbol{x}) = \boldsymbol{x}^T P_h \boldsymbol{x}$ is a valid Lyapunov function candidate. Then, for the forward difference of the Lyapunov function to be negative definite (recall notation introduced at the beginning of this chapter), it is sufficient that

$$\Delta V = V(t+1) - V(t) = \boldsymbol{x}^T(t) \left(A_h^T P_{h(t+1)} A_h - P_h \right) \boldsymbol{x}(t) < 0,$$

which, by Schur complement (Lemma A.1) followed by congruence (Property A.1) with block-diag$(I, P_{h(t+1)})$, is equivalent to:

$$\begin{bmatrix} P_h & (*) \\ P_{h(t+1)} A_h & P_{h(t+1)} \end{bmatrix} > 0. \tag{4.55}$$

LMIs (4.54) can be easily obtained from this expression; no relaxation scheme can be applied since *all* the index combinations between the instant t and $t+1$ must be taken into account. Thus, $V(\boldsymbol{x})$ is proven to be a Lyapunov function for this system and Theorem B.5 can be invoked guaranteeing $\boldsymbol{x} = 0$ is asymptotically stable. As with other stability theorems formerly presented, the fact that any trajectory within the outermost Lyapunov level in a compact set $\mathcal{X}$ of the state-space guaranteeing $z \in \mathcal{C}$ goes asymptotically to $\boldsymbol{x} = 0$, is a consequence of the MFs holding the convex sum property in $\mathcal{C}$. □

The reader should notice that, in this context, the forward difference ΔV has been expressed as $V(t+1) - V(t)$ instead of $V(\boldsymbol{x}(t+1)) - V(\boldsymbol{x}(t))$. This is to underline that, when it comes to fuzzy Lyapunov functions $V(\boldsymbol{x}) = \boldsymbol{x}^T P_h \boldsymbol{x}$, the MFs in P_h depending on $z(t)$ should change accordingly, i.e. $V(t+1) := \boldsymbol{x}^T(t+1) P_{h(t+1)} \boldsymbol{x}(t+1)$, where the forward operator is applied to *any* argument depending on time. Clearly, this was not necessary when quadratic Lyapunov functions $V(\boldsymbol{x}) = \boldsymbol{x}^T P \boldsymbol{x}$ were employed. The relevance of time as an argument suggests including it for expressions which ordinarily did not take it into account, namely, $A_{h(t)} := A_h$ or $P_{h(t)} := P_h$.

Finsler's Lemma A.3 introduces more degrees of freedom to the previous result, as explained hereinafter. With:

$$\begin{aligned} \Delta V &= V(t+1) - V(t) \\ &= \begin{bmatrix} \boldsymbol{x}^T(t) & \boldsymbol{x}^T(t+1) \end{bmatrix} \begin{bmatrix} -P_{h(t)} & 0 \\ 0 & P_{h(t+1)} \end{bmatrix} \begin{bmatrix} \boldsymbol{x}(t) \\ \boldsymbol{x}(t+1) \end{bmatrix} < 0 \end{aligned} \tag{4.56}$$

and the open-loop TS model (4.53) written as an equality constraint:

$$\begin{bmatrix} A_{h(t)} & -I \end{bmatrix} \begin{bmatrix} \boldsymbol{x}(t) \\ \boldsymbol{x}(t+1) \end{bmatrix} = 0,$$

it directly follows from Lemma A.3 that with $M = \begin{bmatrix} \mathcal{U}(\cdot) \\ \mathcal{V}(\cdot) \end{bmatrix}$:

$$\begin{aligned} &\begin{bmatrix} -P_{h(t)} & 0 \\ 0 & P_{h(t+1)} \end{bmatrix} + \begin{bmatrix} \mathcal{U}(\cdot) \\ \mathcal{V}(\cdot) \end{bmatrix} \begin{bmatrix} A_{h(t)} & -I \end{bmatrix} + (*) \\ &= \begin{bmatrix} \mathcal{U}(\cdot) A_{h(t)} + (*) - P_{h(t)} & (*) \\ \mathcal{V}(\cdot) A_{h(t)} - \mathcal{U}^T(\cdot) & P_{h(t+1)} - \mathcal{V}(\cdot) - \mathcal{V}^T(\cdot) \end{bmatrix} < 0. \end{aligned}$$

Appropriate slack variables $\mathcal{U}(\cdot)$ and $\mathcal{V}(\cdot)$ must be chosen for the previous inequality, in the sense that they should contain at least the double convex sum

$$\sum_{i=1}^{r} \sum_{j=1}^{r} h_i(z(t))\, h_j(z(t+1))$$

in order to allow crossed terms enabling a sum relaxation scheme to be applied. $\mathcal{U}_{h(t)h(t+1)}$ and $\mathcal{V}_{h(t)h(t+1)}$ are therefore an appropriate choice, representing a trade-off between the number of LMI constraints and the number of slack variables, resulting in:

$$\begin{bmatrix} \mathcal{U}_{h(t)h(t+1)} A_{h(t)} + (*) - P_{h(t)} & (*) \\ \mathcal{V}_{h(t)h(t+1)} A_{h(t)} - \mathcal{U}^T_{h(t)h(t+1)} & P_{h(t+1)} - \mathcal{V}_{h(t)h(t+1)} - \mathcal{V}^T_{h(t)h(t+1)} \end{bmatrix} < 0. \tag{4.57}$$

LMI conditions can be obtained from the previous result by applying any sum relaxation scheme such as those in [3, 50–52].

Example 4.13 Consider a discrete-time TS system (4.53) with $r = 3$ vertex models associated to matrices

$$A_1 = \begin{bmatrix} 1.15 & a \\ b & -0.37 \end{bmatrix}, \quad A_2 = \begin{bmatrix} -0.07 & -0.57 \\ 0.23 & -0.78 \end{bmatrix}, \quad A_3 = \begin{bmatrix} 0.12 & -0.4 \\ 0.3 & -0.3 \end{bmatrix}$$

with $a = 0.4$ and $b = -0.9$; it is assumed that the MFs h_i, $i \in \{1, 2, 3\}$ hold the convex sum property everywhere.

Using conditions (4.57) with the relaxation scheme in Lemma D.1 as well as $\mathcal{U}_{h(t)h(t+1)} = 0$ and $\mathcal{V}_{h(t)h(t+1)} = H_h$, the following feasible solution has been found:

$$P_1 = \begin{bmatrix} 0.38 & 0.1321 \\ 0.13 & 0.07 \end{bmatrix}, \quad P_2 = \begin{bmatrix} 0.058 & -0.007 \\ -0.007 & 0.30 \end{bmatrix}, \quad P_3 = \begin{bmatrix} 0.16 & 0.02 \\ 0.02 & 0.22 \end{bmatrix},$$
$$H_1 = \begin{bmatrix} 0.36 & 0.19 \\ 0.24 & 0.43 \end{bmatrix}, \quad H_2 = \begin{bmatrix} 0.44 & -0.03 \\ -0.002 & 0.27 \end{bmatrix}, \quad H_3 = \begin{bmatrix} 0.31 & 0.02 \\ 0.026 & 0.25 \end{bmatrix},$$

which means the origin $\boldsymbol{x} = 0$ is asymptotically stable. Of course, any other relaxation scheme can also be used. □

The Lyapunov function candidate (4.2) with $\boldsymbol{P}(\cdot) = P_{h(t-1)}$ is an interesting choice. This Lyapunov function will be discussed in more detail for control and estimation in the next chapter. In this case, expression (4.56) is replaced with:

$$\begin{aligned} \Delta V &= V(t+1) - V(t) \\ &= \begin{bmatrix} \boldsymbol{x}(t) \\ \boldsymbol{x}(t+1) \end{bmatrix}^T \begin{bmatrix} -P_{h(t-1)} & 0 \\ 0 & P_{h(t)} \end{bmatrix} \begin{bmatrix} \boldsymbol{x}(t) \\ \boldsymbol{x}(t+1) \end{bmatrix} < 0. \end{aligned}$$

Following the same outline of the previous proof, we have the previous inequality to be equivalent to the following:

$$\begin{bmatrix} \mathcal{U}(\cdot)A_{h(t)} + (*) - P_{h(t-1)} & (*) \\ \mathcal{V}(\cdot)A_{h(t)} - \mathcal{U}^T(\cdot) & P_{h(t)} - \mathcal{V}(\cdot) - \mathcal{V}^T(\cdot) \end{bmatrix} < 0.$$

Thus, using at least a double convex sum on $h(t-1)$ and $h(t)$ for the slack matrices, a convenient choice representing a trade-off between the number of LMI constraints and the number of slack variables is:

$$\begin{bmatrix} \mathcal{U}_{h(t-1)h(t)}A_{h(t)} + (*) - P_{h(t-1)} & (*) \\ \mathcal{V}_{h(t-1)h(t)}A_{h(t)} - \mathcal{U}^T_{h(t-1)h(t)} & P_{h(t)} - \mathcal{V}_{h(t-1)h(t)} - \mathcal{V}^T_{h(t-1)h(t)} \end{bmatrix} < 0. \quad (4.58)$$

Although results (4.57) and (4.58) look similar, their feasibility sets are different depending upon the system under consideration.

Example 4.14 To exhibit the differences between the aforementioned options for stability, reconsider Example 4.13. Parameters are chosen from a grid in $a \in [0, 2]$

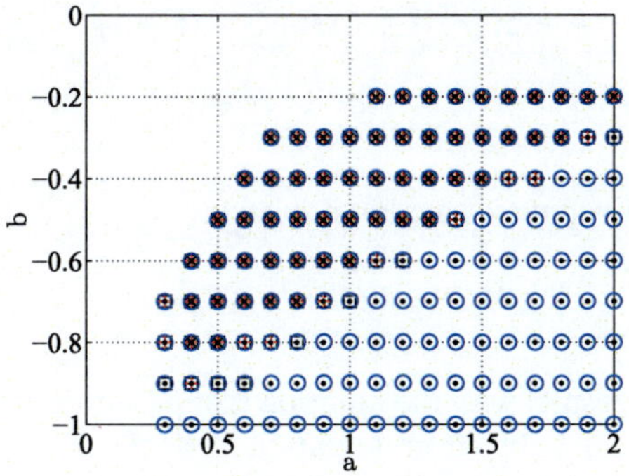

Fig. 4.6 Comparing discrete non-quadratic feasibility domains for choices in Table 4.1

Table 4.1 Comparison of non-quadratic discrete approaches

Case	$\mathcal{U}$	$\mathcal{V}$	Mark
(4.57)	$\mathcal{U}_{h(t)h(t+1)} = 0$	$\mathcal{V}_{h(t)h(t+1)} = H_{h(t)}$	□
(4.58)	$\mathcal{U}_{h(t-1)h(t)} = 0$	$\mathcal{V}_{h(t-1)h(t)} = H_{h(t)}$	×
(4.58)	$\mathcal{U}_{h(t-1)h(t)} = 0$	$\mathcal{V}_{h(t-1)h(t)} = H_{h(t-1)h(t)}$	+
(4.58)	$\mathcal{U}_{h(t-1)h(t)} = G_{h(t)}$	$\mathcal{V}_{h(t-1)h(t)} = H_{h(t-1)h(t)}$	•
(4.58)	$\mathcal{U}_{h(t-1)h(t)} = G_{h(t-1)h(t)}$	$\mathcal{V}_{h(t-1)h(t)} = H_{h(t-1)h(t)}$	○

and $b \in [-1, 0]$ with a step of 0.1; then non-quadratic stability tests (4.57) and (4.58) with Lemma D.1 are tested for each point in the grid with different choices of the slack variables for comparison. Feasibility domains are shown in Fig. 4.6 under the definitions given in Table 4.1.

Recall that, despite this particular example, there is no inclusion between stability approaches (4.57) and (4.58), but of course, there is inclusion among those approaches in (4.58) which use a progressively richer class of indexes. □

Remark 4.3 Note that $V(\boldsymbol{x}) = \boldsymbol{x}^T P_{h(t-1)} \boldsymbol{x}$ does not seem to be a proper Lyapunov function due to the dependence on the past state. However, the LMIs prove actually that for *every* possible value of $h(t-1)$ and every possible value of $h(t)$ the condition:

$$\boldsymbol{x}^T(t) P_{h(t-1)} \boldsymbol{x}(t) - \boldsymbol{x}^T(t+1) P_{h(t)} \boldsymbol{x}(t+1) > 0$$

holds. Hence, we can prove that

$$\begin{aligned}&\min_{h(t)} x^T(t)P_h x(t) - \min_{h(t+1)} x^T(t+1)P_{h(t+1)}x(t+1)\\&\quad\geq \min_{h(t)} x^T(t)P_h x(t) - \max_{h(t+1)} x^T(t+1)P_{h(t+1)}x(t+1) > 0,\end{aligned} \tag{4.59}$$

so the actual formal Lyapunov function proven is

$$V(x) = \min_h x^T(t)P_h x(t) \tag{4.60}$$

which reconciles the results with the intuition that past trajectories should not influence results at current time by sheer definition of state.

A different result, which can be actually applied to any Lyapunov function (4.2) in the discrete-time case, relies on considering the variation of the Lyapunov function during several samples [49, 53, 54]. Up to this point, we imposed that the Lyapunov function decreases in every sample. However, for a discrete-time system, whose origin $x = 0$ is asymptotically stable, the states can converge to zero, even though the Lyapunov function does not decrease every time instant, but only globally. This technique is called α-samples variation.

Lemma 4.1 *Let $V(x(t))$ be a Lyapunov function candidate. If for an $\alpha \in \mathbb{N}$, all $t \in \mathbb{N}$, and any initial conditions $x(0)$ in a domain $\mathcal{D} \subset \mathbb{R}^n$, $\Delta_\alpha V = V(t+\alpha) - V(t) < 0$ holds along the trajectories of* (4.53)*, then the origin of the TS model* (4.53) *is asymptotically stable.*

Proof Consider the function $\bar{V}(x) = \sum_{l=0}^{\alpha-1} V(t+l)$, which is obviously a Lyapunov function candidate since it is a sum of Lyapunov function candidates. Hence, the origin $x = 0$ of (4.53) is asymptotically stable if the forward difference of $\bar{V}(x)$ along the trajectories of (4.53) is negative definite, i.e.

$$\begin{aligned}\Delta\bar{V} &= \bar{V}(t+1) - \bar{V}(t) < 0,\\&= \sum_{l=0}^{\alpha-1} V(t+1+l) - \sum_{l=0}^{\alpha-1} V(t+l)\\&= V(t+\alpha) - V(t) := \Delta_\alpha V,\end{aligned}$$

consequently, if $\Delta_\alpha V < 0$, then $\Delta\bar{V} < 0$, which concludes the proof. □

Therefore, a generic Lyapunov function of the form (4.2) writes

$$\Delta_\alpha V = x^T(t+\alpha)P(\cdot)x(t+\alpha) - x^T(t)P(\cdot)x(t). \tag{4.61}$$

Consider for example: $P(\cdot) = P_{h(t)} > 0$ altogether with the open-loop TS model (4.53), which can be expanded as

$$x(t+\alpha) = \prod_{l=0}^{\alpha-1} A_{h(t+l)}x(t) = A_{h(t+\alpha-1)} \cdots A_{h(t)}x(t). \tag{4.62}$$

Substituting (4.62) in (4.61) yields

$$\Delta_\alpha V = x^T\big(A_{h(t)}^T \cdots A_{h(t+\alpha-1)}^T P_{h(t+\alpha)} A_{h(t+\alpha-1)} \cdots A_{h(t)} - P_{h(t)}\big)x < 0. \tag{4.63}$$

By Schur complement (Lemma A.1), the previous expression holds if and only if

$$\begin{bmatrix} P_{h(t)} & (*) \\ P_{h(t+\alpha)} A_{h(t+\alpha-1)} \cdots A_{h(t)} & P_{h(t+\alpha)} \end{bmatrix} > 0. \tag{4.64}$$

Note that with $\alpha = 1$, we recover the results of a Lyapunov variation on 1 sample (4.55).

As previously, Finsler's Lemma A.3 comes at hand to introduce slack variables. In this case, several options are available; for instance, the Lyapunov inequality can be written as

$$\Delta_\alpha V = \begin{bmatrix} x(t) \\ x(t+\alpha) \end{bmatrix}^T \begin{bmatrix} -P_{h(t)} & 0 \\ 0 & P_{h(t+\alpha)} \end{bmatrix} \begin{bmatrix} x(t) \\ x(t+\alpha) \end{bmatrix}, \tag{4.65}$$

or

$$\Delta_\alpha V = \begin{bmatrix} x(t) \\ \vdots \\ x(t+\alpha) \end{bmatrix}^T \begin{bmatrix} -P_{h(t)} & 0 & \cdots & 0 \\ 0 & 0 & & \vdots \\ \vdots & & \ddots & \vdots \\ 0 & \cdots & \cdots & P_{h(t+\alpha)} \end{bmatrix} \begin{bmatrix} x(t) \\ \vdots \\ x(t+\alpha) \end{bmatrix} < 0. \tag{4.66}$$

Similarly, (4.62) can be written as

$$\begin{bmatrix} A_{h(t+\alpha-1)} \cdots A_{h(t)} & -I \end{bmatrix} \begin{bmatrix} x(t) \\ x(t+\alpha) \end{bmatrix} = 0 \tag{4.67}$$

or

$$\begin{bmatrix} A_{h(t)} & -I & 0 & \cdots & 0 \\ 0 & \ddots & \ddots & & \vdots \\ \vdots & & \ddots & \ddots & 0 \\ 0 & \cdots & 0 & A_{h(t+\alpha-1)} & -I \end{bmatrix} \begin{bmatrix} x(t) \\ x(t+1) \\ \vdots \\ x(t+\alpha) \end{bmatrix} = 0. \tag{4.68}$$

Combining (4.65) and (4.67) and following the same procedure as for 1-sample variation, the following condition is obtained:

$$\begin{bmatrix} \mathcal{U}(\cdot)A_{h(t+\alpha-1)} \cdots A_{h(t)} + (*) - P_{h(t)} & (*) \\ \mathcal{V}(\cdot)A_{h(t+\alpha-1)} \cdots A_{h(t)} - \mathcal{U}^T(\cdot) & P_{h(t+\alpha)} - \mathcal{V}(\cdot) - \mathcal{V}^T(\cdot) \end{bmatrix} < 0. \tag{4.69}$$

A possible choice for $\mathcal{U}(\cdot)$ and $\mathcal{V}(\cdot)$ can be $\mathcal{U}_{h(t+\alpha),\ldots,h(t)}$ and $\mathcal{V}_{h(t+\alpha),\ldots,h(t)}$, respectively.

Similarly, combining (4.66) and (4.68) by Finsler's Lemma A.3 leads to

$$\begin{bmatrix} -P_{h(t)} & 0 & \cdots & 0 \\ 0 & 0 & & \vdots \\ \vdots & & \ddots & \vdots \\ 0 & \cdots & \cdots & P_{h(t+\alpha)} \end{bmatrix} + M \begin{bmatrix} A_{h(t)} & -I & 0 & \cdots & 0 \\ 0 & \ddots & \ddots & & \vdots \\ \vdots & & \ddots & \ddots & 0 \\ 0 & \cdots & 0 & A_{h(t+\alpha-1)} & -I \end{bmatrix} < 0,$$

with $M \in \mathbb{R}^{(n-\alpha)\times(n-\alpha)}$ being a free matrix whose choice can be, for instance, $M_{h(t+\alpha),\ldots,h(t)}$.

Remark 4.4 These approaches are soon faced with such a complexity that memory and numerical solver problems may arise. Therefore, the slack variables have to be chosen according to both the size of the model (number of vertices and order) and the possibilities of the solver.

Example 4.15 This example is taken from [49]; it shows the feasibility advantages of the α-sample approach for non-quadratic stability analysis of TS models. Consider a TS system with two vertex models whose matrices are given by

$$A_1 = \begin{bmatrix} b & -0.5 \\ a & -0.9 \end{bmatrix}, \quad A_2 = \begin{bmatrix} 0.9 & -a \\ -0.6 & 0.8 \end{bmatrix},$$

with $a \in [-0.3, 0.05]$ and $b \in [-0.4, 0.7]$. It is assumed that its MFs $h_1(t)$ and $h_2(t)$ hold the convex sum property everywhere.

Quadratic stability conditions in Theorem 4.4 are unfeasible for any value of a and b in the aforementioned intervals. On the other hand, conditions (4.64) are feasible for several values of a and b as shown in Fig. 4.7 when $\alpha = 2$ (mark "+"), $\alpha = 4$ (mark "$\times$"), and $\alpha = 6$ (mark "o"), . As α grows, so it does the number of LMI conditions and the resulting feasibility set. □

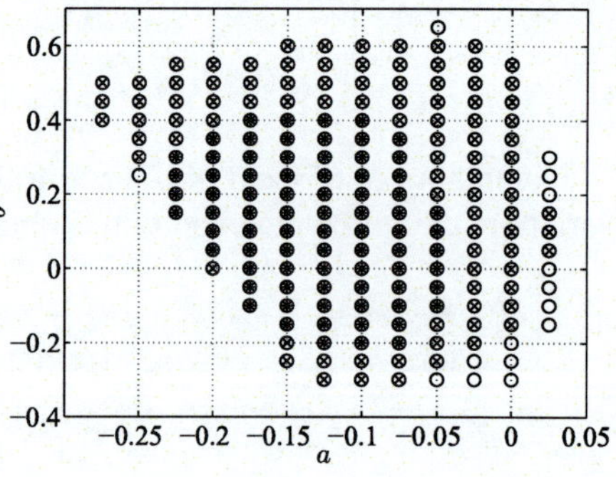

Fig. 4.7 Feasibility sets of the α-sample approach

More generalisations based on the same techniques are available in the literature [55], such as extending (4.2) to

$$V(\boldsymbol{x}) = \bar{\boldsymbol{x}}^T(t)\boldsymbol{P}(\cdot)\bar{\boldsymbol{x}}(t),$$

with $\bar{\boldsymbol{x}}(t) = \left[\boldsymbol{x}^T(t)\ \boldsymbol{x}^T(t+1)\ \cdots\ \boldsymbol{x}^T(t+\alpha)\right]^T$,

$$\boldsymbol{P}(\cdot) = \boldsymbol{P}_{h(t),\dots,h(t+\alpha-1)} = \begin{bmatrix} P^{(0,0)}_{h(t),\dots,h(t+\alpha-1)} & \cdots & P^{(0,\alpha-1)}_{h(t),\dots,h(t+\alpha-1)} \\ \vdots & \ddots & \vdots \\ P^{(\alpha-1,0)}_{h(t),\dots,h(t+\alpha-1)} & \cdots & P^{(\alpha-1,\alpha-1)}_{h(t),\dots,h(t+\alpha-1)} \end{bmatrix}.$$

These variations are beyond the scope of this book and readers are referred to [49, 56] for more details.

Lyapunov Functions Depending on Past or Future Values

Using Lyapunov functions that depend on states or interpolating memberships at several time instants seems, at first glance, counter-intuitive. If the state is all that is needed to predict the future, then how is it possible that we need future or past values of MFs $h_i(z)$ to improve the results? Lyapunov–Krasovskii constructs are needed for delay systems, but they should be useless in delay-free systems.

The answer needs refining the reasoning on several aspects

First, we should consider the conservatism of the Lyapunov function parametrisation: note that being $h(\cdot)$ a function of $\boldsymbol{x}$ then $P_{h(t)}$ encodes an ordinary Lyapunov function $\boldsymbol{x}^T P(\boldsymbol{x}(t))\boldsymbol{x}$. If no robustness/uncertainty comes into account, then $h(t+1)$ may, philosophically speaking, be exactly computed from $\boldsymbol{x}(t)$ by using the prediction model; hence $P_{h(t)h(t+1)}$ would mean using $V(\boldsymbol{x}) = \boldsymbol{x}^T(t)P(\boldsymbol{x}(t), A_{h(t)}\boldsymbol{x}(t))\boldsymbol{x}(t)$. Thus, dependence on future values of the memberships is just a convenient way of getting a richer parametrisation of a usual Lyapunov function. The issue of past values of $h(z)$ might render amenable to a similar reasoning because, basically, all expressions could be thought to be a function of the most-delayed state: for instance, $\boldsymbol{x}^T(t)P_{h(t-1)}\boldsymbol{x}(t)$ can be interpreted as $V(\boldsymbol{x}(t-1)) = (A_{h(t-1)}\boldsymbol{x}(t-1))^T P_{h(t-1)}(A_{h(t-1)}\boldsymbol{x}(t-1))$. So, the expressions in this section are a convenient parametrisation of actual functions of the state at a single point in time. However, the discussion below actually proves that the idea just considered is not actually used at all to justify the developments.

Second, the actual LMIs in this section are shape-independent which means that explicit shape information on $h(\boldsymbol{x})$ is not used and, instead, decreasing of the Lyapunov function is proven for any possible future value of $h(\boldsymbol{x}(t))$, $h(\boldsymbol{x}(t+1))$, etc. This is the same for constructs involving *past* values of the Lyapunov function.

So, generalising the idea in (4.60) following [57, Theorem 3], shape-independence entails that the actually proven Lyapunov function, given some past horizon $m_1 > 0$ and future one $m_2 > 0$, is

$$V(x) = \min_{h(t-1),h(t-2),\ldots,h(t-m_1)} \; \max_{h(0),h(t+1),\ldots,h(t+m_2)} x^T P_{h(t-m_1),\ldots,h(0),\ldots,h(t+m_2)} x, \quad (4.70)$$

where the best case (min) can be chosen for past memberships (as the only data that matters for future system evolution is the present state, the record of past memberships can be freely "invented", at will, to suit our needs) but the worst case (max) is forcefully needed for future states to ensure that our Lyapunov function will decrease whatever the future membership trajectory is.

Reducing Computational Complexity

When the future or past horizons increase, the number of LMIs and decision variables increases exponentially with them; thus in systems with a large number of vertex models, conditions become quickly intractable.

As a first observation, Finsler's Lemma multiplier M is not needed in stability analysis, as the product of vertex matrices in (4.69) can be easily computed in stability analysis (of course, the interest of Finsler Lemma is in control design where the product of *closed-loop* matrices would imply the product of decision variables unsolvable by LMIs, so this is the motivation to add it to the presented results).

A second, more insightful idea lies in the fact that Lemma 4.1 can also be extended to any $V(x)$ that decreases at *irregularly* spaced time intervals (as long as they are upper bounded by, say $\bar{m}$). The formal statement appears in [58], but the idea is simple: if the maximum singular value of A_h is σ, the system state cannot increase at a rate larger than σ, i.e. $\|x(t+\alpha)\| \le \sigma^\alpha \|x(t)\|$. Thus, if we find an irregularly spaced sequence of instants with maximum dwell time $\bar{m}$, so that some $V(x)$ exponentially decreases at these sampling instants, then at intermediate times $V(x)$ will be no larger than $\sigma^{\bar{m}}$ times the value of $V(x)$ at sampling instants: if $V(x)$ tends to zero at sampling instants so it does at inter-sampling ones.

Combining this idea with the generation of a *scenario* (set of vertex sequences of maximum length $\bar{m}$) such that it is complete (understood as fulfilling that, for any arbitrary vertex sequence, there is a leading subsequence in the scenario) then a reduced set of LMIs similar to (4.69) can be set up to prove stability. The reader is referred to [58] for details where, for instance, 88 LMIs suffice to prove the fastest-possible decay rate of a system with a quadratic LF in an example where the full exponentially-increasing scenario would need more than 10^9 LMIs.

Homogeneous Polynomially Parameter-Dependent (HPPD) Lyapunov Functions

The simultaneous development of non-quadratic schemes based on parameter-dependent Lyapunov functions and the sum relaxations discussed in Appendix D, led to the appearance of homogeneous polynomially parameter-dependent (HPPD) Lyapunov functions, especially in the discrete-time context [59]. They usually involve homogeneous polynomials of degree p with respect to a set of matrices $X_{\mathbf{p}}$:

$$X_{\mathbf{p},\mathbf{h}} = \sum_{\mathbf{p}\in\mathcal{K}(p)} \mathbf{h}^{\mathbf{p}} X_{\mathbf{p}},$$

where $X_{\mathbf{p}}$ is a matrix indexed by $\mathbf{p} = p_1 p_2 \cdots p_r$, $\mathbf{h} = h_1^{p_1} h_2^{p_2} \cdots h_r^{p_r}$, with each h_i being an MF associated to some TS representation, and $\mathcal{K}(p)$ is a set of r-tuples containing all possible combinations of nonnegative integers p_i, $i \in \{1, 2, \ldots, r\}$ such that $p_1 + p_2 + \cdots + p_r = p$.

Indeed, multiple convex sums can be seen as homogeneous polynomials of membership functions, a fact which is proved to be directly related to Pòlya's Theorem [38, 60]. Using such nested convex sums for Lyapunov function candidates led to the HPPD approach, based on which a flurry of results have emerged within the discrete-time context (with extensions to time-delay systems) [61–64] and back to continuous-time systems [65, 66].

4.5 Stability of Polynomial Models

Section 3.3 showed how polynomials of arbitrary degree can be used instead of linear functions to bound nonlinearities, so that a polynomial convex representation of the form

$$\begin{aligned} \delta x(t) &= \sum_{i=1}^{r} h_i(\xi)\widehat{f}_i, \\ y(t) &= \sum_{i=1}^{r} h_i(\xi)\widehat{h}_i \end{aligned} \tag{4.71}$$

is obtained, where functions $h_i(\xi)$ hold the convex sum property and capture the system nonlinearities as they did for TS models, while $\widehat{f}_i$ and $\widehat{h}_i$ are vectors of polynomials. If this representation of a nonlinear system arises from the generalisation of the sector nonlinearity methodology described in [67], then, as previously done via TS models, the stability properties of the origin of a nonlinear system can be analyzed via sufficient conditions. In the TS context, these conditions are in the form of linear matrix inequalities; in the current framework, they will be expressed as a sum of squares (SOS) decomposition [68] that can be analyzed via semidefinite programming [69], i.e. the sort of numerical problems the LMIs belong to [1].

Independently of the convex representation of a nonlinear system $\dot{\boldsymbol{x}}(t) = \boldsymbol{f}(\boldsymbol{x})$, its stability can be analyzed by the direct Lyapunov method by replacing the Lyapunov candidate positivity condition by the slightly more conservative condition of being able to prove that it can be expressed as a sum of squares. The same goes for the negative of the time derivative of the Lyapunov function candidate which should meet the requirement of being sum of squares [70], i.e.

$$\begin{aligned} V(\boldsymbol{x}) - \phi(\boldsymbol{x}) &\text{ is SOS,} \\ -\frac{\partial V}{\partial \boldsymbol{x}} f(\boldsymbol{x}) &\text{ is SOS,} \end{aligned} \tag{4.72}$$

where $\phi(\boldsymbol{x})$ is some positive definite polynomial, would be sufficient conditions for guaranteeing that the zero equilibrium point is stable. Since the set of functions $V(\boldsymbol{x})$ satisfying the conditions above is convex, the task of finding such function (or establishing that the problem is infeasible) is a semidefinite programming problem; therefore, it can be computationally solved in an efficient manner, for instance, by free third-party software such as SOS Tools for MATLAB®[71].

What is then the motivation of expressing a nonlinear system as a convex sum of polynomial models? In doing so, via exact convex polynomial representations such as (4.71), we embed the analysis of a nonlinear system into that of a polytope of polynomials, a condition which might be easier to satisfy and, certainly, more systematic. At the same time, the TS/LMI-based analysis will be included as a particular case, which speaks of the generality of the polynomial/SOS-based methods. To summarise, the direct Lyapunov method can be employed for analysis of a nonlinear system via TS models (which leads to LMI conditions) or via convex polynomial models (which leads to SOS conditions); both analyses are semidefinite programming problems, though the polynomial framework covers a wider range of Lyapunov functions than the quadratic/TS one [72].

The notation in [72] preserves the matrix structure of ordinary TS systems, this is to say, a polynomial model is defined as

$$\dot{\boldsymbol{x}}(t) = \sum_{i=1}^{r} h_i(z) A_i(\boldsymbol{x}) \hat{\boldsymbol{x}}(\boldsymbol{x}), \tag{4.73}$$

with $A_i(\boldsymbol{x})$ is a polynomial matrix in $\boldsymbol{x}(t)$, $\hat{\boldsymbol{x}}$ is a column vector of monomials in $\boldsymbol{x}$, and $h_i(z)$, $i \in \{1, 2, \ldots, r\}$, are membership functions holding the convex sum property. Note that this description fits with the fuzzy point of view rather than the nonlinear one, i.e. the nature of the MFs is not of interest as long as they belong to a simplex $\Gamma = \{h_i : \sum_{i=1}^{r} h_i = 1, 0 \le h_i \le 1\}$.

Theorem 4.13 *[72] The origin of system* (4.73) *is stable if there exists a symmetric polynomial matrix* $P(\boldsymbol{x})$ *such that*

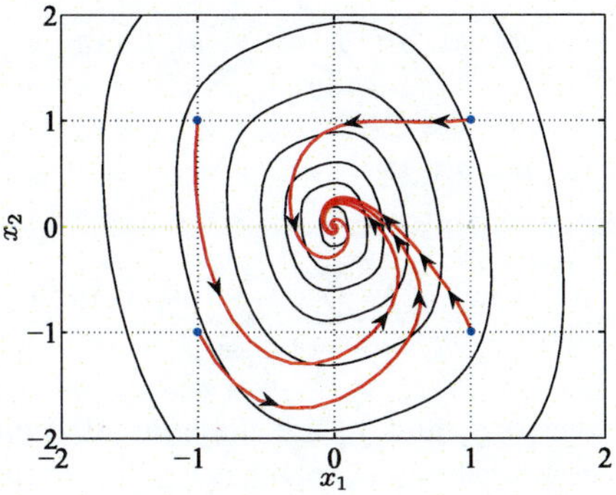

Fig. 4.8 Phase portrait and Lyapunov levels via polynomial models

$$\hat{\boldsymbol{x}}^T\left(P(\boldsymbol{x})-\epsilon_1(\boldsymbol{x})I\right)\hat{\boldsymbol{x}} \text{ is SOS},$$

$$-\hat{\boldsymbol{x}}^T\left(P(\boldsymbol{x})T(\boldsymbol{x})A_i(\boldsymbol{x})+(*)+\sum_{k=1}^{n}\frac{\partial P}{\partial x_k}A_i^k(\boldsymbol{x})\hat{\boldsymbol{x}}+\epsilon_{2i}(\boldsymbol{x})I\right)\hat{\boldsymbol{x}} \text{ is SOS},$$

where $\epsilon_1(\boldsymbol{x})>0$, $\boldsymbol{x}\neq 0$, *and* $\epsilon_{2i}(\boldsymbol{x})\geq 0$ *are nonnegative polynomials,* $\forall i$, $T(\boldsymbol{x})$ *is a matrix whose entries are defined as* $T_{ij}(\boldsymbol{x})=\partial\hat{x}_i/\partial x_j(\boldsymbol{x})$.

Example 4.16 Consider the following nonlinear model in [72]

$$\dot{\boldsymbol{x}}(t)=\begin{bmatrix}-\frac{7}{2}x_1-\frac{3}{2}x_1\sin x_1-4x_2\\ \frac{19}{2}x_1-\frac{21}{2}x_1\sin x_1-2x_2\end{bmatrix}, \tag{4.74}$$

which can be exactly represented by a TS model (which is particular case of a convex polynomial one) with $h_1(x_1)=(1+\sin x_1)/2$, $h_2(x_1)=1-h_1(x_1)$ and system matrices

$$A_1=\begin{bmatrix}-5 & -4\\ -1 & -2\end{bmatrix},\ A_2=\begin{bmatrix}-2 & -4\\ 20 & -2\end{bmatrix}.$$

The stability test in Theorem 4.13 is unfeasible if $V(\boldsymbol{x})$ is assumed to be a 2-nd degree polynomial. Nevertheless, with a 6-th degree polynomial the solution $V(\boldsymbol{x})=46.113x_1^6+22.692x_1^5x_2+33.55x_1^4x_2^2+11.912x_1^3x_2^3+14.018x_1^2x_2^4-1.572x_1x_2^5+1.9735x_2^6$ is obtained via SOSTools of MATLAB®.

Figure 4.8 shows some system trajectories as well as some Lyapunov levels from the obtained 6-th degree polynomial Lyapunov function. □

Theorem 4.13 is associated to $V(\boldsymbol{x})=\hat{\boldsymbol{x}}^T P(\boldsymbol{x})\hat{\boldsymbol{x}}$ as a Lyapunov function; a more general class can be found in [73], where *polynomial fuzzy Lyapunov functions* (PFLF) have been used. To this end, this work considers a tensor-product notation, both for the polynomial model:

$$\dot{x}(t)=\sum_{i_1=0}^{1}\sum_{i_2=0}^{1}\cdots\sum_{i_p=0}^{1}w_{i_1}^1(z_1)w_{i_2}^2(z_2)\cdots w_{i_p}^p(z_p)\boldsymbol{q}_{i_1i_2\ldots i_p}(\boldsymbol{x})=\sum_{\mathbf{i}\in\mathbb{B}^p}\mathbf{w}_{\mathbf{i}}(z)\boldsymbol{q}_{\mathbf{i}}(\boldsymbol{x}) \tag{4.75}$$

and the PFLF:

$$V(\boldsymbol{x})=\sum_{i_1=0}^{1}\sum_{i_2=0}^{2}\cdots\sum_{i_p=0}^{1}w_{i_1}^1(z_1)w_{i_2}^2(z_2)\cdots w_{i_p}^p(z_p)\boldsymbol{p}_{i_1i_2\ldots i_p}(\boldsymbol{x})=\sum_{\mathbf{i}\in\mathbb{B}^p}\mathbf{w}_{\mathbf{i}}(z)\boldsymbol{p}_{\mathbf{i}}(\boldsymbol{x}), \tag{4.76}$$

where $\boldsymbol{q}_{\mathbf{i}}$ and $\boldsymbol{p}_{\mathbf{i}}$ are polynomial vectors, $\mathbf{i}=(i_1,i_2,\ldots,i_p)$, $\mathbb{B}^p$ is the set of all combinations of p binary digits i_j, and $\mathbf{w}_{\mathbf{i}}(z)=w_{i_1}^1(z_1)w_{i_2}^2(z_2)\cdots w_{i_p}^p(z_p)$ is the product of p particular weights (equivalent to MFs $h_i(\boldsymbol{\xi})$ in (4.73)). Clearly, this is a polynomial generalisation of the parameter-dependent (fuzzy) Lyapunov function of Sect. 4.4.

Results in [73], in contrast with those of [72], intend to fit better in the spirit of exact Taylor-based polynomial representations of nonlinear systems [67]. As previously done in [40], the time derivatives of the MFs are regrouped as

$$\dot{V}=\sum_{\mathbf{i}\in\mathbb{B}^p}\mathbf{w}_{\mathbf{i}}\left(\dot{\boldsymbol{p}}_{\mathbf{i}}+\sum_{k=1}^{p}\frac{\partial w_0^k}{\partial z_k}\dot{z}_k\left(\boldsymbol{p}_{\mathbf{i}}-\boldsymbol{p}_{\bar{\mathbf{i}}}\right)\right),$$

with $\bar{\mathbf{i}}$ defined as the p-bit binary index from changing the k-th bit of $\mathbf{i}$ to its complement.

With suitable bounds on the partial derivatives of $w_j^i(z_j)$ (for instance, constant or polynomials in the state), stability conditions generalising those in Sect. 4.4 to the polynomial case can be found. The main idea is to express the time derivatives of the weighting functions as exact polynomial fuzzy models, themselves; assuming the polynomials of this representation are grouped in a matrix

$$\mathbf{R}_{\mathbf{v}}=\begin{bmatrix}\left(\mathbf{r}_{\mathbf{v}_1}^1\right)^T\\ \vdots\\ \left(\mathbf{r}_{\mathbf{v}_p}^p\right)^T\end{bmatrix}\in\mathbb{R}^{p\times n},$$

and the resulting polynomial differences in $\dot{V}(\boldsymbol{x})$ are grouped as

$$\hat{\mathbf{p}}_{\mathbf{i}}=\begin{bmatrix}p_{\mathbf{i}}-p_{\bar{\mathbf{i}}(1)}\\ \vdots\\ p_{\mathbf{i}}-p_{\bar{\mathbf{i}}(p)}\end{bmatrix}\in\mathbb{R}^{p\times 1},$$

the following result follows:

Theorem 4.14 *The origin of the polynomial fuzzy model* (4.75) *with polynomials in* $\mathbf{R}_{\mathbf{v}}$ *from an exact fuzzy model of its MF derivatives, is asymptotically stable if there* *exists polynomials* $p_{\mathbf{i}}(\boldsymbol{x})$ *and nonnegative, radially unbounded polynomials* $\epsilon_1(\boldsymbol{x})$ *and* $\epsilon_2(\boldsymbol{x})$*, such that*

$$p_{\mathbf{l}}(\boldsymbol{x}) - \epsilon_1(\boldsymbol{x}) \text{ is SOS},$$

$$-\left(\frac{\partial p_{\mathbf{i}}}{\partial \boldsymbol{x}}\right)^T \mathbf{q}_{\mathbf{l}} - \hat{\mathbf{p}}_{\mathbf{i}}^T \mathbf{R}_{\mathbf{v}} \mathbf{q}_{\mathbf{l}} - \epsilon_2(\boldsymbol{x}) \text{ is SOS}$$

for all $\mathbf{i}, \mathbf{l} \in \mathbb{B}^p$, $\mathbf{v} \in \mathbb{B}^\sigma$.

4.5.1 Locality Issues in the Polynomial Approach

In Sect. 4.3, it has been shown that the use of the S-procedure (A.2) in the LMI framework helps piecewise Lyapunov functions to include constraints related to the area they operate on. Similarly, as the works of [70, 74] first discussed, the *Positivstellensatz* argumentation extends the use of Lagrange multipliers and S-procedure to the polynomial-SOS case; constraints are thus expressed as SOS conditions.

Consider m polynomial constraints $f_i(\boldsymbol{x}) \geq 0$, $i \in \{1, 2, \ldots, m\}$, that hold in the modelling region Ω and are grouped in a vector $\boldsymbol{\psi}(\boldsymbol{x}) = \{f_1(\boldsymbol{x}), f_2(\boldsymbol{x}), \ldots, f_m(\boldsymbol{x})\} \geq 0$. Then, a sufficient condition for a polynomial $\pi(\boldsymbol{x})$ for being positive in Ω is that there exist multiplier SOS polynomials $q_i(\boldsymbol{x})$, $i = 1, 2, \ldots, n$, such that $\pi(\boldsymbol{x}) - \sum_{i=1}^n q_i(\boldsymbol{x})\phi_i(\boldsymbol{x})$ is SOS, where $\phi_i(\boldsymbol{x})$ are arbitrary fixed polynomials that are composed of products of those in $\boldsymbol{\psi}$. Usually, polynomials $\phi_i(\boldsymbol{x})$ are all the product combinations of restrictions in $\boldsymbol{\psi}$ preserving the same sign up to a certain user-defined degree d_1. Coefficients of q_i are free decision variables.

Results in [67, 73] incorporate these local considerations. The results in [72] can be easily adapted to incorporate local conditions as those discussed.

Example 4.17 We resume the stability analysis of nonlinear model (4.74), which has been proved to be globally stable in Example 4.16. We now focus our attention on its stability properties in $\Omega = \{|x_i| \leq 1\}$. To do so, the nonlinearity $\sin x_1$ is written as a convex sum of polynomials [67], leading to the following polynomial fuzzy model structure:

$$\dot{\boldsymbol{x}}(t) = w_0^1(\boldsymbol{x}) \begin{bmatrix} -\frac{7}{2}x_1 - 4x_2 - \frac{3}{2}x_1 q_0^1(\boldsymbol{x}) \\ \frac{19}{2}x_1 - 2x_2 - \frac{21}{2}x_1 q_0^1(\boldsymbol{x}) \end{bmatrix} + w_1^1(\boldsymbol{x}) \begin{bmatrix} -\frac{7}{2}x_1 - 4x_2 - \frac{3}{2}x_1 q_1^1(\boldsymbol{x}) \\ \frac{19}{2}x_1 - 2x_2 - \frac{21}{2}x_1 q_1^1(\boldsymbol{x}) \end{bmatrix},$$

where $q_0^1(\boldsymbol{x})$, $q_1^1(\boldsymbol{x})$ are polynomials of certain degree, and $w_0^1(\boldsymbol{x})$, $w_1^1(\boldsymbol{x})$ are the corresponding MFs.

A possible option for modelling is $q_0^1(\boldsymbol{x}) = 0.8414$ and $q_0^1(\boldsymbol{x}) = -0.8414$, with $w_0^1(\boldsymbol{x}) = 0.5942 \sin x_1 + 0.5$ and $w_1^1(\boldsymbol{x}) = 1 - w_0^1(\boldsymbol{x})$ as the corresponding MFs. In order to use Theorem 4.14, the expression $\partial w_0^1/\partial x_1 = 0.5942 \cos x_1$ should also be modelled in a polynomial manner; to do so, consider a 0-degree mod-

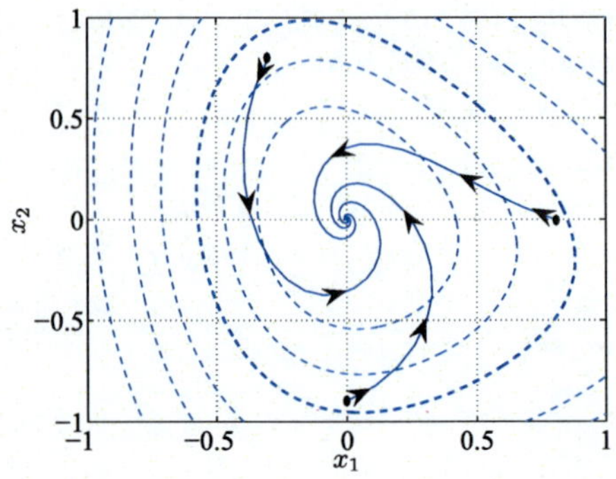

Fig. 4.9 Phase portrait and Lyapunov levels via polynomial models

elling as with bounds $r_0^1(\boldsymbol{x}) = 0.5942$ and $r_1^1(\boldsymbol{x}) = 0.3211$, and MFs $w_0^1(\boldsymbol{x}) = 2.1755 \cos x_1 - 1.1755$ and $w_1^1(\boldsymbol{x}) = 1 - w_0^1(\boldsymbol{x})$.

Using a 2-degree PFLF candidate of the form $V(\boldsymbol{x}) = w_0^1(\boldsymbol{x})p_1(\boldsymbol{x}) + w_1^1(\boldsymbol{x})p_2(\boldsymbol{x})$, the problem is infeasible. Nevertheless, once Lagrange multipliers are considered for local analysis, the problem has a solution. Considering a list of 1- and 2-degree polynomial restrictions such as $\psi = \{x_1^2 - 1 < 0,\ x_2^2 - 1 < 0,\ 1 - x_1 < 0, 1 - x_2 < 0\ -(x_1 + 1) < 0,\ -(x_2 + 1) < 0\}$, from which double products of constraints in the list have been used to construct a 2-nd order polynomial Lagrange multiplier multiplied by the constraints (valid in Ω):

$$
\begin{gathered}
x_1^2 - 1 < 0,\ x_2^2 - 1 < 0,\ -\left(x_1^2 - 1\right)\left(x_2^2 - 1\right) < 0,\ -\left(x_1^2 - 1\right)(x_1 - 1) < 0, \\
\left(x_1^2 - 1\right)(x_1 + 1) < 0,\ -\left(x_2^2 - 1\right)(x_2 - 1) < 0,\ \left(x_2^2 - 1\right)(x_2 + 1) < 0,
\end{gathered}
$$

the solutions $p_1(\boldsymbol{x}) = 3.9106x_1^2 + 1.863x_1x_2 + 4.1858x_2^2$ and $p_2(\boldsymbol{x}) = 10.692x_1^2 + 1.2375x_1x_2 + 2.569x_2^2$ were found. In Fig. 4.9 some level curves of the resulting PFLF are shown in dashed lines while system trajectories are shown in solid lines.

4.6 Stability of Descriptor Models

Let us consider an autonomous nonlinear descriptor model of the form (4.1), exactly represented by a TS descriptor model:

$$
\sum_{j=1}^{r_e} v_j(z)E_j\delta\boldsymbol{x}(t) = \sum_{i=1}^{r} h_i(z)A_i\boldsymbol{x}(t), \tag{4.77}
$$

where, as before, $\delta\boldsymbol{x}(t)$ with $t \in \mathbb{R}$ stands for $\dot{\boldsymbol{x}}(t)$ in the continuous-time case, or $\boldsymbol{x}(t+1)$ with $t \in \mathbb{Z}$ in the discrete-time case, E_j and A_i are constant matrices of appropriate sizes, $v_j(z)$ and $h_i(z)$ with $i \in \{1, 2, \ldots, r\}$ and $j \in \{1, 2, \ldots, r_e\}$ are

MFs depending upon the premise vector z, which in turn may depend on states, time, or exogenous variables. Importantly, the MFs $h_i(z)$ and $v_j(z)$ hold the convex sum property for every $z \in \mathcal{C} \subset \mathbb{R}^p$.

Stability analysis of such models has been performed by rewriting them as a *descriptor redundancy form* for continuous-time models, i.e. [75–77]:

$$\bar{E}\dot{\bar{x}}(t) = \bar{A}_{hv}\bar{x}(t), \tag{4.78}$$

with

$$\bar{x} = \begin{bmatrix} x \\ \dot{x} \end{bmatrix}, \; \bar{E} = \begin{bmatrix} I & 0 \\ 0 & 0 \end{bmatrix}, \; \bar{A}_{hv} = \begin{bmatrix} 0 & I \\ A_h & -E_v \end{bmatrix}, \; A_h = \sum_{i=1}^{r} h_i(z) A_i, \; E_v = \sum_{i=1}^{r_e} v_i(z) E_i;$$

or by means of the following equality constraint for the discrete-time case, first appeared in [78]:

$$\begin{bmatrix} A_h & -E_v \end{bmatrix} \begin{bmatrix} x(t) \\ x(t+1) \end{bmatrix} = 0. \tag{4.79}$$

The representations above can be obtained regardless of the nature of the descriptor, i.e. they hold for descriptors representing singular (differential/difference algebraic equations, DAEs) or non-singular systems (i.e. descriptor that can be directly put as ODEs).

Yet, a more general point of view consists in considering descriptors of the form

$$\bar{E}\dot{\bar{x}} = A_h \bar{x}, \tag{4.80}$$

where $\bar{E} = \text{block-diag}(I_{n-p}, 0_p)$ and A_h is the shorthand for a polytope of matrices of r vertices, as usual. Note that (4.78) can indeed be considered to have that form, as $\bar{A}_{hv}$ is just a polyhedron of matrices, so there is no need, in a formal sense, to keep the distinction between h and v. Note also that A_h has no particular structure as in (4.78).

Most of the stability conditions for descriptor systems via convex optimisation address the non-singular case, as they are infeasible if some E_j is singular; this is the case of the quadratic and non-quadratic results given below.

Naturally, a non-singular descriptor represented by (4.77) will have only full-rank E_j; in such case, it might be treated as a TS model of the ordinary nonlinear system that results from multiplying (4.77) by the inverse of the convex sum on the left-hand side. Nevertheless, the vertex linear systems may *not* be the same, and therefore, not surprisingly, the origin of a system that can be proven stable based on the vertices of its descriptor form may not be provable based on any of its ordinary TS models.

Similarly, non-singular E_j may lead to equivalent LFT representations.

The Quadratic Case

A quadratic Lyapunov function candidate for a continuous-time descriptor model (4.78) is

$$V(\bar{x}) = \bar{x}^T \bar{E}^T P \bar{x},$$

under the conditions $\bar{E}^T P = P^T \bar{E} \geq 0, \det(P) \neq 0$ [79]. Based on it, the following generalisation of [75, Theorem 1] can be made:

Theorem 4.15 *The origin $\mathbf{x} = 0$ of the unforced descriptor model (4.77) with $\delta \mathbf{x}(t) = \dot{\mathbf{x}}(t)$, $t \in \mathbb{R}$, is asymptotically stable if there exists a matrix $P \in \mathbb{R}^{2n\times 2n}$ such that*

$$P = \begin{bmatrix} P_1 & 0 \\ P_2 & P_3 \end{bmatrix}, \ \det P \neq 0,$$

and the following LMIs hold:

$$P_1 = P_1^T > 0, \ \bar{A}_{ij}^T P + P^T \bar{A}_{ij} < 0, \ \bar{A}_{ij} = \begin{bmatrix} 0 & I \\ A_i & -E_j \end{bmatrix}, \tag{4.81}$$

for $i \in \{1, 2, \ldots, r\}$, $j \in \{1, 2, \ldots, r_e\}$. Any trajectory $\mathbf{x}(t)$ starting within the outermost Lyapunov level $V(\mathbf{x}) = \mathbf{x}^T P_1 \mathbf{x} \leq k$, $k > 0$, in a compact set $\mathcal{X}$ of the state-space guaranteeing $z \in \mathcal{C} \subseteq \mathbb{R}^p$ goes asymptotically to $\mathbf{x} = 0$.

Proof From

$$V(\bar{x}) = \bar{x}^T \bar{E}^T P \bar{x} = \mathbf{x}^T P_1 \mathbf{x},$$

it follows that $V(\bar{x})$ is a valid Lyapunov function candidate; its time derivative along the trajectories of the unforced model (4.78) is

$$\begin{aligned} \dot{V}(\bar{x}) &= \bar{x}^T \bar{E}^T P \dot{\bar{x}} + \dot{\bar{x}}^T \bar{E}^T P \bar{x} = \bar{x}^T P^T \bar{E} \dot{\bar{x}} + \dot{\bar{x}}^T \bar{E}^T P \bar{x} \\ &= \bar{x}^T \left(P^T \bar{A}_{hv} + \bar{A}_{hv}^T P \right) \bar{x}, \end{aligned}$$

from which it follows that $P^T \bar{A}_{hv} + A_{hv}^T P < 0$ guarantees $\dot{V}(\bar{x}) < 0, \forall \bar{x} \neq 0$, which is implied by LMI conditions (4.81). Thus, $V(\bar{x})$ a valid Lyapunov function and asymptotic stability of $\mathbf{x} = 0$ is established according to Theorem B.1. Since MFs $h_i(z)$, $v_j(z)$, hold the convex sum property in $\mathcal{C}$, trajectories in the outermost Lyapunov level within $\mathcal{X}$, which, in turn, guarantees $z \in \mathcal{C}$, go asymptotically to $\mathbf{x} = 0$ (see assumption 4.2). □

Similarly, if the descriptor (4.80) is employed instead, along with the same Lyapunov function candidate (with adequate dimensions), we have the following result:

Theorem 4.16 *The origin* $x = 0$ *of the unforced descriptor model* (4.80) *with* $\delta x(t) = \dot{x}(t)$, $t \in \mathbb{R}$, *is asymptotically stable if there exists a matrix* $P \in \mathbb{R}^{n\times n}$ *such that*

$$P = \begin{bmatrix} P_1 & 0 \\ P_2 & P_3 \end{bmatrix}, \ \det P \neq 0,$$

and the following LMIs hold:

$$P_1 = P_1^T > 0, \ \bar{A}_i^T P + P^T \bar{A}_i < 0 \tag{4.82}$$

for $i \in \{1, 2, \ldots, r\}$. *Any trajectory* $x(t)$ *starting within the outermost Lyapunov level* $V(x) = x^T P_1 x \leq k$, $k > 0$, *in a compact set* $\mathcal{X}$ *of the state-space guaranteeing* $z \in \mathcal{C} \subseteq \mathbb{R}^p$ *goes asymptotically to* $x = 0$.

Proof It follows directly from the proof in Theorem 4.15. □

Remark 4.5 Note that P_2 and P_3 may be replaced with premise-dependent P_{2h} and P_{3h} if so wished and Theorem 4.16 would still be valid, because $V(\bar{x})$ does neither actually depend on P_2 nor P_3, so neither it does $\dot{V}(\bar{x})$.

In discrete-time context, the simpler quadratic Lyapunov function candidate $V(x) = x^T P x$ is employed as it is intended to be used along with (4.79)

Theorem 4.17 *The origin* $x = 0$ *of the unforced descriptor model* (4.77) *with* $\delta x(t) = x(t+1)$, $t \in \mathbb{Z}$, *is asymptotically stable if there exists a matrix* $P \in \mathbb{R}^{n\times n}$ *such that*

$$P = P^T > 0, \ \begin{bmatrix} -P & A_i^T P \\ P A_i & -P E_j - E_j^T P + P \end{bmatrix} < 0, \tag{4.83}$$

for $i \in \{1, 2, \ldots, r\}$, $j \in \{1, 2, \ldots, r_e\}$. *Any trajectory* $x(t)$ *starting within the outermost Lyapunov level* $V(x) = x^T P x \leq k$, $k > 0$, *in a compact set* $\mathcal{X}$ *of the state-space guaranteeing* $z \in \mathcal{C} \subseteq \mathbb{R}^p$ *goes asymptotically to* $x = 0$.

Proof From $P = P^T > 0$, it is clear that $V(x) = x^T P x$ is a valid Lyapunov function candidate; taking the forward difference of the Lyapunov function, we have that

$$V(t+1) - V(t) = \begin{bmatrix} x(t) \\ x(t+1) \end{bmatrix}^T \begin{bmatrix} -P & 0 \\ 0 & P \end{bmatrix} \begin{bmatrix} x(t) \\ x(t+1) \end{bmatrix} < 0$$

must hold for the origin $x = 0$ to be asymptotically stable. This can be combined with (4.79) by Finsler's Lemma A.3, yielding the equivalent condition:

$$\begin{bmatrix} -P & 0 \\ 0 & P \end{bmatrix} + \begin{bmatrix} \mathcal{U}(\cdot) \\ \mathcal{V}(\cdot) \end{bmatrix} \begin{bmatrix} A_h & -E_v \end{bmatrix} + \begin{bmatrix} A_h & -E_v \end{bmatrix}^T \begin{bmatrix} \mathcal{U}(\cdot) \\ \mathcal{V}(\cdot) \end{bmatrix}^T < 0,$$

which is guaranteed by LMIs (4.83) once the (conservative) choice $\mathcal{U}(\cdot) = 0$, $\mathcal{V}(\cdot) = P$ is made. This establishes the validity of $V(x)$ as a genuine Lyapunov function, proving asymptotic stability of $\boldsymbol{x} = 0$ according to Theorem B.5. Again, the fact that MFs $h_i(z)$, $v_j(z)$, hold the convex sum property in $\mathcal{C}$, guarantees that trajectories within the outermost Lyapunov level in $\mathcal{X}$ go asymptotically to $\boldsymbol{x} = 0$, which is included in $\mathcal{X}$ (see assumptions 4.1 and 4.2). □

Example 4.18 Consider a continuous-time descriptor model of the form (4.77), i.e. $\delta\boldsymbol{x}(t) = \dot{\boldsymbol{x}}(t)$ and $t \in \mathbb{R}$, with

$$A_1 = \begin{bmatrix} -1 & 0 \\ -0.5 & -2 \end{bmatrix}, \; A_2 = \begin{bmatrix} -1 & 0 \\ 0.5 & -2 \end{bmatrix}, \; E_1 = \begin{bmatrix} 1 & 0.5 \\ 0.5 & 1 \end{bmatrix}, \; E_2 = \begin{bmatrix} 1 & -0.5 \\ -0.5 & 1 \end{bmatrix}.$$

It is assumed that the MFs $h_i(z)$ and $v_j(z)$ hold the convex sum property everywhere.

LMIs (4.81) are feasible with

$$P = \begin{bmatrix} 0.57 & 0 & 0 & 0 \\ 0 & 0.44 & 0 & 0 \\ 0.42 & 0 & 0.57 & 0 \\ 0 & 0.34 & 0 & 0.44 \end{bmatrix} \Rightarrow P_1 = \begin{bmatrix} 0.57 & 0 \\ 0 & 0.44 \end{bmatrix},$$

for all the possible combinations A_i, E_i, $i \in \{1, 2\}$, thus establishing the stability of the TS descriptor model as long as MFs h_i and v_i hold the convex sum property. Note that P has been chosen with two identical 2×2 diagonal blocks, but there is no need for diagonal form or $P_1 = P_3$. This structure is suggested in [7, 75] since it guarantees P to be invertible. □

Theorems 4.15 and 4.17 follow two different paths: one is based on the descriptor extension of the quadratic Lyapunov function $V = \boldsymbol{x}\bar{E}^T P\boldsymbol{x}$, the other one on the Finsler's Lemma A.3. Nevertheless, both versions exist for the continuous- and the discrete-time case.

The Non-Quadratic Case

As with classical TS models, non-quadratic developments for continuous-time descriptor models face the problem of dealing with the time derivatives of the MFs. To avoid duplication, some results circumventing this problem will be later shown in Sect. 5.5, where the stabilisation problem is treated.

Nevertheless, non-quadratic extensions for discrete-time descriptor models are quite straightforward. To see this, consider the following parameter-dependent (fuzzy) Lyapunov function candidate:

$$V(\boldsymbol{x}) = \boldsymbol{x}^T \left(\sum_{i=1}^{r} h_i(z) P_i \right) \boldsymbol{x} = \boldsymbol{x}^T P_{h(t)} \boldsymbol{x},$$

where $P_i = P_i^T > 0, i \in \{1, 2, \ldots, r\}$. Recall that in discrete-time context, notation $P_{h(t)}$ is preferred over P_h as time instances are relevant in the developments.

The forward difference of this function is given by

$$V(t+1) - V(t) = \begin{bmatrix} x(t) \\ x(t+1) \end{bmatrix}^T \begin{bmatrix} -P_{h(t)} & 0 \\ 0 & P_{h(t+1)} \end{bmatrix} \begin{bmatrix} x(t) \\ x(t+1) \end{bmatrix} < 0,$$

which combined via Finsler's Lemma A.3 with the descriptor equation (4.79), yields the equivalent condition

$$\begin{bmatrix} -P_{h(t)} & 0 \\ 0 & P_{h(t+1)} \end{bmatrix} + \begin{bmatrix} \mathcal{U}(\cdot) \\ \mathcal{V}(\cdot) \end{bmatrix} \begin{bmatrix} A_h & -E_v \end{bmatrix} + \begin{bmatrix} A_h & -E_v \end{bmatrix}^T \begin{bmatrix} \mathcal{U}(\cdot) \\ \mathcal{V}(\cdot) \end{bmatrix}^T < 0.$$

Several options for $\mathcal{U}(\cdot)$ and $\mathcal{V}(\cdot)$ are available in order to obtain more relaxed conditions than the quadratic ones, for instance:

1. $\mathcal{U}(\cdot) = 0$, $\mathcal{V}(\cdot) = P_{h(t+1)}$, with conditions

$$\begin{bmatrix} -P_{h(t)} & A_h^T P_{h(t+1)}^T \\ P_{h(t+1)} A_h & -P_{h(t+1)} E_v - E_v^T P_{h(t+1)} + P_{h(t+1)} \end{bmatrix} < 0. \tag{4.84}$$

2. $\mathcal{U}(\cdot) = 0$, $\mathcal{V}(\cdot) = G_{h(t+1)}$, with conditions

$$\begin{bmatrix} -P_{h(t)} & A_h^T G_{h(t+1)}^T \\ G_{h(t+1)} A_h & -G_{h(t+1)} E_v - E_v^T G_{h(t+1)}^T + P_{h(t+1)} \end{bmatrix} < 0. \tag{4.85}$$

3. $\mathcal{U}(\cdot) = 0$, $\mathcal{V}(\cdot) = G_{h(t)h(t+1)}$, with conditions

$$\begin{bmatrix} -P_{h(t)} & A_h^T G_{h(t)h(t+1)}^T \\ G_{h(t)h(t+1)} A_h & -G_{h(t)h(t+1)} E_v - E_v^T G_{h(t)h(t+1)}^T + P_{h(t+1)} \end{bmatrix} < 0. \tag{4.86}$$

Sufficient LMI conditions guaranteeing inequalities (4.84) and (4.85) can be directly obtained by replacing subscripts $h(t)$ (equivalently, h), v, and $h(t+1)$ by indexes i, j, and k, respectively, with $i, k \in \{1, 2, \ldots, r\}$, $j \in \{1, 2, \ldots, r_e\}$; the resulting LMIs have P_i (or P_k) and G_k as decision variables while referring to A_i and E_j in (4.77). As for inequality (4.86), sufficient LMI conditions for it can also be obtained directly replacing $X_{h(t)}$ by X_l, $G_{h(t)h(t+1)}$ by G_{lk}, E_v by E_j, A_h by A_i, and $X_{h(t+1)}$ by X_k, with $i, k, l \in \{1, 2, \ldots, r\}$, $j \in \{1, 2, \ldots, r_e\}$. These substitutions are, of course, the result of dropping the dependency on MFs due to their convexity being guaranteed in a compact set $\mathcal{C}$. In addition, the last inequality allows for algebraic combinations of similar products of MFs leading to less conservative LMIs as double convex sums on $h_i(z)h_l(z)$ appear (see Appendix D).

It can be easily checked that all these conditions include the quadratic case. Moreover, when $E_v = I$, i.e. when a standard TS model is considered instead of

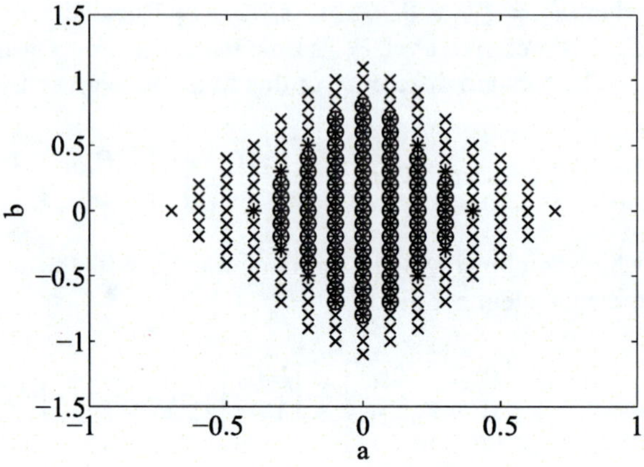

Fig. 4.10 Comparing quadratic (∘), non-quadratic (4.84) (+), and non-quadratic (4.85) (×)

a descriptor one, the non-quadratic conditions for discrete-time TS models (4.55) are recovered, provided that $G(\cdot) = P_{h(t+1)}$.

Example 4.19 Consider a discrete-time descriptor model of the form (4.77), i.e. $\delta x(t) = x(t+1)$ and $t \in \mathbb{Z}$, with:

$$A_1 = \begin{bmatrix} -0.7 & 1.1 \\ -1.3 & 0.5 + a \end{bmatrix}, \; A_2 = \begin{bmatrix} -0.7 & 1.1 \\ -1.3 & 0.5 - a \end{bmatrix},$$
$$E_1 = \begin{bmatrix} 1.9 & 0.3 \\ -0.4 - b & 1.2 \end{bmatrix}, \; E_2 = \begin{bmatrix} 1.9 & 0.3 \\ -0.4 + b & 1.2 \end{bmatrix},$$

where parameters a and b are chosen from the intervals $a \in [-1, 1]$, $b \in [-1.5, 1.5]$ to test the asymptotic stability of the origin $x = 0$. MFs $h_i(z)$ and $v_j(z)$ are assumed to hold the convex sum property everywhere.

As can be seen in Fig. 4.10, the feasibility set of the quadratic approach (∘) can be enlarged by any of the non-quadratic schemes above: that of conditions in (4.84) (+) which is also included in conditions (4.85) (×). □

The Singular Case

The above results can, of course, be applied even if the system under consideration is singular, but in that case they are useless as they always yield infeasible solutions. As discussed in Sect. 3.4, the study of singular systems, even in the linear case, implies dealing with a variety of phenomena such as consistent initialisation, impulsive behaviour, and regularity. This is due to the fact that dynamics are restricted to a subset of the state space for the algebraic restrictions to hold: in contrast with ordinary linear systems, stability properties are no longer global. In order to give a glimpse of this complexity in the linear context, the following definitions and results are given:

These considerations are important as stability properties should be redefined in terms of the subset of the state-space induced by algebraic restrictions.

A regular singular linear system $E\dot{x}(t) = Ax(t)$ is stable if and only if $\sigma(E, A) \subset \mathbb{C}^-$, where $\sigma(E, A) = \{s : \det(sE - A) = 0\}$. This result reduces to stability of linear systems if $E = I$. Stability of regular singular nonlinear systems is still undergoing; the interested reader is referred to [80–82] and the references therein.

4.7 Quadratic Stability of LFT Systems

This section will consider stability analysis of systems described by LFT representations under quadratic Lyapunov functions $V(\boldsymbol{x}) = \boldsymbol{x}^T P\boldsymbol{x}$. Let us consider an LFT model

$$\begin{aligned} \dot{\boldsymbol{x}}(t) &= A\boldsymbol{x}(t) + B_\delta \boldsymbol{w}_\delta \\ \boldsymbol{z}_\delta &= C_\delta \boldsymbol{x} + D_{\delta\delta} \boldsymbol{w}_\delta \end{aligned} \tag{4.87}$$

jointly with an uncertainty equation which, by assumption, can be written as:

$$\begin{bmatrix} \boldsymbol{w}_\delta \\ \boldsymbol{z}_\delta \end{bmatrix} \in \mathcal{S}(\Delta), \tag{4.88}$$

where $\mathcal{S}(\Delta)$ is a particular subspace (depending on Δ). For instance, as usual in most physical modelling cases with uncertain parameters, the equation $\boldsymbol{w}_\delta = \Delta \boldsymbol{z}_\delta$ may be written as

$$\eta \in \mathcal{S}(\Delta) \Leftrightarrow \exists z_\delta : \eta = \begin{bmatrix} \Delta \\ I \end{bmatrix} z_\delta \Leftrightarrow \mathcal{S}(\Delta) = \text{Im}\left(\begin{bmatrix} \Delta \\ I \end{bmatrix}\right) \tag{4.89}$$

but it can, actually, model more complex relationships between $\boldsymbol{w}_\delta$ and $\boldsymbol{z}_\delta$, if so wished (see [83]).

Also, some bounds on the uncertain or nonlinear elements will be assumed to be available, i.e. we will assert that $\Delta \in \mathbb{D}$ for a *known* uncertainty set $\mathbb{D}$.

4.7.1 Non-Augmented LMIs

If we consider the simplified expression $\boldsymbol{z}_\delta = \Delta \boldsymbol{w}_\delta$ instead of (4.88), the first, obvious, approach is eliminating $\boldsymbol{z}_\delta$ and $\boldsymbol{w}_\delta$ by using the full LFT expression (3.73), i.e.

$$\dot{\boldsymbol{x}}(t) = \tilde{A}(\Delta)\boldsymbol{x}(t) = \left(A + B_\delta \Delta (I - D_{\delta\delta}\Delta)^{-1} C_\delta\right) \boldsymbol{x}(t) \tag{4.90}$$

and, considering, say, a quadratic Lyapunov function $V(\boldsymbol{x}) = \boldsymbol{x}^T P \boldsymbol{x}$, posing a Lyapunov inequality as

$$\dot{V}(t) = \boldsymbol{x}^T \left(P\tilde{A}(\Delta) + \tilde{A}^T(\Delta)P \right) \boldsymbol{x} < 0, \tag{4.91}$$

i.e. expressed as a matrix inequality as

$$P\tilde{A}(\Delta) + \tilde{A}^T(\Delta)P < 0 \quad \forall \Delta \in \mathbb{D}. \tag{4.92}$$

However, the rational dependence on Δ in general LFT models usually does not allow posing (4.92) as a *finite* set of LMIs, unless a gridding approach [84]—writing LMIs for many possible values of Δ—is used. Moreover, the gridding approach may be risky if the grid is not dense enough.

In the particular case $D_{\delta\delta} = 0$, (4.92) is linear in Δ so, if Δ can be expressed as the convex sum of some known vertices D_i, with unknown coefficients μ_i, $\sum \mu_i = 1$, $\mu_i > 0$, i.e.

$$\mathbb{D} = \left\{ \Delta \mid \Delta = \sum_{i=1}^{r} \mu_i D_i \right\}, \tag{4.93}$$

then sufficient conditions for (4.92) can be cast as

$$P\tilde{A}(D_i) + \tilde{A}^T(D_i)P < 0 \quad \forall i = 1, 2, \ldots, r. \tag{4.94}$$

Of course, this is *strongly* reminiscent of the TS developments; see Sect. 3.5.1. Note, however, that it does not apply for $D_{\delta\delta} \neq 0$. The developments inspired in the standard LFT generalised plant for linear $\mathcal{H}_2$ and $\mathcal{H}_\infty$, originating in the 1980's [85, 86], gave rise to the results to be discussed next, valid for the general case $D_{\delta\delta} \neq 0$.

4.7.2 Multiplier-Based Conditions

Given the issues arising from rational expressions in the above "condensed" approach, alternative options have been developed in literature. Considering the quadratic Lyapunov function $V(\boldsymbol{x}) = \boldsymbol{x}^T P \boldsymbol{x}$, the Lyapunov stability condition amounts to verifying

$$\frac{dV}{dt} = [\boldsymbol{x}^T \; \dot{\boldsymbol{x}}^T] \begin{bmatrix} 0 & P \\ P & 0 \end{bmatrix} \begin{bmatrix} \boldsymbol{x} \\ \dot{\boldsymbol{x}} \end{bmatrix} < 0. \tag{4.95}$$

Augmenting the left and right vector to $\left[\boldsymbol{x}^T \; \dot{\boldsymbol{x}}^T \; \boldsymbol{w}_\delta^T \; \boldsymbol{z}_\delta^T\right]$, and introducing the plant dynamics, the Lyapunov stability condition can be written as

$$\begin{bmatrix} x \\ \dot{x} \\ w_\delta \\ z_\delta \end{bmatrix}^T \begin{bmatrix} 0 & P & 0 & 0 \\ P & 0 & 0 & 0 \\ 0 & 0 & 0 & 0 \\ 0 & 0 & 0 & 0 \end{bmatrix} \begin{bmatrix} x \\ \dot{x} \\ w_\delta \\ z_\delta \end{bmatrix} = \begin{bmatrix} x \\ w_\delta \end{bmatrix}^T \begin{bmatrix} I & 0 \\ A & B_\delta \\ 0 & I \\ C_\delta & D_{\delta\delta} \end{bmatrix}^T \begin{bmatrix} 0 & P & 0 & 0 \\ P & 0 & 0 & 0 \\ 0 & 0 & 0 & 0 \\ 0 & 0 & 0 & 0 \end{bmatrix} \begin{bmatrix} I & 0 \\ A & B_\delta \\ 0 & I \\ C_\delta & D_{\delta\delta} \end{bmatrix} \begin{bmatrix} x \\ w_\delta \end{bmatrix} < 0,$$

which must hold for all (w_δ, z_δ) in $\mathcal{S}(\Delta)$, so Lagrange/KKT multipliers may be introduced.

Indeed, using now the (full block) S-procedure [83], we can state the following result:

Theorem 4.18 *The origin of the LFT system* (4.87) *is asymptotically stable with Lyapunov function* $V(x) = x^T P x$ *if there exist a multiplier matrix* M *such that*

1. $\eta^T M \eta > 0$ *for all* $\eta \in \mathcal{S}(\Delta)$ *for all* $\Delta \in \mathbb{D}$, *and*
2. *the following LMI is feasible:*

$$\begin{bmatrix} I & 0 \\ A & B_\delta \\ 0 & I \\ C_\delta & D_{\delta\delta} \end{bmatrix}^T \begin{bmatrix} 0 & P & 0 & 0 \\ P & 0 & 0 & 0 \\ 0 & 0 & Q & S \\ 0 & 0 & S^T & R \end{bmatrix} \begin{bmatrix} I & 0 \\ A & B_\delta \\ 0 & I \\ C_\delta & D_{\delta\delta} \end{bmatrix} < 0, \quad M = \begin{bmatrix} Q & S \\ S^T & R \end{bmatrix}. \tag{4.96}$$

Remark 4.6 Under two rather mild assumptions:

- Δ takes values in a compact, path-connected set;
- $\mathcal{S}(\Delta)$ depends continuously on Δ.

The above conditions provide *necessary and sufficient* conditions for *quadratic* stability, see [83]. The "if" part is straightforward, as the left inequality in (4.96) is

$$\begin{bmatrix} x \\ \dot{x} \end{bmatrix}^T \begin{bmatrix} 0 & P \\ P & 0 \end{bmatrix} \begin{bmatrix} x \\ \dot{x} \end{bmatrix} + \begin{bmatrix} w_\delta \\ z_\delta \end{bmatrix}^T M \begin{bmatrix} w_\delta \\ z_\delta \end{bmatrix} < 0; \tag{4.97}$$

so, indeed, if the right term depending on M is positive in $\mathcal{S}(\Delta)$, the Lyapunov function (left term in the sum) decreases there.

The "only-if" part requires more involved discussions available in the cited source [83, Theorem 8], omitted here for brevity.

Remark 4.7 Rewriting (4.96), we get

$$\begin{bmatrix} PA + A^T P + \tilde{R} & PB_\delta + \tilde{S}^T \\ \tilde{S} + B_\delta^T P & \tilde{Q} \end{bmatrix} < 0 \tag{4.98}$$

with $\tilde{R} = C_\delta^T R C_\delta$, $\tilde{S} = S C_\delta + D_{\delta\delta}^T R C_\delta$, $\tilde{Q} = Q + S D_{\delta\delta} + D_{\delta\delta}^T S^T + D_{\delta\delta}^T R D_{\delta\delta}$, which coincides with the concept of $(-R, -S, -Q)$-dissipativity in [87, Lemma2], and has clear links with standard passivity and bounded-real argumentations (see the cited source for details).

4.7.3 *Choice of Multipliers*

The multiplier matrix M may be fully parameterised for arbitrary structures of Δ. The remaining difficult issue is how to write in LMI form with M as decision variable the condition below, required in Theorem 4.18, condition 1.:

$$\forall\, \Delta \in \mathbb{D},\ \eta^T M \eta > 0\, \forall\, \eta \in \mathcal{S}(\Delta). \tag{4.99}$$

For the particular case of block-diagonal Δ arising in many situations, some easy (but conservative) choices can be carried out, inspired in structured-uncertainty and μ-synthesis results derived in the late 1980s.

Block-Diagonal Scalings (Scaled Small Gain)

Let us consider the particular uncertainty description $w_\delta = \Delta z_\delta$, i.e. $\mathcal{S}(\Delta) = (\Delta^T\ I)^T z_\delta$, and an uncertainty bound $\|\Delta\| \le 1$, i.e. $(I - \Delta^T\Delta) \ge 0$.

Consider, too, a matrix L such that $\Delta L = L\Delta$, i.e. L commutes with Δ. The class of such matrices can be easily obtained as follows:

First, it is straightforward that for the case $\Delta = \delta I_{n_1}$, where I_{n_1} denotes the identity matrix of size $n_1 \times n_1$, L can be any arbitrary matrix. Hence, for a general $\Delta =$ block-diag$(\delta_1 I_{n_1}, \delta_2 I_{n_2}, \dots)$, we have $L =$ block-diag$(L_1, L_2, \dots)$, where L_i are arbitrary $n_i \times n_i$ matrices; these scalings come from classical linear robust control and μ-synthesis [86, 88].

So, we can craft a block-diagonal multiplier, because $L^T(I - \Delta^T\Delta)L \ge 0$, so $L^T L - L^T\Delta^T\Delta L = L^T L - \Delta^T L^T L \Delta \ge 0$, and hence letting $\Sigma = L^T L$ for any valid L, we have

$$\begin{bmatrix} \Delta \\ I \end{bmatrix}^T \begin{bmatrix} -\Sigma & 0 \\ 0 & \Sigma \end{bmatrix} \begin{bmatrix} \Delta \\ I \end{bmatrix} \ge 0. \tag{4.100}$$

From the above, Σ can actually be built as a block-diagonal positive semidefinite matrix with the same block-structure as L, the elements of Σ being decision variables in an LMI problem. Indeed, the resulting (sufficient) quadratic stability conditions are

$$\begin{bmatrix} I & 0 \\ A & B_\delta \\ 0 & I \\ C_\delta & D_{\delta\delta} \end{bmatrix}^T \begin{bmatrix} 0 & P & 0 & 0 \\ P & 0 & 0 & 0 \\ 0 & 0 & -\Sigma & 0 \\ 0 & 0 & 0 & \Sigma \end{bmatrix} \begin{bmatrix} I & 0 \\ A & B_\delta \\ 0 & I \\ C_\delta & D_{\delta\delta} \end{bmatrix} < 0, \tag{4.101}$$

which can be written, using (4.98) and applying Schur complement in Lemma A.1, as

$$\begin{bmatrix} PA + A^T P & PB_\delta & C_\delta^T \\ B_\delta^T P & -\Sigma & D_{\delta\delta}^T \\ C_\delta & D_\delta & -\Sigma^{-1} \end{bmatrix} < 0. \tag{4.102}$$

Note that the above expression is the scaled small-gain theorem (Theorem 5.19), which henceforth can be considered a particular case of the generic full-block multiplier theorems from which the above result was derived.

Polytopic $\boldsymbol{\Delta}$

If $\mathcal{S}(\Delta) = (\Delta^T\ I)^T z_\delta$, there are two possible relaxations in the (common) case of polytopic Δ, i.e. $\Delta = \text{co}(\Delta_g)$, Δ_g being a finite generator set.

Approach 1. Adding the extra condition $Q < 0$ to the multiplier structure of M in (4.96), the resulting

$$\Psi(\Delta) = \begin{bmatrix} \Delta \\ I \end{bmatrix}^T \begin{bmatrix} Q & S \\ S^T & R \end{bmatrix} \begin{bmatrix} \Delta \\ I \end{bmatrix} \tag{4.103}$$

is a concave function of Δ so evaluating it in its vertices suffices to set a finite number of LMIs to check the multiplier positiveness on the uncertainty subspace (4.99). A generalisation of the approach when $\Delta_g = \{\sum_{j=1}^N L_j \Delta_j R_j\}$ requiring only $L_j^T Q L_j \leq 0$ appears in [83].

Approach 2. Without adding the definiteness constraint in Q, expressing the convex hull in which Δ takes values as a convex sum $\Delta = \sum_j h_j \Delta_j$, Δ_j being the elements of the generator set Δ_g and $\sum_j h_j = 1$, the multiplier condition (4.99) can be written as:

$$\sum_i \sum_j h_i h_j \begin{bmatrix} \Delta_j \\ I \end{bmatrix}^T \begin{bmatrix} Q & S \\ S^T & R \end{bmatrix} \begin{bmatrix} \Delta_j \\ I \end{bmatrix} > 0, \tag{4.104}$$

which is a double summation on the unit simplex so that standard sum relaxations (say, Pòlya) can be applied to get a finite number of sufficient (asymptotically necessary) LMI conditions on M.

4.7.4 Relation to Polytopic/TS Quadratic Stability Conditions

Let us consider a box polytopic system:

$$\dot{x}(t) = \left(A_0 + \sum_{i=1}^{s} \delta_i A_i\right) x(t), \qquad \delta_i \in \mathbb{R},\ -1 \leq \delta_i \leq 1. \tag{4.105}$$

We will now prove that the multiplier approach encompasses the simpler non-rational quadratic stability conditions (4.5).

Theorem 4.19 *If the standard polytopic LMI conditions at the* 2^s *vertices below hold*

$$P\left(A_0+\sum_{i=1}^{s}\delta_i^* A_i\right)+\left(A_0+\sum_{i=1}^{s}\delta_i^* A_i\right)^T P<0, \quad \delta_i^*\in\{-1,1\} \tag{4.106}$$

there exists a multiplier fulfilling the conditions in Theorem 4.18 for the same P.

Proof It is trivial to write such system in LFT form, as

$$\dot{\boldsymbol{x}}(t)=A_0\boldsymbol{x}(t)+\sum_{i=1}^{s}A_i\boldsymbol{w}_i, \tag{4.107}$$

$$\boldsymbol{z}=\boldsymbol{x}, \qquad \boldsymbol{w}_i=(\delta_i I)\boldsymbol{x} \tag{4.108}$$

with $B_\delta=\begin{bmatrix}A_1 \ A_2 \ \ldots \ A_s\end{bmatrix}$, $C_\delta=I$, and $D_{\delta\delta}=0$.

If (4.106) hold, taking any feasible P, changing signs and reordering, as the inequality is strict, there exists $\delta_0>0$ such that

$$\begin{bmatrix}\boldsymbol{w}_1\\ \ldots\\ \boldsymbol{w}_s\\ \boldsymbol{x}\end{bmatrix}^T\left(\begin{bmatrix}0 & \ldots & 0 & -A_1^T P/s\\ \vdots & \ddots & \vdots & \vdots\\ 0 & \ldots & 0 & -A_s P/s\\ -PA_1/s & \ldots & -PA_s/s & -PA_0-A_0^T P\end{bmatrix}-\delta_0 I\right)\begin{bmatrix}\boldsymbol{w}_1\\ \ldots\\ \boldsymbol{w}_s\\ \boldsymbol{x}\end{bmatrix}>0 \tag{4.109}$$

for $\boldsymbol{w}_i$ being either $\boldsymbol{x}$ or $-\boldsymbol{x}$ (total of 2^s combinations). Indeed, without the term $\delta_0 I$ for the moment being, (4.109) can be written as

$$\begin{aligned}&\boldsymbol{x}^T(PA_0+A_0^T P)\boldsymbol{x}+\frac{1}{s}\sum_{i=1}^{s}\left(\boldsymbol{w}_i^T A_i^T P\boldsymbol{x}+\boldsymbol{x}^T PA_i\boldsymbol{w}_i\right)\\ &\quad=\frac{1}{s}\sum_{i=1}^{s}\left((\boldsymbol{x}^T(A_0^T P)+\boldsymbol{w}_i^T A_i^T P)\boldsymbol{x}+\boldsymbol{x}^T((\delta_0 I+PA_0)\boldsymbol{x}+PA_i\boldsymbol{w}_i)\right)<0,\end{aligned}$$

where negative definiteness comes from the fact that the result is the sum of all combinations in (4.106).

Now, note that conditions (4.109) are *linear* in $\boldsymbol{w}_i$ because the quadratic term on the variables $\boldsymbol{w}$ is zero, hence convex, so it holds also in intermediate points $\boldsymbol{w}_i=\delta_i\boldsymbol{x}$, $-1\le\delta_i\le 1$. Actually, denoting by M the matrix in square brackets in the above expression (4.109), the above linearity argumentation shows that $\boldsymbol{\eta}^T M\boldsymbol{\eta}^T>0$ for all $\boldsymbol{\eta}\in\mathcal{S}(\Delta)$ so we have a valid multiplier fulfilling (4.99). Finally, if we replace M in (4.96) and the model of $\dot{\boldsymbol{x}}$, and operate to obtain (4.97), the resulting expression for (4.97) is $-\delta_0 I<0$ which is, of course, trivially true. □

Regarding TS systems $\dot{\boldsymbol{x}}(t)=\sum_{i=1}^{r}h_i A_i\boldsymbol{x}$, with $\sum_{i=1}^{r}h_i=1$, they can also be written in LFT form with $B_\delta=\begin{bmatrix}A_1 \ \ldots \ A_s\end{bmatrix}$, $C_\delta=I$, $D_{\delta\delta}=0$, and $A_0=0$.

Theorem 4.20 *Given the above TS system, if the standard polytopic LMI conditions at the r vertices below hold*

$$PA_i + A_i^T P < 0, \quad i \in \{1, 2, \ldots, r\}, \tag{4.110}$$

there exists a multiplier fulfilling the conditions in Theorem 4.18 for the same P.

Proof If stability conditions hold, there exists $\delta_0 > 0$ such that (4.109) (with $A_0 = 0$) holds for one of the $\boldsymbol{w}_i$ being equal to $\boldsymbol{x}$ and the rest being zero. The same linearity argument proves that (4.109) holds for all $\boldsymbol{w}_i = h_i\boldsymbol{x}$ with h_i in the standard simplex. The rest of the steps are identical to the box polytopic case. □

Of course, the above developments to find the multiplier can be considered a particular case of the general, constructive, full-block S-procedure argumentation in [83].

Note that the obtained multiplier is *not* diagonal. However, it verifies $Q \leq 0$, so the first approach for polytopic Δ would obtain a non-diagonal LMI solution for polytopic systems when approached in the LFT framework. Basically, the LFT approaches moves the (many) vertex-checking conditions from the main Lyapunov decrement condition to the multiplier one, with decision variables which, a priori, may be different from the Lyapunov function matrices and where system matrices do not intervene, achieving a sort of decoupling.

4.8 Non-Quadratic Stability of LFT Models

We will now develop a parameter-dependent Lyapunov function based on the results in [89]. Let us consider the LFT model (4.87), and use a parameter-dependent Lyapunov function, $V = \boldsymbol{y}_L^T P \boldsymbol{y}_L$ where:

$$\begin{bmatrix} \boldsymbol{y}_L \\ \boldsymbol{z}_L \end{bmatrix} = \begin{bmatrix} C_L & D_L \\ E_L & F_L \end{bmatrix} \begin{bmatrix} \boldsymbol{x} \\ \boldsymbol{w}_L \end{bmatrix}, \quad \boldsymbol{w}_L = \Delta \boldsymbol{z}_L, \tag{4.111}$$

so $\boldsymbol{y}_L$ is LFT-dependent on Δ. Matrices C_L, D_L, E_L, F_L can be arbitrarily[2] chosen, but must not be decision variables in the settings below as they appear multiplied by other Lyapunov/multiplier LMI variables. They must be interpreted as the basis for the search of a Lyapunov function, fixed *a priori*, and P being the Lyapunov decision variables. Note that the size of $\boldsymbol{y}_L$ does not need to be the same as that of the state $\boldsymbol{x}$ and, in fact, augmenting the size can be advantageous in order to increase the generality of the sought Lyapunov function.

For instance, if we had two uncertainties δ_1 and δ_2 choosing

$$C_L = \begin{bmatrix} I \\ 0 \\ 0 \end{bmatrix}, \; D_L = \begin{bmatrix} 0 & 0 \\ I & 0 \\ 0 & I \end{bmatrix}, \; E_L = I \;\; F_L = \begin{bmatrix} 0 & 0 \end{bmatrix}, \; \boldsymbol{w}_L = \begin{bmatrix} \delta_1 I \\ \delta_2 I \end{bmatrix} \boldsymbol{z}_L, \tag{4.112}$$

[2] Some technical well-posedness constraints apply.

we would have $\boldsymbol{y}_L = \left[\delta_1 I \;\; \delta_2 I \;\; I\right]^T \boldsymbol{x}$, so V would be a generic quadratic function of δ_1 and δ_2, which, after suitably partitioning P in 3×3 blocks (symmetric) would be expressed by

$$V = \boldsymbol{x}^T (P_{11}\delta_1^2 + P_{22}\delta_2^2 + (P_{12} + P_{12}^T)\delta_1\delta_2 + (P_{13} + P_{13}^T)\delta_1 + (P_{23} + P_{23}^T)\delta_2 + P_{33})\boldsymbol{x}.$$

Other choices of matrices in (4.111) would lead to different parameterisations of the Lyapunov function (higher degree, rational, etc.), the elements of P being the adjustable decision variables.

In order to ensure that the Lyapunov function is positive for all values of Δ, we set up the following constraint:

$$\begin{bmatrix} \boldsymbol{y}_L \\ \boldsymbol{w}_L \\ \boldsymbol{z}_L \end{bmatrix}^T \begin{bmatrix} P & 0 & 0 \\ 0 & 0 & 0 \\ 0 & 0 & 0 \end{bmatrix} \begin{bmatrix} \boldsymbol{y}_L \\ \boldsymbol{w}_L \\ \boldsymbol{z}_L \end{bmatrix} = \begin{bmatrix} \boldsymbol{x} \\ \boldsymbol{w}_L \end{bmatrix}^T \begin{bmatrix} C_L & D_L \\ 0 & I \\ E_L & F_L \end{bmatrix}^T \begin{bmatrix} P & 0 & 0 \\ 0 & 0 & 0 \\ 0 & 0 & 0 \end{bmatrix} \begin{bmatrix} C_L & D_L \\ 0 & I \\ E_L & F_L \end{bmatrix} \begin{bmatrix} \boldsymbol{x} \\ \boldsymbol{w}_L \end{bmatrix} > 0, \tag{4.113}$$

which must hold for all $\boldsymbol{w}_L, \boldsymbol{z}_L$ in $\mathcal{S}(\Delta)$. So, we must introduce a multiplier for full-block S-procedure and enforce the conditions:

$$\begin{bmatrix} \boldsymbol{x} \\ \boldsymbol{w}_L \end{bmatrix}^T \begin{bmatrix} C_L & D_L \\ 0 & I \\ E_L & F_L \end{bmatrix}^T \begin{bmatrix} P & 0 & 0 \\ 0 & Q_L & S_L \\ 0 & S_L^T & R_L \end{bmatrix} \begin{bmatrix} C_L & D_L \\ 0 & I \\ E_L & F_L \end{bmatrix} \begin{bmatrix} \boldsymbol{x} \\ \boldsymbol{w}_L \end{bmatrix} > 0,$$

$$\eta^T \begin{bmatrix} Q_L & S_L \\ S_L^T & R_L \end{bmatrix} \eta < 0, \; \forall \eta \in \mathcal{S}(\Delta).$$

Now, the Lyapunov-decrement equation is

$$\dot{V} = \begin{bmatrix} \boldsymbol{y}_L \\ \dot{\boldsymbol{y}}_L \end{bmatrix}^T \begin{bmatrix} 0 & P \\ P & 0 \end{bmatrix} \begin{bmatrix} \boldsymbol{y}_L \\ \dot{\boldsymbol{y}}_L \end{bmatrix}. \tag{4.114}$$

Let us develop the vector at the right-hand side in the above expression. We have

$$\begin{bmatrix} \boldsymbol{y}_L \\ \dot{\boldsymbol{y}}_L \end{bmatrix} = \begin{bmatrix} C_L & D_L & 0 & 0 \\ 0 & 0 & D_L & C_L \end{bmatrix} \begin{bmatrix} \boldsymbol{x} \\ \boldsymbol{w}_L \\ \dot{\boldsymbol{w}}_L \\ \dot{\boldsymbol{x}} \end{bmatrix} := \bar{C} \begin{bmatrix} \boldsymbol{x} \\ \boldsymbol{w}_L \\ \dot{\boldsymbol{w}}_L \\ \dot{\boldsymbol{x}} \end{bmatrix}. \tag{4.115}$$

As $\dot{\boldsymbol{w}}_L = \Delta \dot{\boldsymbol{z}}_L + \dot{\Delta} \boldsymbol{z}_L$, and $\dot{\boldsymbol{z}}_L = E_L \dot{\boldsymbol{x}} + D_L \dot{\boldsymbol{w}}_L$, we have:

$$\begin{bmatrix} \boldsymbol{x} \\ \boldsymbol{w}_L \\ \dot{\boldsymbol{w}}_L \\ \dot{\boldsymbol{x}} \end{bmatrix} = \begin{bmatrix} I & 0 & 0 & 0 \\ 0 & 0 & I & 0 \\ 0 & 0 & 0 & I \\ A & B_\delta & 0 & 0 \end{bmatrix} \begin{bmatrix} \boldsymbol{x} \\ \boldsymbol{w}_\delta \\ \boldsymbol{w}_L \\ \dot{\boldsymbol{w}}_L \end{bmatrix} := Q \begin{bmatrix} \boldsymbol{x} \\ \boldsymbol{w}_\delta \\ \boldsymbol{w}_L \\ \dot{\boldsymbol{w}}_L \end{bmatrix}, \tag{4.116}$$

$$\begin{bmatrix} z_L \\ \dot{z}_L \end{bmatrix} = \begin{bmatrix} E_L & F_L & 0 & 0 \\ 0 & 0 & F_L & E_L \end{bmatrix} \begin{bmatrix} x \\ w_L \\ \dot{w}_L \\ \dot{x} \end{bmatrix} := \bar{E} \begin{bmatrix} x \\ w_L \\ \dot{w}_L \\ \dot{x} \end{bmatrix}, \tag{4.117}$$

$$\begin{bmatrix} w_L \\ \dot{w}_L \end{bmatrix} = \begin{bmatrix} \Delta & 0 \\ \dot{\Delta} & \Delta \end{bmatrix} \begin{bmatrix} z_L \\ \dot{z}_L \end{bmatrix} = \begin{bmatrix} I & 0 & 0 \\ 0 & I & I \end{bmatrix} \begin{bmatrix} \Delta & 0 & 0 \\ 0 & \Delta & 0 \\ 0 & 0 & \dot{\Delta} \end{bmatrix} \begin{bmatrix} I & 0 \\ 0 & I \\ I & 0 \end{bmatrix} \begin{bmatrix} z_L \\ \dot{z}_L \end{bmatrix}. \tag{4.118}$$

Then (4.118) can be rewritten by defining temporary pseudo-variables w_L^*, z_L^* such that:

$$\begin{bmatrix} w_L \\ \dot{w}_L \end{bmatrix} = \underbrace{\begin{bmatrix} I & 0 & 0 \\ 0 & I & I \end{bmatrix}}_{T_1} w_L^* := T_1 w_L^*, \tag{4.119}$$

$$z_L^* := \begin{bmatrix} z_L \\ \dot{z}_L \\ z_L \end{bmatrix} = \underbrace{\begin{bmatrix} I & 0 \\ 0 & I \\ I & 0 \end{bmatrix}}_{T_2} \begin{bmatrix} z_L \\ \dot{z}_L \end{bmatrix} := T_2 \begin{bmatrix} z_L \\ \dot{z}_L \end{bmatrix}, \tag{4.120}$$

and

$$w_L^* = \begin{bmatrix} \Delta & 0 & 0 \\ 0 & \Delta & 0 \\ 0 & 0 & \dot{\Delta} \end{bmatrix} z_L^*.$$

Last, we can write, using the conveniently-defined shorthand $\bar{C}$, T_1, T_2, Q, and $\bar{E}$

$$\begin{bmatrix} x \\ w_\delta \\ w_L^* \\ \hline y_L \\ \dot{y}_L \\ \hline z_\delta \\ z_L^* \end{bmatrix} = \begin{bmatrix} I \\ \hline \bar{C} Q \begin{bmatrix} I & 0 \\ 0 & T_1 \end{bmatrix} \\ \hline C_\delta \;\; D_{\delta\delta} \;\; 0 \\ T_2 \bar{E} Q \begin{bmatrix} I & 0 \\ 0 & T_1 \end{bmatrix} \end{bmatrix} \begin{bmatrix} x \\ w_\delta \\ w_L^* \end{bmatrix}; \tag{4.121}$$

so, introducing a suitably partitioned symmetric full-block multiplier M, we get the final result below incorporating positiveness of V and negativeness of $\dot{V}$ in the suitable subspaces

Theorem 4.21 *The origin of the LFT system (4.87) is asymptotically stable with $V = y_L(\Delta) P y_L(\Delta)$ as a non-quadratic Lyapunov function where $y_L(\Delta)$ is defined*

in (4.111)*, if there exist* P*,* M*,* Q_L*,* R_L*,* S_L *such that*

$$\begin{bmatrix} I \\ \bar{C}Q\begin{bmatrix} I & 0 \\ 0 & T_1 \end{bmatrix} \\ C_\delta \;\; D_{\delta\delta} \;\; 0 \\ T_2\bar{E}Q\begin{bmatrix} I & 0 \\ 0 & T_1 \end{bmatrix} \end{bmatrix}^T \begin{bmatrix} 0 & 0 & 0 & 0 & 0 & 0 & 0 \\ 0 & M_{11} & M_{12} & 0 & 0 & M_{13} & M_{14} \\ 0 & M_{12}^T & M_{22} & 0 & 0 & M_{23} & M_{24} \\ 0 & 0 & 0 & 0 & P & 0 & 0 \\ 0 & 0 & 0 & P & 0 & 0 & 0 \\ 0 & M_{13}^T & M_{23}^T & 0 & 0 & M_{33} & M_{34} \\ 0 & M_{14}^T & M_{24}^T & 0 & 0 & M_{34}^T & M_{44} \end{bmatrix} \begin{bmatrix} I \\ \bar{C}Q\begin{bmatrix} I & 0 \\ 0 & T_1 \end{bmatrix} \\ C_\delta \;\; D_{\delta\delta} \;\; 0 \\ T_2\bar{E}Q\begin{bmatrix} I & 0 \\ 0 & T_1 \end{bmatrix} \end{bmatrix} > 0, \tag{4.122}$$

$$\begin{bmatrix} C_L & D_L \\ 0 & I \\ E_L & F_L \end{bmatrix}^T \begin{bmatrix} P & 0 & 0 \\ 0 & Q_L & S_L \\ 0 & S_L^T & R_L \end{bmatrix} \begin{bmatrix} C_L & D_L \\ 0 & I \\ E_L & F_L \end{bmatrix} < 0 \tag{4.123}$$

jointly with the multiplier conditions

$$\begin{bmatrix} \bar{\Delta} \\ I \end{bmatrix}^T M \begin{bmatrix} \bar{\Delta} \\ I \end{bmatrix} > 0 \quad \begin{bmatrix} \Delta \\ I \end{bmatrix}^T \begin{bmatrix} Q_L & S_L \\ S_L^T & R_L \end{bmatrix} \begin{bmatrix} \Delta \\ I \end{bmatrix} < 0 \tag{4.124}$$

for all possible Δ*,* $\bar{\Delta}$ *in a suitable (*a priori *known) compact set, the uncertainty being described as:*

$$\bar{\Delta} = block\text{-}diag(\Delta, \Delta, \Delta, \dot{\Delta}), \quad \begin{bmatrix} \boldsymbol{w}_\delta \\ \boldsymbol{w}_L^* \end{bmatrix} = \bar{\Delta}\begin{bmatrix} z_\delta \\ z_L^* \end{bmatrix}, \quad \boldsymbol{w}_L = \Delta z_L. \tag{4.125}$$

Indeed, $\dot{V} < 0$ plus full-block S-procedure amounts to (4.122) and, in the same way, $V > 0$ ends up in (4.123).

Note that the above theorem requires known bounds on the rate of variation of the parameters and a suitable parametrisation of the multipliers, depending on the uncertainty structure, in order to express (4.124) as a finite number of LMI conditions, as discussed in Sect. 4.7.3.

Remark 4.8 (Discrete-time systems) Similar developments in quadratic and non-quadratic stability analysis would apply to discrete-time systems in LFT form, replacing $\dot{\boldsymbol{x}}$ by $\boldsymbol{x}(t+1)$ in the expressions, and replacing the Lyapunov equation to

$$V(\boldsymbol{x}) = \begin{bmatrix} \boldsymbol{x}(t) \\ \boldsymbol{x}(t+1) \end{bmatrix}^T \begin{bmatrix} -P & 0 \\ 0 & P \end{bmatrix} \begin{bmatrix} \boldsymbol{x}(t) \\ \boldsymbol{x}(t+1) \end{bmatrix} \tag{4.126}$$

or the analogous expressions with $\boldsymbol{y}(t)_L(\Delta)$, $\boldsymbol{y}(t+1)_L(\Delta)$ in the non-quadratic setup. Details are omitted for brevity, as developments are similar to the continuous-time case, and, in fact, a particular case of the control synthesis setup in [89].

4.9 Summary and Conclusions

Stability analysis of convex models (TS or polynomial, descriptor or LFT), is mainly based on quadratic Lyapunov functions. The convex structure helps the analysis to mimic the linear case up to the modelling region. Conservativeness is mainly tackled by using more general classes of Lyapunov functions, such as piecewise or parameter-dependent ones, but the relaxation comes at the price of new difficulties such as continuity and multipliers in piecewise contexts, handling of the time derivatives of the membership functions in parameter-dependent ones. All in all, the different stability analysis techniques constitute the basis for the development of control and observation schemes in the following chapters.

References

1. Boyd, S., El Ghaoui, L., Féron, E., Balakrishnan, V.: Linear Matrix Inequalities in System and Control Theory. Studies in Applied Mathematics, Society for Industrial and Applied Mathematics, Philadelphia, PA (1994)
2. Scherer, C., Weiland, S.: Linear Matrix Inequalities in Control. Delft University, Delft , The Netherlands (2005)
3. Wang, H., Tanaka, K., Griffin, M.: An approach to fuzzy control of nonlinear systems: stability and design issues. IEEE Trans. Fuzzy Syst. **4**(1), 14–23 (1996)
4. Tanaka, K., Wang, H.O.: Fuzzy Control Systems Design and Analysis, analysis Wiley, New York (2001)
5. Johansson, M.: Piecewise linear control systems. PhD thesis, Department of Automatic Control, Lund Institute of Technology, Sweden (1999)
6. Gahinet, P., Nemirovski, A., Laub, A.J., Chilali, M.: LMI control toolbox. The Mathworks Inc (1995)
7. Tanaka, K., Wang, H.O.: Fuzzy Control System Design and Analysis: A Linear Matrix Inequality Approach. Wiley, New York (2001)
8. Bernal, M., Guerra, T.M., Kruszewski, A.: A membership-function-dependent approach for stability analysis and controller synthesis of Takagi–Sugeno models. Fuzzy Sets Syst. **160**(19), 2776–2795 (2009)
9. Lam, H., Leung, F.: LMI-based stability and performance conditions for continuous-time nonlinear systems in Takagi–Sugeno's form. IEEE Trans. Syst. Man Cybern. B Cybern. **37**(5), 1396–1406 (2007)
10. Sala, A., Arino, C.: Relaxed stability and performance conditions for Takagi–Sugeno fuzzy systems with knowledge on membership function overlap. IEEE Transactions on Systems Man and Cybernetics Part B (Cybernetics) **37**(3), 727–732 (2007)
11. Narimani, M., Lam, H.: Relaxed LMI-based stability conditions for Takagi–Sugeno fuzzy control systems using regional-membership-function-shape-dependent analysis approach. IEEE Trans. Fuzzy Syst. **17**(5), 1221–1228 (2009)
12. Sala, A.: Introducing shape-dependent relaxed conditions in fuzzy control of nonlinear systems in Takagi–Sugeno form. In: Fuzzy Systems, 2008. IEEE International Conference on World Congress on Computational Intelligence, pp. 512–517. IEEE, New York (2008)
13. Sala, A., Arino, C.: Relaxed stability and performance LMI conditions for Takagi–Sugeno fuzzy systems with polynomial constraints on membership function shapes. IEEE Trans. Fuzzy Syst. **16**(5), 1328–1336 (2008)
14. Bernal, M., Guerra, T.M.: Generalized non-quadratic stability of continuous-time Takagi–Sugeno models. IEEE Trans. Fuzzy Syst. **18**(4), 815–822 (2010)

15. Lee, D., Kim, D.: Relaxed LMI conditions for local stability and local stabilization of continuous-time Takagi–Sugeno fuzzy systems. IEEE Transactions on Cybernetics **44**(3), 394–405 (2014)
16. Sala, A.: On the conservativeness of fuzzy and fuzzy-polynomial control of nonlinear systems. Annu. Rev. Control. **33**(1), 48–58 (2009)
17. Kruszewski, A., Sala, A., Guerra, T.M., Ariño, C.: A triangulation approach to asymptotically exact conditions for fuzzy summations. IEEE Trans. Fuzzy Syst. **17**(5), 985–994 (2009)
18. Johansson, M., Rantzer, A.: Computation of piecewise quadratic Lyapunov functions for hybrid systems. IEEE Trans. Autom. Control **43**(4), 555–559 (1998)
19. Johansson, M., Rantzer, A., Arzen, K.: Piecewise quadratic stability of fuzzy systems. IEEE Trans. Fuzzy Syst. **7**(6), 713–722 (1999)
20. Feng, G.: Controller synthesis of fuzzy dynamic systems based on piecewise Lyapunov functions. IEEE Trans. Fuzzy Syst. **11**(5), 605–612 (2003)
21. Feng, G.: A survey on analysis and design of model-based fuzzy control systems. IEEE Trans. Fuzzy Syst. **14**(5), 676–697 (2006)
22. Blanco, Y., Perruquetti, W., Borne, P.: Stability and stabilization of nonlinear systems and Takagi–Sugeno's fuzzy models. Math. Probl. Eng. **7**(3), 221–240 (2001)
23. Tanaka, K., Hori, T., Wang, H.: A multiple Lyapunov function approach to stabilization of fuzzy control systems. IEEE Trans. Fuzzy Syst. **11**(4), 582–589 (2003)
24. Arino, C., Perez, E., Sala, A.F.B.: Polytopic invariant and contractive sets for closed-loop discrete fuzzy systems. J. Franklin Inst. **351**(7), 3559–3576 (2014)
25. Guo, H., Qiu, J., Tian, H.H.G.: Fault detection of discrete-time T–S fuzzy affine systems based on piecewise Lyapunov functions. J. Franklin Inst. **351**(7), 3633–3650 (2014)
26. Chen, Y., Tanaka, M., Tanaka, K., Wang, H.: Stability analysis and region-of-attraction estimation using piecewise polynomial Lyapunov functions: Polynomial fuzzy model approach. IEEE Trans. Fuzzy Syst. **23**(4), 1314–1322 (2015)
27. Hedlund, S., Johansson, M.: PWLTLOOL: A matlab toolbox for analysis of piecewise linear systems (2000)
28. Gonzalez, T., Bernal, M.: Progressively better estimates of the domain of attraction for nonlinear systems via piecewise Takagi–Sugeno models: Stability and stabilization issues. Fuzzy Sets Syst. **297**, 73–95 (2016)
29. Gonzalez, T., Sala, A., Bernal, M., Aguiar, B.: Piecewise-Takagi–Sugeno asymptotically exact estimation of the domain of attraction of nonlinear systems. J. Franklin Inst. **354**(3), 1514–1541 (2017)
30. Jarvis-Wloszek, Z., Feeley, R., Tan, W., Sun, K., Packard, A.: Control applications of sum of squares programming. In: Positive Polynomials in Control. Lecture Notes in Control and Information Science, vol. 312, pp. 3–22. Springer, Berlin (2005)
31. Feng, G.: Stability analysis of discrete-time fuzzy dynamic systems based on piecewise Lyapunov functions. IEEE Trans. Fuzzy Syst. **12**(1), 22–28 (2004)
32. Cao, S., Rees, N.W., Feng, G.: Analysis and design of fuzzy control systems using dynamic fuzzy state space models. IEEE Trans. Fuzzy Syst. **7**, 192–200 (1999)
33. Wu, F., Yang, X., Packard, A., Becker, G.: Induced l/sub 2/-norm control for LPV system with bounded parameter variation rates. In: Proceedings of the 1995 American Control Conference, vol. 3, pp. 2379–2383. IEEE, New York (1995)
34. Gahinet, P., Apkarian, P., Chilali, M.: Affine parameter-dependent Lyapunov functions for real parametric uncertainty. IEEE Trans. Autom. Control **41**(3), 436–442 (1996)
35. Guerra, T.M., Vermeiren, L.: LMI-based relaxed nonquadratic stabilization conditions for nonlinear systems in the Takagi–Sugeno's form. Automatica **40**(5), 823–829 (2004)
36. Lee, D., Park, J., Joo, Y.: A fuzzy Lyapunov function approach to estimating the domain of attraction for continuous-time Takagi–Sugeno fuzzy systems. Inf. Sci. **185**(1), 230–248 (2012)
37. Zhou, S., Feng, G.: Generalised H_2 controller synthesis for uncertain discrete-time fuzzy systems via basis-dependent Lyapunov function. IEE Control Theory and Applications **153**(1), 74–80 (2006)

38. Ding, B.: Poly-quadratic stability of discrete-time nonlinear systems in Takagi–Sugeno's form. Asian Journal of Control **11**(6), 610–619 (2009)
39. Wu, F., Packard, A.: LGG control design for LPV systems. In: American Control Conference, Proceedings of the 1995, vol. 6, pp. 4440–4444. IEEE, New York (1995)
40. Guerra, T.M., Bernal, M. A way to escape from the quadratic framework. In: Proceedings of the IEEE International Conference on Fuzzy Systems, pp. 784–789. IEEE, New York (2009)
41. Mozelli, L.A., Palhares, R.M., Souza, F.O., Mendes, E.M.A.M.: Reducing conservativeness in recent stability conditions of TS fuzzy systems. Automatica **45**(6), 1580–1583 (2009)
42. Guerra, T., Bernal, M.: Strategies to exploit non-quadratic local stability analysis. Int. J. Fuzzy Syst. **13**, 372–379 (2012)
43. Rhee, B., Won, S.: A new fuzzy Lyapunov function approach for a Takagi–Sugeno fuzzy control system design. Fuzzy Sets Syst. **157**(9), 1211–1228 (2006)
44. Kalman, R.E.: Lyapunov functions for the problem of Lur'e in automatic control. Proc. Natl. Acad. Sci. U.S.A. **49**(2), 201 (1963)
45. Gonzalez, T., Sala, A., Bernal, M., Robles, R.: A generalisation of line-integral Lyapunov function for Takagi–Sugeno systems. In: Proceedings of the 20th World Congress of the International Federation of Automatic Control (2017c)
46. González, T., Sala, A., Bernal, M.: A generalised integral polynomial Lyapunov function for nonlinear systems. Fuzzy Sets Syst. **356**, 77–91 (2019)
47. Marquez, R., Guerra, T., Kruszewski, A., Bernal, M.: Improvements on non-quadratic stabilization of Takagi–Sugeno models via line-integral Lyapunov functions. In: Proceedings of the 2013 IFAC International Conference on Intelligent Control and Automation Science, pp. 473–478 (2013)
48. Marquez, R., Guerra, T., Kruszewski, A., Bernal, M.: Non-quadratic stabilization of second order continuous Takagi–Sugeno descriptor systems via line-integral Lyapunov function. In: Proceedings of the 2014 IEEE International Conference on Fuzzy Systems, pp. 2451–2456. IEEE, New York (2014)
49. Kruszewski, A., Wang, R., Guerra, T.M.: Nonquadratic stabilization conditions for a class of uncertain nonlinear discrete time TS fuzzy models: A new approach. IEEE Trans. Autom. Control **53**(2), 606–611 (2008)
50. Wang, L.-X., Mendel, J.: Fuzzy basis functions, universal approximation, and orthogonal least-squares learning. IEEE Trans. Neural Networks **3**(5), 807–814 (1992)
51. Tuan, H., Apkarian, P., Narikiyo, T., Yamamoto, Y.: Parameterized linear matrix inequality techniques in fuzzy control system design. IEEE Trans. Fuzzy Syst. **9**(2), 324–332 (2001)
52. Sala, A., Ariño, C.: Asymptotically necessary and sufficient conditions for stability and performance in fuzzy control: Applications of Polya's theorem. Fuzzy Sets Syst. **158**(24), 2671–2686 (2007)
53. Megretski, A.: Integral quadratic constraints derived from the set-theoretic analysis of difference inclusions. In: Proceedings of the 35th IEEE Conference on Decision and Control, vol. 3, pp. 2389–2394. IEEE, New York (1996)
54. Lee, J.-W., Dullerud, G.E.: Uniform stabilization of discrete-time switched and Markovian jump linear systems. Automatica **42**(2), 205–218 (2006)
55. Guerra, T.M., Kruszewski, A., Bernal, M.: Control law proposition for the stabilization of discrete Takagi–Sugeno models. IEEE Transactions of Fuzzy Systems **17**, 724–731 (2009)
56. Lacerda, M.J., Seiler, P.: Stability of uncertain systems using Lyapunov functions with non-monotonic terms. Automatica **82**, 187–193 (2017)
57. Ariño, C., Sala, A., Pérez, E., Bedate, F., Querol, A.: Asymptotically exact stabilisation for constrained discrete Takagi–Sugeno systems via set-invariance. Fuzzy Sets Syst. **316**, 117–138 (2017)
58. Sala, A.: Stability analysis of LPV systems: Scenario approach. Automatica **104**, 233–237 (2019)
59. Ding, B.: Homogeneous polynomially nonquadratic stabilization of discrete-time Takagi–Sugeno systems via nonparallel distributed compensation law. IEEE Transaction on Fuzzy Systems **18**(5), 994–1000 (2010)

60. Polya, G.: Uber positive Darstellung von Polynomen. Vierterterhartschrift d. Naturforschenden Gessellschaft in Zurich **73**, 141–145 (1928)
61. Lendek, Zs., Guerra, T., Lauber, J.: Controller design for TS models using delayed nonquadratic Lyapunov functions. IEEE Transactions on Cybernetics **45**(3), 453–464 (2015)
62. Xie, X., Yue, D., Zhang, H., Xue, Y.: Control synthesis of discrete-time T–S fuzzy systems via a multi-instant homogenous polynomial approach. IEEE Transactions on Cybernetics **46**(3), 630–640 (2016)
63. Xie, X., Yue, D., Zhang, H., Peng, C.: Control synthesis of discrete-time T–S fuzzy systems: reducing the conservatism whilst alleviating the computational burden. IEEE Trans. Cybern **47**(9), 2480–2491 (2017)
64. Lee, D., Joo, Y., Ra, I.-H.: Local stability and local stabilization of discrete-time T–S fuzzy systems with time-delay. Int. J. Control Autom. Syst. **14**(1), 29–38 (2016)
65. Xie, X.-P., Liu, Z.-W., Zhu, X.-L.: An efficient approach for reducing the conservatism of LMI-based stability conditions for continuous-time T–S fuzzy systems. Fuzzy Sets Syst. **263**, 71–81 (2015)
66. Márquez, R., Guerra, T.M., Bernal, M., Kruszewski, A.: Asymptotically necessary and sufficient conditions for Takagi–Sugeno models using generalized non-quadratic parameter-dependent controller design. Fuzzy Sets Syst. **306**, 48–62 (2017)
67. Sala, A., Ariño, C.: Polynomial fuzzy models for nonlinear control: A Taylor series approach. IEEE Trans. Fuzzy Syst. **17**(6), 1284–1295 (2009)
68. Parrilo, P.A.: Structured semidefinite programs and semialgebraic geometry methods in robustness and optimization. PhD thesis, California Institute of Technology (2000)
69. Vandenberghe, L., Boyd, S.: Semidefinite programming. SIAM Rev. **38**(1), 49–95 (1996)
70. Prajna, S., Parrilo, P., Rantzer, A.: Nonlinear control synthesis by convex optimization. IEEE Trans. Autom. Control **49**(2), 310–314 (2004)
71. Prajna, S., Papachristodoulou, A., Seiler, P., Parrilo, P.A.: SOSTOOLS: Sum of Squares Optimization Toolbox for MATLAB. California Institute Technology, Pasadena, CA (2004)
72. Tanaka, K., Yoshida, H., Ohtake, H., Wang, H.: A sum of squares approach to modeling and control of nonlinear dynamical systems with polynomial fuzzy systems. IEEE Trans. Fuzzy Syst. **17**(4), 911–922 (2009)
73. Bernal, M., Sala, A., Jaadari, A., Guerra, T.-M.: Stability analysis of polynomial fuzzy models via polynomial fuzzy Lyapunov functions. Fuzzy Sets Syst. **185**, 5–14 (2011)
74. Parrilo, P.A., Sturmfels, B.: Minimizing polynomial functions. Algorithmic and quantitative real algebraic geometry, DIMACS Series in Discrete Mathematics and Theoretical Computer Science **60**, 83–99 (2003)
75. Taniguchi, T., Tanaka, K., Wang, H.O.: Fuzzy descriptor systems and nonlinear model following control. IEEE Trans. Fuzzy Syst. **8**(4), 442–452 (2000)
76. Fridman, E.: New Lyapunov-Krasovskii functionals for stability of linear retarded and neutral type systems. Syst. Control Lett. **43**, 309–319 (2001)
77. Fridman, E., Shaked, U.: An improved stabilization method for linear time-delay systems. IEEE Trans. Autom. Control **47**(11), 1931–1937 (2002)
78. de Oliveira, M., Skelton, R.: Stability test of constrained linear systems. In: Perspective in Robust Control, pp. 241–257. Springer, Berlin (2001)
79. Duan, G.R.: Analysis and Design of Descriptor Linear Systems. Springer, New York (2010)
80. Arceo, J.C., Sánchez, M., Estrada-Manzo, V., Bernal, M.: Convex stability analysis of nonlinear singular systems via linear matrix inequalities. IEEE Trans. Autom. Control **64**(4), 1740–1745 (2018)
81. Di Franco, P., Scarciotti, G., Astolfi, A.: A globally stable algorithm for the integration of high-index differential-algebraic systems. IEEE Trans. Autom. Control **65**(5), 2107–2122 (2019)
82. Ishihara, J.Y., Terra, M.H.: On the Lyapunov theorem for singular systems. IEEE Trans. Autom. Control **47**(11), 1926–1930 (2002)
83. Scherer, C.W.: LPV control and full block multipliers. Automatica **37**(3), 361–375 (2001)

84. Sala, A.: Computer control under time-varying sampling period: An LMI gridding approach. Automatica **41**(12), 2077–2082 (2005)
85. Zhou, K., Doyle, J.C., Glover, K., et al.: Robust and Optimal Control, vol. 272. Prentice Hall, Englewood Cliffs, NJ (1996)
86. Skogestad, S., Postlethwaite, I.: Multivariable Feedback Control: Analysis and Design, vol. 2. Wiley, New York (2007)
87. Kottenstette, N., McCourt, M., Xia, M., Gupta, V., Antsaklis, P.: On relationships among passivity, positive realness, and dissipativity in linear systems. Automatica **50**(4), 1003–1016 (2014)
88. Bhattacharyya, S., Keel, L., Chapellat, H.: Robust Control: the Parametric Approach. Prentice Hall, New York (1995)
89. Wu, F., Dong, K.: Gain-scheduling control of LFT systems using parameter-dependent Lyapunov functions. Automatica **42**(1), 39–50 (2006)

Chapter 5
State Feedback, Performance, and Robustness

5.1 Introduction

This chapter investigates state feedback design for models with a convex structure such as TS, polynomial, and descriptor; LFT is left for the next chapter. Similar to stability analysis, Lyapunov's direct method and convexity are combined to derive sufficient stabilisation conditions. Based on the assumption that both the plant states and the membership functions are available, a variety of controller structures is presented along with quadratic and non-quadratic Lyapunov functions. The different stabilisation options are presented as instances of a unified "template", which may help the reader to understand recent results as well as envisage future directions, regardless of their complexity or relaxation scheme. Performance specifications such as input/output constraints, $\mathcal{H}_\infty$ disturbance rejection, and robustness are presented. The chapter closes with a real-time application on the wheelchair swing-up problem.

5.2 Quadratic Stabilisation of TS Models

TS models of the form (3.1) (reproduced below in the usual shorthand notation)

$$\delta \boldsymbol{x}(t) = A_h \boldsymbol{x}(t) + B_h \boldsymbol{u}(t) \tag{5.1}$$

will be considered under the usual definitions, along with a "generic" feedback control law

$$\boldsymbol{u}(t) = F(\cdot)H^{-1}(\cdot)\boldsymbol{x}(t), \tag{5.2}$$

where the pair $F(\cdot)$, $H(\cdot)$ will be defined further on according to the Lyapunov function under consideration. Recall that A_h and B_h are convex sums of constant matrices where the interpolating functions $h_i(z)$ hold the convex sum property on a compact set $\mathcal{C} \subset \mathbb{R}^p$, where $z \in \mathcal{C}$ may come from exact modelling of nonlinear

M. Bernal et al., *Analysis and Synthesis of Nonlinear Control Systems*, Studies in Systems, Decision and Control 408, https://doi.org/10.1007/978-3-030-90773-0_5

systems of the form (3.8), i.e. $\delta \boldsymbol{x}(t) = \boldsymbol{f}(\boldsymbol{x}, \boldsymbol{u}, \boldsymbol{d}, \boldsymbol{\theta}, t)$, in order to "capture" bounded expressions that may depend on time t, states $\boldsymbol{x}$, or exogenous (possibly time-varying) signals $\boldsymbol{d}, \boldsymbol{\theta}$. Assumptions 4.1 and 4.2 hold, which means there exists a compact $\mathcal{X}$ in the state space that guarantees $z \in \mathcal{C}$ with $0 \in \mathcal{X}$.

This section is intended as an easy-to-follow introduction to controller design, which may help to understand more complex generalisations to be presented later on. The classical linear state feedback $\boldsymbol{u}(t) = F\boldsymbol{x}(t)$ (corresponding to $F(\cdot) = F$ and $H(\cdot) = I$ in (5.2)) can be used to stabilise a TS model of the form (5.1), in continuous or in discrete time. A more general solution is that of *parallel distributed compensation* (PDC) [1], which is composed of linear state feedback gains blended together using the same MFs $h_i(\cdot)$, $i \in \{1, 2, \ldots, r\}$ of the model; it corresponds to the choice $F(\cdot) = F_h$ and $H(\cdot) = I$ in (5.2):

$$\boldsymbol{u}(t) = \sum_{i=1}^{r} h_i(z) F_i \boldsymbol{x}(t) = F_h \boldsymbol{x}(t). \tag{5.3}$$

There is an advantage in using the same MFs of the TS model for the control law: they can be combined with the MFs of the model during the closed-loop stability analysis, leading to genuine generalisations as they may relax the too-demanding requisite of a common gain. Notice that having a different set of MFs in the control law will not produce the referred relaxation as it will be equivalent to replacing a common *linear* gain state feedback with a common *nonlinear* state feedback.

Since, similar to stability analysis, the resulting conditions are quite different in the continuous and discrete cases, let us consider first the continuous-time case. The TS model is given by

$$\dot{\boldsymbol{x}}(t) = \sum_{i=1}^{r} h_i(z) \left(A_i \boldsymbol{x}(t) + B_i \boldsymbol{u}(t)\right) = A_h \boldsymbol{x}(t) + B_h \boldsymbol{u}(t). \tag{5.4}$$

Substituting the PDC control law (5.3) in the TS model (5.4) gives the closed-loop system

$$\begin{aligned}
\dot{\boldsymbol{x}}(t) &= \sum_{i=1}^{r} h_i(z) \left(A_i + B_i \sum_{j=1}^{r} h_j(z) F_j \right) \boldsymbol{x}(t) \\
&= \sum_{i=1}^{r} h_i(z) \left(\sum_{j=1}^{r} h_j(z) A_i + B_i \sum_{j=1}^{r} h_j(z) F_j \right) \boldsymbol{x}(t) \\
&= \sum_{i=1}^{r} \sum_{j=1}^{r} h_i(z) h_j(z) \left(A_i + B_i F_j\right) \boldsymbol{x}(t) \\
&= (A_h + B_h F_h) \boldsymbol{x}(t),
\end{aligned} \tag{5.5}$$

where the convex sum property of the MFs has been employed in the algebraic manipulations. Note that the resulting closed-loop TS model is composed of r^2 vertex models.

The nested convex sum in (5.5) is carried on into the stability analysis from which the stabilisation conditions are derived, allowing a variety of convex sum relaxations to be applied in order to find sufficient LMI conditions. These schemes are described in Appendix D. In order to preserve the generality of results by not choosing a particular sum relaxation scheme, matrix inequalities still depending on nested convex sums will be presented for most of the theorems. In examples, the corresponding relaxation scheme will be detailed.

As before, quadratic schemes are those based on a quadratic Lyapunov function candidate:

$$V(\boldsymbol{x}) = \boldsymbol{x}^T P \boldsymbol{x}, \quad P = P^T > 0. \tag{5.6}$$

Again, the theoretical justification of the following results lies in Lyapunov Theorems B.1 and B.3 for continuous-time systems as well as Theorems B.5 and B.7 for discrete-time systems, depending on whether the system is time-invariant or time-variant (which, in turn, depends on the scheduling vector z as discussed in the previous chapter). Details can be found in Appendix B. To avoid repetition, the theorems that follow assume the system is time-invariant as adaptations to the time-variant case are straightforward.

Theorem 5.1 *The origin $\boldsymbol{x} = 0$ of the continuous-time TS model (5.4) under the PDC control law (5.3) is asymptotically stable if there exist matrices $X > 0$ and M_j, $j \in \{1, 2, \ldots, r\}$, such that*

$$A_h X + B_h M_h + X A_h^T + M_h^T B_h^T < 0 \tag{5.7}$$

holds. In this case, the PDC gains are $F_j = M_j X^{-1}$, $j \in \{1, 2, \ldots, r\}$, and any trajectory $\boldsymbol{x}(t)$ starting within the outermost Lyapunov level $V(\boldsymbol{x}) = \boldsymbol{x}^T P \boldsymbol{x} \le k$, $P = X^{-1}$, $k > 0$, in a compact set $\mathcal{X}$ of the state space guaranteeing $z \in \mathcal{C} \subseteq \mathbb{R}^p$ goes asymptotically to $\boldsymbol{x} = 0$.

Proof The time derivative of the quadratic Lyapunov function candidate (5.6) along the trajectories of the system resulting from the substitution of (5.3) in (5.4), i.e. the closed-loop system (5.5), is given by

$$\dot{V} = \boldsymbol{x}^T \left((A_h + B_h F_h)^T P + P (A_h + B_h F_h) \right) \boldsymbol{x}.$$

Therefore, $\dot{V} < 0$ is ensured if

$$A_h^T P + F_h^T B_h^T P + P A_h + P B_h F_h < 0. \tag{5.8}$$

Due to the term $P B_h F_h$ and its transpose, a sum relaxation on (5.8) will not produce LMIs, e.g.

$$A_i^T P + F_j^T B_i^T P + PA_i + PB_iF_j < 0,\ i \in \{1, 2, \ldots, r\}, j \in \{1, 2, \ldots, r\}$$

is not a set of LMIs because F_j and P are decision variables that appear multiplied in single terms.

However, (5.8) can be easily adapted to produce conditions allowing LMI expressions once a sum relaxation is applied, by using the change of variables $X = P^{-1}$, $M_h = F_hX$. By congruence (Property A.1) with X, (5.8) can be proven equivalent to (5.7). This ensures $V(\boldsymbol{x})$ is a valid Lyapunov function, thus proving asymptotic stability of the origin $\boldsymbol{x} = 0$ as conditions of Theorem B.1 are fulfilled.

Negative-definiteness of (5.7) is guaranteed regardless of MFs $h_i(z)$ as long as they hold the convex sum property. Since this property is guaranteed as long as $z \in \mathcal{C}$, we can conclude that trajectories within the outermost Lyapunov level $V(\boldsymbol{x}) \leq k$, $k > 0$, in $\mathcal{X}$ go asymptotically to 0 (see assumption 4.2). □

Remark 5.1 As it was done for stability results in Chap. 4, conditions as those above are shown in their most general form: any sum relaxation scheme can be applied to them in order to obtain LMI conditions. A sum relaxation helps associating similar terms in order to derive LMI conditions in the least conservative way. For instance, when the relaxation (D.10) is used for Theorem 5.1, results in [2] are recovered; with (D.11), results in [3] arise; with (D.12), conditions in [4] are obtained. Let us take a closer look at the so-called Tanaka's relaxation [2] which takes into account that $h_ih_j = h_jh_i$:

$$\begin{aligned} & A_hX + B_hM_h + XA_h^T + M_h^TB_h^T < 0 \\ \iff & \sum_{i=1}^{r}\sum_{j=1}^{r} h_i(z)h_j(z)\left(A_iX + B_iM_j + XA_i^T + M_j^TB_i^T\right) < 0 \\ \Longleftarrow & \begin{cases} A_iX + B_iM_j + A_jX + B_jM_i + (*) < 0, & i \neq j, \\ A_iX + B_iM_i + (*) < 0, & i = j. \end{cases} \end{aligned}$$

The LMIs at the bottom imply that conditions in Theorem 5.1 hold; yet, they guarantee so in a less conservative way than directly dropping the MFs as to ask for $A_iX + B_iM_j + XA_i^T + M_j^TB_i^T < 0, i, j \in \{1, 2, \ldots, r\}$.

Remark 5.2 It is straightforward to rewrite conditions in Theorem 5.1 as to guarantee that the origin $\boldsymbol{x} = 0$ of the closed-loop TS model (5.5) is exponentially stable with a decay rate of $\alpha > 0$. Based on what has been done for stability analysis in Theorem 4.3 of Chap. 4, it becomes apparent that replacing (5.7) with

$$A_hX + B_hM_h + XA_h^T + M_h^TB_h^T + 2\alpha X < 0 \tag{5.9}$$

would ensure it.

The application of Theorem 5.1 is illustrated in the following example.

Example 5.1 Consider a continuous-time TS system (5.4) with $r = 2$ and $z = \boldsymbol{x}$, whose membership functions, holding the convex sum property in the whole state space, are given by

$$h_1(z) = \frac{1}{1 + (x_1 + x_2)^2}, \quad h_2(z) = 1 - h_1(z),$$

where vertex models are defined by matrices

$$A_1 = \begin{bmatrix} -2 & 3 \\ 0 & 2 \end{bmatrix}, \; A_2 = \begin{bmatrix} -2 & 5 \\ 0 & -2 \end{bmatrix}, \; B_1 = \begin{bmatrix} 2 \\ 1 \end{bmatrix}, \; B_2 = \begin{bmatrix} 2 \\ 2 \end{bmatrix}.$$

Note that A_1 has an unstable eigenvalue at 2.

Using Theorem 5.1 with conditions (D.11) gives the solution[1]:

$$P = \begin{bmatrix} 3.32 & -5.77 \\ -5.77 & 17.5 \end{bmatrix}, \quad \begin{matrix} F_1 = \begin{bmatrix} 3.03 & -7.93 \end{bmatrix}, \\ F_2 = \begin{bmatrix} -0.26 & 1.03 \end{bmatrix}. \end{matrix}$$

Figure 5.1a presents a result[2] using the gains above, with initial conditions $\boldsymbol{x}(0) = \begin{bmatrix} 1 & 1 \end{bmatrix}^T$. Note that without additional constraints, the performance is rather poor.

In the second round, a decay rate of $\alpha = 0.41$ is added to the constraints using LMIs derived from (5.9). The result is now

$$P = \begin{bmatrix} 2.50 & -4.36 \\ -4.36 & 15.83 \end{bmatrix}, \quad \begin{matrix} F_1 = \begin{bmatrix} 2.20 & -7.08 \end{bmatrix}, \\ F_2 = \begin{bmatrix} -0.20 & 0.64 \end{bmatrix}, \end{matrix}$$

with the corresponding simulation being presented in Fig. 5.1b from the same initial conditions as before. As expected, a better performance is obtained. □

Example 5.2 Consider the continuous-time nonlinear system

$$\begin{aligned} \dot{x}_1(t) &= \frac{-x_2}{1 + x_1^2} - 0.5x_1, \\ \dot{x}_2(t) &= -3x_1^2 - 2x_2^4 - 2x_2 + u(t), \end{aligned}$$

whose origin $\boldsymbol{x} = 0$ is intended to be stabilised for initial conditions in a compact of interest $\mathcal{X} = \{\boldsymbol{x} : |x_i| \leq 5, i \in \{1, 2\}\}$. Note that for $u(t) = 0$, the origin is unstable.

The system equations can be rewritten as

$$\dot{x}(t) = \begin{bmatrix} -0.5 & -1/\left(x_1^2 + 1\right) \\ -3x_1 & -2x_2^3 - 2 \end{bmatrix} x(t) + \begin{bmatrix} 0 \\ 1 \end{bmatrix} u(t),$$

[1] Throughout the chapter, all the numerical values are rounded to two decimal places.

[2] For numerical integration, the *ode45* MATLAB®function has been used.

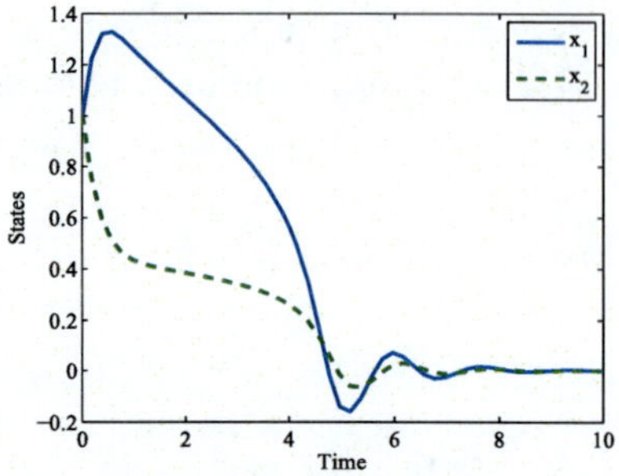

(a) Stabilisation without constraints.

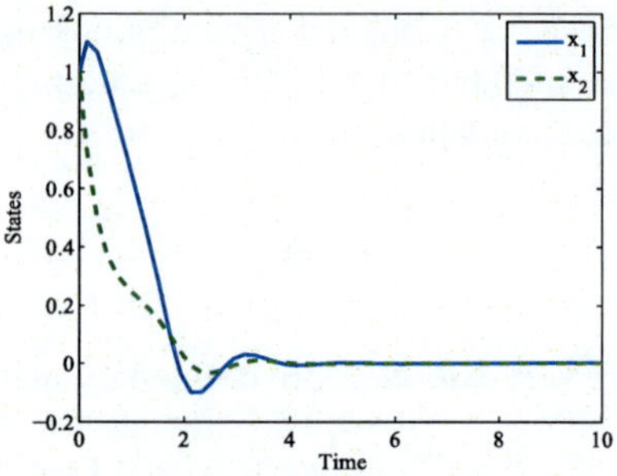

(b) Stabilisation with decay rate $\alpha = 0.41$.

Fig. 5.1 Simulation results for Example 5.1

from which $z_1(x) = 1/(x_1^2 + 1)$, $z_2(x) = x_1$, and $z_3(x) = x_2^3$ can be defined, based on which an equivalent TS model of the form

$$\dot{\boldsymbol{x}}(t) = \sum_{i=1}^{8} h_i(\boldsymbol{z}) \left(A_i \boldsymbol{x}(t) + Bu(t)\right)$$

is obtained with the sector nonlinearity approach, where convexity of MFs $h_i(\boldsymbol{z})$ is ensured in $\mathcal{C} = \{\boldsymbol{z} : z_1 \in [0.0385, 1],\ z_2 \in [-5, 5],\ z_3 \in [-125, 125]\}$, which is compact in $\boldsymbol{z}$ induced by the original $\mathcal{X}$; some matrices of the vertex models are

$$A_1 = \begin{bmatrix} -0.5 & -0.0385 \\ 15 & 248 \end{bmatrix}, \quad A_4 = \begin{bmatrix} -0.5 & -0.0385 \\ -15 & -252 \end{bmatrix}, \quad B = \begin{bmatrix} 0 \\ 1 \end{bmatrix}.$$

Notice that B is constant, which means that a PDC control law will not produce nested convex sums.

Since no double convex sums appear, Theorem 5.1 can be used to synthesise a controller without any sum relaxation; in this way, a feasible solution was found, from which some of the gains are

$$F_1 = \begin{bmatrix} -14.9615 & -248.5 \end{bmatrix}, \quad F_4 = \begin{bmatrix} 15.0385 & 251.5 \end{bmatrix}, \quad F_8 = \begin{bmatrix} 16 & 251.5 \end{bmatrix},$$

and the quadratic Lyapunov matrix is $P = 1 \times 10^{-6}$block-diag$(0.8176, 0.8176)$.

In Fig. 5.2(a), the phase portrait from initial conditions $\boldsymbol{x}(0) = \begin{bmatrix} 3.5 & 3.5 \end{bmatrix}^T, \boldsymbol{x}(0) = \begin{bmatrix} 0 & -4.5 \end{bmatrix}^T$, and $\boldsymbol{x}(0) = \begin{bmatrix} -3 & 4 \end{bmatrix}^T$ is shown, along with the modelling region $\mathcal{X}$ and the outermost Lyapunov level within it.

Consider that the input distribution matrix in the original system is changed to $\begin{bmatrix} 0 & 2 - \sin(x_1) \end{bmatrix}^T$. If the non-constant expressions in $A(\boldsymbol{x})$ and $B(\boldsymbol{x})$ for $\dot{\boldsymbol{x}}(t) = A(\boldsymbol{x})\boldsymbol{x}(t) + B(\boldsymbol{x})u(t)$ are now chosen as $z_1(x) = \sin(x_1)$, $z_2(x) = 1/(x_1^2 + 1)$, $z_3(x) = x_1$, and $z_4(x) = x_2^3$, we have 16 vertex models; some of their matrices are

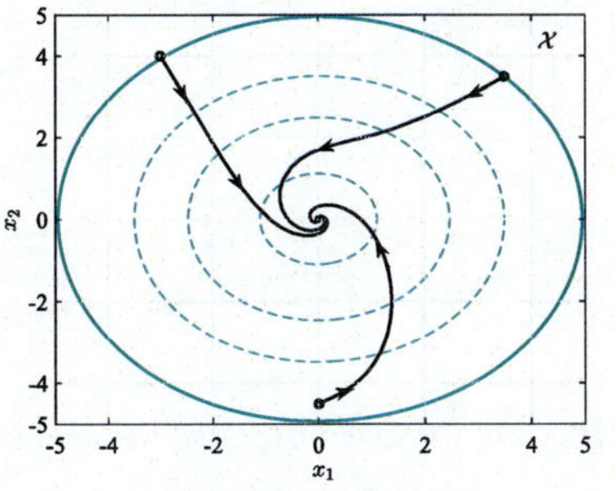

(a) Lyapunov levels.

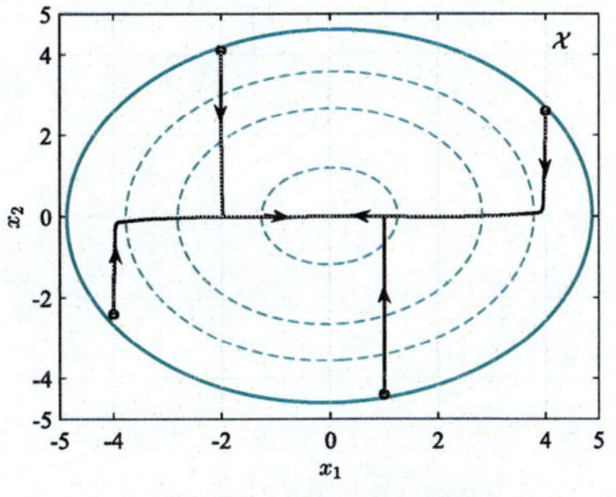

(b) Lyapunov levels.

Fig. 5.2 Simulation results for Example 5.2

$$A_1 = \begin{bmatrix} -0.5 & -0.0385 \\ 15 & 248 \end{bmatrix}, \; A_4 = \begin{bmatrix} -0.5 & -0.0385 \\ -15 & -252 \end{bmatrix}, \; A_{16} = \begin{bmatrix} -0.5 & -1 \\ -15 & -252 \end{bmatrix},$$
$$B_1 = B_2 = \cdots = B_8 = \begin{bmatrix} 0 \\ 3 \end{bmatrix}, \quad B_9 = B_{10} = \cdots = B_{16} = \begin{bmatrix} 0 \\ 1 \end{bmatrix}.$$

Again, Theorem 5.1 yields a feasible solution with sum relaxation in Lemma D.1; some of the results are

$$P = \begin{bmatrix} 0.0063 & -0.0001 \\ -0.0001 & 0.0071 \end{bmatrix}, \quad F_1 = \begin{bmatrix} -6.1728 & -125.4098 \end{bmatrix},$$
$$F_4 = \begin{bmatrix} 5.7342 & 62.8540 \end{bmatrix}, \; F_8 = \begin{bmatrix} 6.0487 & 62.7978 \end{bmatrix}, \; F_{16} = \begin{bmatrix} 6.4206 & 15.2326 \end{bmatrix}.$$

In Fig. 5.2(b), the phase portrait from initial conditions $x(0) = \begin{bmatrix} 4 & 2.6 \end{bmatrix}^T$, $x(0) = \begin{bmatrix} 1 & -4.4 \end{bmatrix}^T$, $x(0) = \begin{bmatrix} -4 & -2.4 \end{bmatrix}^T$, and $x(0) = \begin{bmatrix} -2 & 4.1 \end{bmatrix}^T$ is shown. The modelling region $\mathcal{X}$ and the outermost Lyapunov level within it are also shown.

Let us now consider stabilisation of discrete-time TS models of the form:

$$\boldsymbol{x}(t+1) = \sum_{i=1}^{r} h_i(z) \left(A_i \boldsymbol{x}(t) + B_i \boldsymbol{u}(t) \right) = A_h \boldsymbol{x}(t) + B_h \boldsymbol{u}(t), \tag{5.10}$$

under the usual assumptions for time $t \in \mathbb{Z}$, state $\boldsymbol{x}$, input $\boldsymbol{u}$, and convex nature of matrix functions A_h and B_h. Recall that stabilisation conditions are given without any particular sum relaxation scheme, which is required to derive sufficient LMI conditions for them to hold (see Appendix D).

Using the PDC controller (5.3), i.e. $\boldsymbol{u}(t) = F_h \boldsymbol{x}(t)$, the following generalisation of the classical result in [5] can be established:

Theorem 5.2 *The origin $\boldsymbol{x} = 0$ of the discrete-time TS model (5.10) under the PDC control law (5.3) is asymptotically stable if there exist matrices $X > 0$ and M_j, $j \in \{1, 2, \ldots, r\}$, such that*

$$\begin{bmatrix} X & XA_h^T + M_h^T B_h^T \\ A_h X + B_h M_h & X \end{bmatrix} > 0 \tag{5.11}$$

holds. In this case, the PDC gains are $F_j = M_j X^{-1}$, $j \in \{1, 2, \ldots, r\}$, *and any trajectory* $\boldsymbol{x}(t)$ *starting within the outermost Lyapunov level* $V(\boldsymbol{x}) = \boldsymbol{x}^T P \boldsymbol{x} \leq k$, $P = X^{-1}$, $k > 0$, *in a compact set* $\mathcal{X}$ *of the state space guaranteeing* $z \in \mathcal{C} \subseteq \mathbb{R}^p$, *goes asymptotically to* $\boldsymbol{x} = 0$.

Proof Clearly, $V(\boldsymbol{x}) = \boldsymbol{x}^T P \boldsymbol{x}$, $P = P^T > 0$ is a Lyapunov function candidate for the closed-loop TS model

$$\boldsymbol{x}(t+1) = (A_h + B_h F_h)\, \boldsymbol{x}(t),$$

which, if tested with stability conditions of Theorem 4.4 before any sum relaxation yields

$$\begin{bmatrix} P & (A_h^T + F_h^T B_h^T)P \\ P(A_h + B_h F_h) & P \end{bmatrix} > 0$$

as a sufficient condition for the forward difference $\Delta V = V(t+1) - V(t)$ to be negative-definite. But applying congruence (Property A.1) with block-diag(X, X) in the inequality above as well as performing the change of variables $P = X^{-1}$, $M_h = F_h X$ yields conditions (5.11). As before, negative-definiteness of (5.11) is ensured regardless of the MFs $h_i(z)$ if they hold the convex sum property, which they do for $z \in \mathcal{C}$; therefore, trajectories within the outermost Lyapunov level $V(\boldsymbol{x}) \leq k$,$k > 0$, in $\mathcal{X}$ go asymptotically to $\boldsymbol{x} = 0$ (see assumption 4.2). □

Example 5.3 Consider the discrete-time system

$$\begin{bmatrix} x_1(t+1) \\ x_2(t+1) \end{bmatrix} = \begin{bmatrix} 0.5 - 0.01e^{x_1(t)} & 1 \\ -0.02\theta(t) & -1.5 \end{bmatrix} \begin{bmatrix} x_1(t) \\ x_2(t) \end{bmatrix} + \begin{bmatrix} 0 \\ 5\sin(t) + 30 \end{bmatrix} u(t),$$

where $\boldsymbol{x} = [x_1\ x_2]^T$ is the state, $|x_1| \leq 5$, and $\theta(t) \in [0, 25]$ is an unknown bounded signal.

Convex modelling can be done by defining $z_1(x) = e^{x_1(t)} \in [0.0067, 148.4132]$, $z_2(x) = \theta \in [0, 25]$, and $z_3(x) = 5\sin(t) \in [-5, 5]$, based on which, weights for the TS model $\dot{\boldsymbol{x}}(t) = A_h \boldsymbol{x}(t) + B_h u(t)$ can be constructed (see Sect. 3.2.1); some of its vertex matrices are

$$A_1 = \begin{bmatrix} 0.4999 & 1 \\ 0 & -1.5 \end{bmatrix}, A_4 = \begin{bmatrix} 0.4999 & 1 \\ -0.5 & -1.5 \end{bmatrix}, A_8 = \begin{bmatrix} -0.9841 & 1 \\ -0.5 & -1.5 \end{bmatrix},$$

$$B_1 = \begin{bmatrix} 0 \\ 25 \end{bmatrix}, \quad B_4 = \begin{bmatrix} 0 \\ 35 \end{bmatrix}, \quad B_8 = \begin{bmatrix} 0 \\ 35 \end{bmatrix}.$$

Using Theorem 5.2 with sum relaxation given by Lemma D.1, a solution is found; the Lyapunov matrix and some controller gains are

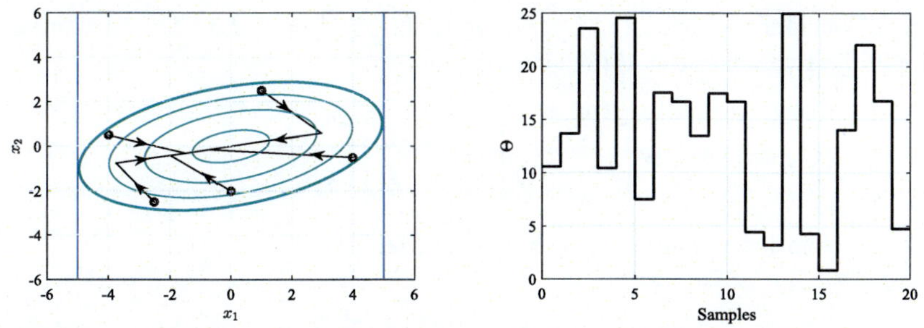

(a) Outermost Lyapunov level within the modelling region.

(b) Signal $\theta(t)$.

Fig. 5.3 Simulation results in Example 5.3

$$P = \begin{bmatrix} 0.0723 & -0.0437 \\ -0.0437 & 0.2126 \end{bmatrix}, \quad F_1 = \begin{bmatrix} 0.0046 \; 0.0666 \end{bmatrix},$$
$$F_4 = \begin{bmatrix} 0.0180 \; 0.0478 \end{bmatrix}, \quad F_8 = \begin{bmatrix} 0.0091 \; 0.0476 \end{bmatrix}.$$

In Fig. 5.3(a), four trajectories are shown from initial conditions $\boldsymbol{x}(0) = \begin{bmatrix} 0 \; -2 \end{bmatrix}^T$, $\boldsymbol{x}(0) = \begin{bmatrix} -2.5 \; -2.5 \end{bmatrix}^T$, $\boldsymbol{x}(0) = \begin{bmatrix} -4 \; 0.5 \end{bmatrix}^T$, $\boldsymbol{x}(0) = \begin{bmatrix} 1 \; 2.5 \end{bmatrix}^T$, and $\boldsymbol{x}(0) = \begin{bmatrix} 4 \; 2 \end{bmatrix}^T$, along with the outermost Lyapunov level within the modelling region $\mathcal{X} = \{\boldsymbol{x} : |x_1| \leq 5\}$. Figure 5.3(b) shows signal $\theta(t)$.

Let us now consider a generalisation of the PDC controller (5.3) that introduces more degrees of freedom:

$$\boldsymbol{u}(t) = F_h H^{-1} \boldsymbol{x}(t), \tag{5.12}$$

which corresponds to the choice $F(\cdot) = F_h$ and $H(\cdot) = H$ in (5.2).

The closed-loop system can be written as

$$\boldsymbol{x}(t+1) = \left(A_h + B_h F_h H^{-1}\right) \boldsymbol{x}(t). \tag{5.13}$$

Based on Finsler's Lemma, the following result can be established:

Theorem 5.3 *The origin $\boldsymbol{x} = 0$ of the discrete-time TS model (5.10) under the control law (5.12) is asymptotically stable if there exist matrices $X = X^T > 0$, H, and F_j, $j \in \{1, 2, \ldots, r\}$, such that*

$$\begin{bmatrix} -H - H^T + X & H^T A_h^T + F_h^T B_h^T \\ A_h H + B_h F_h & -X \end{bmatrix} < 0 \tag{5.14}$$

holds. In this case, any trajectory $\boldsymbol{x}(t)$ starting within the outermost Lyapunov level $V(\boldsymbol{x}) = \boldsymbol{x}^T P \boldsymbol{x} \leq k$, $P = X^{-1}$, $k > 0$, in a compact set $\mathcal{X}$ of the state space guaranteeing $z \in \mathcal{C} \subseteq \mathbb{R}^p$ goes asymptotically to $\boldsymbol{x} = 0$.

Proof Consider the quadratic Lyapunov function $V = \boldsymbol{x}^T P \boldsymbol{x}$. The forward difference of the Lyapunov function can be written as

$$\Delta V = \begin{bmatrix} \boldsymbol{x}(t) \\ \boldsymbol{x}(t+1) \end{bmatrix}^T \begin{bmatrix} -P & 0 \\ 0 & P \end{bmatrix} \begin{bmatrix} \boldsymbol{x}(t) \\ \boldsymbol{x}(t+1) \end{bmatrix}.$$

The closed-loop system (5.13) can be written as

$$\begin{bmatrix} A_h + B_h F_h H^{-1} & -I \end{bmatrix} \begin{bmatrix} \boldsymbol{x}(t) \\ \boldsymbol{x}(t+1) \end{bmatrix} = 0.$$

Using Finsler's Lemma A.3, we have that $\Delta V < 0$ if there exists M so that

$$\begin{bmatrix} -P & 0 \\ 0 & P \end{bmatrix} + M \begin{bmatrix} A_h + B_h F_h H^{-1} & -I \end{bmatrix} + \begin{bmatrix} A_h + B_h F_h H^{-1} & -I \end{bmatrix}^T M^T < 0.$$

Choosing $M = \begin{bmatrix} 0 & P \end{bmatrix}^T$, a sufficient condition for $\Delta V < 0$ is

$$\begin{bmatrix} -P & (*) \\ PA_h + PB_h F_h H^{-1} & -P \end{bmatrix} < 0.$$

Applying the congruence Property A.1 with block-diag(H^T, X), $X = P^{-1}$, and Property A.4 leads to conditions (5.14), which means $V(\boldsymbol{x})$ is a valid Lyapunov function that allows invoking Theorem B.5 to establish asymptotic stability of $\boldsymbol{x} = 0$. Again, convexity of the MFs $h_i(z)$ for $z \in \mathcal{C}$ makes negative-definiteness of (5.14) independent of the particular choice of $h_i(z)$. Thus, any trajectory in the outermost Lyapunov level $V(\boldsymbol{x}) \leq k$, $k > 0$, in $\mathcal{X}$ (see assumption 4.2) goes asymptotically to $\boldsymbol{x} = 0$. □

Of course, any relaxation scheme can be applied to the previous result to derive sufficient LMI conditions, as double convex sums appear (see Remark 5.1).

Example 5.4 Consider a discrete-time TS system (5.10) with $r = 2$ vertex models described by matrices

$$A_1 = \begin{bmatrix} -2.75 & 0.75 \\ -0.05 & -0.1 \end{bmatrix}, \; A_2 = \begin{bmatrix} 0.5 & 1.5 \\ -0.6 & 0.3 \end{bmatrix}, \; B_1 = \begin{bmatrix} 1 \\ 1 \end{bmatrix}, \; B_2 = \begin{bmatrix} 2 \\ 1 \end{bmatrix}.$$

Note that A_1 has an eigenvalue in -2.73. Same as before, consider $z = \boldsymbol{x}$ with MFs

$$h_1(z) = \frac{1}{1 + (x_1 + x_2)^2}, \; h_2(z) = 1 - h_1(z),$$

which hold the convex sum property in the whole state space.

Theorem 5.3 gives the solution

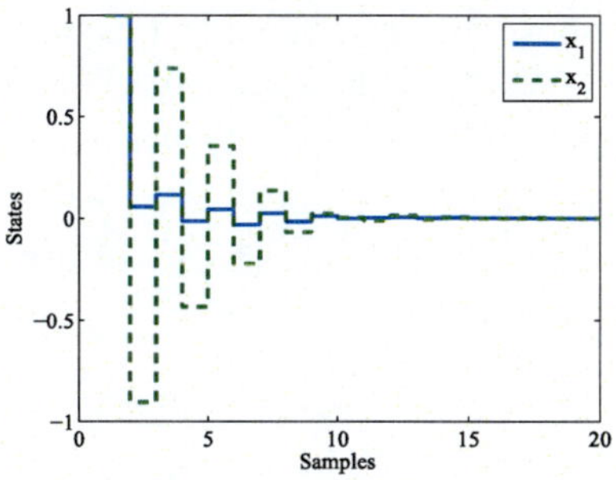

(a) Trajectories of the closed-loop system.

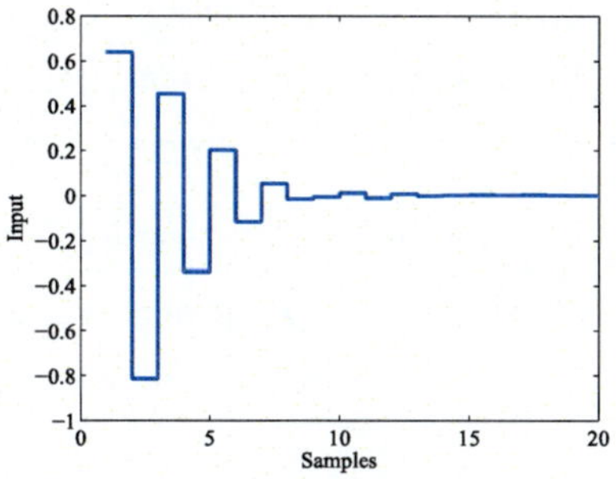

(b) Control input.

Fig. 5.4 Simulation results for Example 5.4

$$X = \begin{bmatrix} 0.045 & 0.07 \\ 0.07 & 0.63 \end{bmatrix}, \qquad H = \begin{bmatrix} 0.065 & 0.09 \\ 0.03 & 0.47 \end{bmatrix},$$
$$F_1 = \begin{bmatrix} 0.14 & -0.18 \end{bmatrix}, \qquad F_2 = \begin{bmatrix} -0.055 & -0.40 \end{bmatrix}.$$

Figure 5.4a presents simulations with the gains above and initial conditions $\boldsymbol{x}(0) = \begin{bmatrix} 1 & 1 \end{bmatrix}^T$, while Fig. 5.4b presents the corresponding control input. □

5.3 Parameter-Dependent Lyapunov Functions

Similar to stability analysis, stabilisation conditions presented in the previous sections are only sufficient, i.e. if they are found infeasible, they do not imply a controller does not exist. Therefore, as in Chap. 4, conservativeness is tackled by using more general classes of Lyapunov functions along with different control schemes. A first option is the use of piecewise Lyapunov functions (PWLFs) [6–9], with the state-space partition based on the activation of the linear models, and the Lyapunov function changing from one region to another. Unfortunately, when applied to control problems, PWLFs lead to bilinear matrix inequalities (BMIs), which require non-convex optimisation problems to be solved [10]; the work in [11] presents a piecewise control synthesis procedure keeping the LMI structure, at the cost of conservatism in some steps; practical applications of piecewise control for affine TS models appear in [12], while an interesting application for non-holonomic mobile robots is reported in [13].

The second alternative is the use of parameter-dependent Lyapunov functions (PDLF) for which a "generic model" will be considered:

$$V(\boldsymbol{x}) = \boldsymbol{x}^T \boldsymbol{P}^{-1}(\cdot)\boldsymbol{x}. \tag{5.15}$$

Remark 5.3 As it will become clearer in the next chapter, Lyapunov functions for stability and observation do not use inverse matrices while those for controller design do; these differences are directly linked to the observer matrix function (to be discussed) and the control matrix function $H(\cdot)$, respectively. This point is very relevant to understanding the "template-like" presentation of the following results.

Of course, if $\boldsymbol{P}(\cdot) = P$ the quadratic case is recovered.

5.3.1 The Continuous-Time Case

The PDLF in [14], where it was referred to as *fuzzy Lyapunov function*, corresponds to the following choice in (5.15):

$$\boldsymbol{P}(\cdot) = \sum_{i=1}^{r} h_i(z) P_i = P_h > 0.$$

The corresponding control law corresponds to $H(\cdot) = P_h$ in (5.2).

As with stability, the designer is faced with the problem of handling the time derivatives of the MFs, but this time the derivatives may include the control signal, thus creating an algebraic loop. Before briefly reviewing the different solutions to this problem, an important result on *local non-quadratic stabilisability* which generalises that in Theorem 4.8 is given.

Theorem 5.4 *[15] The origin of the TS model* (5.1) *under the control law* (5.2) *with* $F(\cdot) = F_h$ *and* $H(\cdot) = P_h$ *is asymptotically stable if there exist matrices* $P_i = P_i^T > 0$, F_i, $i \in \{1, 2, \ldots, r\}$ *such that*

$$P_h A_h + B_h F_h + A_h^T P_h + F_h^T B_h^T < 0. \tag{5.16}$$

This is a "frozen parameter" result, arising from the same ideas as Theorem 4.8, i.e. as the origin is an equilibrium point, when the state is close to it, and derivatives of $\boldsymbol{x}$ and henceforth derivatives of z tend to zero.

Example 5.5 Consider a continuous-time TS system (5.4) with $z = \boldsymbol{x}$ and $r = 2$ vertex models described by

$$A_1 = \begin{bmatrix} -1.30 & 0.85 \\ -1.5 & -1.6 \end{bmatrix}, \qquad A_2 = \begin{bmatrix} -0.05 & 0.3 \\ 2 & -1.4 \end{bmatrix},$$
$$B_1 = \begin{bmatrix} 0 & 1 \end{bmatrix}^T, \qquad B_2 = \begin{bmatrix} 0 & -1 \end{bmatrix}^T.$$

One of the eigenvalues of the second local model A_2 is positive, therefore, the uncontrolled system is not stable. It is assumed the MFs $h_i(z)$ hold the convex sum property in the whole state space.

Using Lemma D.1 to derive LMI conditions guaranteeing inequalities (5.7) in Theorem 5.1 leads to infeasibility: this is the quadratic approach. However, local non-quadratic asymptotic stabilisation can be achieved, by using Lemma D.1 to derive LMI conditions guaranteeing inequality (5.16) in Theorem 5.4, i.e. solving the LMIs

$$\begin{aligned} &P_i > 0, \\ &2(P_iA_i + B_iF_i) + (P_iA_j + B_jF_i) + (P_jA_i + B_iF_j) + (*) < 0, \end{aligned}$$

$i,\ j = 1,\ 2$, from which

$$\begin{aligned} P_1 &= \begin{bmatrix} 0.63 & 0.48 \\ 0.48 & 0.50 \end{bmatrix}, & P_2 &= \begin{bmatrix} 2.03 & 0.21 \\ 0.21 & 0.03 \end{bmatrix}, \\ F_1 &= \begin{bmatrix} -1.46 & -1.27 \end{bmatrix}, & F_2 &= \begin{bmatrix} -4.01 & -0.63 \end{bmatrix} \end{aligned}$$

is obtained. □

In a certain way, Theorem 5.4 is a result of existence; it gives no clue about how large the stabilisability region is; such knowledge requires proper handling of the time derivatives of the MFs. As with stability, the first proposals on this subject that assume bounds $\phi_i \geq 0$ are known such that $|\dot{h}_i| \leq \phi_i$; the LMI conditions above are refined as to take these bounds into account. For TS systems arising from linearisation or fuzzy modelling, i.e. systems where the MFs are assumed to hold the convex sum property *everywhere*, regardless of the state, bounding the time derivatives of the MFs is no different from customary bounds in the LPV literature [16]. Nevertheless, for TS models resulting from the algebraic rewriting of a nonlinear system within a compact set, such bounds proved very conservative and unrealistic for stability analysis, a problem aggravated by the presence of the control law [17].

In the spirit of the proposal in [18], where time derivatives of MFs are split via partial derivation in order to exploit the underlying structure for stability analysis, an extension for controller design was developed in [19]. Without getting into details, this work handles the algebraic loop due to the presence of the control signal by imposing LMI bounds over the input; these bounds can simultaneously be guaranteed and introduced into the stabilisation conditions within an LMI framework. Moreover, the control law can be upgraded to a more general class corresponding to the choice $F(\cdot) = F_h + \delta(z)$, $H(\cdot) = P_h$ in (5.2), i.e.

$$\boldsymbol{u}(t) = (F_h + \delta(z))\, P_h^{-1}\boldsymbol{x}(t),$$

where

$$\delta(z) = \sum_{i=1}^{p} \frac{\partial w_0^k}{\partial z_k} \left(\frac{\partial z_k}{\partial \boldsymbol{x}}^T A_j \boldsymbol{x} \right) G_h^k,$$

G_h^k being additional design gains corresponding to the additional structure due to partial derivation. Guaranteeing $\|\boldsymbol{u}(t)\| \leq \mu$ for a given $\mu > 0$ is a standard LMI task when $u(t)$ has the form (5.2) with $F(\cdot)$ and $H(\cdot)$ having appropriate convex structures, as will be shown in Sect. 5.6.1; thus, the controller can be synthesised without any a priori knowledge on the time solutions of the system. A wide variety of matrix manipulations have helped improving these results; see, for instance, [15].

The work of [20] resumed the assumption of $\dot{h}_i \in [-\phi, \phi]$ which, while combined with a multi-index Lyapunov function of the sort found in [21] as well as a multi-index version of the control law (5.2), produced progressively less conservative results that expanded the domain of attraction of the closed-loop system while preserving the LMI quality of solutions.

The multi-index notation in Sect. 3.2.1 has been employed with slight variations in non-quadratic approaches so as to get richer structures for Lyapunov function candidates and controller proposals. Depending on whether the model arises from an ordinary TS model with MFs h_i, $i \in \{1, 2, \ldots, r\}$, or a tensor-product model where each nonlinearity is rewritten via weighting functions $w^i_{i_j}$, $i_j \in \{0, 1\}$, the indices in the multi-index $\mathbf{i} = (i_1, i_2, \ldots, i_q)$ may take integer values between 1 and r, or between 0 and 1. The generalised PDLF and its corresponding control law in [20] have the following forms:

$$P(\cdot) = \sum_{\mathbf{i}} \mathbf{h_i}(z) P_{\mathbf{i}} = \sum_{i_1=1}^{r} \sum_{i_2=1}^{r} \cdots \sum_{i_q=1}^{r} h_{i_1}(z) h_{i_2}(z) \cdots h_{i_q}(z) P_{(i_1, i_2, \ldots, i_q)},$$

$$F(\cdot) = \sum_{\mathbf{i}} \mathbf{h_i}(z) F_{\mathbf{i}}, \quad H(\cdot) = \sum_{\mathbf{i}} \mathbf{h_i}(z) P_{\mathbf{i}},$$

since they employ MFs h_i in the TS model. Similar arrangements can be done when a tensor-product model is employed instead.

First suggested in [22], where the idea of feeding back the time derivatives of the MFs via a robust differentiator of the sort of [23] has been proposed, the work in [24] employed the multi-index structure of a tensor-product model

$$\dot{x}(t) = A_{w} x(t) + B_{w} u(t),$$

with

$$A_{w} = \sum_{\mathbf{i} \in \mathbb{B}^p} \mathbf{w_i} A_{\mathbf{i}} = \sum_{i_1=0}^{1} \sum_{i_2=0}^{1} \cdots \sum_{i_p=0}^{1} w^1_{i_1} w^2_{i_2} \cdots w^p_{i_p} A_{(i_1, i_2, \ldots, i_p)},$$

$$B_{w} = \sum_{\mathbf{i} \in \mathbb{B}^p} \mathbf{w_i} B_{\mathbf{i}} = \sum_{i_1=0}^{1} \sum_{i_2=0}^{1} \cdots \sum_{i_p=0}^{1} w^1_{i_1} w^2_{i_2} \cdots w^p_{i_p} B_{(i_1, i_2, \ldots, i_p)},$$

for the following proposals in the Lyapunov function and the control law (arguments omitted):

$$P(\cdot) = P_{\mathbf{w}} = \left(\sum_{i_1^1=0}^{1} w_{i_1^1}^1 \cdots \sum_{i_1^{c_1}=0}^{1} w_{i_1^{c_1}}^1\right) \cdots \left(\sum_{i_p^1=0}^{1} w_{i_p^1}^p \cdots \sum_{i_p^{c_p}=0}^{1} w_{i_p^{c_p}}^p\right) P_{b_1 b_2 \cdots b_p},$$
$$F(\cdot) = F_{\mathbf{w}} + \dot{G}_{\mathbf{w}}, \quad H(\cdot) = P_{\mathbf{w}},$$

for $b_i, c_i \in \mathbb{N}$, where the already-discussed algebraic loop due to the control law can be solved to cancel out the effect of the time derivatives of the weighting functions in $P_{\mathbf{w}}$ by means of the time derivatives fed back through the term $\dot{G}_{\mathbf{w}}$. Despite the progress in terms of feasibility domains and larger stabilisability regions, this proposal has a cumbersome notation and has little practical interest.

Finally, stabilisation via line-integral Lyapunov functions of the sort first appeared in [25] have remained, unfortunately, limited to special cases [26].

5.3.2 The Discrete-Time Case

For the discrete-time case, the previous choices are also applicable, but there is a huge variety of Lyapunov functions that fit the generic form (5.15), for instance:

1. The proposal in [27] uses
$$P^{-1}(\cdot) = H_h^{-T} P_h H_h^{-1} > 0, \tag{5.17}$$
along with a control law (5.2) where $H(\cdot) = H_h$.
2. The structure first described in [28] chooses
$$P(\cdot) = \sum_{i=1}^{r} h_i\,(z(t-1))\, P_i = P_{h(t-1)} > 0, \tag{5.18}$$
with $H(\cdot)$ in (5.2) not directly related to $P_{h(t-1)}$; this point will be discussed further on.
3. The proposal in [29] is
$$P(\cdot) = P_{k,h}^{-1} > 0,$$
with $P_{k,h}^{-1}$ being homogeneous polynomials of degree k with respect to matrix P and the membership functions h_i; this proposal generalises what (5.15) initiated for a double sum P_{hh}^{-1}.
4. Further enhancements in [30, 31] are based on proposals such as
$$P(\cdot) = x^T \mathbb{H}_{G_0^H}^{-T} \mathbb{P}_{G_0^P} \mathbb{H}_{G_0^H}^{-1} x,$$
with G_0^H, G_0^P being multisets of delays; these candidates generalise (5.17).

5. Extensions of all forms of Lyapunov functions can also be done via the so-called α-sample approach described in Chap. 4 [32].

In order to deliver a clear way of doing from these numerous possibilities, in the next sections we will focus on the developments based on Finsler's Lemma A.3 [33]. Some proofs come at hand easily and therefore can be mimicked for more complex Lyapunov functions without difficulty. Recall the notation for convex nested sums is extensively used and not to be confused with constant matrices, i.e. any expression containing subscripts of the sort $h(t)$ (e.g. A_h or $P_{h(t+1)}$) is a *convex sum* weighted by MFs $h_i(z)$, with $z \in \mathcal{C} \subseteq \mathbb{R}^p$ depending on time, states, and/or exogenous variables.

An Approach with No Nested Convex Sums in the Lyapunov Function

Consider first the generic Lyapunov function candidate (5.15) with $\boldsymbol{P}^{-1}(\cdot) = P_{h(t)}^{-1}$ and the generic control law (5.2) with $F(\cdot) = F_{h(t)}$ and $H(\cdot) = P_{h(t)}$, i.e. $\boldsymbol{u}(t) = F_{h(t)} P_{h(t)}^{-1} \boldsymbol{x}(t)$.

Theorem 5.5 *The origin $\boldsymbol{x} = 0$ of the discrete-time TS model (5.10) $\boldsymbol{x}(t+1) = A_h \boldsymbol{x}(t) + B_h \boldsymbol{u}(t)$ under the control law $\boldsymbol{u}(t) = F_{h(t)} P_{h(t)}^{-1} \boldsymbol{x}(t)$ is asymptotically stable if there exist matrices $P_i = P_i^T > 0$, F_i, $i \in \{1, 2, \ldots, r\}$, such that*

$$\begin{bmatrix} -P_{h(t)} & P_{h(t)} A_{h(t)}^T + F_{h(t)}^T B_{h(t)}^T \\ A_{h(t)} P_{h(t)} + B_{h(t)} F_{h(t)} & -P_{h(t+1)} \end{bmatrix} < 0 \tag{5.19}$$

holds. In this case, any trajectory $\boldsymbol{x}(t)$ starting within the outermost Lyapunov level $V(\boldsymbol{x}) = \boldsymbol{x}^T P_{h(t)}^{-1} \boldsymbol{x} \leq k$, $k > 0$, in a compact set $\mathcal{X}$ of the state space guaranteeing $z \in \mathcal{C} \subseteq \mathbb{R}^p$ goes asymptotically to $\boldsymbol{x} = 0$.

Proof The fact that $P_i = P_i^T > 0$, $i \in \{1, 2, \ldots, r\}$, guarantees $V(\boldsymbol{x}) = \boldsymbol{x}^T P_{h(t)}^{-1} \boldsymbol{x}$ is a valid Lyapunov function candidate for the closed-loop system resulting from the substitution of $\boldsymbol{u}(t) = F_{h(t)} P_{h(t)}^{-1} \boldsymbol{x}(t)$ in $\boldsymbol{x}(t+1) = A_h \boldsymbol{x}(t) + B_h \boldsymbol{u}(t)$, i.e.

$$\begin{bmatrix} A_{h(t)} + B_{h(t)} F_{h(t)} P_{h(t)}^{-1} & -I \end{bmatrix} \begin{bmatrix} \boldsymbol{x}(t) \\ \boldsymbol{x}(t+1) \end{bmatrix} = 0. \tag{5.20}$$

For $V(\boldsymbol{x})$ to be a genuine Lyapunov function, its forward difference should satisfy

$$\begin{bmatrix} \boldsymbol{x}(t) \\ \boldsymbol{x}(t+1) \end{bmatrix}^T \begin{bmatrix} -P_{h(t)}^{-1} & 0 \\ 0 & P_{h(t+1)}^{-1} \end{bmatrix} \begin{bmatrix} \boldsymbol{x}(t) \\ \boldsymbol{x}(t+1) \end{bmatrix} < 0, \tag{5.21}$$

which can be combined with (5.20) via Finsler's Lemma A.3 to yield the equivalent condition

$$\begin{bmatrix}\mathcal{U}(\cdot)\\ \mathcal{V}(\cdot)\end{bmatrix}\left[A_{h(t)}+B_{h(t)}F_{h(t)}P_{h(t)}^{-1}\ \ -I\right]+(*)+\begin{bmatrix}-P_{h(t)}^{-1} & 0\\ 0 & P_{h(t+1)}^{-1}\end{bmatrix}<0. \tag{5.22}$$

The property of congruence (A.1) with block-diag$(P_{h(t)}, P_{h(t+1)})$ gives

$$\begin{bmatrix}P_{h(t)}\mathcal{U}(\cdot)\\ P_{h(t+1)}\mathcal{V}(\cdot)\end{bmatrix}\left[A_{h(t)}P_{h(t)}+B_{h(t)}F_{h(t)}\ \ -P_{h(t+1)}\right]+(*)+\begin{bmatrix}-P_{h(t)} & 0\\ 0 & P_{h(t+1)}\end{bmatrix}<0.$$

A choice for the slack variables is $\mathcal{U}(\cdot) = 0$ and $\mathcal{V}(\cdot) = P_{h(t+1)}^{-1}$, which leads to sufficient conditions (5.19) guaranteeing the previous inequality. This establishes $V(\boldsymbol{x})$ as a valid Lyapunov function, which, in turn, proves $\boldsymbol{x} = 0$ to be asymptotically stable. From assumptions 4.1 and 4.2, any trajectory in the outermost Lyapunov level within $\mathcal{X}$ goes asymptotically to $\boldsymbol{x} = 0$, as lying in $\mathcal{X}$ guarantees $z \in \mathcal{C}$, which is where MFs $h_i(z)$ hold the convex sum property as required by the steps above. □

Right before the choice of slack variables is made, the expression contained a triple convex sum on $h(t)h(t)h(t+1)$; therefore, instead of $F_{h(t)}$ we may have used a "minimal" choice of $F_{h(t)h(t+1)}$ to fully exploit combinations. Unfortunately, the future value $h(t+1)$ is not available, which obliges the control law to be restricted to our original option.

An Approach with Nested Convex Sums in the Lyapunov Function

The second generic Lyapunov function will be now considered, namely that in (5.17), which asks for $\boldsymbol{P}^{-1}(\cdot) = H_h^{-T}P_hH_h^{-1} > 0$ in (5.15); the same control law will be used, i.e. $F(\cdot) = F_{h(t)}$ and $H(\cdot) = P_{h(t)}$ in (5.2), which yields $\boldsymbol{u}(t) = F_{h(t)}P_{h(t)}^{-1}\boldsymbol{x}(t)$.

Theorem 5.6 *The origin $\boldsymbol{x} = 0$ of the discrete-time TS model (5.10) $\boldsymbol{x}(t+1) = A_h\boldsymbol{x}(t) + B_h\boldsymbol{u}(t)$ under the control law $\boldsymbol{u}(t) = F_{h(t)}P_{h(t)}^{-1}\boldsymbol{x}(t)$ is asymptotically stable if there exist matrices $P_i = P_i^T > 0$, non-singular H_i, F_i, and $i \in \{1, 2, \ldots, r\}$, such that*

$$\begin{bmatrix}-P_{h(t)} & H_{h(t)}^TA_{h(t)}^T+F_{h(t)}^TB_{h(t)}^T\\ A_{h(t)}H_{h(t)}+B_{h(t)}F_{h(t)} & P_{h(t+1)}-H_{h(t+1)}-H_{h(t+1)}^T\end{bmatrix}<0 \tag{5.23}$$

holds. In that case, any trajectory $\boldsymbol{x}(t)$ starting within the outermost Lyapunov level $V(\boldsymbol{x}) = \boldsymbol{x}^TH_{h(t)}^{-T}P_{h(t)}H_{h(t)}^{-1}\boldsymbol{x} \leq k$, $k > 0$, in a compact set $\mathcal{X}$ of the state space guaranteeing $z \in \mathcal{C} \subseteq \mathbb{R}^p$ goes asymptotically to $\boldsymbol{x} = 0$.

Proof From $P_i = P_i^T > 0$, it follows that $H_h^{-T}P_hH_h^{-1} > 0$ as matrices H_i are non-singular; therefore, $V(\boldsymbol{x})$ is a valid Lyapunov function candidate, whose forward difference is

$$\Delta V = \begin{bmatrix} x(t) \\ x(t+1) \end{bmatrix}^T \begin{bmatrix} -H_{h(t)}^{-T} P_{h(t)} H_{h(t)}^{-1} & 0 \\ 0 & H_{h(t+1)}^{-T} P_{h(t+1)} H_{h(t+1)}^{-1} \end{bmatrix} \begin{bmatrix} x(t) \\ x(t+1) \end{bmatrix}. \tag{5.24}$$

The dynamics of the closed-loop system with $u(t) = F_{h(t)} H_{h(t)}^{-1} x(t)$ are

$$\begin{bmatrix} A_{h(t)} + B_{h(t)} F_{h(t)} H_{h(t)}^{-1} & -I \end{bmatrix} \begin{bmatrix} x(t) \\ x(t+1) \end{bmatrix} = 0. \tag{5.25}$$

Again, using Finsler's Lemma A.3 with a multiplier choice of

$$\begin{bmatrix} 0 \\ H_{h(t+1)}^{-T} \end{bmatrix},$$

we can guarantee negative-definiteness of (5.24) taking into account restriction (5.25) as to get

$$\begin{aligned} &\begin{bmatrix} 0 \\ H_{h(t+1)}^{-T} \end{bmatrix} \begin{bmatrix} A_{h(t)} + B_{h(t)} F_{h(t)} H_{h(t)}^{-1} & -I \end{bmatrix} + (*) \\ &+ \begin{bmatrix} -H_{h(t)}^{-T} P_{h(t)} H_{h(t)}^{-1} & 0 \\ 0 & H_{h(t+1)}^{-T} P_{h(t+1)} H_{h(t+1)}^{-1} \end{bmatrix} \\ &= \begin{bmatrix} -H_{h(t)}^{-T} P_{h(t)} H_{h(t)}^{-1} & (*) \\ H_{h(t+1)}^{-T} \left(A_{h(t)} + B_{h(t)} F_{h(t)} H_{h(t)}^{-1} \right) & \begin{matrix} -H_{h(t+1)}^{-T} - H_{h(t+1)}^{-1} \\ + H_{h(t+1)}^{-T} P_{h(t+1)} H_{h(t+1)}^{-1} \end{matrix} \end{bmatrix} < 0. \end{aligned}$$

Applying congruence in Property A.1 with block-diag$(H_{h(t)}^T, H_{h(t+1)}^T)$ allows rewriting the inequality as (5.23), thus establishing the validity of $V(x)$ as a Lyapunov function that proves $x = 0$ is asymptotically stable. Since conditions are based on the MFs $h_i(z)$ holding the convex sum property in $z \in \mathcal{C}$, trajectories within the outermost Lyapunov level within $\mathcal{X}$ are guaranteed to go asymptotically to $x = 0$ by assumptions 4.1 and 4.2. □

Example 5.6 Consider a TS system with two vertex models described by matrices

$$A_1 = \begin{bmatrix} 1.2 & -0.55 \\ 0.45 & 0.88 \end{bmatrix}, \qquad A_2 = \begin{bmatrix} -1.24 & 0.08 \\ -0.91 & -1.73 \end{bmatrix},$$
$$B_1 = \begin{bmatrix} 0 \\ 1 \end{bmatrix}, \qquad B_2 = \begin{bmatrix} 0 \\ -1 \end{bmatrix},$$

where MFs $h_i(z)$ are assumed to hold the convex sum property everywhere in the state space.

Using the Relaxation Lemma D.1 to derive LMI conditions from those in Theorem 5.2 (based on a quadratic Lyapunov function), no solution can be found as the LMIs are infeasible. However, LMI conditions derived from (5.23) in Theorem 5.6 with

Relaxation Lemma D.1 (based on a non-quadratic Lyapunov function) yield the following feasible solution:

$$P_1 = \begin{bmatrix} 0.06 & 0.12 \\ 0.12 & 0.50 \end{bmatrix}, \qquad P_2 = \begin{bmatrix} 0.05 & 0.15 \\ 0.15 & 0.52 \end{bmatrix},$$
$$H_1 = \begin{bmatrix} 0.06 & 0.14 \\ 0.15 & 0.51 \end{bmatrix}, \qquad H_2 = \begin{bmatrix} 0.05 & 0.15 \\ 0.16 & 0.60 \end{bmatrix},$$
$$F_1 = \begin{bmatrix} -0.12 & -0.76 \end{bmatrix}, \quad F_2 = \begin{bmatrix} -0.16 & -0.76 \end{bmatrix},$$

which illustrates the conservativeness reduction of non-quadratic approaches over their quadratic counterparts: more TS models can be stabilised at the price of higher computational burden. □

An Approach using Past Values of the Scheduling Vector

Consider now the third generic Lyapunov function candidate as in (5.18), namely $\boldsymbol{P}^{-1}(\cdot) = P_{h(t-1)} > 0$ in (5.15), along with $\boldsymbol{u}(t) = F_{h(t)h(t-1)} H^{-1}_{h(t)h(t-1)} \boldsymbol{x}(t)$ as the control law. The reasons for these choices will become apparent in the sequel.

Theorem 5.7 *The origin $\boldsymbol{x} = 0$ of the discrete-time TS model (5.10) $\boldsymbol{x}(t+1) = A_h \boldsymbol{x}(t) + B_h \boldsymbol{u}(t)$ under the control law $\boldsymbol{u}(t) = F_{h(t)h(t-1)} H^{-1}_{h(t)h(t-1)} \boldsymbol{x}(t)$ is asymptotically stable if there exist matrices $P_i = P_i^T > 0$, H_{ij}, and F_{ij}, $i \in \{1, 2, \ldots, r\}$, and $j \in \{1, 2, \ldots, r\}$, such that*

$$\begin{bmatrix} P_{h(t-1)} - H^T_{h(t)h(t-1)} - H_{h(t)h(t-1)} & (*) \\ A_{h(t)} H_{h(t)h(t-1)} + B_{h(t)} F_{h(t)h(t-1)} & -P_{h(t)} \end{bmatrix} < 0 \tag{5.26}$$

holds. In this case, any trajectory $\boldsymbol{x}(t)$ starting within the outermost Lyapunov level $V(\boldsymbol{x}) = \boldsymbol{x}^T P^{-1}_{h(t-1)} \boldsymbol{x} \le k$, $k > 0$, in a compact set $\mathcal{X}$ of the state space guaranteeing $z \in \mathcal{C} \subseteq \mathbb{R}^p$ goes asymptotically to $\boldsymbol{x} = 0$.

Proof From $P_i = P_i^T > 0$, $i \in \{1, 2, \ldots, r\}$, $V(\boldsymbol{x})$ is a valid Lyapunov function candidate. Instead of considering the proposed control law, let us take a step back to the generic form $\boldsymbol{u}(t) = F(\cdot) H^{-1}(\cdot) \boldsymbol{x}(t)$ as to show how the proposed choice follows "naturally" from the development. Following the same procedure as before, i.e. combining via Finsler's Lemma A.3 the closed-loop system formulation with the inequality derived from the Lyapunov forward difference, it is easy to show that

$$\begin{bmatrix} -P^{-1}_{h(t-1)} & 0 \\ 0 & P^{-1}_{h(t)} \end{bmatrix} + \begin{bmatrix} \mathcal{U}(\cdot) \\ \mathcal{V}(\cdot) \end{bmatrix} \begin{bmatrix} A_{h(t)} + B_{h(t)} F(\cdot) H^{-1}(\cdot) & -I \end{bmatrix} + (*) < 0$$

is a sufficient condition to guarantee the forward difference of the Lyapunov function to be negative-definite.

Congruence via Property A.1 with block-diag$(H^T(\cdot), P_{h(t)})$ gives

$$\begin{bmatrix} H^T(\cdot)\mathcal{U}(\cdot) \\ P_{h(t)}\mathcal{V}(\cdot) \end{bmatrix} \begin{bmatrix} A_{h(t)}H(\cdot) + B_{h(t)}F(\cdot) & -P_{h(t)} \end{bmatrix} + (*) \\ + \begin{bmatrix} -H^T(\cdot)P_{h(t-1)}^{-1}H(\cdot) & 0 \\ 0 & P_{h(t)} \end{bmatrix} < 0. \tag{5.27}$$

Choosing $\mathcal{U}(\cdot) = 0$ and $\mathcal{V}(\cdot) = P_{h(t)}^{-1}$ turns (5.27) into

$$\begin{bmatrix} -H^T(\cdot)P_{h(t-1)}^{-1}H(\cdot) & (*) \\ A_{h(t)}H(\cdot) + B_{h(t)}F(\cdot) & -P_{h(t)} \end{bmatrix} < 0.$$

Finally, with Property A.4, sufficient condition

$$\begin{bmatrix} P_{h(t-1)} - H^T(\cdot) - H(\cdot) & (*) \\ A_{h(t)}H(\cdot) + B_{h(t)}F(\cdot) & -P_{h(t)} \end{bmatrix} < 0 \tag{5.28}$$

is obtained. It can be seen that (5.28) corresponds to a double sum $z(t)z(t-1)$, which implies that a "minimal" choice for $H(\cdot)$ and $F(\cdot)$ (considering keeping the instant t) is $H_{h(t)h(t-1)}$ and $F_{h(t)h(t-1)}$, respectively, which leads to sufficient conditions (5.26). This choice allows exploiting all the possible combinations without adding extra convex sums.

Thus, $\boldsymbol{x} = 0$ is proved asymptotically stable via the Lyapunov function $V(\boldsymbol{x})$ and, since the MFs are required to hold the convex sum property to sustain the previous reasoning, trajectories within the outermost Lyapunov level within $\mathcal{X}$ are guaranteed to go asymptotically to $\boldsymbol{x} = 0$ by assumptions 4.1 and 4.2. □

Notice that invertibility of $H_{h(t)h(t-1)}$ is ensured from the first block of (5.26): indeed, if (5.26) holds then

$$H_{h(t)h(t-1)}^T + H_{h(t)h(t-1)} > P_{h(t-1)} > 0.$$

The interest of (5.26) compared to (5.19) is the possibility of introducing the past values of the premise variables in the control law. Clearly, this approach is easily generalised to more past samples.

The previous results are summarised in the following table. The similarity and relationships between the different options justify the "template-like" presentation of the generic Lyapunov function candidate (5.15) and the generic control law (5.2).

Remark 5.4 The second row in Table 5.1 with $H_{h(t)} = P_{h(t)}$ gives the first one, thus proving that it is a direct generalisation; for all three of them, the quadratic conditions are recovered with $H(\cdot) = P(\cdot) = P$.

Generalisations: Taking a look at Table 5.1, where the idea was to keep triple sums, the generalisation $\boldsymbol{P}(\cdot) = P_{h(t)h(t-1)}^{-1}$ can be obtained by combining the first and third rows. Therefore, the controller gain will be the same as in the third row, i.e. $F_{h(t)h(t-1)}H_{h(t)h(t-1)}^{-1}$, and using the same procedure based on Finsler's Lemma

Table 5.1 Non-quadratic controller design for discrete-time TS models

Lyapunov function: $V(\boldsymbol{x}(t)) = \boldsymbol{x}^T(t)P_{h(t)}^{-1}\boldsymbol{x}(t)$. Controller: $\boldsymbol{u}(t) = F_{h(t)}P_{h(t)}^{-1}\boldsymbol{x}(t)$. Conditions: $\begin{bmatrix} -P_{h(t)} & (*) \\ A_{h(t)}P_{h(t)} + B_{h(t)}F_{h(t)} & -P_{h(t+1)} \end{bmatrix} < 0.$
Lyapunov function: $V(\boldsymbol{x}(t)) = \boldsymbol{x}^T(t)H_{h(t)}^{-1}P_{h(t)}H_{h(t)}^{-T}\boldsymbol{x}(t)$. Controller: $\boldsymbol{u}(t) = F_{h(t)}H_{h(t)}^{-1}\boldsymbol{x}(t)$. Conditions: $\begin{bmatrix} -P_{h(t)} & (*) \\ A_{h(t)}H_{h(t)} + B_{h(t)}F_{h(t)} & P_{h(t+1)} - H_{h(t+1)} - H_{h(t+1)}^T \end{bmatrix} < 0.$
Lyapunov function: $V(\boldsymbol{x}(t)) = \boldsymbol{x}^T(t)P_{h(t-1)}^{-1}\boldsymbol{x}(t)$. Controller: $\boldsymbol{u}(t) = F_{h(t)h(t-1)}H_{h(t)h(t-1)}^{-1}\boldsymbol{x}(t)$. Conditions: $\begin{bmatrix} P_{h(t-1)} - H_{h(t)h(t-1)}^T - H_{h(t)h(t-1)} & (*) \\ A_{h(t)}H_{h(t)h(t-1)} + B_{h(t)}F_{h(t)h(t-1)} & -P_{h(t)} \end{bmatrix} < 0.$

A.3 the following condition for the forward difference of the Lyapunov function candidate is obtained:

$$\begin{bmatrix} P_{h(t)h(t-1)} - H_{h(t)h(t-1)}^T - H_{h(t)h(t-1)} & (*) \\ A_{h(t)}H_{h(t)h(t-1)} + B_{h(t)}F_{h(t)h(t-1)} & -P_{h(t+1)h(t)} \end{bmatrix} < 0.$$

The previous approach has the downside of having four nested convex sums on $h(t+1)h(t)h(t)h(t-1)$; on the other hand, it has an advantage in terms of generalisation since $P_{h(t)h(t-1)} = P_{h(t-1)}$ directly gives the third row, while choices $P_{h(t)h(t-1)} = P_{h(t)}$ and $F_{h(t)h(t-1)} = F_{h(t)}$ lead to the first one. As mentioned before, many more options for further relaxation are possible: Pólya associations, homogeneous Lyapunov functions, multiple delays, α-samples, etc.

As we have discussed at the end of Sect. 4.4.3, working on Lyapunov functions that depend on past/future values of the membership seems somehow counterintuitive. Yet, for a reformulation with no intervention of past/future variables, see [34, Thm.3] as well as [35, 36].

5.4 Stabilisation of Polynomial Models

As expected, stabilisation via polynomial models intends to go further than the one based on Takagi–Sugeno models. To do so, instead of a PDC control law, the following generalisation has been proposed [37]:

$$u(t) = \sum_{j=1}^{r} h_j(z) F_j(x) \hat{x}(x), \tag{5.29}$$

where $\hat{x}(x)$ is a vector of monomials in $x(t)$, following the notation in Sect. 4.5. For the closed-loop polynomial model in the matrix form

$$\dot{x}(t) = \sum_{i=1}^{r} \sum_{i=1}^{r} h_i(z) h_j(z) \left(A_i(x) - B_i(x) F_j(x)\right) \hat{x}(x), \tag{5.30}$$

the following result, due to [37], provides SOS controller design conditions:

Theorem 5.8 *[37] The origin of system (5.30) is asymptotically stable if there exist a symmetric polynomial matrix $X(\tilde{x})$ and polynomial matrices $M_i(x)$ of adequate dimensions, such that*

$$\begin{aligned}
&v^T \left(X(\tilde{x}) - \epsilon_1(x) I\right) v \text{ is SOS,} \\
&- v^T \Big(T(x) A_i(x) X(\tilde{x}) + T(x) B_i(x) M_j(x) + T(x) A_j(x) X(\tilde{x}) \\
&\quad + T(x) B_j(x) M_i(x) + (*) - \sum_{k \in K} \frac{\partial X}{\partial x_k} \left(A_i^k(x) + A_j^k(x)\right) \hat{x} + \epsilon_{2ij}(x) I \Big) v \\
&\quad \text{is SOS}
\end{aligned}$$

hold for nonnegative polynomials $\epsilon_1(x) > 0$, $x \neq 0$, and $\epsilon_{2ij}(x) \geq 0$, $\forall i \leq j$, $T(x)$ is a matrix whose entries are defined as $T_{ij}(x) = \partial \hat{x}_i / \partial x_j(x)$, K denotes the set of row indices of $B_i(x)$ equal to 0, $A_i^k(x)$ denotes the kth row of $A_i(x)$, and v are vectors which are independent of x. The stabilising gains in (5.29) are thus determined as $F_i(x) = M_i(x) X^{-1}(\tilde{x})$, with the Lyapunov function given by $V(x) = \hat{x}^T(x) X^{-1}(\tilde{x}) \hat{x}(x)$.

Example 5.7 Consider the following polynomial system:

$$\begin{aligned}
\dot{x}_1 &= -x_1 + 0.25 x_2 - 0.1 x_1 x_2, \\
\dot{x}_2 &= 3 x_1 - 2 x_2 - 0.25 x_1^2 x_2 - 0.2 x_2^2 + 0.1 u x_2 + 7u,
\end{aligned}$$

which can be written in the form $\dot{x}(t) = A(x) x(t) + B(x) u(t)$, with matrices

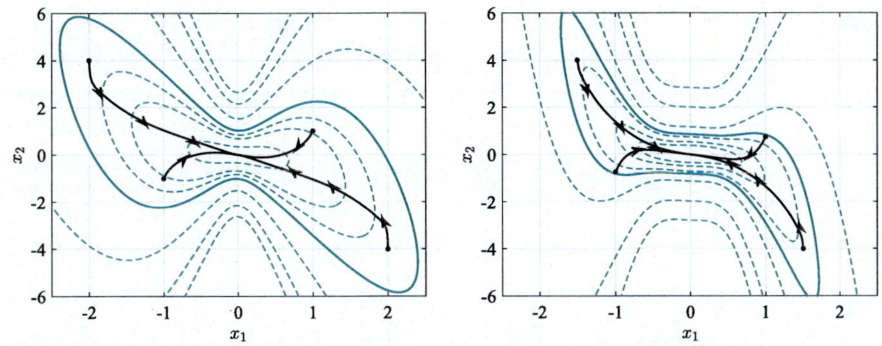

(a) Phase portrait and Lyapunov levels, polynomial matrix $X(\tilde{x})$ of degree 2.

(b) Phase portrait and Lyapunov levels, polynomial matrix $X(\tilde{x})$ of degree 4.

Fig. 5.5 Phase portraits for Example 5.7

$$A(\boldsymbol{x}) = \begin{bmatrix} -1 - 0.1x_2 & 0.25 \\ 3 - 0.25x_1x_2 & -2 - 0.2x_2 \end{bmatrix}, \quad B(\boldsymbol{x}) = \begin{bmatrix} 0 \\ 7 + 0.1x_2 \end{bmatrix}.$$

By means of the sector nonlinearity approach and considering the nonlinearities $z_1 = x_1$ and $z_2 = x_2$, the system is rewritten as a TS model with $r = 4$. Conditions in Theorem 5.8 hold considering a polynomial matrix $X(\tilde{x})$ for the case when its entries are of degree 2 and 4. It is considered that $\epsilon_1(\boldsymbol{x}) = \epsilon_{2ij}(\boldsymbol{x}) = 0.001\left(x_1^2 + x_2^2\right)$, $\hat{x} = x$, i.e. $T(\boldsymbol{x}) = \partial\hat{\boldsymbol{x}}/\partial\boldsymbol{x} = I$. In this example, $K = \{1\}$ is the set of rows in $B(\boldsymbol{x})$ equal to 0, which implies $\tilde{x} = x_1$; $A_i^k(\boldsymbol{x})$ denotes the kth row of A_i.

In Fig. 5.5(a) and 5.5(b), Lyapunov levels are shown in dashed lines while system trajectories are shown in solid lines for $X(\tilde{x})$ having polynomial entries of degree 2 and 4, respectively.

This approach is not pursued in this book as its development has not reached the maturity of those based on LMI conditions. Notice the difficulty in handling $B(\boldsymbol{x})$ in order to avoid algebraic loops with the control input $\boldsymbol{u}$, which is the reason behind limiting the number of rows of $B(\boldsymbol{x})$ that is different from 0.

5.5 Stabilisation of Descriptor Models

Let us consider nonlinear descriptor models of the form (3.57) exactly represented by a TS descriptor model, i.e. an extension of (4.77) for the non-autonomous case:

$$\sum_{i=1}^{r_e} v_i(z)E_i\delta\boldsymbol{x}(t) = \sum_{i=1}^{r} h_i(z)\left(A_i\boldsymbol{x}(t) + B_i\boldsymbol{u}(t)\right), \tag{5.31}$$

where, as before, $\delta x(t)$ stands for $\dot{x}(t)$ with $t \in \mathbb{R}$ in the continuous-time case or $x(t+1)$ with $t \in \mathbb{Z}$ in the discrete-time case, E_i, $i \in \{1, 2, \ldots, r_e\}$, and A_i, B_i, $i \in \{1, 2, \ldots, r\}$, are constant matrices of proper sizes, $v_i(\cdot)$ and $h_i(\cdot)$ are MFs depending upon the premise vector z, which in turn may depend on states, time, or exogenous variables.

As for stability analysis, the *descriptor redundancy form* is adopted for continuous-time models [38]:

$$\bar{E}\dot{\bar{x}}(t) = \bar{A}_{hv}\bar{x}(t) + \bar{B}_h u(t) \tag{5.32}$$

with

$$\bar{x} = \begin{bmatrix} x \\ \dot{x} \end{bmatrix}, \ \bar{E} = \begin{bmatrix} I & 0 \\ 0 & 0 \end{bmatrix}, \ \bar{A}_{hv} = \begin{bmatrix} 0 & I \\ A_h & -E_v \end{bmatrix}, \ \bar{B}_h = \begin{bmatrix} 0 \\ B_h \end{bmatrix},$$

$$A_h = \sum_{i=1}^{r} h_i(z)A_i, \ B_h = \sum_{i=1}^{r} h_i(z)B_i, \ E_v = \sum_{i=1}^{r_e} v_i(z(t))E_i,$$

while the following equality constraint is used for the discrete-time case [39] along with Finsler's Lemma A.3:

$$\begin{bmatrix} A_h + B_h F(\cdot) & -E_v \end{bmatrix} \begin{bmatrix} x(t) \\ x(t+1) \end{bmatrix} = 0, \tag{5.33}$$

where $F(\cdot)$ will be defined according to a specific control law.

Stabilisation of continuous-time descriptor TS models has begun with the seminal paper of [38], which is based on [40]; the results make use of the quadratic Lyapunov function described in Sect. 4.6:

$$V(x) = \bar{x}^T \bar{E} X \bar{x} = \begin{bmatrix} x \\ \dot{x} \end{bmatrix}^T \begin{bmatrix} I & 0 \\ 0 & 0 \end{bmatrix} \begin{bmatrix} X_1 & 0 \\ X_3 & X_4 \end{bmatrix} \begin{bmatrix} x \\ \dot{x} \end{bmatrix}, \ X \equiv \begin{bmatrix} X_1 & 0 \\ X_3 & X_4 \end{bmatrix}.$$

Later on, a semi-quadratic extension towards a non-quadratic approach has been proposed in [41]: it allowed the use of convex sums in those entries of the Lyapunov matrix that, due to the structure, are not going to be derived, i.e. X_3 and X_4 above are replaced by functions depending on the scheduling vector z, namely $X_3(z)$ and $X_4(z)$. Indeed, notice that

$$\dot{V}(x) = \dot{\bar{x}}^T \bar{E} X(z) \bar{x} + \bar{x}^T \bar{E} X(z) \dot{\bar{x}} + \bar{x}^T \bar{E} \dot{X}(z) \bar{x},$$

where

$$\bar{E}\dot{X}(z) = \begin{bmatrix} I & 0 \\ 0 & 0 \end{bmatrix} \begin{bmatrix} \dot{X}_1 & 0 \\ \dot{X}_3(z) & \dot{X}_4(z) \end{bmatrix} = \begin{bmatrix} 0 & 0 \\ 0 & 0 \end{bmatrix}$$

if X_1 remains constant despite $X_3(z)$ and $X_4(z)$ being non-constant. This relaxation was introduced in [41] and avoids dealing with $\dot{h}_i$ while introducing some convex sums in the Lyapunov function, hence the semi-quadratic characteristic.

Otherwise, the use of non-quadratic Lyapunov functions faces the same problems as with ordinary TS models: dealing with the time derivative of the MFs leading to local approaches [42].

For the discrete-time case, [43, 44] have considered both quadratic and non-quadratic results in a single framework using Finsler's Lemma A.3. As has been previously seen, this important lemma establishes the equivalence between an inequality subject to an equality condition and a single inequality which includes a Lagrange multiplier. This equivalence is usually broken due to conservative assumptions, but even so it has been proved advantageous, especially in the discrete-time case.

Theorem 5.9 *The origin $\boldsymbol{x} = 0$ of the continuous-time descriptor model (5.31) with $\delta\boldsymbol{x}(t) = \dot{\boldsymbol{x}}(t)$, $t \in \mathbb{R}$, under the control law $\boldsymbol{u}(t) = F_{hv}X_1\boldsymbol{x}(t)$ is asymptotically stable if there exists matrices $P_1 = P_1^T > 0$, P_{3ik}, P_{4ik} and F_{ij}, $i, k \in \{1, 2, \ldots, r\}$, $j \in \{1, 2, \ldots, r_e\}$ such that*

$$\begin{bmatrix} P_{3hh}+P_{3hh}^T & P_1A_h^T+F_{hv}^TB_h^T+P_{4hh}-P_{3hh}^TE_v^T \\ A_hP_1+B_hF_{hv}+P_{4hh}^T-E_vP_{3hh} & -E_vP_{4hh}-P_{4hh}^TE_v^T \end{bmatrix} < 0 \qquad (5.34)$$

hold. In that case, any trajectory $\boldsymbol{x}(t)$ starting within the outermost Lyapunov level $V(\boldsymbol{x}) = \boldsymbol{x}^TX_1\boldsymbol{x} \leq k$, $X_1 = P_1^{-1}$, $k > 0$, in a compact set $\mathcal{X}$ of the state space guaranteeing $z \in \mathcal{C} \subseteq \mathbb{R}^p$ goes asymptotically to $\boldsymbol{x} = 0$.

Proof The control law $\boldsymbol{u}(t) = F_{hv}X_1x(t)$ can be rewritten as

$$\boldsymbol{u}(t) = \bar{F}_{hv}X(z)\bar{\boldsymbol{x}}(t), \quad \bar{F}_{hv} = \begin{bmatrix} F_{hv} & 0 \end{bmatrix},$$

with

$$X(z) = \begin{bmatrix} X_1 & 0 \\ X_3(z) & X_4(z) \end{bmatrix},$$

such that the closed-loop system is

$$\bar{E}\dot{\bar{\boldsymbol{x}}}(t) = \left(\bar{A}_{hv} + \bar{B}_h\bar{F}_{hv}X(z)\right)\bar{\boldsymbol{x}}(t).$$

Since $P_1 = P_1^T > 0$, $V(\boldsymbol{x}) = \boldsymbol{x}^TP_1^{-1}\boldsymbol{x} = \boldsymbol{x}^TX_1\boldsymbol{x} = \bar{\boldsymbol{x}}^T\bar{E}^TX(z)\bar{\boldsymbol{x}}$ is a valid Lyapunov function candidate.

The time derivative of the Lyapunov function is

$$\dot{V} = \bar{\boldsymbol{x}}^T\left(\left(\bar{A}_{hv} + \bar{B}_h\bar{F}_{hv}X(z)\right)^TX(z) + X^T(z)\left(\bar{A}_{hv} + \bar{B}_h\bar{F}_{hv}X(z)\right)\right)\bar{\boldsymbol{x}},$$

which means that

$$\left(\bar{A}_{hv} + \bar{B}_h\bar{F}_{hv}X(z)\right)^TX(z) + X^T(z)\left(\bar{A}_{hv} + \bar{B}_h\bar{F}_{hv}X(z)\right) < 0$$

guarantees $\dot{V} < 0$.

This condition can be transformed into (5.34) if congruence (Property A.1) is applied to it with

$$P_{hh} = \begin{bmatrix} P_1 & 0 \\ P_{3hh} & P_{4hh} \end{bmatrix},$$

where $P_{hh} \equiv X^{-1}(z)$, obliging to $P_1 = X_1^{-1}$, $P_{4hh} = X_4^{-1}(z)$, and $P_{3hh} = -X_4^{-1}(z) X_3(z)X_1^{-1}$. Thus, $V(x)$ is proven a valid Lyapunov function establishing asymptotic stability of $x = 0$ as long as MFs $h_i(z)$, $v_j(z)$ hold the convex sum property, which they do in $\mathcal{C}$. Therefore, any trajectory contained in the outermost Lyapunov level within $\mathcal{X}$ goes asymptotically to $x = 0$ by assumptions 4.1 and 4.2. □

Remark 5.5 Quadratic conditions can be obtained if $X(z)$ is chosen constant. Interestingly, the control law $u(t) = F_{hv}X_1x(t)$ remains unaffected in such case, which means that the reduction of conservativeness achieved by the semi-quadratic approach is entirely due to the richer choices associated with the slack terms $X_3(z)$ and $X_4(z)$ in the Lyapunov function.

As before in discrete-time contexts, whenever convenient, the time dependence of the convex sums will be omitted or shown, e.g. P_h or $P_{h(t)}$ will be used indistinctively.

Theorem 5.10 *The origin $x = 0$ of the discrete-time descriptor model (5.31) with $\delta x(t) = x(t+1)$, $t \in \mathbb{Z}$, under the control law $u(t) = F_{h(t)v(t)}H^{-1}_{h(t)v(t)}x(t)$ is asymptotically stable if there exist matrices $P_i = P_i^T > 0$, H_{ij}, and F_{ij}, $i \in \{1, 2, \ldots, r\}$, $j \in \{1, 2, \ldots, r_e\}$, such that*

$$\begin{bmatrix} P_{h(t)} - H_{hv} - H_{hv}^T & H_{hv}^T A_h^T + F_{hv}^T B_h^T \\ A_h H_{hv} + B_h F_{hv} & -E_v P_{h(t+1)} - P_{h(t+1)}^T E_v^T + P_{h(t+1)} \end{bmatrix} < 0 \tag{5.35}$$

holds. In this case, any trajectory $x(t)$ starting within the outermost Lyapunov level $V(x) = x^T X_{h(t)} x \le k$, $X_{h(t)} = P^{-1}_{h(t)}$, $k > 0$, in a compact set $\mathcal{X}$ of the state space guaranteeing $z \in \mathcal{C} \subseteq \mathbb{R}^p$ goes asymptotically to $x = 0$.

Proof Since $P_i = P_i^T > 0$, $V(x) = x^T X_{h(t)} x$ is a valid Lyapunov function candidate with $X_{h(t)} = P^{-1}_{h(t)}$.

The closed loop can be rewritten as an equality constraint:

$$\begin{bmatrix} A_h + B_h F_{hv} H_{hv}^{-1} & -E_v \end{bmatrix} \begin{bmatrix} x(t) \\ x(t+1) \end{bmatrix} = 0. \tag{5.36}$$

The forward difference of the Lyapunov function is

$$\begin{aligned} \Delta V(t) =& x^T(t+1) X_{h(t+1)} x(t+1) - x^T(t+1) X_{h(t)} x(t+1) \\ =& \begin{bmatrix} x(t) \\ x(t+1) \end{bmatrix}^T \begin{bmatrix} -X_{h(t)} & 0 \\ 0 & X_{h(t+1)} \end{bmatrix} \begin{bmatrix} x(t) \\ x(t+1) \end{bmatrix} < 0. \end{aligned}$$

Via Finsler's Lemma A.3, equality (5.36) under the previous inequality is equivalent to

$$\begin{bmatrix} \mathcal{U}(\cdot) \\ \mathcal{V}(\cdot) \end{bmatrix} \begin{bmatrix} A_h + B_h F_{hv} H_{hv}^{-1} & -E_v \end{bmatrix} + (*) + \begin{bmatrix} -X_{h(t)} & 0 \\ 0 & X_{h(t+1)} \end{bmatrix} < 0, \tag{5.37}$$

where matrices $\mathcal{U}(\cdot) \in \mathbb{R}^{n\times n}$ and $\mathcal{V}(\cdot) \in \mathbb{R}^{n\times n}$ are free matrices.

Using the congruence (Property A.1) with the full-rank matrix block-diag$(H_{hv}^T, P_{h(t+1)})$ on (5.37) yields

$$\begin{aligned} &\begin{bmatrix} H_{hv}^T \mathcal{U}(\cdot) \\ P_{h(t+1)} \mathcal{V}(\cdot) \end{bmatrix} \begin{bmatrix} A_h H_{hv} + B_h F_{hv} & -E_v P_{h(t+1)} \end{bmatrix} + (*) \\ &+ \begin{bmatrix} -H_{hv}^T X_{h(t)} H_{hv} & 0 \\ 0 & P_{h(t+1)} X_{h(t+1)} P_{h(t+1)} \end{bmatrix} < 0. \end{aligned} \tag{5.38}$$

Taking $\mathcal{U}(\cdot) = 0$ and $\mathcal{V}(\cdot) = X_{h(t+1)}$, (5.38) is rewritten as

$$\begin{bmatrix} -H_{hv}^T X_{h(t)} H_{hv} & H_{hv}^T A_h^T + F_{hv}^T B_h^T \\ A_h H_{hv} + B_h F_{hv} & P_{h(t+1)} - E_v P_{h(t+1)} - P_{h(t+1)}^T E_v^T \end{bmatrix} < 0.$$

Finally, applying Property A.4 $(Y - X)^T X^{-1} (Y - X) \geq 0 \Leftrightarrow Y^T X^{-1} Y \geq Y + Y^T - X$ gives the desired conditions (5.35), guaranteeing $V(\boldsymbol{x})$ is a valid Lyapunov function and $\boldsymbol{x} = 0$ is asymptotically stable.

As before, this development is based on the MFs holding the convex sum property, which is only valid in $\mathcal{C}$. But ensuring $z \in \mathcal{C}$ can be done by assumptions 4.1 and 4.2, thus guaranteeing that any trajectory in the outermost Lyapunov level within $\mathcal{X}$ goes asymptotically to $\boldsymbol{x} = 0$. □

Remark 5.6 Different sufficient conditions from those in the previous theorem can be found by slightly changing the choices. For instance, using the congruence Property A.1 on (5.37) with block-diag$(H_{hv}^T, G_{h(t)h(t)h(t+1)}^T)$ gives

$$\begin{aligned} &\begin{bmatrix} H_{hv}^T \mathcal{U}(\cdot) \\ G_{h(t)h(t)h(t+1)}^T \mathcal{V}(\cdot) \end{bmatrix} \begin{bmatrix} A_h H_{hv} + B_h F_{hv} & -E_v G_{h(t)h(t)h(t+1)} \end{bmatrix} + (*) \\ &+ \begin{bmatrix} -H_{hv}^T X_{h(t)} H_{hv} & 0 \\ 0 & G_{h(t)h(t)h(t+1)}^T X_{h(t+1)} G_{h(t)h(t)h(t+1)} \end{bmatrix} < 0. \end{aligned}$$

Note that a new matrix $G_{h(t)h(t)h(t+1)}$ is introduced, thus adding extra degrees of freedom to the inequality. Choosing $\mathcal{U}(\cdot) = 0$ and $\mathcal{V}(\cdot) = G_{h(t)h(t)h(t+1)}^{-T}$ yields

$$\begin{bmatrix} -H_{hv}^T X_{h(t)} H_{hv} & (*) \\ A_h H_{hv} + B_h F_{hv} & \begin{matrix} -E_v G_{h(t)h(t)h(t+1)} - (*) \\ +G_{h(t)h(t)h(t+1)}^T X_{h(t+1)} G_{h(t)h(t)h(t+1)} \end{matrix} \end{bmatrix} < 0. \tag{5.39}$$

Employing Property A.1 and Lemma A.1 (Schur complement), (5.39) gives

$$\begin{bmatrix} -H_{hv} - H_{hv}^T + P_h & (*) & (*) \\ A_h H_{hv} + B_h F_{hv} & -E_v G_{h(t)h(t)h(t+1)} + (*) & (*) \\ 0 & G_{h(t)h(t)h(t+1)} & -P_{h(t+1)} \end{bmatrix} < 0, \tag{5.40}$$

which is a new set of conditions that can be used instead of (5.38).

Remark 5.7 The quadratic case is recovered by choosing $G_{h(t)h(t)h(t+1)} = H_{hv} = P_h = P$ in inequality (5.40) and employing the Schur complement in Lemma A.1:

$$\begin{bmatrix} -P & PA_h^T + F_{hv}^T B_h^T \\ A_h P + B_h F_{hv} & P - E_v P - P E_v^T \end{bmatrix} < 0. \tag{5.41}$$

Note that conditions for the classical TS model are recovered if $E_v = I$.

Example 5.8 Consider a continuous-time descriptor model of the form (5.31) with $r = 2$ right-hand side vertices and $r_e = 2$ left-hand side vertices, where

$$A_1 = \begin{bmatrix} 1 & 0.2 \\ 2\beta & -2 \end{bmatrix}, \; A_2 = \begin{bmatrix} 1 & 0.2 \\ -2\beta & -2 \end{bmatrix}, \; B_1 = \begin{bmatrix} 0.4 \\ 1 \end{bmatrix}, \; B_2 = \begin{bmatrix} 0.6 \\ 1 \end{bmatrix},$$
$$E_1 = \begin{bmatrix} 1 & -0.7 \\ -0.7 & 0.9 \end{bmatrix}, \; E_2 = \begin{bmatrix} 1 & 0.7 \\ 0.7 & 0.9 \end{bmatrix},$$

where $\beta > 0$ is a parameter to be maximised while searching for a controller. With $z = x$, MFs are defined as $h_1(z) = (1 + \cos x_1)/2$, $h_2(z) = (1 - \cos x_1)/2$, $v_1(z) = (1 + \cos x_2)/2$, and $v_2(z) = (1 - \cos x_2)/2$. Clearly, they hold the convex sum property in the whole state space.

Controllers via the corresponding continuous quadratic conditions above were found up to the value $\beta = 1.37$, whereas for the semi-quadratic conditions this value can be pushed up to $\beta = 1.98$. Wherever a double nested convex sum appeared, Tuan's relaxation D.1 was used to obtain LMI conditions to guarantee the matrix inequalities in Theorem 5.9. This test allows appreciating the advantages of the non-quadratic approaches over the quadratic ones when it comes to controller design.

We analyse the behaviour of the system with parameter $\beta = 1.98$, both for the open-loop case ($u(t) = 0$) and under the corresponding semi-quadratic controller obtained from the previous setup. The open-loop behaviour in Fig. 5.6a shows that the system is unstable. The closed-loop behaviour in Fig. 5.6b shows that the semi-quadratic controller does indeed stabilise the origin $x = 0$ of the TS descriptor model. Simulations are run from initial conditions $x(0) = \begin{bmatrix} 0.5 & -0.7 \end{bmatrix}^T$. □

Example 5.9 Conditions (5.35) in Theorem 5.10 are compared against conditions (5.41) in Remark 5.7. To this end, consider a setup of the form (5.31) with

$$A_1 = \begin{bmatrix} -1 & 1+a \\ -1.5 & 0.5 \end{bmatrix}, \; A_2 = \begin{bmatrix} -1 & 1-a \\ -1.5 & 0.5 \end{bmatrix}, \; B_1 = \begin{bmatrix} 1-b \\ 0.1 \end{bmatrix}, \; B_2 = \begin{bmatrix} 1+b \\ 0.1 \end{bmatrix},$$
$$E_1 = \begin{bmatrix} 0.9 & 0.1+a \\ -0.4-b & 1.1 \end{bmatrix}, \; E_2 = \begin{bmatrix} 0.9 & 0.1-a \\ -0.4+b & 1.1 \end{bmatrix},$$

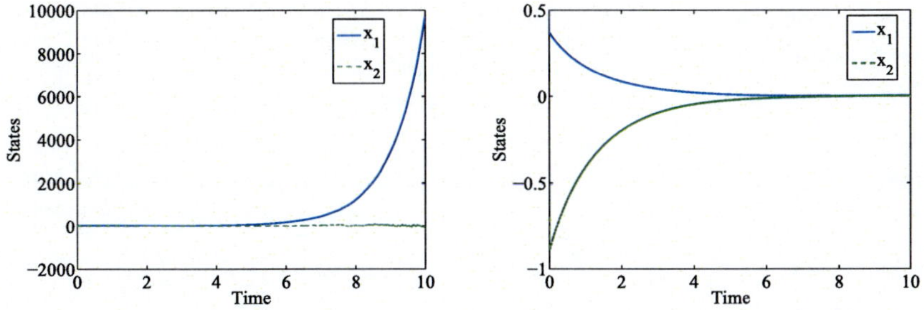

(a) Open-loop behaviour. (b) Closed-loop behaviour.

Fig. 5.6 Simulation results for Example 5.8

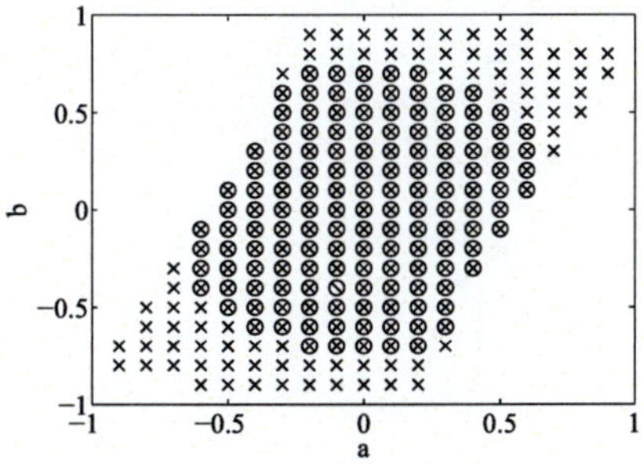

Fig. 5.7 Feasibility sets for discrete TS descriptors: quadratic (○) versus non-quadratic (×)

where a and b are parameters to be chosen from the interval $[-1, 1]$.

As expected, results in Fig. 5.7 show that the feasibility set of the non-quadratic conditions (5.35) marked with (×) contains that of the quadratic ones (5.41) marked with (○). □

Table 5.2 offers an overview of the previous results on the stabilisation of TS descriptor models up to any sum relaxation scheme; it includes quadratic as well as non-quadratic approaches.

A Word on LFT Stabilisation

LFT models are not considered in this chapter for several reasons: first, the fact that descriptor models encompass LFT ones; second, for brevity, given that output feedback for LFT systems, in Chap. 6, naturally includes state feedback as a particular case.

Table 5.2 Stabilisation of descriptor TS models

Lyapunov function and control law	Conditions
Continuous quadratic $V(\bar{x}) = \bar{x}^T \bar{E}^T X \bar{x}$, $X = \begin{bmatrix} X_1 & 0 \\ X_3 & X_4 \end{bmatrix}$, $u(t) = \bar{F}_{hv} X \bar{x}(t)$ $\bar{F}_{hv} = \begin{bmatrix} F_{hv} & 0 \end{bmatrix}$.	$P_1 = P_1^T > 0$, $\begin{bmatrix} P_3 + (*) & (*) \\ A_h P_1 + B_h F_{hv} + P_4^T - E_v P_3 & -E_v P_4 + (*) \end{bmatrix} < 0$, $X = P^{-1} = \begin{bmatrix} P_1 & 0 \\ P_3 & P_4 \end{bmatrix}^{-1}$.
Continuous semi-quadratic $V(\bar{x}) = \bar{x}^T \bar{E}^T X(z) \bar{x}$, $X(z) = \begin{bmatrix} X_1 & 0 \\ X_3(z) & X_4(z) \end{bmatrix}$, $u(t) = \bar{F}_{hv} X(z) \bar{x}(t)$, $\bar{F}_{hv} = \begin{bmatrix} F_{hv} & 0 \end{bmatrix}$.	$P_1 = P_1^T > 0$, $\begin{bmatrix} P_{3hh} + (*) & (*) \\ A_h P_1 + B_h F_{hv} - E_v P_{3hh} + (P_{4hh})^T & -E_v P_{4hh} + (*) \end{bmatrix} < 0$, $X(z) = P_{hh}^{-1} = \begin{bmatrix} P_1 & 0, \\ P_{3hh} & P_{4hh} \end{bmatrix}^{-1}$.
Discrete quadratic $V(x) = x^T X x$, $u(t) = F_{hv} X x(t)$.	$P = P^T > 0$, $\begin{bmatrix} -P & (*) \\ A_h P + B_h F_{hv} & 0.5P - E_v P + (*) \end{bmatrix} < 0$, $X = P^{-1}$.
Discrete non-quadratic $V(x) = x^T X_h x$, $u(t) = F_{hv} H_{hv}^{-1} x(t)$.	$P_h = P_h^T > 0$, $\begin{bmatrix} 0.5P_h - H_{hv} + (*) & (*) \\ A_h H_{hv} + B_h F_{hv} & (0.5I - E_v) P_{h(t+1)} + (*) \end{bmatrix} < 0$, $X_h = P_h^{-1}$.

5.6 Performance Specifications and Robustness

Among the numerous choices for performance indices that can be adapted to the different stability and stabilisation schemes of the previous chapters, this section focuses on giving the main idea of two of them for TS models: conditions for imposing bounds on the input and output vectors, and $\mathcal{H}_\infty$ disturbance attenuation. Examples illustrating these results are presented.

5.6.1 *Constraints on the System Input and Output*

It is easy to understand that, in order to constrain the input or output norm of a system, initial conditions should be restricted. To incorporate such restrictions into the Lyapunov-based analysis and synthesis results of the previous chapters, the following result comes at hand.

Lemma 5.1 *Let us assume that the initial condition* $\boldsymbol{x}(t_0) = \boldsymbol{x}_0$ *of a given nonlinear system is unknown, but bounded as* $\|\boldsymbol{x}_0\| \leq \sigma^2$*. If* $V(\boldsymbol{x}) = \boldsymbol{x}^T P^{-1}\boldsymbol{x}$*,* $P = P^T > 0$*, is a Lyapunov function candidate, then*

$$P \geq \sigma^2 I \Rightarrow \begin{bmatrix} 1 & \boldsymbol{x}_0^T \\ \boldsymbol{x}_0 & P \end{bmatrix} \geq 0. \tag{5.42}$$

Proof The fact that $\|\boldsymbol{x}_0\| \leq \sigma^2$ is equivalent to $\boldsymbol{x}_0^T\boldsymbol{x}_0 \leq \sigma^2$; therefore,

$$P^{-1} \leq \frac{1}{\sigma^2} I \Rightarrow \boldsymbol{x}_0^T P^{-1}\boldsymbol{x}_0 \leq \frac{1}{\sigma^2}\boldsymbol{x}_0^T\boldsymbol{x}_0 \leq \sigma^2\frac{1}{\sigma^2} \leq 1,$$

which, by Schur complement (A.1), yields the desired result. □

Consider a nonlinear output which has already been modelled via a convex structure shared with the MFs of the TS model of a system, namely

$$\boldsymbol{y}(t) = \sum_{i=1}^{r} h_i(z) C_i \boldsymbol{x}(t) = C_h \boldsymbol{x}(t), \tag{5.43}$$

where $C_i, i \in \{1, 2, \ldots, r\}$, are known matrices corresponding to the vertex models.

Under the same assumptions on the Lyapunov function candidate in Lemma 5.1, the following result provides conditions for imposing a norm bound on the output (5.43).

Theorem 5.11 *Let us assume that the initial condition* $\boldsymbol{x}_0$ *of a given nonlinear system is unknown, but bounded as* $\|\boldsymbol{x}_0\| \leq \sigma^2$*. If* $V(\boldsymbol{x}) = \boldsymbol{x}^T P^{-1}\boldsymbol{x}$*,* $P = P^T > \sigma^2 I$*, is a Lyapunov function candidate, and* (5.43) *is the output of the system, then*

$$\begin{bmatrix} P & PC_i^T \\ C_i P & \lambda^2 I \end{bmatrix} > 0,\ i \in \{1, 2, \ldots, r\} \tag{5.44}$$

guarantees $\|\boldsymbol{y}(t)\| \leq \lambda$*,* $\forall t \geq t_0$*.*

Proof If (5.44) hold, then convexity of MFs $h_i(z)$ guarantees

$$\sum_{i=1}^{r} h_i(z) \begin{bmatrix} P & PC_i^T \\ C_i P & \lambda^2 I \end{bmatrix} = \begin{bmatrix} P & PC_h^T \\ C_h P & \lambda^2 I \end{bmatrix} > 0,$$

which, by Schur complement (A.1), is equivalent to

$$P - \frac{1}{\lambda^2} P C_h^T C_h P \geq 0 \Rightarrow \frac{1}{\lambda^2} C^T C \leq P^{-1},$$

which, after pre- and post-multiplying by $\boldsymbol{x}^T(t)$ and $\boldsymbol{x}(t)$, respectively, yields

$$\frac{1}{\lambda^2}x^T(t)C^TCx(t) = \frac{1}{\lambda^2}\|y(t)\|^2 \leq x^T(t)P^{-1}x(t) = V(x) \leq V(x_0) \leq 1,$$

where the last inequality is guaranteed by Lemma 5.1 as its hypotheses are met.

Taking the second and last terms in the inequality chain, it follows that $\|y(t)\| \leq \lambda$ as desired. □

Remark 5.8 LMIs (5.42) in Lemma 5.1 and (5.44) in Theorem 5.11 are supposed to be added to any analysis or design conditions in the previous chapters to guarantee the desired output constraint. Incorporating these LMIs may require adjustments; for instance, performing operations to use $X = P^{-1}$ as a decision variable instead of P or considering outputs that include direct connections with the input.

Let us now consider the case when the input has to be bounded, a situation that frequently occurs in real-time applications. Our objective is to develop conditions so that the input is norm-bounded, i.e. $\|u(t)\| \leq \mu$, which, depending on the control structure, may lead to slight variations on the conditions. Let us assume a control law similar to (5.12) is employed, i.e.

$$u(t) = \left(\sum_{i=1}^{r} h_i(z)F_i\right)\left(\sum_{i=1}^{r} h_i(z)H_i\right)^{-1} x(t) = F_h H_h^{-1} x(t), \tag{5.45}$$

which includes the PDC as a particular case. Assuming the Lyapunov function candidate in Lemma 5.1 is used with the hypotheses therein, the following result provides conditions for guaranteeing a norm bound on the input (5.45).

Theorem 5.12 *Let us assume that the initial condition x_0 of a given nonlinear system is unknown, but bounded as $\|x_0\| \leq \sigma^2$. If $V(x) = x^T P^{-1} x$, $P = P^T > \sigma^2 I$, is a Lyapunov function candidate and the control law (5.45) is employed, then*

$$\begin{bmatrix} H_i + H_i^T - P & F_i^T \\ F_i & \mu^2 I \end{bmatrix} \geq 0 \tag{5.46}$$

guarantees $\|u(t)\| \leq \mu$, $\forall t \geq t_0$.

Proof Convexity of MFs $h_i(z)$ along with (5.46) ensures that

$$\sum_{i=1}^{r} h_i(z) \begin{bmatrix} H_i + H_i^T - P & F_i^T \\ F_i & \mu^2 I \end{bmatrix} = \begin{bmatrix} H_h + H_h^T - P & F_h^T \\ F_h & \mu^2 I \end{bmatrix} \geq 0,$$

which, by Schur complement (A.1) and Property A.4, implies

$$H_h^T P^{-1} H_h - \frac{1}{\mu^2} F_h^T F_h \geq 0.$$

Applying the congruence (Property A.1) with H^{-T} and H^{-1} leads to

$$\frac{1}{\mu^2} H_h^{-T} F_h^T F_h H_h^{-1} \leq P^{-1} \iff \frac{1}{\mu^2} (F_h H_h^{-1})^T (F_h H_h^{-1}) \leq P^{-1},$$

which, after pre- and post-multiplying by $\boldsymbol{x}^T(t)$ and $\boldsymbol{x}(t)$, respectively, yields

$$\begin{aligned}\frac{1}{\mu^2}\boldsymbol{x}^T(t)(F_h H_h^{-1})^T(F_h H_h^{-1})\boldsymbol{x}(t) =& \frac{1}{\mu^2}\boldsymbol{u}^T(t)\boldsymbol{u}(t)\\ \leq& \boldsymbol{x}^T(t)P^{-1}\boldsymbol{x}(t) = V(\boldsymbol{x}) \leq V(\boldsymbol{x}_0) \leq 1,\end{aligned}$$

where the last inequality is ensured by Lemma 5.1 whose hypotheses are met.

Thus, taking the second and last terms in the inequality chain above yields the desired result $\|\boldsymbol{u}(t)\| \leq \mu^2$. □

Remark 5.9 Again, LMIs (5.42) in Lemma 5.1 and (5.46) in Theorem 5.12 should be added to any analysis or design conditions in the previous chapters to guarantee the desired input constraint. Moreover, since convex conditions can be combined in a single optimisation problem without further adjustments, conditions for input/output constraints can be combined too; these conditions do not change if the TS model is continuous- or discrete-time. Naturally, if the Lyapunov function changes, conditions may change accordingly.

Before we proceed to the next two sections, we would like to stress the fact that more general forms of TS models can be found in the literature, capturing further characteristics, e.g.: uncertain systems [5, 45] for which $\mathcal{H}_2$ and $\mathcal{H}_\infty$ specifications are made [46–48], delayed systems [49, 50], systems with bounded external disturbances [51], periodic systems [52], switching systems [53–55], and so on [5, 9, 56].

In the next two sections, uncertain TS models with external disturbances will be considered, i.e.

$$\begin{aligned}\delta\boldsymbol{x}(t) &= (A_h + \Delta A_h)\,\boldsymbol{x}(t) + (B_h + \Delta B_h)\,\boldsymbol{u}_k + B_{dh}\boldsymbol{d}(t)\\ \boldsymbol{y}(t) &= (C_h + \Delta C_h)\,\boldsymbol{x}(t) + D_{dh}\boldsymbol{d}(t)\end{aligned} \tag{5.47}$$

under the usual definitions for operator $\delta\boldsymbol{x}(t)$, state $\boldsymbol{x}$, input $\boldsymbol{u}$, output $\boldsymbol{y}$, and disturbance $\boldsymbol{d}$; with A_i, B_i, C_i, B_{di}, and D_{di}, $i \in \{1, 2, \ldots, r\}$, being vertex matrices of appropriate dimensions which, if combined in convex sums via the MFs $h_i(z)$, form A_h, B_h, C_h, B_{dh}, and D_h, respectively. We detail those which are used for the first time:

$$B_{dh} = \sum_{i=1}^{r} h_i(z)B_{di}, \; D_{dh} = \sum_{i=1}^{r} h_i(z)D_{di},$$

$$\Delta A_h = \sum_{i=1}^{r} h_i(z)\Delta A_i, \; \Delta B_h = \sum_{i=1}^{r} h_i(z)\Delta B_i, \; \Delta C_h = \sum_{i=1}^{r} h_i(z)\Delta C_i.$$

The first two convex sums above are the distribution matrices for the disturbance vector $\boldsymbol{d}(t)$ with respect to the state and output equations, respectively; the rest of them gather the parameter uncertainties, which are usually written as in [5]: $\Delta A_i = H_a \Delta_a E_{ai}$, $\Delta B_i = H_b \Delta_b E_{bi}$, $\Delta C_i = H_c \Delta_c E_{ci}$ with H_a, H_b, H_c, E_{ai}, E_{bi}, and E_{ci} being constant matrices, and Δ_a, Δ_b, and Δ_c time-varying[3] expressions that satisfy conditions

$$\Delta_a^T \Delta_a \leq I \quad \Delta_b^T \Delta_b \leq I \quad \Delta_c^T \Delta_c \leq I. \tag{5.48}$$

Depending on what parts of the model (5.47) are considered and what the desired goals are, numerous results exist. We focus on two important cases: $\mathcal{H}_\infty$ disturbance rejection and robust stabilisation conditions in the presence of uncertainties.

5.6.2 $\mathcal{H}_\infty$ Attenuation

We would like to recall the origin of the $\mathcal{H}_\infty$ denomination, which has been borrowed from the field of linear systems:

Definition 5.1 [57] The $\mathcal{H}_\infty$ norm of the transfer function $H(s)$ is

$$\|H\|_\infty \equiv \sup\{\|H(s)\| : \mathrm{Re}(s) > 0\}.$$

This has the same meaning as in (A.48) in Appendix A.

Definition 5.2 [57] A linear system $\dot{\boldsymbol{x}}(t) = A\boldsymbol{x}(t) + B\boldsymbol{u}(t)$, $\boldsymbol{y}(t) = C\boldsymbol{x}(t) + D\boldsymbol{u}(t)$ is *non-expansive* if

$$\int_0^T \boldsymbol{y}^T(t)\boldsymbol{y}(t)dt \leq \int_0^T \boldsymbol{u}^T(t)\boldsymbol{u}(t)dt.$$

Non-expansivity of a linear system is equivalent to $\|H\|_\infty \leq 1$ with $H(s)$ being its transfer function. Naturally, as transfer functions do not apply in nonlinear contexts, the name has only historical roots, relying on the non-expansivity concept to provide a clear meaning with regard to disturbances $\boldsymbol{d}(t)$ as inputs and their effect on the system output $\boldsymbol{y}(t)$ [58].

Definition 5.3 A nonlinear system subject to a disturbance signal $\boldsymbol{d}(t)$ is said to have an $\mathcal{H}_\infty$ attenuation level $\gamma > 0$ if

$$\sup_{\boldsymbol{d}(t)\neq 0} \frac{\|\boldsymbol{y}(t)\|_2}{\|\boldsymbol{d}(t)\|_2} \leq \gamma \tag{5.49}$$

holds with respect to its output $\boldsymbol{y}(t)$, where $\|\cdot\|_2$ stands for the $\mathcal{L}_2$ signal norm in (A.45) in Appendix A.

---

[3] For the ease of notation, the explicit time dependence is omitted.

Guaranteeing an $\mathcal{H}_\infty$ attenuation level $\gamma > 0$ is also known as $\mathcal{H}_\infty$ disturbance rejection. It will become apparent that having an $\mathcal{H}_\infty$ attenuation level $\gamma > 0$ for a linear system reduces to non-expansivity between the disturbance $\boldsymbol{d}(t)$ (considered as an input) and the system output $\boldsymbol{y}(t)$, over a time horizon from 0 to ∞.

We are now ready to discuss $\mathcal{H}_\infty$ controller design for nonlinear systems via equivalent TS models. To this end, consider first a continuous-time disturbed TS model

$$\begin{aligned}\dot{\boldsymbol{x}}(t) =& A_h\boldsymbol{x}(t) + B_h\boldsymbol{u}(t) + B_{dh}\boldsymbol{d}(t),\\ \boldsymbol{y}(t) =& C_h\boldsymbol{x}(t),\end{aligned} \tag{5.50}$$

with initial conditions $\boldsymbol{x}(0) = 0$, which may arise from rewriting a nonlinear system like (3.8) within a compact set $\mathcal{C} \subset \mathbb{R}^p$ of the scheduling variables z.

The goal is to obtain the best $\mathcal{L}_2 \to \mathcal{L}_2$ gain via a PDC control law $\boldsymbol{u}(t) = F_h\boldsymbol{x}(t)$, i.e. to minimise the worst case (5.49).

Theorem 5.13 *The TS model (5.50) under the PDC control law (5.3) $\boldsymbol{u}(t) = F_h x(t)$ has an $\mathcal{H}_\infty$ attenuation level $\gamma > 0$ for any trajectory in the outermost ellipsoid $\boldsymbol{x}^T P\boldsymbol{x} \le k$, $P = X^{-1}$, $k > 0$, within a compact $\mathcal{X}$ of the state space guaranteeing $z \in \mathcal{C} \subseteq \mathbb{R}^p$, if there exist matrices $X = X^T > 0$, M_i, $i \in \{1, 2, \ldots, r\}$, such that*

$$\begin{bmatrix} XA_h^T + A_hX + M_h^TB_h^T + B_hM_h & B_{dh} & XC_h^T \\ B_{dh}^T & -\gamma^2 I & 0 \\ C_hX & 0 & -I \end{bmatrix} < 0 \tag{5.51}$$

holds. If this is the case, the PDC gains are given by $F_i = M_iX^{-1}$, $i \in \{1, 2, \ldots, r\}$.

Proof Based on the fact that $X = X^T > 0$ and $P = X^{-1}$, consider a quadratic function $V = \boldsymbol{x}^T P\boldsymbol{x}$. It is well known that if

$$\dot{V} + \boldsymbol{y}^T\boldsymbol{y} - \gamma^2\boldsymbol{d}^T\boldsymbol{d} \le 0 \tag{5.52}$$

holds, then (5.49) holds too [57, 59]. To see this, consider the integration of this expression over $[0, \infty)$:

$$V(\boldsymbol{x}(\infty)) - V(\boldsymbol{x}(0)) \le \int_0^\infty \left(\gamma^2\boldsymbol{d}^T\boldsymbol{d} - \boldsymbol{y}^T\boldsymbol{y}\right) dt.$$

Since the origin is assumed to be asymptotically stable, $\boldsymbol{x}(\infty) = 0$ and with initial conditions $\boldsymbol{x}(0) = 0$, we obtain

$$0 < \int_0^\infty \left(\gamma^2\boldsymbol{d}^T\boldsymbol{d} - \boldsymbol{y}^T\boldsymbol{y}\right) dt,$$

which is equivalent to

$$\int_0^\infty \boldsymbol{y}^T \boldsymbol{y} dt < \gamma^2 \int_0^\infty \boldsymbol{d}^T \boldsymbol{d} dt \Leftrightarrow \|\boldsymbol{y}\|_2 < \gamma \|\boldsymbol{d}\|_2$$

and γ satisfies (5.49).

Condition (5.52) translates into the following:

$$\begin{aligned}\dot{V}+\boldsymbol{y}^T\boldsymbol{y}-\gamma^2\boldsymbol{d}^T\boldsymbol{d} &= 2\boldsymbol{x}^T P\left((A_h+B_hF_h)\boldsymbol{x}+B_{dh}\boldsymbol{d}\right)+\boldsymbol{x}^T C_h^T C_h\boldsymbol{x}-\gamma^2\boldsymbol{d}^T\boldsymbol{d} \\ &= \begin{bmatrix}\boldsymbol{x}\\ \boldsymbol{d}\end{bmatrix}^T \begin{bmatrix} P(A_h+B_hF_h)+(*)+C_h^TC_h & (*) \\ B_{dh}^T P & -\gamma^2 I\end{bmatrix}\begin{bmatrix}\boldsymbol{x}\\ \boldsymbol{d}\end{bmatrix} < 0.\end{aligned}$$

Using the Schur complement in Lemma A.1, the last inequality is guaranteed if

$$\begin{bmatrix} P(A_h+B_hF_h)+(A_h+B_hF_h)^T P & PB_{dh} & C_h^T \\ B_{dh}^T P & -\gamma^2 I & 0 \\ C_h & 0 & -I \end{bmatrix} < 0,$$

which, by means of the congruence (Property A.1) with block-diag(X, I, I) and substitutions $X = P^{-1}$ and $M_h = F_h X, i \in \{1, 2, \ldots, r\}$, yields the desired conditions (5.51). Assumptions 4.1 and 4.2 guarantee there is a compact $\mathcal{X}$ ensuring convexity of MFs for $\boldsymbol{z} \in \mathcal{C}$; therefore, the reasoning above holds for trajectories within the outermost ellipsoid $\boldsymbol{x}^T P\boldsymbol{x} \le k$, $P = X^{-1}$, $k > 0$, contained in such $\mathcal{X}$. □

If the output equation has the form

$$\boldsymbol{y}(t) = C_h\boldsymbol{x}(t) + D_{dh}\boldsymbol{d}(t),$$

instead of $\boldsymbol{y}(t) = C_h\boldsymbol{x}(t)$ in (5.50), i.e. the disturbance is also affecting the output, a similar result can be found following the outline of the previous proof [4, 47], where

$$\begin{bmatrix} XA_h^T + A_hX + M_h^T B_h^T + B_hM_h & B_{dh} & XC_h^T \\ B_{dh}^T & -\gamma^2 I & D_{dh}^T \\ C_hX & D_{dh} & -I \end{bmatrix} < 0 \tag{5.53}$$

replaces (5.51).

Remark 5.10 As we have done in previous chapters, results are given as nested convex sums of matrix expressions. Sufficient LMI conditions guaranteeing them can be obtained in a variety of ways as shown in Appendix D. Moreover, minimising $\gamma > 0$ in (5.51) or (5.53), which means maximising disturbance attenuation with respect to the output in an $\mathcal{L}_2 - \mathcal{L}_2$-gain sense, is also an LMI problem, more specifically, a *generalised eigenvalue problem* (GEVP) which can be efficiently solved by most of the LMI toolboxes. See Appendix C for details.

The design of a controller with disturbance attenuation is illustrated in the following example.

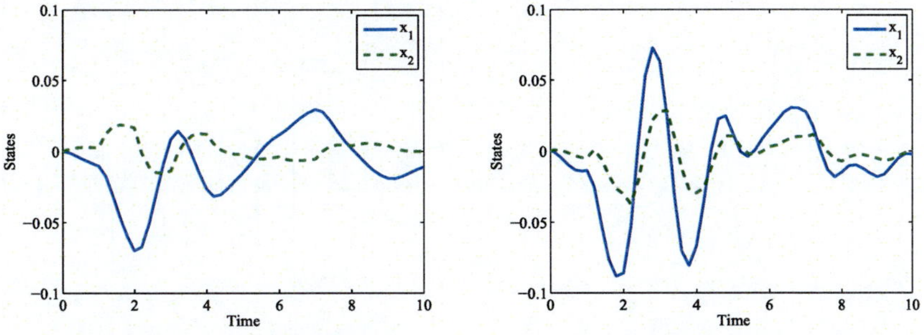

(a) Stabilisation with attenuation.

(b) Stabilisation without attenuation.

Fig. 5.8 Simulation results for Example 5.10

Example 5.10 Consider a TS fuzzy system of the form (5.50), with $r = 2$ vertex models described by

$$A_1 = \begin{bmatrix} -2.75 & 0.75 \\ -0.05 & -0.1 \end{bmatrix}, \ B_1 = \begin{bmatrix} 1 \\ 1 \end{bmatrix}, \ B_{d1} = \begin{bmatrix} 0 \\ 0.1 \end{bmatrix}, \ C_1 = \begin{bmatrix} -1 & 0 \end{bmatrix},$$

$$A_2 = \begin{bmatrix} 0.5 & 1.5 \\ -0.6 & 0.3 \end{bmatrix}, \ B_2 = \begin{bmatrix} 2 \\ 1 \end{bmatrix}, \ B_{d2} = \begin{bmatrix} -0.1 \\ 0.1 \end{bmatrix}, \ C_2 = \begin{bmatrix} -1 & 1 \end{bmatrix},$$

and membership functions $h_1(z) = 1/(1 + (x_1 + x_2)^2)$ and $h_2(z) = 1 - h_1(z)$, $z = \boldsymbol{x}$, which hold the convex sum property everywhere in the state space.

The highest $\mathcal{H}_\infty$ attenuation, i.e. the smallest γ, is $\gamma_{\min}^2 = 0.01$. For an attenuation of $\gamma^2 = 0.05$, the results are presented in Fig. 5.8a, while Fig. 5.9a,b presents the control signal and the generated disturbance signal d (band-limited white noise with power 0.1 and sample time 1). For the sake of comparison, Fig. 5.8b presents the results without considering $\mathcal{H}_\infty$ attenuation, with the same gains as those computed in Example 5.1 without the decay rate. □

Let us now consider $\mathcal{H}_\infty$ controller design of discrete-time TS models of the form:

$$\begin{aligned} \boldsymbol{x}(t+1) &= A_h \boldsymbol{x}(t) + B_h \boldsymbol{u}(t) + B_{dh} \boldsymbol{d}(t), \\ \boldsymbol{y}(t) &= C_h \boldsymbol{x}(t) + D_{dh} \boldsymbol{u}(t) + K_{dh} \boldsymbol{d}(t), \end{aligned} \tag{5.54}$$

under the usual definitions of time, states, inputs, disturbances, outputs, and vertex matrices. The formerly used controller proposal

$$\boldsymbol{u}(t) = F_h H^{-1} \boldsymbol{x}(t) \tag{5.55}$$

will be considered along with initial condition $\boldsymbol{x}(0) = 0$.

Theorem 5.14 *The TS model (5.54) under the control law (5.55) has an $\mathcal{H}_\infty$ attenuation level $\gamma > 0$ for any trajectory in the outermost ellipsoid $\boldsymbol{x}^T H^{-T} P H^{-1} \boldsymbol{x} \le k$,*

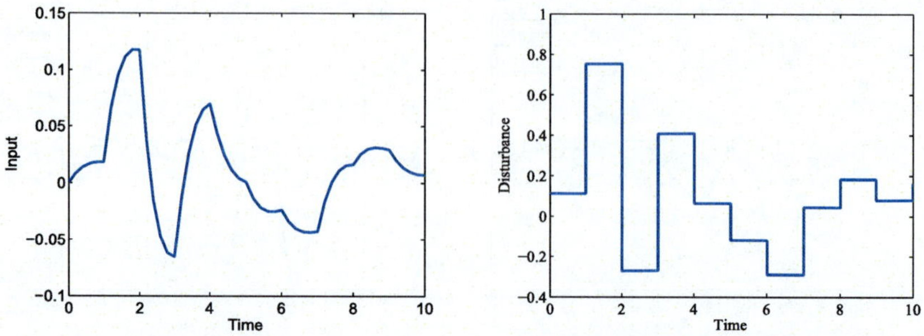

(a) Control signal u. (b) Disturbance signal d.

Fig. 5.9 Control and disturbance signals for Example 5.10

$k > 0$, within a compact $\mathcal{X}$ of the state space guaranteeing $z \in \mathcal{C} \subseteq \mathbb{R}^p$, if there exist matrices $P = P^T > 0$, F_i, $i \in \{1, 2, \ldots, r\}$, and H such that

$$\begin{bmatrix} -P & 0 & H^T A_h^T + F_h^T B_h^T & H^T C_h^T + F_h^T D_{dh}^T \\ 0 & -\gamma I & B_{dh}^T & K_{dh}^T \\ A_h H + B_h F_h & B_{dh} & -H - H^T + P & 0 \\ C_h H + D_{dh} F_h & K_{dh} & 0 & -\gamma I \end{bmatrix} < 0. \tag{5.56}$$

Proof From $P = P^T > 0$, it follows that $V = \boldsymbol{x}(t)^T H^{-T} P H^{-1} \boldsymbol{x}(t)$ is positive-definite and quadratic. For such functions, it is well known that condition

$$\Delta V + \boldsymbol{y}^T \boldsymbol{y} - \gamma^2 \boldsymbol{d}^T \boldsymbol{d} \leq 0 \tag{5.57}$$

guarantees (5.49) in the discrete-time case [59]. To see this, it is enough to perform a summation from 0 to ∞ on both sides of (5.57); the arguments being the same as those in the first part of the proof of Theorem 5.13.

The closed-loop system resulting from the substitution of (5.55) in (5.54) is

$$\begin{aligned} \boldsymbol{x}(t+1) &= \left(A_h + B_h F_h H^{-1}\right) \boldsymbol{x}(t) + B_{dh} \boldsymbol{d}(t), \\ \boldsymbol{y}(t) &= \left(C_h + D_{dh} F_h H^{-1}\right) \boldsymbol{x}(t) + K_{dh} \boldsymbol{d}(t). \end{aligned} \tag{5.58}$$

It will be substituted in the forward difference of $V = \boldsymbol{x}(t)^T H^{-T} P H^{-1} \boldsymbol{x}(t)$ (arguments omitted when convenient):

$$\begin{aligned} \Delta V =& V(t+1) - V(t) \\ =& \boldsymbol{x}(t+1)^T H^{-T} P H^{-1} \boldsymbol{x}(t+1) - \boldsymbol{x}(t)^T H^{-T} P H^{-1} \boldsymbol{x}(t) \\ =& \begin{bmatrix} \boldsymbol{x} \\ \boldsymbol{d} \end{bmatrix}^T \begin{bmatrix} \Omega - H^{-T} P H^{-1} & (*) \\ B_{dh}^T H^{-T} P H^{-1} (A_h + B_h F_h H^{-1}) & B_{dh}^T H^{-T} P H^{-1} B_{dh} \end{bmatrix} \begin{bmatrix} \boldsymbol{x} \\ \boldsymbol{d} \end{bmatrix}, \end{aligned}$$

where $\Omega = (A_h + B_h F_h H^{-1})^T H^{-T} P H^{-1}(A_h + B_h F_h H^{-1})$.

On the other hand, $\boldsymbol{y}(t)^T \boldsymbol{y}(t) - \gamma^2 \boldsymbol{d}(t)^T \boldsymbol{d}(t)$ can be written as

$$\begin{bmatrix} \boldsymbol{x}(t) \\ \boldsymbol{d}(t) \end{bmatrix}^T \begin{bmatrix} (C_h + D_{dh} F_h H^{-1})^T (C_h + D_{dh} F_h H^{-1}) & (*) \\ K_{dh}^T (C_h + D_{dh} F_h H^{-1}) & K_{dh}^T K_{dh} - \gamma^2 I \end{bmatrix} \begin{bmatrix} \boldsymbol{x}(t) \\ \boldsymbol{d}(t) \end{bmatrix}.$$

With the S-procedure in Property A.2, (5.57) is ensured if

$$\begin{aligned}
&\begin{bmatrix} \Omega - H^{-T} P H^{-1} & (*) \\ B_{dh}^T H^{-T} P H^{-1} (A_h + B_h F_h H^{-1}) & B_{dh}^T H^{-T} P H^{-1} B_{dh} \end{bmatrix} \\
&+ \begin{bmatrix} \gamma^{-1} (C_h + D_{dh} F_h H^{-1})^T (C_h + D_{dh} F_h H^{-1}) & (*) \\ \gamma^{-1} K_{dh}^T (C_h + D_{dh} F_h H^{-1}) & \gamma^{-1} K_{dh}^T K_{dh} - \gamma I \end{bmatrix} \\
&= \begin{bmatrix} (A_h + B_h F_h H^{-1})^T \\ B_{dh}^T \end{bmatrix} H^{-T} P H^{-1} \begin{bmatrix} A_h + B_h F_h H^{-1} & B_{dh} \end{bmatrix} \\
&+ \gamma^{-1} \begin{bmatrix} (C_h + D_{dh} F_h H^{-1})^T \\ K_{dh}^T \end{bmatrix} \begin{bmatrix} C_h + D_{dh} F_h H^{-1} & K_{dh} \end{bmatrix} \\
&- \begin{bmatrix} H^{-T} P H^{-1} & 0 \\ 0 & \gamma I \end{bmatrix} \\
&= \begin{bmatrix} A_h + B_h F_h H^{-1} & B_{dh} \\ C_h + D_{dh} F_h H^{-1} & K_{dh} \end{bmatrix}^T \begin{bmatrix} H^{-T} P H^{-1} & 0 \\ 0 & \gamma^{-1} I \end{bmatrix} \begin{bmatrix} A_h + B_h F_h H^{-1} & B_{dh} \\ C_h + D_{dh} F_h H^{-1} & K_{dh} \end{bmatrix} \\
&- \begin{bmatrix} H^{-T} P H^{-1} & 0 \\ 0 & \gamma I \end{bmatrix} < 0.
\end{aligned}$$

Using the congruence (Property A.1) with block-diag(H^T, I), the previous inequality turns into

$$(*) \begin{bmatrix} H^{-T} P H^{-1} & 0 \\ 0 & \gamma^{-1} I \end{bmatrix} \begin{bmatrix} A_h H + B_h F_h & B_{dh} \\ C_h H + D_{dh} F_h & K_{dh} \end{bmatrix} - \begin{bmatrix} P & 0 \\ 0 & \gamma I \end{bmatrix} < 0,$$

which, by Schur complement in Lemma A.1 and Property A.4, results in conditions (5.56).

The reasoning above depends on MFs $h_i(z)$ holding the convex sum property, which is ensured for $z \in \mathcal{C}$. This, in turn, is ensured for trajectories contained in the outermost ellipsoid $\boldsymbol{x}^T H^{-T} P H^{-1} \boldsymbol{x} \leq k$, $k > 0$, in $\mathcal{X}$, as deduced from assumptions 4.1 and 4.2. □

Naturally, different results can be obtained if another quadratic function is employed. For instance, using a simpler one like $V = \boldsymbol{x}(t)^T P^{-1} \boldsymbol{x}(t)$, with $P = P^T > 0$ yields the following result:

Theorem 5.15 *The TS model* (5.54) *under the control law* (5.55) *has an $\mathcal{H}_\infty$ attenuation level $\gamma > 0$ for any trajectory in the outermost ellipsoid $\boldsymbol{x}^T P^{-1} \boldsymbol{x} \leq k$, $k > 0$, within a compact set $\mathcal{X}$ of the state space guaranteeing $z \in \mathcal{C} \subseteq \mathbb{R}^p$, if there exist matrices $P = P^T$, F_i, $i \in \{1, 2, \ldots, r\}$, and H, such that*

$$\begin{bmatrix} -H - H^T + P & 0 & H^T A_h^T + F_h^T B_h^T & H^T C_h^T + F_h^T D_{dh}^T \\ 0 & -\gamma I & B_{dh}^T & K_{dh}^T \\ A_h H + B_h F_h & B_{dh} & -P & 0 \\ C_h H + D_{dh} F_h & K_{dh} & 0 & -\gamma I \end{bmatrix} < 0. \tag{5.59}$$

Proof Due to $P = P^T > 0$, $V = \boldsymbol{x}(t)^T P^{-1} \boldsymbol{x}(t)$ is a positive-definite quadratic function whose forward difference is

$$\begin{aligned} \Delta V =& V(t+1) - V(t) \\ =& \boldsymbol{x}(t+1)^T P^{-1} \boldsymbol{x}(t+1) - \boldsymbol{x}(t)^T P^{-1} \boldsymbol{x}(t) \\ =& \begin{bmatrix} \boldsymbol{x}(t) \\ \boldsymbol{d}(t) \end{bmatrix}^T \begin{bmatrix} (A_h + B_h F_h H^{-1})^T P^{-1} (*) - P^{-1} & (*) \\ B_{dh}^T P^{-1} (A_h + B_h F_h H^{-1}) & B_{dh}^T P^{-1} B_{dh} \end{bmatrix} \begin{bmatrix} \boldsymbol{x}(t) \\ \boldsymbol{d}(t) \end{bmatrix}. \end{aligned}$$

Again, $\boldsymbol{y}(t)^T \boldsymbol{y}(t) - \gamma^2 \boldsymbol{d}(t)^T \boldsymbol{d}(t) < 0$ can be written as

$$\begin{bmatrix} \boldsymbol{x}(t) \\ \boldsymbol{d}(t) \end{bmatrix}^T \begin{bmatrix} (C_h + D_{dh} F_h H^{-1})^T (*) & (*) \\ K_{dh}^T (C_h + D_{dh} F_h H^{-1}) & K_{dh}^T K_{dh} - \gamma^2 I \end{bmatrix} \begin{bmatrix} \boldsymbol{x}(t) \\ \boldsymbol{d}(t) \end{bmatrix} < 0,$$

which, by the S-procedure in Property A.2, turns inequality (5.57) into

$$\begin{aligned} & \begin{bmatrix} (A_h + B_h F_h H^{-1})^T P^{-1} (*) - P^{-1} & (*) \\ B_{dh}^T P^{-1} (A_h + B_h L_h H^{-1}) & B_{dh}^T P^{-1} B_{dh} \end{bmatrix} \\ & + \begin{bmatrix} \gamma^{-1} (C_h + D_{dh} F_h H^{-1})^T (*) & (*) \\ \gamma^{-1} K_{dh}^T (C_h + D_{dh} F_h H^{-1}) & \gamma^{-1} K_{dh}^T K_{dh} - \gamma I \end{bmatrix} \\ & = (*) \begin{bmatrix} P^{-1} & 0 \\ 0 & \gamma^{-1} I \end{bmatrix} \begin{bmatrix} A_h + B_h F_h H^{-1} & B_{dh} \\ C_h + D_{dh} F_h H^{-1} & K_{dh} \end{bmatrix} - \begin{bmatrix} P^{-1} & 0 \\ 0 & \gamma I \end{bmatrix}. \end{aligned}$$

Applying the congruence (Property A.1) with block-diag(H^T, I) gives

$$(*) \begin{bmatrix} P^{-1} & 0 \\ 0 & \gamma^{-1} I \end{bmatrix} \begin{bmatrix} A_h H + B_h F_h & B_{dh} \\ C_h H + D_{dh} F_h & K_{dh} \end{bmatrix} - \begin{bmatrix} H^T P^{-1} H & 0 \\ 0 & \gamma I \end{bmatrix} < 0,$$

on which using the Schur complement in Lemma A.1 and Property A.4 yields the desired result (5.59).

Trajectories in the outermost ellipsoid $\boldsymbol{x}^T P^{-1} \boldsymbol{x} \leq k$, $k > 0$, within a compact set $\mathcal{X}$ of the state space guaranteeing $z \in \mathcal{C} \subseteq \mathbb{R}^p$ (which, by assumptions 4.1 and 4.2 must exist), must hold the specified $\mathcal{H}_\infty$ attenuation level as $z \in \mathcal{C}$ guarantees the MFs $h_i(z)$ employed in the steps above hold the convex sum property. □

Remark 5.11 As with conditions in Theorem 5.13, any relaxation scheme can be applied to derive LMI conditions guaranteeing (5.56) in Theorem 5.14 or (5.59) in Theorem 5.15; see Appendix D. The minimum $\gamma > 0$ satisfying (5.56) can also be found via a GEVP; see Appendix C for details.

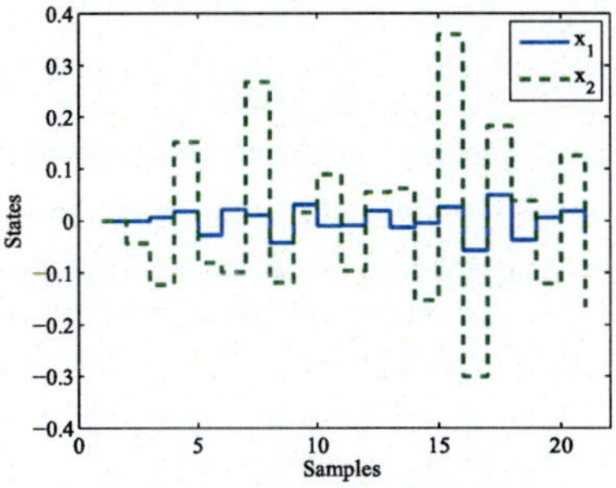

(a) Stabilisation with attenuation.

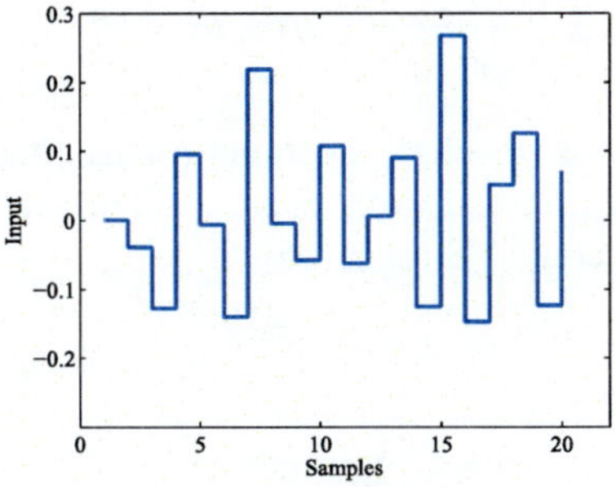

(b) Control signal u.

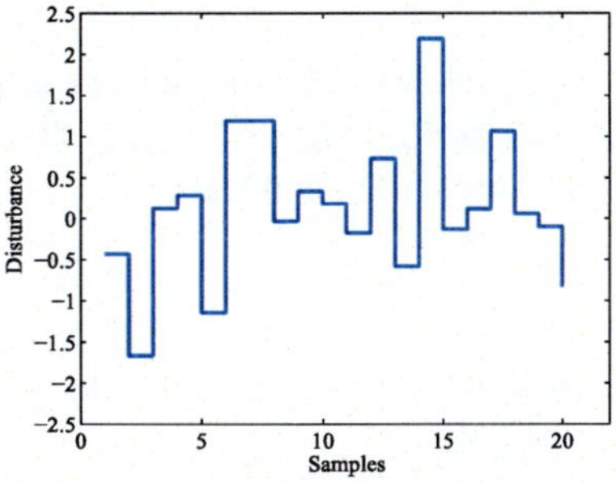

(c) Disturbance signal d.

Fig. 5.10 State, control, and disturbance signals for Example 5.11

Stabilisation of a discrete-time system with $\mathcal{H}_\infty$ attenuation is illustrated in the following example.

Example 5.11 Consider the TS system from Example 5.4 with the $r = 2$ vertex models described by

$$A_1 = \begin{bmatrix} -2.75 & 0.75 \\ -0.05 & -0.1 \end{bmatrix}, \; B_1 = \begin{bmatrix} 1 \\ 1 \end{bmatrix}, \; B_{d1} = \begin{bmatrix} 0 \\ 0.1 \end{bmatrix}, \; \begin{matrix} C_1 = \begin{bmatrix} -1 & 0 \end{bmatrix}, \\ K_{d1} = D_{d1} = 0, \end{matrix}$$

$$A_2 = \begin{bmatrix} 0.5 & 1.5 \\ -0.6 & 0.3 \end{bmatrix}, \; B_2 = \begin{bmatrix} 2 \\ 1 \end{bmatrix}, \; B_{d2} = \begin{bmatrix} -0.1 \\ 0.1 \end{bmatrix}, \; \begin{matrix} C_2 = \begin{bmatrix} -1 & 1 \end{bmatrix}, \\ K_{d2} = D_{d2} = 0. \end{matrix}$$

Again, membership functions $h_1(z) = 1/(1 + (x_1 + x_2)^2)$ and $h_2(z) = 1 - h_1(z)$, globally holding the convex sum property, will be considered.

Using Theorem 5.15, the highest $\mathcal{H}_\infty$ attenuation, i.e. the smallest $\gamma > 0$, is $\gamma = 0.47$. For this attenuation, the results are presented in Fig. 5.10a, while Fig. 5.10b,c presents the control signal $u(t)$ and the generated disturbance signal $d(t)$ (band-limited white noise with power 0.1 and sample time 1). □

5.6.3 Robust Control

Several results concerning robust control of TS systems can be found in the literature. In order to show the main principle, we consider the very basic norm-bounded uncertain TS model [5]:

$$\dot{\boldsymbol{x}}(t) = (A_h + \Delta A_h)\,\boldsymbol{x}(t) + (B_h + \Delta B_h)\,\boldsymbol{u}(t) \tag{5.60}$$

in continuous time, and

$$\boldsymbol{x}(t+1) = (A_h + \Delta A_h)\,\boldsymbol{x}(t) + (B_h + \Delta B_h)\,\boldsymbol{u}(t) \tag{5.61}$$

in discrete time, with the parameter uncertainties written as in [5]: $\Delta A_i = H_a \Delta_a E_{ai}$, $\Delta B_i = H_b \Delta_b E_{bi}$, where the matrices H_a, H_b, E_{ai}, and E_{bi} are constants, and Δa and Δb are time-varying and satisfy the conditions

$$\Delta_a^T \Delta_a \leq I \quad \Delta_b^T \Delta_b \leq I. \tag{5.62}$$

The goal is to derive stability conditions for the closed-loop system for every variation of the uncertainties in their domain.

Let us consider first the continuous-time case with the PDC law (5.3), i.e. $\boldsymbol{u}(t) = F_h \boldsymbol{x}(t)$. The closed-loop dynamics are

$$\dot{\boldsymbol{x}}(t) = (A_h + B_h F_h + \Delta A_h + \Delta B_h F_h)\,\boldsymbol{x}(t).$$

For this, the following result can be formulated.

Theorem 5.16 *The origin of the continuous TS model* (5.60) *with the PDC control law $\boldsymbol{u}(t) = F_h \boldsymbol{x}(t)$ is robustly asymptotically stable, i.e. whenever the uncertainties satisfy the boundary conditions* (5.62)*, if there exist matrices $X = X^T > 0$, and M_i, $i \in \{1, 2, \ldots, r\}$, and scalars $\tau_a > 0$ and $\tau_b > 0$, such that*

$$\begin{bmatrix} A_h X + B_h M_h + (*) + \tau_a H_a H_a^T + \tau_b H_b H_b^T & X E_{ah}^T & M_j^T E_{bh}^T \\ E_{ah} X & -\tau_a I & 0 \\ E_{bh} M_h & 0 & -\tau_b I \end{bmatrix} < 0 \tag{5.63}$$

holds. In that case, the PDC gains are given by $F_i = M_i X^{-1}$, $i \in \{1, 2, \ldots, r\}$.

Proof Using the quadratic Lyapunov function candidate $V = \boldsymbol{x}^T P \boldsymbol{x}$ with $P = X^{-1}$, we have that

$$\dot{V} = 2\boldsymbol{x}^T P\,(A_h + B_h F_h + \Delta A_h + \Delta B_h F_h)\,\boldsymbol{x}.$$

Then, $\dot{V} < 0$ if

$$P A_h + P B_h F_h + P \Delta A_h + P \Delta B_h F_h + (*) < 0.$$

Using congruence (Property A.1) with $X = P^{-1}$ and making the change of variables $M_h = F_h X$ gives

$$A_h X + B_h M_h + \Delta A_h X + \Delta B_h M_h + (*) < 0. \tag{5.64}$$

Using the definitions of the uncertain matrices, it is clear that

$$\begin{aligned}\Theta \equiv& (\Delta A_h)X + (\Delta B_h)M_h + (*)\\ =& \begin{bmatrix} H_a & H_b \end{bmatrix}\begin{bmatrix} \Delta_a & 0 \\ 0 & \Delta_b \end{bmatrix}\begin{bmatrix} E_{ah}X \\ E_{bh}M_h \end{bmatrix} + \begin{bmatrix} XE_{ah}^T & M_h^T E_{bh}^T \end{bmatrix}\begin{bmatrix} \Delta_a^T & 0 \\ 0 & \Delta_b^T \end{bmatrix}\begin{bmatrix} H_a^T \\ H_b^T \end{bmatrix}.\end{aligned}$$

Using Property A.3 (completion of squares) with $Q = \text{block-diag}(\tau_a I, \tau_b I)$ gives the bound

$$\begin{aligned}\Theta \le& \begin{bmatrix} H_a & H_b \end{bmatrix}\begin{bmatrix} \Delta_a & 0 \\ 0 & \Delta_b \end{bmatrix}\begin{bmatrix} \tau_a I & 0 \\ 0 & \tau_b I \end{bmatrix}\begin{bmatrix} \Delta_a^T & 0 \\ 0 & \Delta_b^T \end{bmatrix}\begin{bmatrix} H_a^T \\ H_b^T \end{bmatrix}\\ &+ \begin{bmatrix} XE_{ah}^T & M_h^T E_{bh}^T \end{bmatrix}\begin{bmatrix} \tau_a^{-1} I & 0 \\ 0 & \tau_b^{-1} I \end{bmatrix}\begin{bmatrix} E_{ah}X \\ E_{bh}M_h \end{bmatrix}.\end{aligned}$$

With the uncertainty bounds (5.48), we have

$$\begin{aligned}\Theta \le& \begin{bmatrix} H_a & H_b \end{bmatrix}\begin{bmatrix} \tau_a I & 0 \\ 0 & \tau_b I \end{bmatrix}\begin{bmatrix} H_a^T \\ H_b^T \end{bmatrix}\\ &+ \begin{bmatrix} XE_{ah}^T & M_h^T E_{bh}^T \end{bmatrix}\begin{bmatrix} \tau_a^{-1} I & 0 \\ 0 & \tau_b^{-1} I \end{bmatrix}\begin{bmatrix} E_{ah}X \\ E_{bh}M_h \end{bmatrix},\end{aligned}$$

and therefore (5.64) is satisfied if

$$\begin{aligned}A_h X + B_h M_h + (*) + \begin{bmatrix} H_a & H_b \end{bmatrix}\begin{bmatrix} \tau_a I & 0 \\ 0 & \tau_b I \end{bmatrix}\begin{bmatrix} H_a^T \\ H_b^T \end{bmatrix}\\ + \begin{bmatrix} XE_{ah}^T & M_h^T E_{bh}^T \end{bmatrix}\begin{bmatrix} \tau_a^{-1} I & 0 \\ 0 & \tau_b^{-1} I \end{bmatrix}\begin{bmatrix} E_{ah}X \\ E_{bh}M_h \end{bmatrix} < 0.\end{aligned}$$

With the Schur complement in Lemma A.1, the last inequality can be written as we obtain (5.63), which completes the proof. □

Example 5.12 Consider a TS system of the form (5.60), with $r = 2$ vertex nominal models described by

$$A_1 = \begin{bmatrix} -2.75 & 0.75 \\ -0.05 & -0.1 \end{bmatrix},\ B_1 = \begin{bmatrix} 1 \\ 1 \end{bmatrix},\ A_2 = \begin{bmatrix} 0.5 & 1.5 \\ -0.6 & 0.3 \end{bmatrix},\ B_2 = \begin{bmatrix} 2 \\ 1 \end{bmatrix},$$

uncertainty matrices

$$H_a = \begin{bmatrix} 0 \\ 1 \end{bmatrix},\ H_b = 1,\ E_{a1} = \begin{bmatrix} 0 & 0.5 \end{bmatrix},\ E_{a2} = \begin{bmatrix} -0.5 & 0.5 \end{bmatrix},\ E_{b1} = E_{b2} = \begin{bmatrix} 0.5 \end{bmatrix},$$

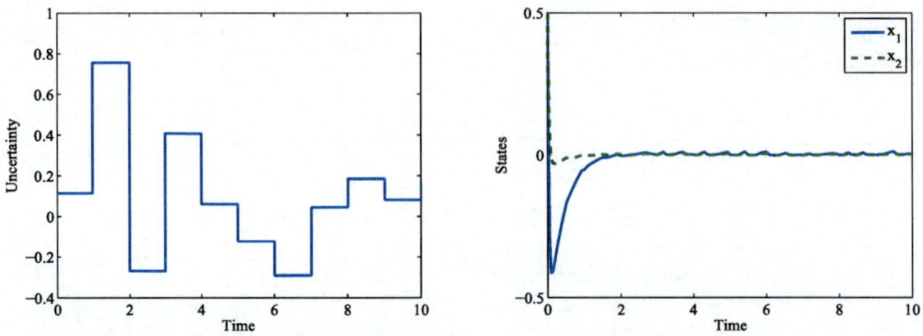

(a) Uncertainty signals ($\Delta a = \Delta b$). (b) Stabilisation results.

Fig. 5.11 Simulation results for Example 5.12

and membership functions with $z = x$, $h_1(z) = 1/(1 + (x_1 + x_2)^2)$, and $h_2(z) = 1 - h_1(z)$ that globally hold the convex sum property.

Using Theorem 5.16, the solution

$$P = \begin{bmatrix} 2.98 & -8.20 \\ -8.20 & 133.93 \end{bmatrix}, \quad \begin{array}{l} F_1 = \begin{bmatrix} 2.87 & -37.70 \end{bmatrix}, \\ F_2 = \begin{bmatrix} 1.78 & -24.58 \end{bmatrix} \end{array}$$

is obtained.

Consider also the uncertainty signals $\Delta a(t)$ and $\Delta b(t)$, with $\Delta a(t) = \Delta b(t)$, as a band-limited white noise of power 0.1 with sample time $1s$, given in Fig. 5.11a. The simulation results with the gains above from the initial states $\begin{bmatrix} 1 & 1 \end{bmatrix}^T$ are presented in Fig. 5.11(b). □

Let us now consider the discrete-time case, i.e. stabilisation of (5.61) using the controller (5.55), i.e. $\boldsymbol{u}(t) = F_h H^{-1}\boldsymbol{x}(t)$. The closed-loop system is expressed as

$$\boldsymbol{x}(t+1) = (A_h + \Delta A_h + B_h F_h H^{-1} + \Delta B_h F_h H^{-1})\boldsymbol{x}(t), \tag{5.65}$$

and the following result can be stated.

Theorem 5.17 *The origin of the closed-loop system* (5.65) *is robustly asymptotically stable, if there exist $P = P^T > 0$, full-rank H, F_i, $i \in \{1, 2, \ldots, r\}$, $S_a = S_a^T > 0$, and $S_b = S_b^T > 0$, such that*

$$\begin{bmatrix} -P & H^T E_{ah}^T & F_h^T E_{bh}^T & H^T A_h^T + F_h^T B_h^T \\ E_{ah} H & -S_a & 0 & 0 \\ E_{bh} F_h & 0 & -S_b & 0 \\ A_h H + B_h F_h & 0 & 0 & \begin{array}{c} -H - H^T + P \\ +H_a S_a H_a^T + H_b S_b H_b^T \end{array} \end{bmatrix} < 0. \tag{5.66}$$

Proof Using the Lyapunov function candidate $V(\boldsymbol{x}) = \boldsymbol{x}^T H^{-T} P H^{-1} \boldsymbol{x}$ for system (5.65), the forward difference in the Lyapunov function $\Delta V = V(t+1) - V(t)$

is negative if

$$\begin{bmatrix} -P & (*) \\ A_hH + B_hF_h + \Delta A_hH + \Delta B_hF_h & -H - H^T + P \end{bmatrix} < 0,$$

which can be obtained by the Schur complement in Lemma A.1 on

$$x^T(t+1)H^{-T}PH^{-1}x(t+1) - x^T(t)H^{-T}PH^{-1}x(t),$$

once the closed-loop equation (5.65) is substituted.

Using the fact that

$$\begin{aligned} \begin{bmatrix} 0 & (*) \\ \Delta A_hH & 0 \end{bmatrix} &= \begin{bmatrix} 0 \\ H_a \end{bmatrix} \Delta_a \begin{bmatrix} E_{ah}H & 0 \end{bmatrix} + (*) \\ &\leq \begin{bmatrix} 0 \\ H_a \end{bmatrix} S_a \begin{bmatrix} 0 \\ H_a \end{bmatrix}^T + \begin{bmatrix} E_{ah}H & 0 \end{bmatrix}^T S_a^{-1} \begin{bmatrix} E_{ah}H & 0 \end{bmatrix} \end{aligned}$$

for $S_a = S_a^T > 0$, and similarly for

$$\begin{bmatrix} 0 & (*) \\ \Delta B_hF_h & 0 \end{bmatrix},$$

we obtain directly (5.66) once the Schur complement in Lemma A.1 is applied again twice. □

Using the Lyapunov function candidate $V(x) = x^TP^{-1}x$, $P = P^T > 0$, instead of the previous one $V(x) = x^TH^{-T}PH^{-1}x$, the following result can be formulated:

Theorem 5.18 *The origin of the closed-loop system (5.65) is robustly asymptotically stable, if there exist $P = P^T$, H, F_i, $S_a = S_a^T > 0$, and $S_b = S_b^T > 0$, such that*

$$\begin{bmatrix} -H - H^T + P & H^TE_{ah}^T & F_h^TE_{bh}^T & H^TA_h^T + F_h^TB_h^T \\ E_{ah}H & -S_a & 0 & 0 \\ E_{bh}F_h & 0 & -S_{bh} & 0 \\ A_hH + B_hF_h & 0 & 0 & \begin{matrix} -P + H_aS_aH_a \\ +H_bS_bH_b \end{matrix} \end{bmatrix} < 0. \tag{5.67}$$

Proof It follows the same outline as the proof of Theorem 5.17. □

Example 5.13 Consider a TS system of the form (5.61), with $r = 2$ nominal vertex models described by

$$A_1 = \begin{bmatrix} -2.75 & 0.75 \\ -0.05 & -0.1 \end{bmatrix}, \; A_2 = \begin{bmatrix} 0.5 & 1.5 \\ -0.6 & 0.3 \end{bmatrix}, \; B_1 = \begin{bmatrix} 1 \\ 1 \end{bmatrix}, \; B_2 = \begin{bmatrix} 2 \\ 1 \end{bmatrix},$$

the uncertainty matrices

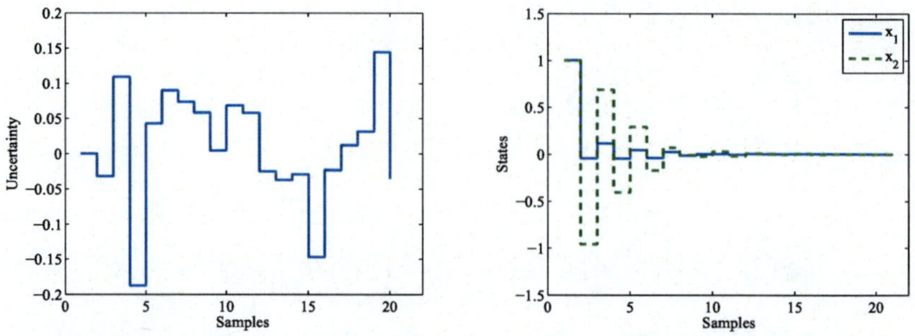

(a) Uncertainty signals ($\Delta a = \Delta b$). (b) Stabilisation results.

Fig. 5.12 Simulation results for Example 5.13

$$H_a = \begin{bmatrix} 0 \\ 0.5 \end{bmatrix}, \; H_b = 0.5, \; E_{a1} = \begin{bmatrix} 0 & 0.5 \end{bmatrix}, \; E_{a2} = \begin{bmatrix} -0.1 & 0.1 \end{bmatrix}, \; E_{b1} = E_{b2} = \begin{bmatrix} 0.1 \end{bmatrix},$$

and membership functions which hold the convex sum property in the whole state space $h_1(z) = 1/(1 + (x_1 + x_2)^2)$ and $h_2(z) = 1 - h_1(z)$, with $z = \boldsymbol{x}$.

Using Theorem 5.17, the solution

$$P = \begin{bmatrix} 0.04 & 0.10 \\ 0.10 & 0.57 \end{bmatrix}, \qquad H = \begin{bmatrix} 0.04 & 0.11 \\ 0.10 & 0.60 \end{bmatrix},$$
$$F_1 = \begin{bmatrix} 0.027 & -0.24 \end{bmatrix}, \qquad F_2 = \begin{bmatrix} -0.10 & -0.54 \end{bmatrix}$$

is obtained.

Consider the uncertainty signals $\Delta a(t)$ and $\Delta b(t)$, with $\Delta a(t) = \Delta b(t)$, given in Fig. 5.12a. The simulation results from initial states $\begin{bmatrix} 1 & 1 \end{bmatrix}^T$ are presented in Fig. 5.12b. □

Structured Uncertainty and the Scaled Small-Gain Theorem

The *scaled small-gain theorem* (Theorem 5.19) can be used as the basic tool to incorporate "structured" uncertainty in stability, performance, and stabilisation problems.

Consider an uncertainty which might take values onto a set of square matrices $\mathcal{D}$, and an associated set of matrices $\mathcal{S}$ such that:

1. $\|\Delta\|_\infty < 1, \forall \Delta \in \mathcal{D}$.
2. $S\Delta = \Delta S, \forall (\Delta, S) \in \mathcal{D} \times \mathcal{S}$.

The matrix S is denoted in some references as "multiplier", and the set $\mathcal{S}$ is a set of matrices that commute with uncertainty Δ.

For instance,

- if Δ is a full matrix, without any known structure, then $\mathcal{S} = \{\gamma I : \gamma \in \mathbb{R}\}$ is a valid set of multipliers;
- if Δ is a block-diagonal construct (say, with m block elements), each of its block matrices at the diagonal being full unstructured matrices, then $\mathcal{S} = \{$block-diag $(\gamma_1 I, \ldots, \gamma_m I) : \gamma_i \in \mathbb{R}\}$ is a valid set of multipliers;
- if Δ is built from a repeated time-varying real parameter, i.e. $\Delta = \delta I$, then $\mathcal{S}$ is the set of full matrices with compatible dimensions.

Consider now an uncertain function of the state:

$$\boldsymbol{q}(\boldsymbol{x}, \Delta) = (M_{11} + M_{12}\Delta(I - M_{22}\Delta)^{-1}M_{21})\boldsymbol{x} \tag{5.68}$$

and an LMI condition such as

$$\begin{bmatrix} \boldsymbol{x} \\ \boldsymbol{q}(\boldsymbol{x}, \Delta) \end{bmatrix}^T \tilde{P} \begin{bmatrix} \boldsymbol{x} \\ \boldsymbol{q}(\boldsymbol{x}, \Delta) \end{bmatrix} \geq 0. \tag{5.69}$$

Theorem 5.19 *(Scaled small-gain) There exists a matrix $\tilde{P}$ fulfilling (5.69) for all $\Delta \in \mathcal{D}$ if the following matrix inequality is feasible for a decision variable $Q = S^T S$, with $S \in \mathcal{S}$*

$$\begin{bmatrix} \Theta^T \tilde{P} \Theta + \begin{bmatrix} 0 & 0 \\ 0 & Q \end{bmatrix} & \begin{matrix} M_{21}^T \\ M_{22}^T \end{matrix} \\ M_{21} \; M_{22} & Q^{-1} \end{bmatrix} \geq 0, \quad \Theta = \begin{bmatrix} I & 0 \\ M_{11} & M_{12} \end{bmatrix}. \tag{5.70}$$

Proof Use the fact that $\boldsymbol{q}(\boldsymbol{x}, \Delta)$ can be expressed as the LFT

$$\begin{bmatrix} \boldsymbol{q} \\ z \end{bmatrix} = \begin{bmatrix} M_{11} & M_{12} \\ M_{21} & M_{22} \end{bmatrix} \begin{bmatrix} \boldsymbol{x} \\ \boldsymbol{w} \end{bmatrix}, \quad \boldsymbol{w} = \Delta z, \tag{5.71}$$

and, given the fact that $\forall \Delta \in \mathcal{D}$

$$\boldsymbol{w}^T Q \boldsymbol{w} = \boldsymbol{w}^T S^T S \boldsymbol{w} = z^T \Delta^T S^T S \Delta z = z^T S^T \Delta^T \Delta S z \leq z^T S^T S z = z^T Q z, \tag{5.72}$$

use the S-procedure in Property A.2 rewriting (5.72) as

$$\begin{bmatrix} \boldsymbol{x} \\ \boldsymbol{w} \end{bmatrix}^T \begin{bmatrix} 0 & 0 \\ 0 & Q \end{bmatrix} \begin{bmatrix} \boldsymbol{x} \\ \boldsymbol{w} \end{bmatrix} - \begin{bmatrix} \boldsymbol{x} \\ \boldsymbol{w} \end{bmatrix}^T [M_{21} \; M_{22}] \, Q \begin{bmatrix} M_{21}^T \\ M_{22}^T \end{bmatrix} \begin{bmatrix} \boldsymbol{x} \\ \boldsymbol{w} \end{bmatrix} \leq 0,$$

followed by Schur complement in Lemma A.1. □

Actually, regarding computational complexity, depending on the presence or absence of further LMI decision variables on M_{21} and M_{22} the problem is BMI or LMI, respectively (in the latter case, it follows from applying congruence (Property A.1) with block-diag(I, I, Q)).

Note also that the scaled small-gain theorem was derived from an alternative approach in (4.102).

For instance, the stability analysis of an uncertain continuous-time system

$$\dot{x}(t) = q(x, \Delta) = (A + H\Delta E)x(t) \tag{5.73}$$

would amount to selecting

$$M_{11} = A,\; M_{12} = H,\; M_{21} = E,\; M_{22} = 0,\; \tilde{P} = \begin{bmatrix} 0 & P \\ P & 0 \end{bmatrix}.$$

Details are left to the reader. The discrete-time case can also be written with $\tilde{P} =$ block-diag(P, P).

Indeed, applying the above theorem, if Δ is a fully unknown uncertain (unstructured) square matrix, then Q can be only a positive scalar, $\epsilon > 0$, multiplied by the identity matrix with compatible dimensions, resulting in

$$\begin{bmatrix} A^T P + PA & PH & E^T \\ H^T P & -\epsilon I & 0 \\ E & 0 & -\epsilon^{-1} I \end{bmatrix} < 0. \tag{5.74}$$

Note that a Schur complement A.1 would transform the above to

$$\begin{bmatrix} A^T P + PA + \epsilon E^T E & PH \\ H^T P & -\epsilon I \end{bmatrix} < 0, \tag{5.75}$$

and, if feasible, so it will be with $P' = P/\epsilon$, so (5.75) would end up being the condition on $\mathcal{H}_\infty$ norm (or, more precisely, induced $\mathcal{L}_{2\to 2}$ norm, as explained in the previous section) being less than 1 of the system $\dot{x} = Ax + Ew$, $z = Hx$. Hence, the scaled small-gain conditions reduce to the non-scaled small-gain ones under unstructured Δ.

In synthesis problems with a term Bu added to the right-hand side of (5.73), congruence with $X = P^{-1}$ is usually needed after (5.74), rendering, instead of (5.75),

$$\begin{bmatrix} XA^T + AX + \epsilon^{-1} HH^T & XE^T \\ EX & -\epsilon^{-1} I \end{bmatrix} < 0. \tag{5.76}$$

If Δ had some structure, for instance, if it results from uncertain parameters in physical modelling, then ϵI would be replaced by Q with a suitable structure in order to be less conservative.

Note that in all the developments, matrix Δ is assumed to be a static time-varying uncertainty. If Δ were unmodelled dynamics, S-procedure in Property A.2 applied time-by-time would not apply. In such a case, we need to prove the so-called small-gain theorem by a different setup.

TS Systems and LFT Uncertainty

In case a parametric/nonlinear component is modelled in a polytopic/TS setup but further uncertainty is unmodelled, we may replace (5.68) by

$$\boldsymbol{q}(\boldsymbol{x}, \Delta) = (M_{11,h} + M_{12,h}\Delta(I - M_{22,h}\Delta)^{-1}M_{21,h})\boldsymbol{x}, \tag{5.77}$$

so Theorem 5.19 can be restated using the h-dependent matrices. Note that, in such a case, the term $\Theta^T \tilde{P}\Theta$ might require further Schur complements or sum relaxations, depending on the particular problem to be solved.

As an example, the generalisation of (5.73) to:

$$\dot{\boldsymbol{x}}(t) = \left([A_h \; B_h] + \begin{bmatrix} H_a & H_b \end{bmatrix} \begin{bmatrix} \Delta_a & 0 \\ 0 & \Delta_b \end{bmatrix} \begin{bmatrix} E_a \\ E_b \end{bmatrix}\right) \begin{bmatrix} \boldsymbol{x} \\ \boldsymbol{u} \end{bmatrix} \tag{5.78}$$

using the compatible multiplier to such block-diagonal uncertainty, is left to the reader. It can be accomplished by considering some decision variables $\tau_a \geq 0, \tau_b \geq 0$, building:

$$Q = \begin{bmatrix} \tau_a I & 0 \\ 0 & \tau_b I \end{bmatrix};$$

then Theorem 5.19 can be used to prove Theorem 5.16, noting the analogy with (5.76). The scaled small-gain framework would allow for other memoryless uncertainty structures, if needed in a particular application.

5.7 Application Case Study: Wheelchair Swing-Up

This work has been done in collaboration between the LAMIH UMR CNRS 8201, Polytechnic University Hauts-de-France in Valenciennes and the company Autonomad Mobility. The objective is to swing up a wheelchair using an electrical motorisation toolkit. The company is selling such kits, where the wheels are equipped with a brushless motor, Fig. 5.13. The goal is to increase the mobility of the disabled, as this swing-up (also called self-balancing mode) allows them to cross over obstacles and to navigate on rough terrains (grass, sand, and stony ground). The main issues are ensuring a safe and secure swing-up (as it is an unstable dangerous position), having smooth transitions between the modes (standard and self-balancing).

Using a simplified model and unknown parameters such as the mass of the disabled, we define $\dot{y}$ as the wheelchair speed, ψ as the angle from the vertical axis, and $\dot{\psi}$ as the angular velocity. The model is put into a descriptor form and is derived from Lagrange–Euler equations [60]:

$$E_c(\psi)\dot{\boldsymbol{x}}(t) = A_c(\psi, \dot{\psi})\boldsymbol{x}(t) + B_c\boldsymbol{u}(t), \tag{5.79}$$

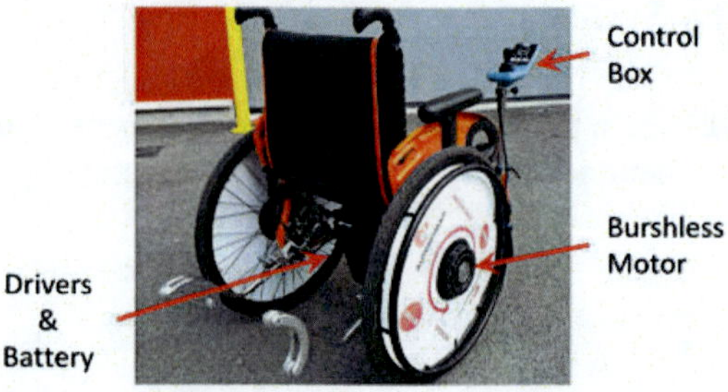

Fig. 5.13 The wheelchair system

where

$$x = \begin{bmatrix} \dot{y} \\ \dot{\psi} \\ \psi \end{bmatrix}, \; E_c(\psi) = \begin{bmatrix} M_b + 2M_w + 2J_w/(r^2) & M_b l\cos(\psi) & 0 \\ M_b l\cos(\psi) & M_b l^2 + J_b & 0 \\ 0 & 0 & 1 \end{bmatrix}, \; B_c = \begin{bmatrix} K_t/r \\ -K_t \\ 0 \end{bmatrix},$$

$$A_c(\psi, \dot{\psi}) = \begin{bmatrix} -\mu_{wg} - 2\mu_m/(r^2) & 2\mu_m/r & M_b l\dot{\psi}^2 \sin(\psi/\pi)/(\psi/\pi) \\ 2\mu_m/r & -2\mu_m & M_b g l \sin(\psi/\pi)/(\psi/\pi) \\ 0 & 1 & 0 \end{bmatrix},$$

where M_b is the mass of user and wheelchair frame, M_w is the mass of each push-wheel, J_b is the inertia of the user and wheelchair frame, J_w is the inertia of each push-wheel, μ_m and μ_{wg} are the viscous friction coefficients of the motor and the push-wheel ground interface, K_t is the torque constant of the brushless motor, l is the distance from push-wheel axle to centre of mass M_b, r is the radius of the push-wheels, and g is the acceleration of gravity.

The first step consists in getting a convex descriptor model [61] from (5.79) for the following compact set of interest:

$$\mathcal{X} = \left\{ (\psi, \dot{\psi}) \in \mathbb{R}^2 : |\psi| \le \bar{\psi}, \; |\dot{\psi}| \le \bar{\dot{\psi}} \right\} \tag{5.80}$$

with $\bar{\psi} = 0.4$rad and $\bar{\dot{\psi}} = 0.4$rad/s for the application. The sector nonlinearity approach in Sect. 3.2.1 can be directly applied to (5.79) with 3 nonlinearities, namely $\dot{\psi}^2$, $\cos(\psi)$, and $\sin(\psi/\pi)/(\psi/\pi)$, and may produce an 8-vertex convex descriptor model.

Nevertheless, the number of vertex models can be reduced if functions $\cos(\psi)$ and $\sin(\psi/\pi)/(\psi/\pi)$ are related as in [62], leaving only 2 nonlinearities to be taken into account:

$$z_1(x) = \dot{\psi}^2, \quad z_2(x) = \cos(\psi).$$

Thus, we consider the membership functions

$$w_0^1(z) = w_0^1(\dot{\psi}) = \frac{\bar{\dot{\psi}}^2 - \dot{\psi}^2}{\bar{\dot{\psi}}^2} = \frac{z_1^1 - z_1}{z_1^1 - z_1^0},$$
$$v_0^1(z) = v_0^1(\psi) = \frac{1 - \cos(\psi)}{1 - \cos(\bar{\psi})} = \frac{z_2^1 - z_2}{z_2^1 - z_2^0},$$

where $z_1^1 = \bar{\dot{\psi}}^2$, $z_1^0 = 0$, $z_2^1 = 1$, $z_2^0 = \cos(\bar{\psi})$, $w_1^1(z) = 1 - w_0^1(z)$, and $v_1^1(z) = 1 - v_0^1(z)$.

The resulting convex descriptor model is

$$\sum_{i=0}^{1} v_i(\psi) E_{ci} \dot{x}(t) = \sum_{i=0}^{1} \sum_{j=0}^{1} v_i(\psi) w_j(\dot{\psi}) \left(A_{cij} x(t) + Bu(t)\right), \tag{5.81}$$

where $E_{c0} = E_c(\bar{\psi})$, $E_{c1} = E_c(0)$, $A_{c00} = A_c(\bar{\psi}, 0)$, $A_{c01} = A_c(\bar{\psi}, \bar{\dot{\psi}})$, $A_{c10} = A_c(0, 0)$, $A_{c11} = A_c(0, \bar{\dot{\psi}})$, and $B = B_c$. This is an exact convex rewriting of model (5.79) in $\mathcal{X}$.

The second step is to derive a discrete-time representation of (5.81), for which we use $t \in \mathbb{Z}$ as the discrete time and T_s as the sampling period to perform Euler discretisation, i.e. to approximate $\dot{x}$ as

$$\dot{x}(t) \approx \frac{x(t+1) - x(t)}{T_s},$$

which results in

$$\sum_{i=0}^{1} v_i(\psi(t)) E_{ci} x(t+1) \tag{5.82}$$
$$= \sum_{i=0}^{1} \sum_{j=0}^{1} v_i(\psi(t)) \left(w_j(\dot{\psi}(t)) T_s A_{cij} + E_{ci}\right) x(t) + T_s B_c u(t). \tag{5.83}$$

This can be expressed in compact notation for convex sums as

$$E_v x(t+1) = A_{vw} x(t) + Bu(t), \tag{5.84}$$

where convex sums are expressed as in tensor-product form, i.e. weights w_i and v_j are not merged together as $h_i(z)$.

As the motorisation kit is adaptable to any wheelchair and has to run for any user, mass uncertainty is an important issue that has been solved using the robust control technique (see Sect. 5.6.3). In order to introduce the mass uncertainty, we transform the definition of the matrices in (5.79) as

$$\tilde{E}_c(\psi) = E_c(\psi) + \Delta M_b \begin{bmatrix} 1 & l\cos(\psi) & 0 \\ l\cos(\psi) & l^2 & 0 \\ 0 & 0 & 0 \end{bmatrix},$$

$$\tilde{A}_c(\psi, \dot{\psi}) = A_c(\psi, \dot{\psi}) + \Delta M_b \begin{bmatrix} 0 & 0 & l\dot{\psi}^2 \sin(\psi/\pi)/(\psi/\pi) \\ 0 & 0 & gl\sin(\psi/\pi)/(\psi/\pi) \\ 0 & 0 & 0 \end{bmatrix},$$

where ΔM_b represents the maximum acceptable mass variation. Using the convex form (5.81), we have

$$\tilde{E}_v = E_v + \Delta M_b \begin{bmatrix} 1 & 0 \\ 0 & 1 \\ 0 & 0 \end{bmatrix} \Upsilon_v, \quad \tilde{A}_{vw} = A_{vw} + \Delta M_b \begin{bmatrix} 1 & 0 \\ 0 & 1 \\ 0 & 0 \end{bmatrix} \Omega_{vw},$$

$$\Upsilon_v = \begin{bmatrix} 1 & l\cos(\psi(t)) & 0 \\ l\cos(\psi(t)) & l^2 & 0 \end{bmatrix} = v_0(\psi(t))\Upsilon_0 + (1 - v_0(\psi(t)))\,\Upsilon_1,$$

$$\Omega_{vw} = \begin{bmatrix} 1 & l\cos(\psi) & T_s l\dot{\psi}\sin(\psi/\pi)/(\psi/\pi) \\ l\cos(\psi) & l^2 & T_s gl\sin(\psi/\pi)/(\psi/\pi) \end{bmatrix}$$

$$= \sum_{i=0}^{1}\sum_{j=0}^{1} v_i(\psi(t))w_j(\dot{\psi}(t))\Omega_{cij}, \quad \Omega_{cij} = \Omega_{vw}|_{v_i=w_j=1}.$$

A stable trajectory generation is obtained using a simplified linearised version of (5.84), namely

$$E_r x_r(t+1) = A_{rBF} x_r(t), \quad A_{rBF} \equiv A_r + BF_r, \tag{5.85}$$

where F_r can be obtained via any classical linear technique. The control law for the swinging-up depends on both the trajectory to follow $x_r(t)$ and the trajectory error $e(t) = x(t) - x_r(t)$, i.e.:

$$u(t) = \begin{bmatrix} F(\cdot) & G(\cdot) \end{bmatrix} H_{(\cdot)}^{-1} \begin{bmatrix} x_r(t) \\ e(t) \end{bmatrix}, \tag{5.86}$$

where $(\cdot)$ denotes a dependence on variables to be fixed once the LMI formulation has been obtained. Thus, the global model under consideration is

$$\bar{E}_v \bar{x}(t+1) = \left(\bar{A}_{vw} + \bar{B}\bar{F}(\cdot)\right)\bar{x}(t) \tag{5.87}$$

with

$$\bar{x} = \begin{bmatrix} x_r \\ e \end{bmatrix}, \; \bar{E}_v = \begin{bmatrix} E_r & 0 \\ 0 & E_v \end{bmatrix}, \; \bar{A}_{vw} = \begin{bmatrix} A_{rBF} & 0 \\ \tilde{A}_{vw} + \tilde{E}_v E_r^{-1} A_{rBF} & \tilde{A}_{vw} \end{bmatrix}, \; \bar{B} = \begin{bmatrix} 0 \\ B \end{bmatrix},$$

and $\bar{F} = \begin{bmatrix} F(\cdot) & G(\cdot) \end{bmatrix} H_{(\cdot)}^{-1}$.

After some standard manipulations based on the discrete-time methodologies in Sect. 5.5 and including the parameter uncertainties [44], we end up with the control problem formulated as

$$\begin{bmatrix} -H_{(\cdot)} - H_{(\cdot)}^T + P_{(\cdot)} & (*) & 0 & (*) \\ \bar{A}_{vw} H_{(\cdot)} + \bar{B}\begin{bmatrix} F_{(\cdot)} & G_{(\cdot)} \end{bmatrix} & -\begin{bmatrix} E_r & 0 \\ 0 & E_v \end{bmatrix} J_{(\cdot)} - (*) + \begin{bmatrix} 0 & 0 & 0 \\ 0 & m^2 S & 0 \\ 0 & 0 & 0 \end{bmatrix} & (*) & (*) \\ 0 & J_{(\cdot)} & -P_{(\cdot)} & 0 \\ \begin{bmatrix} \Omega_{vw} - \Upsilon_v E_r^{-1} A_{rBF} & \Omega_{vw} \end{bmatrix} H_{(\cdot)} & -\begin{bmatrix} 0 & \Upsilon_v \end{bmatrix} J_{(\cdot)} & 0 & -S \end{bmatrix} < 0. \tag{5.88}$$

The last step is to determine the dependence of the variables $(\cdot)$. One goal is to derive conditions with the minimum number of LMI constraints. Since the input matrix B is constant, we can avoid cross terms in (5.88). Consider a quadratic Lyapunov function $P_{(\cdot)} = P$; thus, a good choice for the free variables is $F_{(\cdot)} = F_{vw}$, $J_{(\cdot)} = J_w$, and $H_{(\cdot)} = H$ as to get double convex sums from which sufficient LMI conditions can be obtained following any relaxation scheme in Appendix D. Therefore, the control will be

$$u(t) = \bar{F}_{vw}\bar{x}(t) = \begin{bmatrix} F_{vw} & G_{vw} \end{bmatrix} H^{-1}\bar{x}(t).$$

Some extra performances have to be considered such as acting on the norm of the gains (see Section 5.6). Therefore, the following constraints are added:

$$\begin{bmatrix} F_{vw} & G_{vw} \end{bmatrix} P^{-1} \begin{bmatrix} F_{vw}^T \\ G_{vw}^T \end{bmatrix} < \lambda I, \quad \lambda P > I,$$

where the first one is transformed via a Schur complement in Lemma A.1 into the following constraints:

$$\begin{bmatrix} -\lambda I & \begin{bmatrix} F_{vw} & G_{vw} \end{bmatrix} \\ \begin{bmatrix} F_{vw}^T \\ G_{vw}^T \end{bmatrix} & -P \end{bmatrix} < 0, \quad \lambda P > I, \tag{5.89}$$

with $\lambda > 0$ a parameter allowing to act on the norm of the control gains.

The following values are used for the tests: $M_b = 90$kg, $M_w = 5$kg, $J_b = 9$kgm^2, $J_w = 0.23$kgm^2, $\mu_m = 0.1$kg/s, $\mu_{wg} = 0.2$ kg/s, $r = l = 0.3$m, and $K_t = 1.5$Nm/A. Considering $\max(\Delta M_b) = 20$kg and the free parameter $\lambda = 230$ yields

$$\begin{aligned} F_{00} &= \begin{bmatrix} 0.54 & -1.22 & 0.64 \end{bmatrix}, & F_{10} &= \begin{bmatrix} -0.2 & 0.19 & 0.56 \end{bmatrix}, \\ F_{01} &= \begin{bmatrix} -0.23 & 1.27 & 0.23 \end{bmatrix}, & F_{11} &= \begin{bmatrix} -0.47 & 1.13 & 0.41 \end{bmatrix}, \\ G_{00} &= \begin{bmatrix} -1.86 & 1.58 & 0.88 \end{bmatrix}, & G_{10} &= \begin{bmatrix} -1.12 & 0.04 & 1 \end{bmatrix}, \\ G_{01} &= \begin{bmatrix} -0.88 & 0.65 & 0.85 \end{bmatrix}, & G_{11} &= \begin{bmatrix} -0.73 & 1.48 & 0.5 \end{bmatrix}. \end{aligned}$$

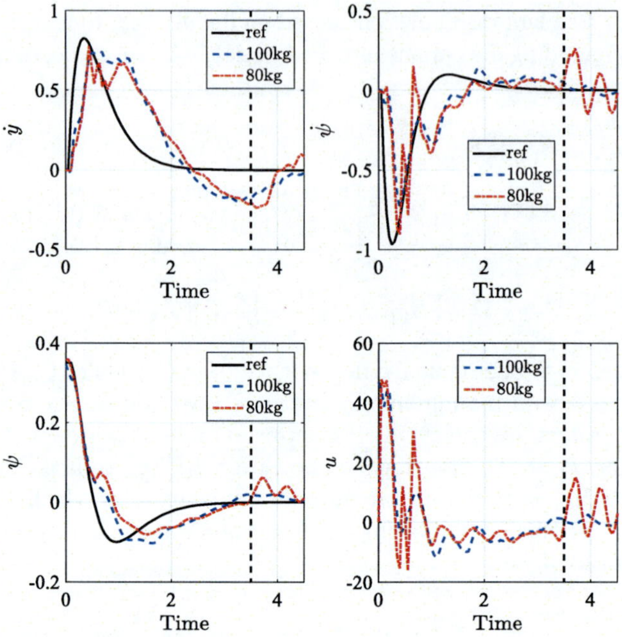

Fig. 5.14 Real-time results for $\dot{y}$, $\dot{\psi}$, ψ, and u for the 100kg individual (dashed line) and 80kg individual (red line) compared with the references (black lines)

Real-time experiments are provided using the Autonomad Mobility kit[4]. Two different users have tested the unique control law with masses corresponding to $M_b = \{80, 100\}$kg. Figure 5.14 shows the swing-up and robustness tests. Trajectory generation (5.85) is in black; user 1 results in red, user 2 in blue. The wheelchair speed $\dot{y}$ is top-left, the angular velocity $\dot{\psi}$ top-right, the angle ψ bottom-left, and the control u bottom-right. The real-time experiments show a very good behaviour according to the swing-up of both users (Fig. 5.14 before $t = 3.5$s) as well as excellent robustness to disturbances. Effectively, after $t = 3.5$s, i.e. indicated with the dashed vertical line, the users begin to voluntarily (strongly) move inside the chair, making oscillations appear while maintaining a stable position (the reader can see it clearly on the video). The other smaller oscillations are due to both the encoder quantisation error present on $\dot{y}$ and the gyroscopic measurement $\dot{\psi}$ noise.

[4] A full video of the session is available at https://pod.uphf.fr/video/1583-discrete-reference-tracking-control-to-swing-up-an-electric-wheelchair/

5.8 Summary and Conclusions

Stabilisation of nonlinear systems based on convex models requires choosing an appropriate controller structure, which may depend on the sort of Lyapunov function to be employed. All the schemes, including the simplest which is parallel distributed compensation, employ the membership functions of the model in order to enrich the controller choices as well as take advantage of the convexity properties. Nested convex sums appear from which LMI or SOS conditions can be derived by dropping the membership functions in a variety of ways, some less conservative—but more computationally demanding—than others. LFT controller design has been left for the next chapter as this approach includes observers and output feedback, which, along with performance and robustness, are the subject of the final part of the book.

References

1. Wang, H., Tanaka, K., Griffin, M.: An approach to fuzzy control of nonlinear systems: stability and design issues. IEEE Transactions on Fuzzy Systems **4**(1), 14–23 (1996)
2. Tanaka, K., Ikeda, T., Wang, H.: Fuzzy regulators and fuzzy observers: relaxed stability conditions and LMI-based designs. IEEE Transactions on Fuzzy Systems **6**(2), 250–265 (1998)
3. Tuan, H., Apkarian, P., Narikiyo, T., Yamamoto, Y.: Parameterized linear matrix inequality techniques in fuzzy control system design. IEEE Transactions on Fuzzy Systems **9**(2), 324–332 (2001)
4. Liu, X., Zhang, H.: Stability analysis of uncertain fuzzy large-scale system. Chaos, Solitons & Fractals **25**(5), 1107–1122 (2005)
5. Tanaka, K., Wang, H.O.: Fuzzy Control System Design and Analysis: A Linear Matrix Inequality Approach. Wiley, New York (2001)
6. Johansson, M., Rantzer, A.: Computation of piecewise quadratic Lyapunov functions for hybrid systems. IEEE Transactions on Automatic Control **43**(4), 555–559 (1998)
7. Johansson, M., Rantzer, A., Arzen, K.: Piecewise quadratic stability of fuzzy systems. IEEE Transactions on Fuzzy Systems **7**(6), 713–722 (1999)
8. Feng, G.: Controller synthesis of fuzzy dynamic systems based on piecewise Lyapunov functions. IEEE Transactions on Fuzzy Systems **11**(5), 605–612 (2003)
9. Feng, G.: A survey on analysis and design of model-based fuzzy control systems. IEEE Transactions on Fuzzy Systems **14**(5), 676–697 (2006)
10. Hu, J., Han, Y., Zhao, L.: Synchronizing chaotic systems using control based on a special matrix structure and extending to fractional chaotic systems. Communications in Nonlinear Science and Numerical Simulation **15**(1), 115–123 (2010)
11. Qiu, J., Feng, G., Gao, H.: Static-output-feedback H_∞ control of continuous-time T–S fuzzy affine systems via piecewise Lyapunov functions. IEEE Transactions on Fuzzy Systems **21**(2), 245–261 (2013)
12. Schulte, H., Hahn, H.: Fuzzy state feedback gain scheduling control of servo-pneumatic actuators. Control Engineering Practice **12**, 639–650 (2004)
13. Cuesta, F., Ollero, A.: Fuzzy control of reactive navigation with stability analysis based on conicity and Lyapunov theory. Control Engineering Practice **12**(5), 625–638 (2004)
14. Blanco, Y., Perruquetti, W., Borne, P.: Stability and stabilization of nonlinear systems and Takagi–Sugeno's fuzzy models. Mathematical Problems in Engineering **7**(3), 221–240 (2001)

15. Pan, J.-T., Guerra, T.M., Fei, S.-M., Jaadari, A.: Nonquadratic stabilization of continuous T–S fuzzy models: LMI solution for a local approach. IEEE Transactions of Fuzzy Systems **20**, 594–602 (2012)
16. Mozelli, L., Palhares, R., Avellar, G.: A systematic approach to improve multiple Lyapunov function stability and stabilization conditions for fuzzy systems. Information Sciences **179**(8), 1149–1162 (2009)
17. Guelton, K., Guerra, T., Bernal, M., Bouarar, T., Manamanni, N.: Comments on fuzzy control systems design via fuzzy Lyapunov functions. IEEE Transactions on Systems, Man, and Cybernetics, Part B: Cybernetics **40**(3), 970–972 (2010)
18. Guerra, T.M., Bernal, M.: A way to escape from the quadratic framework. In: Proceedings of the IEEE International Conference on Fuzzy Systems, pp. 784–789. IEEE, New York (2009)
19. Guerra, T., Bernal, M., Guelton, K., Labiod, S.: Non-quadratic local stabilization for continuous-time Takagi–Sugeno models. Fuzzy Sets and Systems **201**, 40–54 (2012)
20. Lee, D., Kim, D.: Relaxed LMI conditions for local stability and local stabilization of continuous-time Takagi–Sugeno fuzzy systems. IEEE Transactions on Cybernetics **44**(3), 394–405 (2014)
21. Bernal, M., Guerra, T.M.: Generalized non-quadratic stability of continuous-time Takagi–Sugeno models. IEEE Transactions on Fuzzy Systems **18**(4), 815–822 (2010)
22. Aguiar, B., Márquez, R., Bernal, M.: Towards a global nonquadratic controller design for nonlinear systems via robust differentiators and Takagi–Sugeno models. In: 2015 IEEE International Conference on Fuzzy Systems (FUZZ-IEEE), pp. 1–7. IEEE, New York (2015)
23. Levant, A.: Robust exact differentiation via sliding mode technique. Automatica **34**(3), 379–384 (1998)
24. Gonzalez, T., Bernal, M., Sala, A., Aguiar, B.: Cancellation-based nonquadratic controller design for nonlinear systems via Takagi–Sugeno models. IEEE Transactions on Cybernetics **47**(9), 2628–2638 (2017)
25. Rhee, B., Won, S.: A new fuzzy Lyapunov function approach for a Takagi–Sugeno fuzzy control system design. Fuzzy Sets and Systems **157**(9), 1211–1228 (2006)
26. Marquez, R., Guerra, T., Kruszewski, A., Bernal, M.: Improvements on non-quadratic stabilization of Takagi–Sugeno models via line-integral Lyapunov functions. In: Proceedings of the 2013 IFAC International Conference on Intelligent Control and Automation Science, pp. 473–478 (2013)
27. Guerra, T.M., Vermeiren, L.: LMI-based relaxed nonquadratic stabilization conditions for nonlinear systems in the Takagi–Sugeno's form. Automatica **40**(5), 823–829 (2004)
28. Guerra, T.M., Kerkeni, H., Lauber, J., Vermeiren, L.: An efficient Lyapunov function for discrete T–S models: observer design. IEEE Transactions of Fuzzy Systems **20**(1), 187–192 (2012)
29. Ding, B., Sun, H., Yang, P.: Further studies on LMI-based relaxed stabilization conditions for nonlinear systems in Takagi–Sugeno's form. Automatica **42**(3), 503–508 (2006)
30. Lendek, Z., Guerra, T., Lauber, J.: Construction of extended Lyapunov functions and control laws for discrete-time TS systems. In: Proceedings of the 2012 IEEE World Congress on Computational Intelligence, IEEE International Conference on Fuzzy Systems, pp. 286–291. IEEE, New York (2012)
31. Lendek, Zs., Guerra, T., Lauber, J.: Controller design for TS models using delayed non-quadratic Lyapunov functions. IEEE Transactions on Cybernetics **45**(3), 453–464 (2015)
32. Guerra, T.M., Kruszewski, A., Bernal, M.: Control law proposition for the stabilization of discrete Takagi–Sugeno models. IEEE Transactions of Fuzzy Systems **17**, 724–731 (2009)
33. Skelton, R.E., Iwasaki, T., Grigoriadis, K.: A Unified Approach to Linear Control Design. Taylor & Francis, London (1998)
34. Ariño, C., Sala, A., Pérez, E., Bedate, F., Querol, A.: Asymptotically exact stabilization for constrained discrete Takagi–Sugeno systems via set-invariance. Fuzzy Sets and Systems **316**, 117–138 (2017)
35. Hu, T., Lin, Z.: Composite quadratic Lyapunov functions for constrained control systems. IEEE Transactions on Automatic Control **48**(3), 440–450 (2003)

36. Hu, T., Blanchini, F.: Non-conservative matrix inequality conditions for stability/stabilizability of linear differential inclusions. Automatica **46**(1), 190–196 (2010)
37. Tanaka, K., Yoshida, H., Ohtake, H., Wang, H.: A sum of squares approach to modeling and control of nonlinear dynamical systems with polynomial fuzzy systems. IEEE Transactions on Fuzzy Systems **17**(4), 911–922 (2009)
38. Taniguchi, T., Tanaka, K., Wang, H.O.: Fuzzy descriptor systems and nonlinear model following control. IEEE Transactions on Fuzzy Systems **8**(4), 442–452 (2000)
39. de Oliveira, M., Skelton, R.: Stability test of constrained linear systems. In: Perspective in Robust Control, pp. 241–257. Springer, Berlin (2001)
40. Taniguchi, T., Tanaka, K., Yamafuji, K., Wang, H.O.: Fuzzy descriptor systems: stability analysis and design via LMIs. In: Proceedings of the American Control Conference, pp. 1827–1831 (1999)
41. Guerra, T.M., Bernal, M., Kruszewski, A., Afroun, M.: A way to improve results for the stabilization of continuous-time fuzzy descriptor models. In: 46th IEEE Conference on Decision and Control, pp. 5960–5964. IEEE, New York (2007)
42. Bouarar, T., Guelton, K., Manamanni, N.: Robust fuzzy Lyapunov stabilization for uncertain and disturbed Takagi–Sugeno descriptors. ISA Transactions **49**(4), 447–461 (2010)
43. Estrada-Manzo, V., Guerra, T.M., Lendek, Z., Pudlo, P.: Discrete-time Takagi–Sugeno descriptor mdoels: controller design. In: Proceedings of the IEEE International Conference on Fuzzy Systems, pp. 1–5. IEEE, New York (2014b)
44. Estrada-Manzo, V., Lendek, Z., Guerra, T.M., Pudlo, P.: Controller design for discrete-time descriptor models: a systematic LMI approach. IEEE Transactions on Fuzzy Systems **23**(5), 1608–1621 (2015)
45. Xu, S., Song, B., Lu, J., Lam, J.: Robust stability of uncertain discrete-time singular fuzzy systems. Fuzzy Sets and Systems **158**(20), 2306–2316 (2007)
46. Liu, X., Zhang, Q.: New approaches to H_∞ controller designs based on fuzzy observers for T–S fuzzy systems via LMI. Automatica **39**(9), 1571–1582 (2003)
47. Delmotte, F., Guerra, T.M., Ksontini, M.: Continuous Takagi–Sugeno's models: Reduction of the number of LMI conditions in various fuzzy control design technics. IEEE Transactions on Fuzzy Systems **15**(3), 426–438 (2007)
48. Wu, H.-N.: Robust H_2 fuzzy output feedback control for discrete-time nonlinear systems with parametric uncertainties. International Journal of Approximate Reasoning **46**(1), 151–165 (2007)
49. Yoneyama, J.: Robust stability and stabilization for uncertain Takagi–Sugeno fuzzy time-delay systems. Fuzzy Sets and Systems **158**(2), 115–134 (2007)
50. Liu, F., Wu, M., He, Y., Yokoyama, R.: New delay-dependent stability criteria for T–S fuzzy systems with time-varying delay. Fuzzy Sets and Systems **161**(15), 2033–2042 (2010)
51. Zhou, S., Feng, G.: Generalised H_2 controller synthesis for uncertain discrete-time fuzzy systems via basis-dependent Lyapunov function. IEE Control Theory and Applications **153**(1), 74–80 (2006)
52. Kruszewski, A., Guerra, T.M.: Stabilization of a class of nonlinear model with periodic parameters in the Takagi–Sugeno form. In: Proceedings of the IFAC Workshop Periodic Control Systems, pp. 1–6 (2007)
53. Dong, J., Yang, G.: Dynamic output feedback H_∞ control synthesis for discrete-time T–S fuzzy systems via switching fuzzy controllers. Fuzzy Sets and Systems **160**(19), 482–499 (2009)
54. Wang, T., Qu, L.-J.: Robust decentralized control for T–S uncertain fuzzy interconnected systems with time-delay. In: Proceedings of the Sixth International Conference on Machine Learning and Cybernetics, pp. 631–636 (2007)
55. Choi, D., Park, P.: Guaranteed cost controller design for discrete-time switching fuzzy systems. IEEE Transactions on Systems, Man, and Cybernetics, Part B: Cybernetics **34**(1), 110–119 (2004)
56. Sala, A., Guerra, T.M., Babuška, R.: Perspectives of fuzzy systems and control. Fuzzy Sets and Systems **156**(3), 432–444 (2005)

57. Boyd, S., El Ghaoui, L., Féron, E., Balakrishnan, V.: Linear Matrix Inequalities in System and Control Theory. Studies in Applied Mathematics, Society for Industrial and Applied Mathematics, Philadelphia, PA (1994)
58. Van Der Schaft, A.J.: L2-gain analysis of nonlinear systems and nonlinear state feedback h_∞ control. IEEE Transactions on Automatic Control **37**(6), 770–784 (1992)
59. Tanaka, K., Wang, H.O.: Fuzzy Control Systems Design and Analysis, analysis Wiley, New York (2001)
60. Gray, M., Guerra, T., Delprat, S., Mohammed, S.: Control of an automated wheelchair. In: 21st IFAC World Congress (2020)
61. Ohtake, H., Tanaka, K., Wang, H.: Fuzzy modeling via sector nonlinearity concept. In: Proceedings of the Joint 9th IFSA World Congress and 20th NAFIPS International Conference, vol. 1, pp. 127–132 (2001)
62. Guerra, T., Bernal, M., Blandeau, M.: Reducing the number of vertices in some Takagi–Sugeno models: example in the mechanical field. IFAC-PapersOnLine **51**(10), 133–138 (2018)

Chapter 6
Observation and Output Feedback

6.1 Introduction

This chapter considers more complex control problems—mainly observer design and output feedback—for which not every convex structure considered in this book has found a mature answer. As expected, the simplest modelling choices—ordinary TS models, LFT representations—along with the simplest Lyapunov functions—quadratic—have reached a wider set of easy-to-express results at the price of being conservative. For more up-to-date results on these subjects, usually with more involved methodologies, the reader is referred to research reports whenever appropriate. Two real-time applications are presented too: sitting control for people living with spinal cord injury as well as observer-based trajectory tracking for power-assisted wheelchairs.

6.2 Observer Design

Most of the control schemes of the last chapter assume all the states are available. Nevertheless, in real-life applications often not all variables of interest of a given system can be measured. In such cases, an observer has to be designed to estimate the unmeasured variables—usually states—based on the system model and the available input-output data. Expectedly, observer design based on convex models reproduces the convex structure of the observed system. This means that the membership functions from the Takagi–Sugeno, descriptor, or polynomial model are used in the observer structure, thus leading to an important implementation question: are these membership functions available?

In traditional fuzzy TS contexts, membership functions are nothing less than exogenous signals which hold the convex sum property; thus, they are not available for the observer, which has to deal with the estimation task relying on a single observer gain to drive the observer states to those of the plant. Yet, model-based approaches

M. Bernal et al., *Analysis and Synthesis of Nonlinear Control Systems*, Studies in Systems, Decision and Control 408, https://doi.org/10.1007/978-3-030-90773-0_6

as those presented in this book lead to convex models whose membership functions embed nonlinearities, uncertainties, and other signals, some of which might be known, some of which might not. As it turns out, the availability of the membership functions constitutes a crucial distinction amongst observer design proposals. The next two Sects. 6.2.1 and 6.2.2 will only treat the case where the membership functions (grouped in what is usually referred to as scheduling vector) are available; the last Sect. 6.2.3 will consider the case where these functions, also known as premises, are not available.

6.2.1 Observer Design for TS Models

Consider a TS model of the form (3.1) (reproduced below)

$$\begin{aligned}\delta\boldsymbol{x}(t) &= \sum_{i=1}^{r} h_i(z)(A_i\boldsymbol{x}(t) + B_i\boldsymbol{u}(t) + \boldsymbol{a}_i),\\ \boldsymbol{y}(t) &= \sum_{i=1}^{r} h_i(z)(C_i\boldsymbol{x}(t) + \boldsymbol{c}_i),\end{aligned} \tag{6.1}$$

under the usual meanings for operator $\delta\boldsymbol{x}(t)$, state $\boldsymbol{x}$, input $\boldsymbol{u}$, output $\boldsymbol{y}$, scheduling vector $z \in \mathcal{C}$, and constant matrices and affine terms possibly resulting from modelling in $\mathcal{C} \subseteq \mathbb{R}^p$.

Since the scheduling vector z is assumed available, it can be used in the observer proposal, namely

$$\begin{aligned}\delta\widehat{\boldsymbol{x}}(t) &= \sum_{i=1}^{r} h_i(z)(A_i\widehat{\boldsymbol{x}}(t) + B_i\boldsymbol{u}(t) + \boldsymbol{a}_i + K_i(\boldsymbol{y}(t) - \widehat{\boldsymbol{y}}(t))),\\ \widehat{\boldsymbol{y}}(t) &= \sum_{i=1}^{r} h_i(z)(C_i\widehat{\boldsymbol{x}}(t) + \boldsymbol{c}_i),\end{aligned} \tag{6.2}$$

where $\widehat{\boldsymbol{x}}(t)$ and $\widehat{\boldsymbol{y}}$ are the observer state and the observer output, respectively; K_i observer gains to be designed.

Observer design is based on the dynamics of the estimation error $\boldsymbol{e} = \boldsymbol{x} - \widehat{\boldsymbol{x}}$. Assumptions 4.1 and 4.2 should be updated as follows:

Assumption 6.1 Given a convex model of the error dynamics, its origin $\boldsymbol{e} = 0$ is an equilibrium point $\forall z \in \mathcal{C}$.

Assumption 6.2 Given a convex model of the error dynamics, there exists a compact set $\mathcal{E}$ of the error state space containing the origin $\boldsymbol{e} = 0$ such that $\forall \boldsymbol{e} \in \mathcal{E}$ then $z \in \mathcal{C}$.

The Continuous Quadratic Case

Let us focus on the continuous-time case, i.e. $\delta x(t) = \dot{x}(t)$, $\delta\widehat{x}(t) = \dot{\widehat{x}}(t)$, $t \in \mathbb{R}$, in (6.1)–(6.2); the error dynamics can be written as

$$\begin{aligned}\dot{e}(t) &= \sum_{i=1}^{r}\sum_{j=1}^{r} h_i(z)h_j(z)(A_i - K_iC_j)e(t) \\ &= (A_h - K_hC_h)e(t),\end{aligned} \tag{6.3}$$

with $e = x - \widehat{x}$ as the estimation error. Note that, if the scheduling vector is known, the (also known) input u, the input matrices B_i, $i \in \{1, 2, \ldots, r\}$, and the affine terms a_i and c_i, $i \in \{1, 2, \ldots, r\}$, respectively, do not influence the observer design conditions.

The error dynamics (6.3) is itself a TS system. Therefore, sufficient conditions for observer design are extensions of the stability conditions presented in Chap. 4 and the dual of those described in Chap. 5.

The next result provides sufficient conditions to prove asymptotic stability of the origin $e = 0$ of the error dynamics (6.3), by means of a quadratic Lyapunov function candidate $V(e) = e^TPe$; it generalises the formulation in [1]:

Theorem 6.1 *The origin $e = 0$ of the continuous-time TS estimation error model (6.3), resulting from the TS model (6.1) and the TS observer (6.2) with $\delta x(t) = \dot{x}(t)$, $\delta\widehat{x}(t) = \dot{\widehat{x}}(t)$, $t \in \mathbb{R}$, is asymptotically stable if there exist matrices $P = P^T > 0$ and N_j, $j \in \{1, 2, \ldots, r\}$, such that*

$$PA_h - N_hC_h + A_h^TP - C_h^TN_h^T < 0 \tag{6.4}$$

holds. In this case, the observer gains are $K_j = P^{-1}N_j$, $j \in \{1, 2, \ldots, r\}$, and any trajectory $e(t)$ starting within the outermost Lyapunov level $V(e) = e^TPe \le k$, $k > 0$, in a compact set $\mathcal{E}$ of the error state space guaranteeing $z \in \mathcal{C} \subseteq \mathbb{R}^p$ goes asymptotically to $e = 0$.

Proof From $P = P^T > 0$ we know that $V(e) = e^TPe$ is a valid Lyapunov function candidate for the estimation error model (6.3); its time derivative is

$$\dot{V} = 2e^TP\dot{e} = e^T\left(P(A_h - K_hC_h) + (A_h - K_hC_h)^TP\right)e,$$

from which it follows that

$$PA_h - PK_hC_h + A_h^TP - C_h^TK_h^TP < 0$$

guarantees $\dot{V} < 0$.

Condition (6.4) follows from the inequality above by performing the substitution $N_h = PK_h$; this, in turn, depends on MFs $h_i(z)$ holding the convex sum property, which by modelling of the TS model (6.1) is guaranteed in $\mathcal{C}$. Therefore, any trajectory starting within the outermost Lyapunov level $V(\boldsymbol{e}) = \boldsymbol{e}^T P \boldsymbol{e} \leq k, k > 0$, in a compact set $\mathcal{E}$ of the error state space holding Assumptions 6.1 and 6.2 goes asymptotically to $\boldsymbol{e} = 0$, as these assumptions guarantee $z \in \mathcal{C}$, which, in turn, ensures the convexity of MFs $h_i(z)$ on which the steps above are based. □

Remark 6.1 As with most results in this book involving nested convex sums, inequality (6.4) in Theorem 6.1 can be the subject of any relaxation scheme in Appendix D to find sufficient LMIs on P and N_j guaranteeing them. Examples of such instances can be found in [2, 3] by gathering expressions in bigger LMIs, in [4, 5] by using Lemma D.1, and in [6] for fuzzy TS models where not all the vertices are active at the same time.

The design of an observer using the conditions of Theorem 6.1 with Tanaka's relaxation scheme in (D.10) is illustrated in the following example.

Example 6.1 Consider a TS system with $r = 2$ vertex models whose matrices are given by

$$A_1 = \begin{bmatrix} -7 & 1.25 \\ 0.85 & -1.6 \end{bmatrix}, \; A_2 = \begin{bmatrix} -1.45 & -0.4 \\ 0.57 & 0.7 \end{bmatrix}, \; B_1 = B_2 = \begin{bmatrix} 0 \\ 1 \end{bmatrix}, \; \begin{matrix} C_1 = \begin{bmatrix} 1 & 0 \end{bmatrix}, \\ C_2 = \begin{bmatrix} 0 & 1 \end{bmatrix}, \end{matrix}$$

with $x_1, x_2 \in [-1, 1]$, $u \in \mathbb{R}$ and membership functions $h_1(z) = (1 - z)/2$, $h_2(z) = 1 - h_1$, where $z \in [-1, 1]$ is an exogenous measurable signal.

Conditions (6.4) are satisfied with

$$P = \begin{bmatrix} 0.1087 & 0.0268 \\ 0.0268 & 0.3298 \end{bmatrix}, \; N_1 = \begin{bmatrix} -0.4362 \\ 0.3096 \end{bmatrix}, \; N_2 = \begin{bmatrix} -0.0018 \\ 0.5814 \end{bmatrix},$$

from which the observer gains can be calculated as

$$K_1 = P^{-1} N_1 = \begin{bmatrix} -4.33 \\ 1.29 \end{bmatrix}, \qquad K_2 = P^{-1} N_2 = \begin{bmatrix} -0.46 \\ 1.80 \end{bmatrix}.$$

A trajectory of the estimation error using the observer gains above is presented in Fig. 6.1. For this particular trajectory, the true initial states were $[1 \; 1]^T$, and the estimated initial states were $[0 \; 0]^T$, the input signal $u = \sin x_1$ was employed. As it can be seen, the estimation error converges to zero. □

Remark 6.2 The results in Theorem 6.1 can be modified as to guarantee the origin $\boldsymbol{e} = 0$ of the error dynamics (6.3) is exponentially stable with decay rate of $\alpha > 0$, by using the concepts presented in Chap. 4. This is quite common in the implementation of observers as it affects the convergence rate of the estimation error, which might be critical in many applications. To this end, it is straightforward to prove that conditions (6.4) should be replaced by

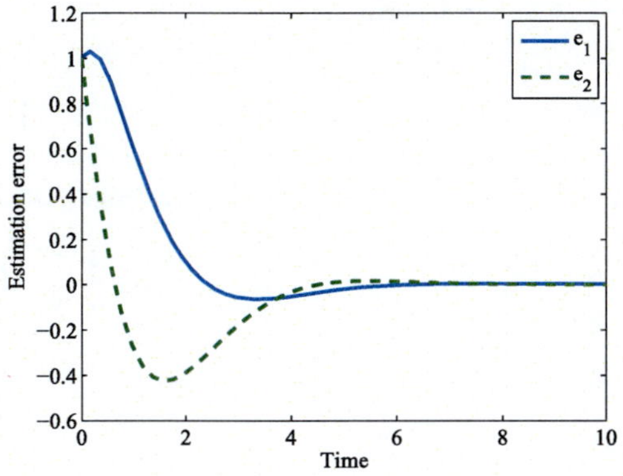

Fig. 6.1 Simulation results for Example 6.1

$$PA_h - N_h C_h + A_h^T P - C_h^T N_h^T + 2\alpha P < 0 \tag{6.5}$$

in order to guarantee a decay rate of $\alpha > 0$.

The observer design such that a desired convergence rate is obtained is illustrated in the following example.

Example 6.2 Consider the TS system in Example 6.1. Using Tanaka's relaxation in (D.10) an observer with decay rate $\alpha > 0$ can be designed by solving the LMI problem of finding $P = P^T > 0$, $M_i, i, j = 1, 2, 3, 4, i \leq j$, so that

$$\begin{aligned} &(PA_i - M_i C_i) + (*) + 2\alpha P < 0, \\ &(PA_i + PA_j - M_i C_j - M_j C_i) + (*) + 4\alpha P \leq 0, \end{aligned}$$

which, of course, are sufficient to guarantee (6.5).

Solving these LMIs for a decay rate $\alpha = 1.5$, the following observer gains are found

$$K_1 = \begin{bmatrix} -2.77 \\ 1.75 \end{bmatrix}, \quad K_2 = \begin{bmatrix} 2.35 \\ 2.65 \end{bmatrix}.$$

A trajectory of the estimation error using these gains is presented in Fig. 6.2. For this particular trajectory, the true initial states were $[1\ 1]^T$, and the estimated initial states were $[0\ 0]^T$; the input signal was $u = \sin x_1$. As it can be seen, the estimation error converges to $\boldsymbol{e} = 0$ much quicker than in the previous example. □

Example 6.3 Consider the following system, inspired in [7, Example 2.3]:

$$\begin{bmatrix} \dot{x}_1 \\ \dot{x}_2 \end{bmatrix} = \begin{bmatrix} -3x_1^2 & x_1 - 2 \\ 1 & -x_1^2 \end{bmatrix} \begin{bmatrix} x_1 \\ x_2 \end{bmatrix} + \begin{bmatrix} e^{x_1} \\ 1 \end{bmatrix} u, \quad y = x_1.$$

Defining $z_1 = x_1 \in [-1, 1.5]$, $z_2 = x_1^2 \in [0, 2.25]$, and $z_3 = e^{x_1} \in [e^{-1}, e^{1.5}]$, and using the sector nonlinearity approach, an equivalent TS model can be obtained with

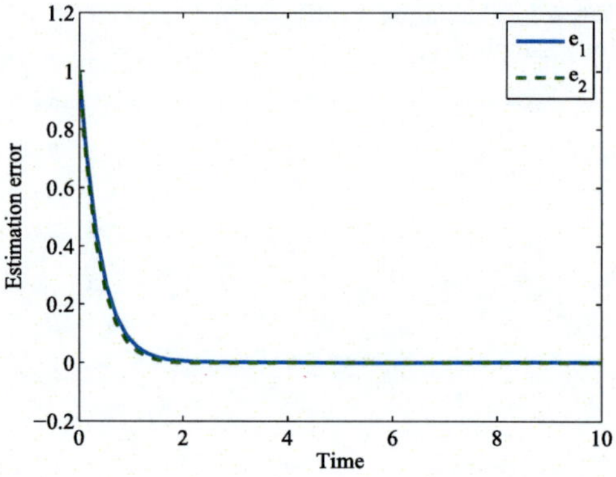

Fig. 6.2 Simulation results for Example 6.2

weighting functions $w_0^j(z_j) = (z_j^1 - z_j)/(z_j^1 - z_j^0)$, $w_1^j(z_j) = 1 - w_0^j$, $j = 1, 2, 3$, (consequently, 8 different MFs $h_i(z)$ with $z = [z_1\ z_2\ z_3]^T$), and vertex matrices

$$A_1 = A_2 = \begin{bmatrix} 0 & -3 \\ 1 & 0 \end{bmatrix}, \quad A_3 = A_4 = \begin{bmatrix} -\frac{27}{4} & -3 \\ 1 & -\frac{9}{4} \end{bmatrix},$$
$$A_5 = A_6 = \begin{bmatrix} 0 & -\frac{1}{2} \\ 1 & 0 \end{bmatrix}, \quad A_7 = A_8 = \begin{bmatrix} -\frac{27}{4} & -\frac{1}{2} \\ 1 & -\frac{9}{4} \end{bmatrix},$$

$$B_1 = B_3 = B_5 = B_7 = \begin{bmatrix} e^{-1} \\ 1 \end{bmatrix}, \quad B_2 = B_4 = B_6 = B_8 = \begin{bmatrix} e^{1.5} \\ 1 \end{bmatrix},$$

and $C_i = \begin{bmatrix} 1 & 0 \end{bmatrix}$ for $i \in \{1, 2, \ldots, 8\}$. Conditions in Theorem 6.1 are feasible with the sum relaxation in (D.10); they yield

$$P = \begin{bmatrix} 11.71 & 1 \\ 1 & 0.84 \end{bmatrix}, \ K_1 = K_2 = \begin{bmatrix} 4.29 \\ -45.97 \end{bmatrix}, \ K_3 = K_4 = \begin{bmatrix} -2.2 \\ -43.96 \end{bmatrix},$$
$$K_5 = K_6 = \begin{bmatrix} 0.96 \\ -7.12 \end{bmatrix}, \ K_7 = K_8 = \begin{bmatrix} -5.53 \\ -10.12 \end{bmatrix}.$$

Simulation results are depicted in Fig. 6.3. The phase portrait shows the trajectories from initial conditions (squares): $x(0) = [-0.3 \quad 1.5]^T$ and $\hat{x}(0) = [1.1 \quad -0.5]^T$; $x(0) = [0.5 \quad 1.5]^T$ and $\hat{x}(0) = [1 \quad -3.9]^T$; $x(0) = [0.5 \quad 1.5]^T$ and $\hat{x}(0) = [0 \quad -2.6]^T$; $x(0) = [0.5 \quad -3]^T$ and $\hat{x}(0) = [0 \quad 2.4]^T$; $x(0) = [1.45 \quad -2]^T$ and $\hat{x}(0) = [0 \quad 0]^T$, under the input signal $u(t) = 0.25 \sin t + 0.25 \cos 10t + 0.5 \sin 20t$.

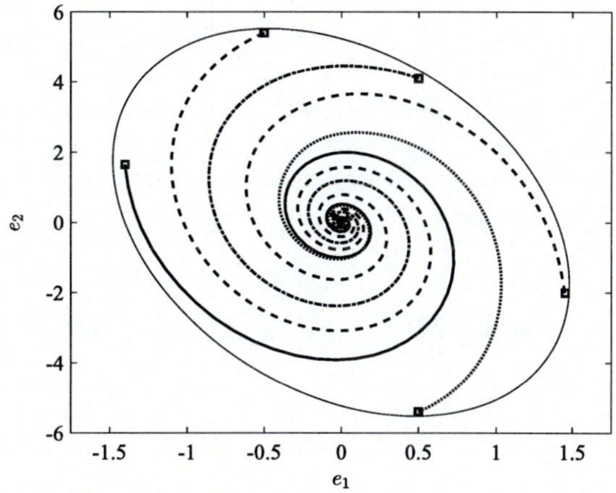

Fig. 6.3 Phase portrait from 5 initial conditions in Example 6.3

The Discrete Quadratic Case

We now turn our attention to the discrete-time case, where the error dynamics, assuming that the scheduling vector z is available, is

$$e(t+1) = (A_h - K_h C_h)e(t), \tag{6.6}$$

as results from TS model (6.1) and observer (6.2), with $\delta x(t) = x(t+1)$, $\delta\widehat{x}(t) = \widehat{x}(t+1)$, $t \in \mathbb{Z}$.

Theorem 6.2 *The origin $e = 0$ of the discrete-time TS estimation error model (6.6), resulting from the TS model (6.1) and the TS observer (6.2) with $\delta x(t) = x(t+1)$, $\delta\widehat{x}(t) = \widehat{x}(t+1)$, $t \in \mathbb{Z}$, is asymptotically stable if there exist matrices $P = P^T > 0$ and N_j, $j \in \{1, 2, \ldots, r\}$, such that*

$$\begin{bmatrix} -P & A_h^T P - C_h^T N_h^T \\ PA_h - N_h C_h & -P \end{bmatrix} < 0 \tag{6.7}$$

holds. In that case, the observer gains are $K_j = P^{-1}N_j$, $j \in \{1, 2, \ldots, r\}$, and any trajectory $e(t)$ starting within the outermost Lyapunov level $V(e) = e^T Pe \leq k$, $k > 0$, in a compact set $\mathcal{E}$ of the error state space guaranteeing $z \in \mathcal{C} \subseteq \mathbb{R}^p$ goes asymptotically to $e = 0$.

Proof Given that $P = P^T > 0$, $V(e) = e^T(t)Pe(t)$ is a valid Lyapunov function candidate. The forward difference of this candidate can be written as

$$\Delta V = \begin{bmatrix} e(t) \\ e(t+1) \end{bmatrix}^T \begin{bmatrix} -P & 0 \\ 0 & P \end{bmatrix} \begin{bmatrix} e(t) \\ e(t+1) \end{bmatrix},$$

while the error dynamics can be expressed as

$$\begin{bmatrix} A_h - K_h C_h & -I \end{bmatrix} \begin{bmatrix} e(t) \\ e(t+1) \end{bmatrix} = 0.$$

Using Finsler's Lemma A.3, we have that $\Delta V < 0$ if there exists a multiplier M such that

$$\begin{bmatrix} -P & 0 \\ 0 & P \end{bmatrix} + M \begin{bmatrix} A_h - K_h C_h & -I \end{bmatrix} + \begin{bmatrix} A_h - K_h C_h & -I \end{bmatrix}^T M^T < 0.$$

Choosing $M = \begin{bmatrix} 0 & P \end{bmatrix}^T$, a sufficient condition for $\Delta V < 0$ is

$$\begin{bmatrix} -P & A_h^T P - C_h^T K_h^T P \\ P A_h - P K_h C_h & -P \end{bmatrix} < 0,$$

from which inequality (6.7) follows by substituting K_j by $P^{-1} N_j$. This ensures $V(\boldsymbol{e})$ is a Lyapunov function guaranteeing asymptotic stability of $\boldsymbol{e} = 0$. Moreover, since the steps above are valid as long as MFs $h_i(z)$ hold the convex sum property, which they do for any $z \in \mathcal{C}$, it follows that any trajectory starting in the outermost Lyapunov level $V(\boldsymbol{e}) = \boldsymbol{e}^T P \boldsymbol{e} \leq k$, $k > 0$, in a compact set $\mathcal{E}$ of the error state space, should go asymptotically to $\boldsymbol{e} = 0$ (see Assumptions 6.1 and 6.2). □

To illustrate the application of the above conditions, consider the following example.

Example 6.4 Consider a TS system with $r = 2$ vertex models described by

$$A_1 = \begin{bmatrix} -7 & 1.25 \\ 0.85 & -1.6 \end{bmatrix}, \qquad A_2 = \begin{bmatrix} -1.45 & -0.4 \\ 0.57 & 0.7 \end{bmatrix},$$
$$C_1 = \begin{bmatrix} 1 & 0 \end{bmatrix}, \qquad C_2 = \begin{bmatrix} 2 & 0 \end{bmatrix},$$

where the membership functions are assumed to be exogenous signals always holding the convex sum property; one of them, $h_1(t)$, is illustrated in Fig. 6.4a; the other one is of course $h_2(t) = 1 - h_1(t)$. In this case, $z = t$ is the scheduling vector.

Using Theorem 6.2 with the sum relaxation in Lemma D.1, amounts to solving LMI conditions

$$\frac{2}{r-1}\Gamma_{ii} + \Gamma_{ij} + \Gamma_{ji} < 0,$$

for $i, j \in \{1, 2\}$, with

$$\Gamma_{ij} = \begin{bmatrix} -P & A_i^T P - C_i^T N_j^T \\ P A_i - N_j C_i & -P \end{bmatrix},$$

for decision variables $P = P^T > 0$, N_i, $i \in \{1, 2, \ldots, r\}$. These LMIs are sufficient to guarantee (6.7). A solution has been found with the Lyapunov matrix and observer gains as follows:

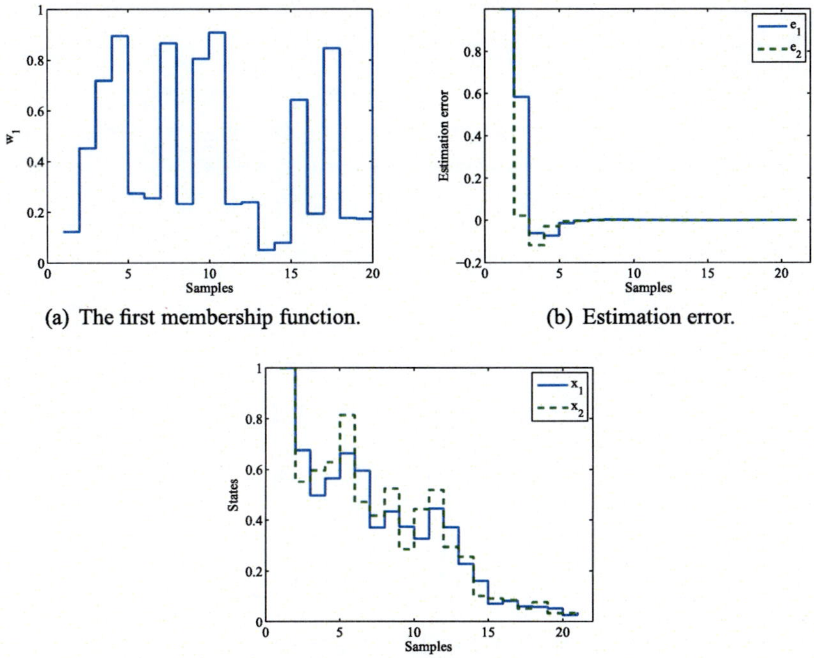

(a) The first membership function.

(b) Estimation error.

(c) True states.

Fig. 6.4 Simulation results for Example 6.4

$$P = \begin{bmatrix} 0.42 & -0.06 \\ -0.06 & 0.64 \end{bmatrix}, \; K_1 = \begin{bmatrix} 0.38 \\ 0.82 \end{bmatrix}, \; K_2 = \begin{bmatrix} 0.015 \\ 0.27 \end{bmatrix}.$$

Figure 6.4b presents the estimation error when the true initial states are $\boldsymbol{x}(0) = \begin{bmatrix} 1 & 1 \end{bmatrix}^T$ and the estimated initial states are $\widehat{\boldsymbol{x}}_0 = 0$. As expected, the estimation error converges to zero. The evolution of the true states is presented in Fig. 6.4c. □

Example 6.5 Consider the discrete-time nonlinear system given by the equations [8]

$$\begin{bmatrix} x_1(t+1) \\ x_2(t+1) \\ x_3(t+1) \end{bmatrix} = \begin{bmatrix} 0 & -x_3 & -1 \\ 1 & 0 & 0 \\ 0 & 1 & 0 \end{bmatrix} \begin{bmatrix} x_1(t) \\ x_2(t) \\ x_3(t) \end{bmatrix} + \begin{bmatrix} 1 & 0 \\ 0 & 0 \\ 0 & 1 \end{bmatrix} \begin{bmatrix} u_1(t) \\ u_2(t) \end{bmatrix},$$

$$y(t) = \begin{bmatrix} 0 & 2+x_3 & 0 \\ 0 & 0 & 1 \end{bmatrix} \begin{bmatrix} x_1(t) \\ x_2(t) \\ x_3(t) \end{bmatrix}.$$

The system above can be represented as a TS one with $r = 2$ vertex models described by matrices

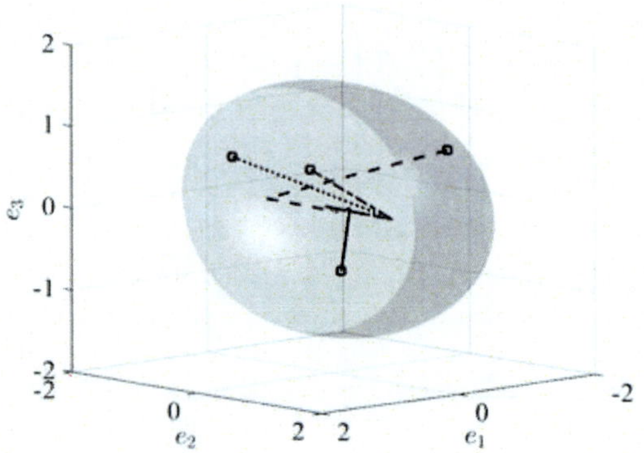

Fig. 6.5 Phase portrait from 4 initial conditions in Example 6.5

$$A_1 = \begin{bmatrix} 0 & 1 & -1 \\ 1 & 0 & 0 \\ 0 & 1 & 0 \end{bmatrix}, \; A_2 = \begin{bmatrix} 0 & -1 & -1 \\ 1 & 0 & 0 \\ 0 & 1 & 0 \end{bmatrix}, \; B_1 = B_2 = \begin{bmatrix} 1 & 0 \\ 0 & 0 \\ 0 & 1 \end{bmatrix},$$
$$C_1 = \begin{bmatrix} 0 & 1 & 0 \\ 0 & 0 & 1 \end{bmatrix}, \; C_2 = \begin{bmatrix} 0 & 3 & 0 \\ 0 & 0 & 1 \end{bmatrix},$$

and membership functions $h_1(x_3) = 0.5 - 0.5x_3$ and $h_2(x_3) = 0.5x_3 + 0.5$. The LMI conditions in Theorem 6.2 with relaxation (D.10) yield the solution

$$P = \begin{bmatrix} 0.79 & 0 & -0.066 \\ 0 & 0.47 & 0 \\ -0.066 & 0 & 0.75 \end{bmatrix}, \; K_1 = \begin{bmatrix} 0.34 & -1 \\ 0 & 0.06 \\ 0.67 & 0 \end{bmatrix}, \; K_2 = \begin{bmatrix} -0.36 & -1 \\ 0 & 0.064 \\ 0.32 & 0 \end{bmatrix}.$$

The simulation results are depicted in Fig. 6.5 from 4 initial conditions in the system state: $x(0) = [1 \; -0.5 \; 0.7]^T$, $x(0) = [-1 \; 0.75 \; 0.7]^T$, $x(0) = [1 \; 0.75 \; 0.7]^T$, and $x(0) = [0.5 \; 0.7 \; -0.6]^T$. The initial condition for the observer in all cases is $\hat{x}(0) = [0 \; 0 \; 0]^T$. The control input given in [8] is applied, namely

$$u(t) = \begin{bmatrix} 0.0437 - 0.265x_3 & 0.168x_3 + 0.278 \\ -0.121x_3 - 0.284 & 0.0658x_3 - 0.499 \end{bmatrix} y + \begin{bmatrix} -0.1 \sin 10t \\ 0.1 \cos 10t \end{bmatrix}.$$

Similar to the continuous-time case, performance criteria, such as decay rate can be added to the conditions.

The Continuous Non-Quadratic Case

We will now take a look at generalisations such as non-quadratic schemes, under the same assumption on the availability of the scheduling vector z. To this end, consider a generic TS observer model of the form:

$$\begin{aligned}\delta\widehat{\boldsymbol{x}}(t) &= A_{h(t)}\hat{\boldsymbol{x}}(t) + B_{h(t)}\boldsymbol{u}(t) + \boldsymbol{a}_{h(t)} + S^{-1}(\cdot)K(\cdot)\left(\boldsymbol{y}(t) - \hat{\boldsymbol{y}}(t)\right), \\ \widehat{\boldsymbol{y}}(t) &= C_{h(t)}\widehat{\boldsymbol{x}}(t) + \boldsymbol{c}_{h(t)}\end{aligned} \tag{6.8}$$

to be used along with a TS model of the form (6.1), where $K(\cdot)$ and $S(\cdot)$ are design matrices to be defined according to the Lyapunov function under consideration.

A generic non-quadratic Lyapunov function candidate of the form

$$V(\boldsymbol{e}) = \boldsymbol{e}^T(t)\boldsymbol{P}(\cdot)\boldsymbol{e}(t) \tag{6.9}$$

will be considered for estimation.

Common options are $S(\cdot) = \boldsymbol{P}(\cdot) = P_{h(t)}$ and $S(\cdot) = H_{h(t)}$.

Remark 6.3 Notice that, as said before in Remark 5.3, the Lyapunov function candidate for observation (as that for stability) does not use an inverse of the involved matrix $\boldsymbol{P}(\cdot)$; the controller, on the other hand, does.

Let us consider first the continuous-time case. As shown before, the input and the affine terms do not affect the design when the scheduling variables are available. Therefore, the error dynamics resulting from the TS model (6.1) and the observer (6.8) with $\delta\boldsymbol{x}(t) = \dot{\boldsymbol{x}}(t)$, $\delta\widehat{\boldsymbol{x}}(t) = \dot{\widehat{\boldsymbol{x}}}(t)$, $t \in \mathbb{R}$, and choices $S(\cdot) = \boldsymbol{P}(\cdot) = P_{h(t)}$, $K(\cdot) = K_{h(t)}$, is given by

$$\dot{\boldsymbol{e}}(t) = (A_h - P_h^{-1}K_hC_h)\boldsymbol{e}(t). \tag{6.10}$$

Theorem 6.3 *[Local observability] The origin $\boldsymbol{e} = 0$ of the continuous-time TS estimation error model (6.10), resulting from the TS model (6.1) and the TS observer (6.2) with $\delta\boldsymbol{x}(t) = \dot{\boldsymbol{x}}(t)$, $\delta\widehat{\boldsymbol{x}}(t) = \dot{\widehat{\boldsymbol{x}}}(t)$, $t \in \mathbb{R}$, is asymptotically stable in a domain $\mathcal{D}$ including the origin $\boldsymbol{e} = 0$ if there exist matrices $P_i = P_i^T > 0$, $i \in \{1, 2, \ldots, r\}$, and K_j, $j \in \{1, 2, \ldots, r\}$, such that the following holds:*

$$P_hA_h - K_hC_h + A_h^TP_h - C_h^TK_h^T < 0. \tag{6.11}$$

Proof Clearly, $V(\boldsymbol{e}) = \boldsymbol{e}^TP_h\boldsymbol{e}$ is a valid Lyapunov function candidate because $P_i = P_i^T > 0$, $i \in \{1, 2, \ldots, r\}$. Its time derivative is

$$\dot{V} = 2\boldsymbol{e}^T\left(P_h\left(A_h - P_h^{-1}K_hC_h\right)\right)\boldsymbol{e} + \boldsymbol{e}^T\dot{P}_h\boldsymbol{e},$$

which is negative-definite for sufficiently small $\lambda > 0$ in a domain $\mathcal{D} = \{\boldsymbol{e} : \|\dot{P}_h\| < \lambda\}$, which means condition (6.11) holds, validating $V(\boldsymbol{e})$ as a Lyapunov function guaranteeing asymptotic stability of $\boldsymbol{e} = 0$. □

Remark 6.4 This effectively means that if the condition of Theorem 6.3 is satisfied, and if the estimated initial state is close enough to the true state, then the estimation error will converge to zero. However, no information is available on what "close enough" means, i.e. on the region in which the estimated states will converge to the true states. To obtain information on this, one has to use knowledge on the membership functions, specifically on the derivatives of the membership functions. For instance, given a bound on the derivatives of the membership functions, the result of [9, 10] can be properly modified. If the TS model has been obtained by using the sector nonlinearity approach, the results of [11] can be applied. Again, obtaining sufficient LMI conditions to guarantee (6.11) can be made with any relaxation scheme in Appendix D.

Example 6.6 Consider a continuous-time TS system with $r = 2$ vertex models described by

$$A_1 = \begin{bmatrix} -1.75 & 0.8 \\ 0.7 & 0.6 \end{bmatrix}, \qquad A_2 = \begin{bmatrix} 1.3 & -0.7 \\ 0.3 & -0.15 \end{bmatrix},$$
$$C_1 = \begin{bmatrix} 1 & 0 \end{bmatrix}, \qquad C_2 = \begin{bmatrix} 0 & 1 \end{bmatrix}.$$

The LMIs using a quadratic Lyapunov function and Lemma D.1 are infeasible. However, local asymptotic stability of the estimation error can be achieved, by using Lemma D.1 with the conditions of Theorem 6.3, i.e. solving the LMIs

$$P_i > 0,$$
$$2(P_i A_i - K_i C_i) + (P_i A_j - K_i C_j) + (P_j A_i - K_j C_i) + (*) < 0,$$

$i, j = 1, 2$, as to obtain

$$P_1 = \begin{bmatrix} 0.45 & -0.43 \\ -0.43 & 0.5 \end{bmatrix}, \qquad P_2 = \begin{bmatrix} 0.02 & -0.17 \\ -0.17 & 1.62 \end{bmatrix},$$
$$K_1 = \begin{bmatrix} -0.36 & 1.37 \end{bmatrix}^T, \qquad K_2 = \begin{bmatrix} 0.46 & 0.93 \end{bmatrix}^T,$$

which proves there is a domain around $\boldsymbol{e} = 0$ where the corresponding estimation error dynamics go asymptotically to the origin. □

The Discrete Non-Quadratic Case

Let us now consider the discrete-time case, i.e. a TS model (6.1) and an observer (6.8) with $\delta \boldsymbol{x}(t) = \boldsymbol{x}(t+1)$, $\delta \widehat{\boldsymbol{x}}(t) = \widehat{\boldsymbol{x}}(t+1)$, $t \in \mathbb{Z}$, and choices $S(\cdot) = H_{h(t)}$, $K(\cdot) = K_{h(t)}$. The error dynamics are given by

$$\boldsymbol{e}(t+1) = (A_h - H_h^{-1} K_h C_h)\boldsymbol{e}(t). \tag{6.12}$$

Theorem 6.4 *The origin $\boldsymbol{e} = 0$ of the discrete-time TS estimation error model (6.12), resulting from the TS model (6.1) and the TS observer (6.2) with $\delta \boldsymbol{x}(t) = \boldsymbol{x}(t+1)$, $\delta \widehat{\boldsymbol{x}}(t) = \widehat{\boldsymbol{x}}(t+1)$, $t \in \mathbb{Z}$, $S(\cdot) = H_{h(t)}$, and $K(\cdot) = K_{h(t)}$, is asymptotically stable if there exist matrices $P_i = P_i^T > 0$, H_i, K_i, $i \in \{1, 2, \ldots, r\}$, such that*

$$\begin{bmatrix} -P_h & A_h^T H_h^T - C_h^T K_h^T \\ H_h A_h - K_h C_h & -H_h - H_h^T + P_{h(t+1)} \end{bmatrix} < 0 \tag{6.13}$$

holds. In that case, any trajectory $\boldsymbol{e}(t)$ starting within the outermost Lyapunov level $V(\boldsymbol{e}) = \boldsymbol{e}^T P \boldsymbol{e} \leq k$, $k > 0$, in a compact set $\mathcal{E}$ of the error state space implying $z \in \mathcal{C} \subseteq \mathbb{R}^p$, goes asymptotically to $\boldsymbol{e} = 0$.

Proof Using the fact that $P_i = P_i^T > 0$, $i \in \{1, 2, \ldots, r\}$, consider the Lyapunov function candidate $V = \boldsymbol{e}^T(t) P_h \boldsymbol{e}(t)$, which corresponds to the choice $\boldsymbol{P}(\cdot) = P_h$ in (6.9); its forward difference is

$$\begin{aligned} \Delta V &= \boldsymbol{e}^T(t+1) P_{h(t+1)} \boldsymbol{e}(t+1) - \boldsymbol{e}^T(t) P_h \boldsymbol{e}(t) \\ &= \begin{bmatrix} \boldsymbol{e}(t) \\ \boldsymbol{e}(t+1) \end{bmatrix}^T \begin{bmatrix} -P_h & 0 \\ 0 & P_{h(t+1)} \end{bmatrix} \begin{bmatrix} \boldsymbol{e}(t) \\ \boldsymbol{e}(t+1) \end{bmatrix}. \end{aligned}$$

The error dynamics can be written as

$$\begin{bmatrix} A_h - H_h^{-1} K_h C_h & -I \end{bmatrix} \begin{bmatrix} \boldsymbol{e}(t) \\ \boldsymbol{e}(t+1) \end{bmatrix} = 0.$$

Using Lemma A.3 with a multiplier $M = \begin{bmatrix} 0 & H_h^T \end{bmatrix}^T$, we guarantee that $\Delta V < 0$ subject to the error dynamics above if

$$\begin{bmatrix} 0 \\ H_h \end{bmatrix} \begin{bmatrix} A_h - H_h^{-1} K_h C_h & -I \end{bmatrix} + (*) + \begin{bmatrix} -P_h & 0 \\ 0 & P_{h(t+1)}^{-1} \end{bmatrix} < 0,$$

which is equivalent to condition (6.13), thus establishing asymptotic stability of $\boldsymbol{e} = 0$ since $V(\boldsymbol{e})$ is thus proved to be a valid Lyapunov function. As this developments depend on MFs $h_i(z)$ holding the convex sum property, trajectories starting within the outermost Lyapunov level $V(\boldsymbol{e}) = \boldsymbol{e}^T P \boldsymbol{e} \leq k$, $k > 0$, in a compact set $\mathcal{E}$ of the error state space holding Assumptions 6.1 and 6.2, go asymptotically to the origin, because these assumptions guarantee $z \in \mathcal{C}$, which is enough for MFs $h_i(z)$ to hold the convex sum property. □

Remark 6.5 Due to $P_{h(t+1)}$, condition (6.13) in the theorem above is a triple convex sum; sufficient LMI conditions guaranteeing it can be found via any relaxation scheme in Appendix D. Note that the quadratic case is recovered if $H_i = P_i = P$.

Let us illustrate the use of the design conditions in the following example.

Example 6.7 Consider a TS system with two $r = 2$ vertex models described by matrices

$$A_1 = \begin{bmatrix} -0.07 & 1.37 \\ 0.27 & 0.17 \end{bmatrix}, \qquad A_2 = \begin{bmatrix} -0.54 & 0.82 \\ 1.63 & 0.23 \end{bmatrix},$$
$$C_1 = \begin{bmatrix} 1 & 0 \end{bmatrix}, \qquad C_2 = \begin{bmatrix} 0 & -1 \end{bmatrix}.$$

Using a quadratic Lyapunov function with the relaxation Lemma D.1, the resulting LMIs are infeasible. However, using Theorem 6.4 with Lemma D.1 a feasible solution is found, namely

$$P_1 = \begin{bmatrix} 0.22 & 0.04 \\ 0.04 & 0.27 \end{bmatrix}, \; H_1 = \begin{bmatrix} 0.34 & 0.21 \\ -0.10 & 0.19 \end{bmatrix}, \; K_1 = \begin{bmatrix} -0.27 \\ 0.02 \end{bmatrix},$$
$$P_2 = \begin{bmatrix} 0.25 & 0.04 \\ 0.04 & 0.25 \end{bmatrix}, \; H_2 = \begin{bmatrix} 0.16 & -0.06 \\ 0.04 & 0.39 \end{bmatrix}, \; K_2 = \begin{bmatrix} -0.36 \\ 0.19 \end{bmatrix};$$

thus illustrating the feasibility advantages of the non-quadratic framework over the quadratic one. □

The use of delayed Lyapunov functions $V(\boldsymbol{e}) = \boldsymbol{e}^T P_{h(t-1)} \boldsymbol{e}$, corresponding to $\boldsymbol{P}(\cdot) = P_{h(t-1)}$ in (6.9), is now explored. First proposed in [12] for state estimation, they improve over the previous conditions without increasing the number of LMIs to be solved.

Consider a discrete-time TS model (6.1) with $\delta \boldsymbol{x}(t) = \boldsymbol{x}(t+1)$ and an observer (6.8) with $\delta\widehat{\boldsymbol{x}}(t) = \widehat{\boldsymbol{x}}(t+1), t \in \mathbb{Z}, S(\cdot) = H_{h(t)h(t-1)}$, and $K(\cdot) = K_{h(t)}$, reproduced below for convenience without terms $B_{h(t)}\boldsymbol{u}(t)$, $\boldsymbol{a}_{h(t)}$, and $\boldsymbol{c}_{h(t)}$ as they play no role in the error dynamics:

$$\begin{aligned} \widehat{\boldsymbol{x}}(t+1) &= A_h \widehat{\boldsymbol{x}}(t) + H^{-1}_{h(t)h(t-1)} K_h (\boldsymbol{y}(t) - \widehat{\boldsymbol{y}}(t)), \\ \widehat{\boldsymbol{y}}(t) &= C_h \widehat{\boldsymbol{x}}(t). \end{aligned}$$

The error dynamics are given by

$$\boldsymbol{e}(t+1) = \left(A_h - H^{-1}_{h(t)h(t-1)} K_h C_h\right) \boldsymbol{e}(t), \tag{6.14}$$

based on which we have the following result:

Theorem 6.5 *The origin $\boldsymbol{e} = 0$ of the discrete-time TS estimation error model (6.14), resulting from the TS model (6.1) with $\delta \boldsymbol{x}(t) = \boldsymbol{x}(t+1)$ and the TS observer (6.8) with $\delta\widehat{\boldsymbol{x}}(t) = \widehat{\boldsymbol{x}}(t+1), t \in \mathbb{Z}, S(\cdot) = H_{h(t)h(t-1)}$, and $K(\cdot) = K_{h(t)}$, is asymptotically stable if there exist matrices $P_i = P_i^T > 0$, K_i, H_{ij}, $i, j\{1, 2, \ldots, r\}$, such that*

$$\begin{bmatrix} -P_{h(t-1)} & (*) \\ H_{h(t)h(t-1)}A_h - K_h C_h & -H_{h(t)h(t-1)} - H^T_{h(t)h(t-1)} + P_h \end{bmatrix} < 0 \tag{6.15}$$

holds. In that case, any trajectory $\boldsymbol{e}(t)$ starting within the outermost Lyapunov level $V(\boldsymbol{e}) = \boldsymbol{e}^T P_{h(t-1)} \boldsymbol{e} \leq k$, $k > 0$ in a compact set $\mathcal{E}$ of the error state space guaranteeing $\boldsymbol{z} \in \mathcal{C} \subseteq \mathbb{R}^p$ goes asymptotically to $\boldsymbol{e} = 0$.

Proof Since $P_i = P_i^T > 0$, $i \in \{1, 2, \ldots, r\}$, it follows that $V(\boldsymbol{e}) = \boldsymbol{e}^T P_{h(t-1)} \boldsymbol{e}$ is a valid Lyapunov function candidate whose forward difference is

$$\begin{aligned}\Delta V &= e^T(t+1)P_h e(t+1) - e^T(t)P_{h(t-1)}e(t) \\ &= \begin{bmatrix} e(t) \\ e(t+1) \end{bmatrix}^T \begin{bmatrix} -P_{h(t-1)} & 0 \\ 0 & P_h \end{bmatrix} \begin{bmatrix} e(t) \\ e(t+1) \end{bmatrix},\end{aligned}$$

which, along with the error dynamics written as

$$\left[A_h - H^{-1}_{h(t)h(t-1)}K_h C_h \ -I\right] \begin{bmatrix} e(t) \\ e(t+1) \end{bmatrix} = 0,$$

allows Finsler's Lemma A.3 to be used to guarantee $\Delta V < 0$ with a multiplier $M = \left[0 \ H^T_{h(t)h(t-1)}\right]^T$ if

$$\begin{bmatrix} 0 \\ H_{h(t)h(t-1)} \end{bmatrix} \left[A_h - H^{-1}_{h(t)h(t-1)}K_h C_h \ -I\right] + (*) + \begin{bmatrix} -P_{h(t-1)} & 0 \\ 0 & P_h^{-1} \end{bmatrix} < 0$$

holds, which is equivalent to inequality (6.15). Thus, $V(e)$ is a valid Lyapunov function that establishes asymptotic stability of $e = 0$; any trajectory starting within the outermost Lyapunov level $V(e) = e^T P_{h(t-1)} e \leq k$, $k > 0$, in a compact set $\mathcal{E}$ of the error state space holding assumptions 6.1 and 6.2, goes asymptotically to $e = 0$, because these assumptions guarantee $z \in \mathcal{C}$, where the MFs $h_i(z)$ hold the convex sum property. □

Remark 6.6 As stated repeatedly, sufficient LMI conditions guaranteeing (6.15) can be found by means of any relaxation scheme in Appendix D. Performance specifications such as convergence rate, robustness, $\mathcal{H}_\infty$ disturbance rejection, etcetera, can be designed following the corresponding outline, both for quadratic and non-quadratic cases. Moreover, for the discrete-time observer design, α-sample variations and multiple delays with Lyapunov function candidates like $V(e) = e^T P_{H_0^P} e$, where H_0^P is a multiset of delays, have also been employed [13].

Table 6.1 summarises some results for non-quadratic observer design in the discrete-time case that can be obtained following the outline of the previous theorems. The second row corresponds to results in Theorem 6.4. The observers do not include the affine terms as they play no role in the observer design.

Remark 6.7 As for controller design, the second row with $H_{h(t)} = P_{h(t)}$ reduces to the first one, whereas the three rows with $H(\cdot) = P(\cdot) = P$ reduce to the quadratic case.

6.2.2 *Observer Design for Descriptor Models*

Estimation for continuous-time descriptor TS models can be found in [14–16] (in the quadratic framework), and in [17] (in the semi-quadratic configuration of [18]). In

Table 6.1 Non-quadratic observer design for discrete-time systems

Lyapunov function: $V(e) = e^T P_{h(t)} e$. Observer: $\hat{x}(t+1) = A_{h(t)}\hat{x}(t) + B_{h(t)}u(t) + P_{h(t)}^{-1} K_{h(t)} \left(y(t) - \hat{y}(t)\right)$. Conditions: $\begin{bmatrix} -P_{h(t)} & (*) \\ P_{h(t)}A_{h(t)} - K_{h(t)}C_{h(t)} & P_{h(t+1)} - 2P_{h(t)} \end{bmatrix} < 0.$
Lyapunov function: $V(e) = e^T P_{h(t)} e$. Observer: $\hat{x}(t+1) = A_{h(t)}\hat{x}(t) + B_{h(t)}u(t) + H_{h(t)}^{-1} K_{h(t)} \left(y(t) - \hat{y}(t)\right)$. Conditions: $\begin{bmatrix} -P_{h(t)} & (*) \\ H_{h(t)}A_{h(t)} - K_{h(t)}C_{h(t)} & P_{h(t+1)} - H_{h(t)} - H_{h(t)}^T \end{bmatrix} < 0.$
Lyapunov function: $V(e) = e^T P_{h(t-1)} e$. Observer: $\hat{x}(t+1) = A_{h(t)}\hat{x}(t) + B_{h(t)}u(t) + H_{h(t)h(t-1)}^{-1} K_{h(t)h(t-1)} \left(y(t) - \hat{y}(t)\right)$. Conditions: $\begin{bmatrix} -P_{h(t-1)} & (*) \\ H_{h(t)h(t-1)}A_{h(t)} - K_{h(t)h(t-1)}C_{h(t)} & P_{h(t)} - H_{h(t)h(t-1)} - H_{h(t)h(t-1)}^T \end{bmatrix} < 0.$

the discrete-time case [19, 20] (for state estimation) have considered both quadratic and non-quadratic results in a single framework which, not surprisingly, makes use of Finsler's Lemma A.3. The convex descriptor model for which observers will be designed in this section is

$$
\begin{aligned}
E_v \delta x(t) &= \sum_{i=1}^{r_e} v_i(z) E_i \delta x(t) = \sum_{i=1}^{r} h_i(z) \left(A_i x(t) + B_i u(t)\right) = A_h x(t) + B_h u(t), \\
y(t) &= \sum_{i=1}^{r} h_i(z) C_i x(t) = C_h x(t),
\end{aligned}
\tag{6.16}
$$

where the different terms and notations have the same meaning as in previous chapters, with $x \in \mathbb{R}^n$, $u \in \mathbb{R}^m$, and $y \in \mathbb{R}^q$. Recall the scheduling variables $z \in \mathcal{C} \subseteq \mathbb{R}^p$ are assumed measurable, i.e. available to be used in the observer.

The Continuous Quadratic Case

In [14, 15], the following observer for (6.16) was proposed:

$$\begin{aligned} \overline{E}\dot{\hat{\overline{x}}}(t) &= \overline{A}_{hv}\hat{\overline{x}}(t) + \overline{B}_h \boldsymbol{u}(t) + \overline{K}_{hv}\left(\boldsymbol{y}(t) - \hat{\boldsymbol{y}}(t)\right), \\ \hat{\boldsymbol{y}}(t) &= \overline{C}_h \hat{\overline{x}}(t), \end{aligned} \tag{6.17}$$

with

$$\hat{\overline{x}} = [\hat{x}\ \dot{\hat{x}}],\ \overline{E} = \begin{bmatrix} I & 0 \\ 0 & 0 \end{bmatrix},\ \overline{A}_h v = \begin{bmatrix} 0 & I \\ A_h & -E_v \end{bmatrix},\ \overline{B}_h = \begin{bmatrix} 0 \\ B_h \end{bmatrix},\ \overline{C} = [C_h\ 0],$$

$$\overline{K}_{hv} = \begin{bmatrix} 0 \\ K_{hv} \end{bmatrix},\ K_{hv} = \sum_{j=1}^{r_e}\sum_{i=1}^{r} v_j(z)h_i(z)K_{ij},\ K_{ij} \in \mathbb{R}^{n\times q}.$$

Our objective is to determine the observer gains K_{ij} such that the estimation error $\boldsymbol{e} = \boldsymbol{x} - \hat{\boldsymbol{x}}$ goes asymptotically to zero when $t \to \infty$. For this purpose, an extended estimation error is defined:

$$\overline{e} = \overline{x} - \hat{\overline{x}} = \begin{bmatrix} \boldsymbol{x} - \hat{\boldsymbol{x}} \\ \dot{\boldsymbol{x}} - \dot{\hat{\boldsymbol{x}}} \end{bmatrix}, \quad \text{with } \overline{x} = \begin{bmatrix} \boldsymbol{x} \\ \dot{\boldsymbol{x}} \end{bmatrix},$$

based on which the extended error dynamics can be obtained from (6.16) and (6.17) as

$$\begin{aligned} \overline{E}\dot{\overline{e}}(t) &= \overline{A}_{hv}\overline{x} + \overline{B}_h \boldsymbol{u} - \left(\overline{A}_{hv}\hat{\overline{x}} + \overline{B}_h \boldsymbol{u} + \overline{K}_{hv}\left(\boldsymbol{y} - \hat{\boldsymbol{y}}\right)\right) \\ &= \overline{A}_{hv}\left(\overline{x} - \hat{\overline{x}}\right) - \overline{K}_{hv}\overline{C}_h\left(\overline{x} - \hat{\overline{x}}\right) \\ &= \left(\overline{A}_{hv} - \overline{K}_{hv}\overline{C}_h\right)\overline{e}(t). \end{aligned} \tag{6.18}$$

Theorem 6.6 *The origin* $\boldsymbol{e} = 0$ *of the continuous-time descriptor estimation error model* (6.18)*, resulting from the descriptor model* (6.16) *with* $\delta\boldsymbol{x}(t) = \dot{\boldsymbol{x}}(t)$, $t \in \mathbb{R}$ *and the descriptor observer* (6.17)*, is asymptotically stable if there exist matrices* $P_1 = P_1^T > 0$, P_3, M_{ij}, $i \in \{1, 2, \ldots, r\}$, $j \in \{1, 2, \ldots, r_e\}$*, such that*

$$\begin{bmatrix} P_3^T A_h - M_{hv}C_h + (*) & (*) \\ P_3^T A_h - M_{hv}C_h + P_1 - E_v^T P_3 & -E_v^T P_3 - P_3^T E_v \end{bmatrix} < 0 \tag{6.19}$$

holds. In this case, the observer gains are $K_{ij} = P_3^{-T} M_{ij}$, $i \in \{1, 2, \ldots, r\}$, $j \in \{1, 2, \ldots, r_e\}$*, and any trajectory* $\boldsymbol{e}(t)$ *starting within the outermost Lyapunov level* $V(\boldsymbol{e}) = \boldsymbol{e}^T P_1 \boldsymbol{e} \le k$, $k > 0$*, in a compact set* $\mathcal{E}$ *of the error state space guaranteeing* $z \in \mathcal{C} \subseteq \mathbb{R}^p$ *goes asymptotically to* $\boldsymbol{e} = 0$.

Proof Since $P_1 = P_1^T > 0$, it is clear that

$$V(\overline{e}) = \overline{e}^T \overline{E}^T P \overline{e} = e^T P_1 e, \quad \overline{E}^T P = P^T \overline{E} \geq 0, \tag{6.20}$$

is a quadratic Lyapunov function candidate with

$$P = \begin{bmatrix} P_1 & 0 \\ P_3 & P_4 \end{bmatrix}. \tag{6.21}$$

The time derivative of (6.20) is

$$\dot{V} = \dot{\overline{e}}^T \overline{E} P \overline{e} + \overline{e} P^T \overline{E} \dot{\overline{e}} = \overline{e}^T P^T \left(\overline{A}_{hv} - \overline{K}_{hv} \overline{C}_h \right) \overline{e} + (*).$$

To ensure that $\dot{V}(\overline{e}) < 0$, $e \neq 0$, a sufficient condition is

$$P^T \overline{A}_{hv} - P^T \overline{K}_{hv} \overline{C}_h + \overline{A}_{hv}^T P - \overline{C}_h^T \overline{K}_{hv}^T P < 0,$$

which is equivalent to

$$\begin{bmatrix} P_3^T A_h - P_3^T K_{hv} C_h + (*) & (*) \\ P_4^T A_h - P_4^T K_{hv} C_h + P_1 - E_v^T P_3 & -E_v^T P_4 - P_4^T E_v \end{bmatrix} < 0.$$

Dropping MFs via a relaxation scheme of Appendix D on the previous expression does not produce LMIs, because the presence of $P_3^T K_{hv} C_h$ and $P_4^T K_{hv} C_h$ makes the possibility of replacing $P_3^T K_{hv}$ and $P_4^T K_{hv}$ by a single decision variable univocally defining K_{hv}, impossible. Therefore, we set $P_3 = P_4$ and define $M_{hv} = P_3^T K_{hv}$ as in [14] to obtain inequality (6.19). Thus, $V(e)$ is a valid Lyapunov function proving $e = 0$ is asymptotically stable as long as MFs $h_i(z)$ and $v_j(z)$ hold the convex sum property, which is guaranteed for $z \in \mathcal{C}$. Therefore, any trajectory starting in the outermost Lyapunov level $V(e) = e^T P_1 e \leq k$, $k > 0$, in a compact set $\mathcal{E}$ of the error state space implying $z \in \mathcal{C}$—i.e. holding Assumptions 6.1 and 6.2—goes asymptotically to $e = 0$. □

Remark 6.8 In [15] the authors suggest a pole placement technique instead of using $P_3 = P_4$ in the expression above. Once the gain K_{hv} is designed, the corresponding expression is used to verify the estimation error convergence. As with other design schemes involving nested convex sums, sufficient LMI conditions guaranteeing (6.19) can be obtained applying any relaxation scheme in Appendix D.

An observer with more degrees of flexibility for the observer gain is now considered. To this end, the following observer is proposed using a new observer state, namely, $\hat{\overline{x}} = \left[\hat{x}^T \ \alpha^T \right]^T$, i.e.

$$\begin{aligned} \overline{E} \dot{\hat{\overline{x}}}(t) &= \overline{A}_{hv} \hat{\overline{x}}(t) + \overline{B}_h u(t) + P^{-T} \overline{K}_{hv} \left(y(t) - \hat{y}(t) \right), \\ \hat{y}(t) &= \overline{C}_h \hat{\overline{x}}(t), \end{aligned} \tag{6.22}$$

where all the terms are defined as in the previous observer scheme, except that $\overline{K}_{hv} = \left[K_{1hv}^T\ K_{2hv}^T\right]^T$, i.e. it is fully exploited, and P is defined as in (6.21) with $P_1 = P_1^T > 0$, P_3 and P_4 free variables, i.e. P_3 is no longer required to be equal to P_4.

As before, an extended estimation error is defined

$$\overline{e} = \overline{x} - \hat{\overline{x}} = \begin{bmatrix} x - \hat{x} \\ \dot{x} - \alpha \end{bmatrix},$$

and its dynamics can be thus expressed as follows:

$$\overline{E}\dot{\overline{e}}(t) = \left(\overline{A}_{hv} - P^{-T}\overline{K}_{hv}\overline{C}_h\right)\overline{e}(t). \tag{6.23}$$

Theorem 6.7 *The origin $e = 0$ of the continuous-time descriptor estimation error model* (6.23)*, resulting from the descriptor model* (6.16) *with $\delta x(t) = \dot{x}(t)$, $t \in \mathbb{R}$ and the descriptor observer* (6.22)*, is asymptotically stable if there exist matrices $P_1 = P_1^T > 0$, P_3, P_4, K_{1ij}, K_{2ij}, $i \in \{1, 2, \dots, r\}$, $j \in \{1, 2, \dots, r_e\}$, such that*

$$\begin{bmatrix} P_3^T A_h - K_{1hv}C_h + (*) & (*) \\ P_4^T A_h - K_{2hv}C_h + P_1 - E_v^T P_3 & -P_4^T E_v - E_v^T P_4 \end{bmatrix} < 0 \tag{6.24}$$

holds. In that case, any trajectory $e(t)$ starting within the outermost Lyapunov level $V(e) = e^T P_1 e \leq k$, $k > 0$ in a compact set $\mathcal{E}$ of the error state space induced by $\mathcal{C} \subseteq \mathbb{R}^p$ goes asymptotically to $e = 0$.

Proof It has already been established that $V(e) = \overline{e}^T\overline{E}^T P\overline{e}$ is a Lyapunov function candidate with $\overline{E}^T P = P^T\overline{E} \geq 0$ and $P_1 = P_1^T > 0$; its time derivative is

$$\dot{V} = \dot{\overline{e}}^T\overline{E}P\overline{e} + \overline{e}^T P^T\overline{E}\dot{\overline{e}} = \overline{e}^T P^T\left(\overline{A}_{hv} - P^{-T}\overline{K}_{hv}\overline{C}_h\right)\overline{e} + (*).$$

A sufficient condition to guarantee $\dot{V} < 0$, $e \neq 0$ is

$$P^T\overline{A}_{hv} - \overline{K}_{hv}\overline{C}_h + \overline{A}_{hv}^T P - \overline{C}_h^T\overline{K}_{hv}^T < 0,$$

which after substitution of P, $\overline{A}_{hv}$, $\overline{K}_{hv}$, and $\overline{C}_h$ yields conditions (6.24), thus establishing that $V(e)$ is a valid Lyapunov function that proves asymptotic stability of $e = 0$.

Since the steps above depend on MFs $h_i(z)$ and $v_j(z)$ holding the convex sum property, it is clear that trajectories starting within the outermost Lyapunov level $V(e) = e^T P_1 e \leq k$, $k > 0$, in a compact set $\mathcal{E}$ of the error state space holding Assumptions 6.1 and 6.2, go asymptotically to $e = 0$, since these assumptions imply $z \in \mathcal{C}$, where the MFs are guaranteed to hold the convex sum property. □

Remark 6.9 The small change in the extended state vector $\hat{\overline{x}}$ introducing α allows rewriting the final form of the observer with a new structure of $\overline{K}_{hv}$. Indeed, (6.22) can be rewritten as

$$\begin{bmatrix} I & 0 \\ 0 & 0 \end{bmatrix} \begin{bmatrix} \dot{\hat{x}} \\ \dot{\alpha} \end{bmatrix} = \begin{bmatrix} 0 & I \\ A_h & -E_h \end{bmatrix} \begin{bmatrix} \hat{x} \\ \alpha \end{bmatrix} + \begin{bmatrix} 0 \\ B_h \end{bmatrix} u + P^{-T} \begin{bmatrix} K_{1hv} \\ K_{2hv} \end{bmatrix} (y - \hat{y}),$$

which, using the notation

$$\begin{aligned} \begin{bmatrix} \mathcal{K}_{1hv} \\ \mathcal{K}_{2hv} \end{bmatrix} &= P^{-T} \begin{bmatrix} K_{1hv} \\ K_{2hv} \end{bmatrix} = \begin{bmatrix} P_1^{-1} & -P_1^{-1} P_3^T P_4^{-T} \\ 0 & P_4^{-T} \end{bmatrix} \begin{bmatrix} K_{1hv} \\ K_{2hv} \end{bmatrix} \\ &= \begin{bmatrix} P_1^{-1} \left(K_{1hv} - P_3^T P_4^{-T} K_{2hv} \right) \\ P_4^{-1} K_{2hv} \end{bmatrix}, \end{aligned} \tag{6.25}$$

yields

$$\begin{bmatrix} I & 0 \\ 0 & 0 \end{bmatrix} \begin{bmatrix} \dot{\hat{x}} \\ \dot{\alpha} \end{bmatrix} = \begin{bmatrix} 0 & I \\ A_h & -E_v \end{bmatrix} \begin{bmatrix} \hat{x} \\ \alpha \end{bmatrix} + \begin{bmatrix} 0 \\ B_h \end{bmatrix} u + \begin{bmatrix} \mathcal{K}_{1hv} C_h \\ \mathcal{K}_{2hv} C_h \end{bmatrix} (x - \hat{x}),$$

or equivalently

$$\begin{aligned} \dot{\hat{x}} &= \alpha + \mathcal{K}_{1hv} C_h (x - \hat{x}), \\ E_v \alpha &= A_h \hat{x} + B_h u + \mathcal{K}_{2hv} C_h (x - \hat{x}). \end{aligned} \tag{6.26}$$

Defining $\alpha = \dot{\hat{x}} - \mathcal{K}_{1hv} C_h (x - \hat{x})$, allows rewriting (6.26) as

$$\begin{aligned} E_v \dot{\hat{x}} &= A_h \hat{x} + B_h u + E_v \mathcal{K}_{1hv} C_h (x - \hat{x}) + \mathcal{K}_{2hv} C_h (x - \hat{x}) \\ &= A_h \hat{x} + B_h u + \begin{bmatrix} E_v & I \end{bmatrix} \begin{bmatrix} \mathcal{K}_{1hv} \\ \mathcal{K}_{2hv} \end{bmatrix} (y - \hat{y}), \end{aligned} \tag{6.27}$$

which, using (6.25) and (6.27), corresponds to the final observer form

$$\begin{aligned} E_v \dot{\hat{x}} &= A_h \hat{x} + B_h u + \begin{bmatrix} E_v & I \end{bmatrix} P^{-T} \begin{bmatrix} K_{1hv} \\ K_{2hv} \end{bmatrix} (y - \hat{y}), \\ \hat{y} &= C_h \hat{x}. \end{aligned}$$

Note that without the new auxiliary variable α, expression (6.26) does not hold unless a perfect observation is obtained. Moreover, if the LMI problem is feasible, it means that $x - \hat{x} \to 0$ and $\dot{x} - \alpha \to 0$ as time goes to infinity.

Example 6.8 Consider a family of TS descriptor models with $u(t) = 0, r = r_e = 2$, and its vertex models described by [16]:

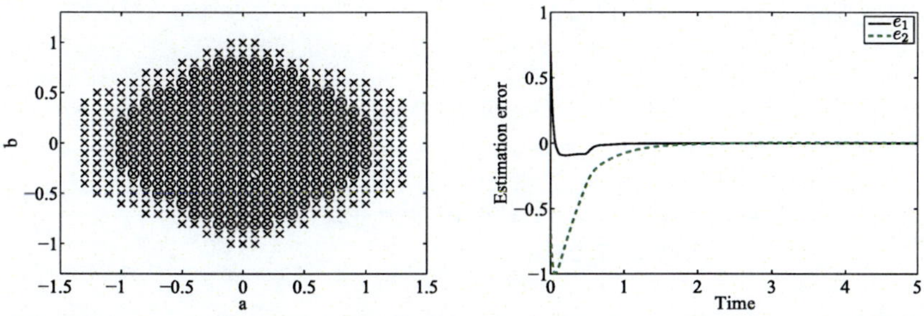

(a) Feasibility sets: (○) for (Guerra et al., 2004); (×) for (Estrada-Manzo et al., 2014a).

(b) Estimation error for $(a, b) = (-1.2, -0.3)$.

Fig. 6.6 Observer design for Example 6.8

$$A_1 = \begin{bmatrix} -0.35 & 1.24 + b \\ -0.74 & 0.5 \end{bmatrix},\ A_2 = \begin{bmatrix} -0.35 & 1.24 - b \\ -0.74 & 0.5 \end{bmatrix},$$
$$E_1 = \begin{bmatrix} 0.37 & 0.43 - a \\ 0.23 & 1.15 \end{bmatrix},\ E_2 = \begin{bmatrix} 0.37 & 0.43 + a \\ 0.23 & 1.15 \end{bmatrix},$$
$$C_1 = \begin{bmatrix} 1.5 + a \\ 0 \end{bmatrix}^T,\ C_2 = \begin{bmatrix} 1.5 - a \\ 0 \end{bmatrix}^T,$$

with parameters $a \in [-1.5,\ 1.5]$ and $b \in [-1.5,\ 1.5]$. This family is investigated to find out whether an observer can be designed or not for a pair of values (a, b). The LMI conditions in [14] corresponding to Theorem 6.6 are compared with the more recent in [16] corresponding to Theorem 6.7, to determine whether they are feasible for a given pair (a, b) or not. The results are illustrated in Fig. 6.6a.

Note that when $a = -1.2$ and $b = -0.3$, there is no solution for LMI conditions in [14] corresponding to Theorem 6.6; nevertheless, conditions in [16] corresponding to Theorem 6.7, provide the following solution:

$$P_1 = \begin{bmatrix} 0.14 & -0.12 \\ -0.12 & 0.21 \end{bmatrix},\ P_3 = \begin{bmatrix} 0.36 & -1 \\ 0.92 & 0.93 \end{bmatrix},\ P_4 = \begin{bmatrix} 0.31 & -0.16 \\ -0.16 & 1.60 \end{bmatrix},$$
$$K_{11} = \begin{bmatrix} 1.80 \\ -0.04 \\ -0.35 \\ -4.89 \end{bmatrix},\ K_{12} = \begin{bmatrix} -0.04 \\ 0.86 \\ -0.29 \\ -2.78 \end{bmatrix},\ K_{21} = \begin{bmatrix} 0.21 \\ 0.36 \\ -0.06 \\ -0.80 \end{bmatrix},\ K_{22} = \begin{bmatrix} 0.25 \\ 0.17 \\ -0.05 \\ -0.57 \end{bmatrix},$$

whose performance is shown in Fig. 6.6b with MFs holding the convex sum property for the whole state space with $z = \boldsymbol{x}$, namely $v_1(z) = 1/\left(1 + x_1^2\right)$, $v_2(z) = 1 - v_1(z)$, $h_1(z) = (\cos x_1 + 1)/2$, and $h_2(z) = 1 - h_1(z)$ and initial conditions $\boldsymbol{x}(0) = \left[0.7\ -0.7\right]^T$, $\hat{\boldsymbol{x}}(0) = \left[0\ 0\right]^T$. □

The Discrete Quadratic Case

An observer for the discrete-time descriptor model (6.16) with $\delta \boldsymbol{x}(t) = \boldsymbol{x}(t+1)$, $t \in \mathbb{R}$, is given by:

$$\begin{aligned} E_v \hat{\boldsymbol{x}}(t+1) &= A_h \hat{\boldsymbol{x}}(t) + B_h \boldsymbol{u}(t) + K_{hv} \left(\boldsymbol{y}(t) - \hat{\boldsymbol{y}}(t) \right) \\ \hat{\boldsymbol{y}}(t) &= C_h \hat{\boldsymbol{x}}(t), \end{aligned} \tag{6.28}$$

where all the terms are defined as in (6.16).

The estimation error $\boldsymbol{e}(t) = \boldsymbol{x}(t) - \hat{\boldsymbol{x}}(t)$ has the following dynamics:

$$E_v \boldsymbol{e}(t+1) = (A_h - K_{hv} C_h)\, \boldsymbol{e}(t), \tag{6.29}$$

which can be equivalently written as

$$\begin{bmatrix} A_h - K_{hv} C_h & -E_v \end{bmatrix} \begin{bmatrix} \boldsymbol{e}(t) \\ \boldsymbol{e}(t+1) \end{bmatrix} = 0. \tag{6.30}$$

Theorem 6.8 *The origin $\boldsymbol{e} = 0$ of the discrete-time descriptor estimation error model (6.29), resulting from the descriptor model (6.16) with $\delta \boldsymbol{x}(t) = \boldsymbol{x}(t+1)$, $t \in \mathbb{Z}$, and the descriptor observer (6.28), is asymptotically stable if there exist matrices $P = P^T > 0$, M_{ij}, $i \in \{1, 2, \ldots, r\}$, $j \in \{1, 2, \ldots, r_e\}$, such that*

$$\begin{bmatrix} -P & A_{h(t)}^T P - C_{h(t)}^T M_{h(t)v(t)}^T \\ P A_{h(t)} - M_{h(t)v(t)} C_{h(t)} & -P E_{v(t)} - E_{v(t)}^T P + P \end{bmatrix} < 0 \tag{6.31}$$

holds. In that case, the observer gains are $K_{ij} = P^{-1} M_{ij}$, $i \in \{1, 2, \ldots, r\}$, $j \in \{1, 2, \ldots, r_e\}$, and any trajectory $\boldsymbol{e}(t)$ starting within the outermost Lyapunov level $V(\boldsymbol{e}) = \boldsymbol{e}^T P \boldsymbol{e} \le k$, $k > 0$, in a compact set $\mathcal{E}$ of the error state space guaranteeing $z \in \mathcal{C} \subseteq \mathbb{R}^p$ goes asymptotically to $\boldsymbol{e} = 0$.

Proof It is clear that $V(\boldsymbol{e}) = \boldsymbol{e}^T P \boldsymbol{e}$ is a Lyapunov function candidate because $P = P^T > 0$; its forward difference is

$$\begin{aligned} \Delta V &= \boldsymbol{e}^T(t+1) P \boldsymbol{e}(t+1) - \boldsymbol{e}^T(t) P \boldsymbol{e}(t) \\ &= \begin{bmatrix} \boldsymbol{e}(t) \\ \boldsymbol{e}(t+1) \end{bmatrix}^T \begin{bmatrix} -P & 0 \\ 0 & P \end{bmatrix} \begin{bmatrix} \boldsymbol{e}(t) \\ \boldsymbol{e}(t+1) \end{bmatrix} < 0. \end{aligned} \tag{6.32}$$

As in many other cases above, Finsler's Lemma A.3 comes at hand to combine the Lyapunov variation inequality above with the error dynamics (equality) in (6.30) as to obtain sufficient conditions guaranteeing $\Delta V < 0$, namely

$$\begin{bmatrix} \mathcal{U}(\cdot) \\ \mathcal{V}(\cdot) \end{bmatrix} \begin{bmatrix} A_h - K_{hv} C_h & -E_v \end{bmatrix} + (*) + \begin{bmatrix} -P & 0 \\ 0 & P \end{bmatrix} < 0.$$

Choosing $\mathcal{U}(\cdot) = 0$ and $\mathcal{V}(\cdot) = P$ yields (6.31) if the substitution $M_{hv} = PK_{hv}$ is made, which proves $V(\boldsymbol{e})$ is a Lyapunov function establishing asymptotic stability of $\boldsymbol{e} = 0$ as long as the MFs $h_i(z)$ and $v_j(z)$ hold the convex sum property. Since the latter is guaranteed in $\mathcal{C}$, it follows that trajectories starting within the outermost Lyapunov level $V(\boldsymbol{e}) = \boldsymbol{e}^T P \boldsymbol{e} \leq k, k > 0$, in a compact set $\mathcal{E}$ of the error state space holding Assumptions 6.1 and 6.2 go asymptotically to $\boldsymbol{e} = 0$. □

Remark 6.10 Relaxation schemes in Appendix D are necessary to obtain sufficient LMIs guaranteeing (6.31). If the output is linear, no relaxation is needed as it yields the same results as directly dropping the MFs.

The Discrete Non-Quadratic Case

The next observer proposal uses Finsler's Lemma to decouple the observer gains from the Lyapunov function, allowing extra slack variables to be included:

$$\begin{aligned} E_v \hat{\boldsymbol{x}}(t+1) &= A_h \hat{\boldsymbol{x}}(t) + B_h \boldsymbol{u}(t) + H_h^{-1} K_{hv} \left(\boldsymbol{y}(t) - \hat{\boldsymbol{y}}(t) \right), \\ \hat{\boldsymbol{y}}(t) &= C_h \boldsymbol{x}(t), \end{aligned} \tag{6.33}$$

where all the terms have the same meanings as before with the addition of H_h which is a convex sum of matrices $H_i, i \in \{1, 2, \ldots, r\}$.

The estimation error $\boldsymbol{e}(t) = \boldsymbol{x}(t) - \hat{\boldsymbol{x}}(t)$ and its dynamics are written as

$$\begin{bmatrix} A_h - H_h^{-1} K_{hv} C_h & -E_v \end{bmatrix} \begin{bmatrix} \boldsymbol{e}(t) \\ \boldsymbol{e}(t+1) \end{bmatrix} = 0. \tag{6.34}$$

Theorem 6.9 *The origin $\boldsymbol{e} = 0$ of the discrete-time descriptor estimation error model (6.34), resulting from the descriptor model (6.16) with $\delta \boldsymbol{x}(t) = \boldsymbol{x}(t+1)$, $t \in \mathbb{Z}$, and the descriptor observer (6.33), is asymptotically stable if there exist matrices $P_i = P_i^T > 0$, H_i, K_{ij} $i \in \{1, 2, \ldots, r\}$, $j \in \{1, 2, \ldots, r_e\}$, such that*

$$\begin{bmatrix} -P_{h(t)} & (*) \\ H_{h(t)} A_{h(t)} - K_{h(t)v(t)} C_{h(t)} & -H_{h(t)} E_{v(t)} - E_{v(t)}^T H_{h(t)}^T + P_{h(t+1)} \end{bmatrix} < 0 \tag{6.35}$$

holds. In that case, any trajectory $\boldsymbol{e}(t)$ starting within the outermost Lyapunov level $V(\boldsymbol{e}) = \boldsymbol{e}^T P_h \boldsymbol{e} \leq k$, $k > 0$ in a compact set $\mathcal{E}$ of the error state space guaranteeing $z \in \mathcal{C} \subseteq \mathbb{R}^p$ goes asymptotically to $\boldsymbol{e} = 0$.

Proof Consider the non-quadratic Lyapunov function candidate:

$$V(\boldsymbol{e}(t)) = \boldsymbol{e}(t)^T P_h \boldsymbol{e}(t), \; P_h = \sum_{i=1}^{r} h_i(z(t)) P_i,$$

where $P_i = P_i^T > 0$ is guaranteed by hypothesis; its forward difference is given by

$$\begin{aligned}\Delta V &= e^T(t+1)P_{h(t+1)}e(t+1) - e^T(t)P_{h(t)}e(t) \\ &= \begin{bmatrix} e(t) \\ e(t+1) \end{bmatrix}^T \begin{bmatrix} -P_{h(t)} & 0 \\ 0 & P_{h(t+1)} \end{bmatrix} \begin{bmatrix} e(t) \\ e(t+1) \end{bmatrix} < 0.\end{aligned} \tag{6.36}$$

Again, both the variation of the non-quadratic Lyapunov function (6.36) and the error dynamics (6.34) can be combined through Finsler's Lemma A.3, to yield the following equivalent condition

$$\begin{bmatrix} \mathcal{U}(\cdot) \\ \mathcal{V}(\cdot) \end{bmatrix} \begin{bmatrix} A_h - H_h^{-1}K_{hv}C_h & -E_v \end{bmatrix} + (*) + \begin{bmatrix} -P_{h(t)} & 0 \\ 0 & P_{h(t+1)} \end{bmatrix} < 0.$$

Assigning $\mathcal{U}(\cdot) = 0$ and $\mathcal{V}(\cdot) = H_h$ yields (6.35), thus proving $V(e)$ is a Lyapunov function guaranteeing asymptotic stability of $e = 0$ as long as MFs $h_i(z)$ and $v_j(z)$ hold the convex sum property, which is ensured for $z \in \mathcal{C}$. Therefore, any trajectory starting within the outermost Lyapunov level $V(e) = e^T P_h e \leq k, k > 0$, in a compact set $\mathcal{E}$ of the error state space holding Assumptions 6.1 and 6.2 goes asymptotically to the origin. □

Remark 6.11 Sufficient LMI conditions for inequality (6.35) can be obtained using any relaxation scheme in Appendix D. Notice that the non-quadratic approach reduces to the quadratic one if $P_h = H_h = P$.

Example 6.9 Consider a discrete-time descriptor model with $r = r_e = 2$ and system matrices:

$$A_1 = \begin{bmatrix} -1 & 1+a \\ -1.5 & 0.5 \end{bmatrix}, A_2 = \begin{bmatrix} -1 & 1-a \\ -1.5 & 0.5 \end{bmatrix}, C_1 = \begin{bmatrix} 0 \\ 1-b \end{bmatrix}^T, C_2 = \begin{bmatrix} 0 \\ 1+b \end{bmatrix},$$
$$E_1 = \begin{bmatrix} 0.9 & 0.1+a \\ -0.4-b & 1.1 \end{bmatrix}, E_2 = \begin{bmatrix} 0.9 & 0.1-a \\ -0.4+b & 1.1 \end{bmatrix},$$

with parameters $a \in [-1, 1]$ and $b \in [-1, 1]$.

The feasibility sets of observer design conditions in Theorems 6.8 and 6.9 are compared in Fig. 6.7a for the system above, with (∘) for the quadratic approach and (×) for the non-quadratic. As expected, the former feasibility set is included in the latter.

For illustration purposes, consider the case where $a = -0.5$ and $b = -0.5$, which has no solution in the quadratic framework. Using the non-quadratic conditions above, the following solution was found:

$$P_1 = \begin{bmatrix} 0.51 & -0.35 \\ -0.35 & 0.26 \end{bmatrix}, P_2 = \begin{bmatrix} 0.49 & -0.34 \\ -0.34 & 0.26 \end{bmatrix},$$

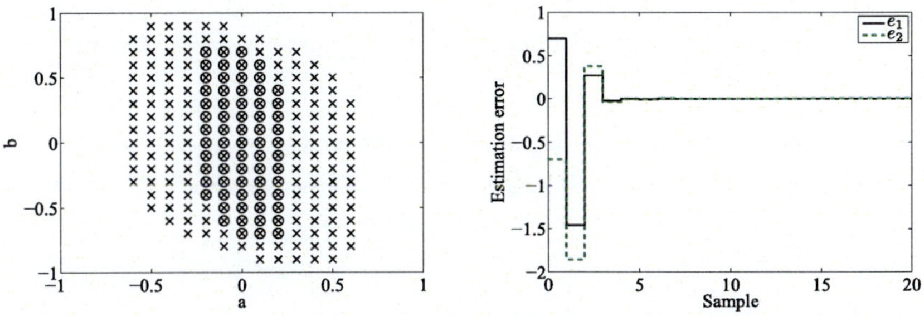

(a) Feasibility sets: (o) for quadratic; (×) for non-quadratic. (b) Estimation error for $(a, b) = (-0.5, -0.5)$.

Fig. 6.7 Observer design for Example 6.9

$$H_1 = \begin{bmatrix} 0.64 & -0.38 \\ -0.24 & 0.36 \end{bmatrix}, \; H_2 = \begin{bmatrix} 0.32 & -0.16 \\ -0.07 & 0.26 \end{bmatrix},$$

$$K_{11} = \begin{bmatrix} 0.0003 \\ -0.0466 \end{bmatrix}, \; K_{12} = \begin{bmatrix} 0.0215 \\ -0.0887 \end{bmatrix},$$

$$K_{21} = \begin{bmatrix} 0.6724 \\ -0.4504 \end{bmatrix}, \; K_{22} = \begin{bmatrix} 0.6654 \\ -0.3708 \end{bmatrix}.$$

The time evolution of the estimation error is plotted in Fig. 6.7 for initial conditions: $\boldsymbol{x}(0) = \begin{bmatrix} 0.7 & -0.7 \end{bmatrix}^T$ and $\hat{\boldsymbol{x}}(0) = \begin{bmatrix} 0 & 0 \end{bmatrix}^T$. MFs $v_1(z) = 1/\left(1 + x_1^2\right)$, $v_2(z) = 1 - v_1(z)$, $h_1(z) = (\cos x_1 + 1)/2$, and $h_2(z) = 1 - h_1(z)$ were used; they hold the convex sum property for all $z = \boldsymbol{x}$. □

State estimation schemes for descriptor models are summarised in Table 6.2. All of them, except the third one which can be proven as an exercise, have been detailed in this section. Recall that these results assume that the premises are measurable.

6.2.3 Observer Design: Unmeasurable Membership Functions

We now turn our attention to the problem of designing an observer for a TS model whose membership functions are not measurable, i.e. the structure of $h_i(\cdot)$ is known, but z is not available. Since the observers presented in the previous sections used the same MFs as the TS models whose states they intended to observe, the usual approach to the non-availability of z was, for a long time, to use an estimate $\hat{z}$ while mimicking the whole convex structure of the TS model. Then, observer design was carried out by adding Lipschitz-like assumptions on a variety of terms that emerge

Table 6.2 Estimation of descriptor TS models

Continuous quadratic

Lyapunov function: $V(\bar{e}(t)) = \bar{e}^T(t)\bar{E}^T P\bar{e}(t)$, $\bar{e}(t) = \begin{bmatrix} e^T & \dot{e}^T \end{bmatrix}^T$.

Observer: $E_v\dot{\hat{x}} = A_h\hat{x} + B_h u + K_{hv}\left(y - \hat{y}\right)$, $\hat{y} = C_h\hat{x}$.

Conditions: $$\begin{bmatrix} P_3^T A_h - M_{hv}C_h + (*) & (*) \\ P_3^T A_h - M_{hv}C_h + P_1 - E_v^T P_3 & -E_v^T P_3 + (*) \end{bmatrix} < 0,$$

$$P_1 = P_1^T > 0,\ P = \begin{bmatrix} P_1 & 0 \\ P_3 & P_3 \end{bmatrix},\ K_{hv} = P_3^{-T} M_{hv}.$$

Continuous quadratic (full-gain)

Lyapunov function: $V(\bar{e}(t)) = \bar{e}^T(t)\bar{E}^T P\bar{e}(t)$, $\bar{e}(t) = \begin{bmatrix} e^T & (\dot{x} - \alpha)^T \end{bmatrix}^T$.

Observer: $E_v\dot{\hat{x}} = A_h\hat{x} + B_h u + \begin{bmatrix} E_v & I \end{bmatrix} P^{-T} \begin{bmatrix} K_{1hv} \\ K_{2hv} \end{bmatrix} \left(y - \hat{y}\right)$, $\hat{y} = C_h\hat{x}$.

Conditions: $$\begin{bmatrix} P_3^T A_h - K_{1hv}C_h + (*) & (*) \\ P_4^T A_h - K_{2hv}C_h + P_1 - E_v^T P_3 & -E_v^T P_4 + (*) \end{bmatrix} < 0,$$

$$P_1 = P_1^T > 0,\ P = \begin{bmatrix} P_1 & 0 \\ P_3 & P_4 \end{bmatrix}.$$

Continuous semi-quadratic (full-gain)

Lyapunov function: $V(\bar{e}(t)) = \bar{e}^T(t)\bar{E}^T P_h\bar{e}(t)$, $\bar{e}(t) = \begin{bmatrix} e^T & (\dot{x} - \alpha)^T \end{bmatrix}^T$.

Observer: $E_v\dot{\hat{x}} = A_h\hat{x} + B_h u + \begin{bmatrix} E_v & I \end{bmatrix} P_h^{-T} \begin{bmatrix} K_{1hv} \\ K_{2hv} \end{bmatrix} \left(y - \hat{y}\right)$, $\hat{y} = C_h\hat{x}$.

Conditions: $$\begin{bmatrix} P_{3h}^T A_h - K_{1hv}C_h + (*) & (*) \\ P_{4h}^T A_h - K_{2hv}C_h + P_1 - E_v^T P_{3h} & -E_v^T P_{4h} + (*) \end{bmatrix} < 0,$$

$$P_1 = P_1^T > 0,\ P_h = \begin{bmatrix} P_1 & 0 \\ P_{3h} & P_{4h} \end{bmatrix}.$$

Discrete quadratic. Lyapunov function: $V(e(t)) = e^T P e$.

Observer: $E_v\hat{x}(t+1) = A_h\hat{x} + B_h u + K_{hv}\left(y - \hat{y}\right)$, $\hat{y} = C_h\hat{x}$.

Conditions: $$\begin{bmatrix} -P & (*) \\ PA_h - M_{hv}C_h & P(0.5I - E_v) + (*) \end{bmatrix} < 0, \quad \begin{matrix} P = P^T > 0, \\ K_{hv} = P^{-1}M_{hv}. \end{matrix}$$

Discrete non-quadratic. Lyapunov function: $V(e(t)) = e^T P_h e$.

Observer: $E_v\hat{x}(t+1) = A_h\hat{x} + B_h u + H_h^{-1} K_{hv}\left(y - \hat{y}\right)$, $\hat{y} = C_h\hat{x}$.

Conditions: $$P_h = P_h^T > 0, \quad \begin{bmatrix} -P_h & (*) \\ H_h A_h - K_{hv}C_h & 0.5P_{h(t+1)} - H_h E_v + (*) \end{bmatrix} < 0.$$

---

during the observer development. This section explains the problem and points out some solutions, referring the reader to the corresponding references.

Consider again an affine TS model of interest whose states are to be observed[1]:

$$\begin{aligned}\delta x(t) &= \sum_{i=1}^{r} h_i(z)(A_i x(t) + B_i u(t) + a_i) = A_h x(t) + B_h u(t) + a_h, \\ y(t) &= \sum_{i=1}^{r} h_i(z)(C_i x(t) + c_i) = C_h x(t) + c_h,\end{aligned} \tag{6.37}$$

where all the terms have the usual meanings, $\delta x(t) = \dot{x}(t)$ and $t \in \mathbb{R}$ for the continuous-time case, $\delta x = x(t+1)$ and $t \in \mathbb{Z}$ for the discrete-time case.

Several solutions ignoring the MFs in (6.37) are available in the literature for observer design, e.g. linear [21] and sliding-mode [22, 23]. Nevertheless, as we are interested in convex solutions, consider the following Thau-Luenberger observer [24, 25], which is the dual of the PDC controller and, in a sense, generalises the observers in the preceding sections:

$$\begin{aligned}\delta \widehat{x}(t) &= \sum_{i=1}^{r} h_i(\widehat{z})(A_i \widehat{x}(t) + B_i u(t) + a_i + K_i(y(t) - \widehat{y}(t))) \\ &= A_{\hat{h}} \widehat{x}(t) + B_{\hat{h}} u(t) + a_{\hat{h}} + K_{\hat{h}}(y(t) - \widehat{y}(t)), \\ \widehat{y}(t) &= \sum_{i=1}^{r} h_i(\widehat{z})(C_i \widehat{x}(t) + c_i) = C_{\hat{h}} \widehat{x} + c_{\hat{h}},\end{aligned} \tag{6.38}$$

where all the terms have the usual meanings with some important differences: $\hat{z}$ clearly indicates that the scheduling variable is an estimate of z and, consequently, the shorthand notation for convex sums should be adapted to reflect this fact, e.g.:

$$A_{\hat{h}} = \sum_{i=1}^{r} h_i(\widehat{z}) A_i, \quad B_{\hat{h}} = \sum_{i=1}^{r} h_i(\widehat{z}) B_i, \quad C_{\hat{h}} = \sum_{i=1}^{r} h_i(\widehat{z}) C_i, \quad K_{\hat{h}} = \sum_{i=1}^{r} h_i(\widehat{z}) K_i.$$

The dynamics of the estimation error when using observer (6.38) for system (6.37) in the continuous-time case can be derived as (arguments omitted):

[1] Recall that, although for controller design affine systems are rarely considered due to the arising technical problems, for observer design they do not present additional difficulties.

$$\begin{aligned}\dot{\boldsymbol{e}}(t) &= \dot{\boldsymbol{x}}(t) - \dot{\widehat{\boldsymbol{x}}}(t)\\ &= A_h\boldsymbol{x} + B_h\boldsymbol{u} + \boldsymbol{a}_h - A_{\hat{h}}\widehat{\boldsymbol{x}} - B_{\hat{h}}\boldsymbol{u} - \boldsymbol{a}_{\hat{h}} - K_{\hat{h}}(\boldsymbol{y} - \widehat{\boldsymbol{y}})\\ &= A_{\hat{h}}\boldsymbol{x} + B_{\hat{h}}\boldsymbol{u} + \boldsymbol{a}_{\hat{h}} - A_{\hat{h}}\boldsymbol{x} - B_{\hat{h}}\boldsymbol{u} - \boldsymbol{a}_{\hat{h}}\\ &\quad + A_h\boldsymbol{x} + B_h\boldsymbol{u} + \boldsymbol{a}_h - A_{\hat{h}}\widehat{\boldsymbol{x}} - B_{\hat{h}}\boldsymbol{u} - \boldsymbol{a}_{\hat{h}} - K_{\hat{h}}(\boldsymbol{y} - \widehat{\boldsymbol{y}})\\ &= A_{\hat{h}}\boldsymbol{e} - K_{\hat{h}}(\boldsymbol{y} - \widehat{\boldsymbol{y}}) + \sum_{i=1}^{r}(h_i(z) - h_i(\widehat{z}))(A_i\boldsymbol{x} + B_i\boldsymbol{u} + \boldsymbol{a}_i).\end{aligned}$$

But noticing that

$$\begin{aligned}\boldsymbol{y} - \widehat{\boldsymbol{y}} &= C_h\boldsymbol{x} + \boldsymbol{c}_h - C_{\hat{h}}\widehat{\boldsymbol{x}} - \boldsymbol{c}_{\hat{h}}\\ &= C_h\boldsymbol{x} + \boldsymbol{c}_h - C_{\hat{h}}\boldsymbol{x} - \boldsymbol{c}_{\hat{h}} + C_{\hat{h}}\boldsymbol{x} + \boldsymbol{c}_{\hat{h}} - C_{\hat{h}}\widehat{\boldsymbol{x}} - \boldsymbol{c}_{\hat{h}}\\ &= C_{\hat{h}}\boldsymbol{e} + \sum_{j=1}^{r}(h_i(z) - h_i(\widehat{z}))(C_j\boldsymbol{x} + c_j),\end{aligned}$$

we can rewrite $\dot{\boldsymbol{e}}(t)$ as

$$\begin{aligned}\dot{\boldsymbol{e}}(t) =& A_{\hat{h}}\boldsymbol{e} - K_{\hat{h}}\left(C_{\hat{h}}\boldsymbol{e} + \sum_{j=1}^{r}(h_i(z) - h_i(\widehat{z}))(C_j\boldsymbol{x} + c_j)\right)\\ &+ \sum_{i=1}^{r}(h_i(z) - h_i(\widehat{z}))(A_i\boldsymbol{x} + B_i\boldsymbol{u} + \boldsymbol{a}_i)\\ =& \left(A_{\hat{h}} - K_{\hat{h}}C_{\hat{h}}\right)\boldsymbol{e} - K_{\hat{h}}\sum_{j=1}^{r}(h_i(z) - h_i(\widehat{z}))(C_j\boldsymbol{x} + c_j)\\ &+ \sum_{i=1}^{r}(h_i(z) - h_i(\widehat{z}))(A_i\boldsymbol{x} + B_i\boldsymbol{u} + \boldsymbol{a}_i).\end{aligned} \tag{6.39}$$

For the discrete-time case, the derivation is similar and is therefore not repeated here. The error dynamics are finally given by

$$\begin{aligned}\boldsymbol{e}(t+1) =& \sum_{i=1}^{r}h_i(z)\sum_{j=1}^{r}h_j(\widehat{z})(A_i - K_iC_j)\boldsymbol{e}(t)\\ &+ \sum_{i=1}^{r}h_i(z)K_i\sum_{j=1}^{r}(h_j(z) - h_j(\widehat{z}))(C_j\boldsymbol{x}(t) + \boldsymbol{c}_j)\\ &+ \sum_{i=1}^{r}(h_i(z) - h_i(\widehat{z}))(A_i\boldsymbol{x}(t) + B_i\boldsymbol{u}(t) + \boldsymbol{a}_i).\end{aligned}$$

Based on this formulation, several works in the literature make Lipschitz assumption on the difference $h_i(z) - h_i(\widehat{z})$. An instance of these approaches can be found in [25] for linear output $C_h = C$:

Theorem 6.10 *Assume the error system (6.39) has $C_h = C$ and that the Lipschitz-like condition*

$$\left\| \sum_{i=1}^{r} (h_i(z) - h_i(\widehat{z}))(A_i \boldsymbol{x} + B_i \boldsymbol{u} + \boldsymbol{a}_i) \right\| \leq \mu \|\boldsymbol{e}\| \tag{6.40}$$

holds for some $\mu > 0$. The origin $\boldsymbol{e} = 0$ of the error system (6.39) is asymptotically stable if there exist $P = P^T > 0$, $Q = Q^T > 0$, K_i, $i \in \{1, 2, \ldots, r\}$, such that

$$PA_h - N_h C \leq -Q,$$
$$\begin{bmatrix} Q - \mu^2 I & P \\ P & I \end{bmatrix} > 0$$

holds. In that case, the observer gains are $K_i = P^{-1} N_i$, $i \in \{1, 2, \ldots, r\}$, and every trajectory $\boldsymbol{e}(t)$ starting in the outermost Lyapunov level $V(\boldsymbol{e}) = \boldsymbol{e}^T P \boldsymbol{e} \leq k$, $k > 0$, within a compact set $\mathcal{E}$ of the error space guaranteeing $z \in \mathcal{C}$, goes asymptotically to $\boldsymbol{e} = 0$.

Remark 6.12 Note that the origin $\boldsymbol{e} = 0$ of (6.39) is not necessarily an equilibrium point; therefore, condition (6.40) enforces that guaranteeing asymptotic stability of the origin.

More conditions based on Lipschitz bounds for observer design with unmeasurable scheduling vector can be found in [26, 27]: these works assume the membership functions hold the convex sum property globally. The differential mean value theorem (DMVT) has been used in [28] to refine the Lipschitz bounds and split measurable from unmeasurable signals while in [29] an explicit form of the DMVT has been used to eliminate the need of Lipschitz bounds. These approaches are not pursued here.

6.3 Observer-Based Stabilisation

Observer-based stabilisation is in general considered in the case when the scheduling variables z do not depend on states that have to be estimated. This is because in this case, the separation principle applies, i.e. the observer and the controller can be designed separately and then combined. In case of scheduling variables that have to be estimated or disturbances affecting the system, the separation principle no longer holds and in general a two-step procedure is employed [9, 30, 31]. In what follows, we briefly describe observer-based stabilisation for the case when the scheduling vector is measured. For the general case when the scheduling vector depends on states to be estimated, the interested reader is referred to [9, 30–33].

Continuous-Time Case

For observer-based stabilisation in continuous time, consider again the TS model

$$\begin{aligned}\dot{\boldsymbol{x}}(t) &= \sum_{i=1}^{r} h_i(z)\left(A_i\boldsymbol{x}(t) + B_i\boldsymbol{u}(t)\right) = A_h\boldsymbol{x}(t) + B_h\boldsymbol{u}(t),\\ \boldsymbol{y}(t) &= \sum_{i=1}^{r} h_i(z)C_i\boldsymbol{x}(t) = C_h\boldsymbol{x}(t),\end{aligned} \tag{6.41}$$

where the scheduling vector z depends only on known (measurable) variables and, moreover, it does not depend on the input $\boldsymbol{u}$ to avoid the problem of solving implicit equations.

The most basic TS observer form is considered:

$$\begin{aligned}\dot{\widehat{\boldsymbol{x}}}(t) &= \sum_{i=1}^{r} h_i(z)\left(A_i\widehat{\boldsymbol{x}}(t) + B_i\boldsymbol{u}(t) + K_i\left(\boldsymbol{y}(t) - \widehat{\boldsymbol{y}}(t)\right)\right)\\ &= A_h\widehat{\boldsymbol{x}}(t) + B_h\boldsymbol{u}(t) + K_h\left(\boldsymbol{y}(t) - \widehat{\boldsymbol{y}}(t)\right),\\ \widehat{\boldsymbol{y}}(t) &= \sum_{i=1}^{r} h_i(z)C_i\widehat{\boldsymbol{x}}(t) = C_h\widehat{\boldsymbol{x}}(t),\end{aligned} \tag{6.42}$$

along with the simplest PDC control law, now based on the estimated states:

$$\boldsymbol{u}(t) = \sum_{i=1}^{r} h_i(z)F_i\widehat{\boldsymbol{x}}(t) = F_h\widehat{\boldsymbol{x}}(t). \tag{6.43}$$

The estimation error dynamics with $\boldsymbol{e} = \boldsymbol{x} - \widehat{\boldsymbol{x}}$ is obtained as (6.3), repeated here for convenience:

$$\begin{aligned}\dot{\boldsymbol{e}}(t) &= \sum_{i=1}^{r}\sum_{j=1}^{r} h_i(z)h_j(z)\left(A_i - K_jC_i\right)\boldsymbol{e}(t)\\ &= (A_h - K_hC_h)\boldsymbol{e}(t).\end{aligned}$$

The closed-loop dynamics using the estimate-based control law is

$$\begin{aligned}\dot{x}(t) &= \sum_{i=1}^{r} h_i(z)\left(A_i x(t) + B_i \sum_{j=1}^{r} h_j(z) F_j \widehat{x}(t)\right)\\ &= \sum_{i=1}^{r}\sum_{j=1}^{r} h_i(z)h_j(z)\left((A_i + B_i F_j)x(t) - B_i F_j e(t)\right)\\ &= (A_h + B_h F_h)x(t) - B_h F_h e(t).\end{aligned} \tag{6.44}$$

Combining the dynamics of the estimation error and the state, we obtain

$$\begin{bmatrix}\dot{x}(t)\\ \dot{e}(t)\end{bmatrix} = \begin{bmatrix}A_h + B_h F_h & -B_h F_h\\ 0 & A_h - K_h C_h\end{bmatrix}\begin{bmatrix}x(t)\\ e(t)\end{bmatrix}. \tag{6.45}$$

Then, the combined observer and controller design problem consists in finding gains K_i and F_i, $i \in \{1, 2, \ldots, r\}$, such that the system (6.45) is asymptotically stable. Obviously, gains K_i, $i \in \{1, 2, \ldots, r\}$ can be directly found using the methodologies in the previous section because the observer error dynamics are decoupled from the state dynamics: this is an LMI problem. Nevertheless, despite this solution guaranteeing $e \to 0$ as $t \to \infty$, the term $-B_h F_h e$ still acts as an external disturbance on the state equation above. Therefore, proving that the separation principle holds as stated in [9, 34, 35], requires more involved arguments as shown in the result below adapted from [36]:

Theorem 6.11 *The origin $[x^T\ e^T]^T = 0$ of the closed-loop dynamics (6.45), resulting from the PDC control law (6.43) which employs the states of the observer (6.42), is asymptotically stable if there exist $X = X^T > 0$, $P = P^T > 0$, M_i, N_i, $i \in \{1, 2, \ldots, r\}$, such that*

$$\begin{aligned}A_h X + B_h M_h + X A_h^T + M_h^T B_h^T &< 0,\\ P A_h - N_h C_h + A_h^T P - C_h^T N_h^T &< 0\end{aligned} \tag{6.46}$$

holds. In that case, the PDC gains are $F_i = M_i X^{-1}$ and the observer gains are $K_i = P^{-1} N_i$, $i \in \{1, 2, \ldots, r\}$, and there exists $\lambda > 0$ such that any trajectory starting within the outermost Lyapunov level

$$V(x, e) = \begin{bmatrix}x\\ e\end{bmatrix}^T \begin{bmatrix}\lambda X^{-1} & 0\\ 0 & P\end{bmatrix}\begin{bmatrix}x\\ e\end{bmatrix} \le k, \quad k > 0,$$

in a compact set $\mathcal{X} \times \mathcal{E}$ of the error state space guaranteeing $z \in \mathcal{C} \subseteq \mathbb{R}^p$ goes asymptotically to $[x^T\ e^T]^T = 0$.

Proof If there is a quadratic Lyapunov function

$$\tilde{V}(x, e) = \begin{bmatrix}x\\ e\end{bmatrix}^T \begin{bmatrix}P_{11} & P_{12}\\ P_{12}^T & P_{22}\end{bmatrix}\begin{bmatrix}x\\ e\end{bmatrix},$$

proving stability of (6.45), then

$$\begin{bmatrix} P_{11} & P_{12} \\ P_{12}^T & P_{22} \end{bmatrix} \begin{bmatrix} A_h + B_h F_h & -B_h F_h \\ 0 & A_h - K_h C_h \end{bmatrix} + (*) < 0, \tag{6.47}$$

which means that

$$P_{11}(A_h + B_h F_h) + (A_h + B_h F_h)^T P_{11} < 0,$$

which, by congruence (Property A.1) and substitutions $X = P_{11}^{-1}$, $M_h = F_h P_{11}^{-1}$, is equivalent to the first equation in (6.46).

If congruence (Property A.1) is applied to (6.47) with

$$Q \equiv \begin{bmatrix} P_{11} & P_{12} \\ P_{12}^T & P_{22} \end{bmatrix}^{-1} = \begin{bmatrix} Q_{11} & Q_{12} \\ Q_{12}^T & Q_{22} \end{bmatrix},$$

we have that

$$\begin{bmatrix} A_h + B_h F_h & -B_h F_h \\ 0 & A_h - K_h C_h \end{bmatrix} \begin{bmatrix} Q_{11} & Q_{12} \\ Q_{12}^T & Q_{22} \end{bmatrix} + (*) < 0,$$

which implies

$$(A_h - K_h C_h) Q_{22} + Q_{22} (A_h - K_h C_h)^T < 0.$$

By congruence (Property A.1) and substitutions $P = Q_{22}^{-1}$, $N_h = Q_{22}^{-1} K_h$, the previous inequality is equivalent to the second one in (6.46).

Therefore, if conditions (6.46) hold, the Lyapunov function candidate $V(\boldsymbol{x}, \boldsymbol{e}) = [\boldsymbol{x}^T \ \boldsymbol{e}^T]\text{block-diag}(\lambda X^{-1}, P)[\boldsymbol{x}^T \ \boldsymbol{e}^T]^T$, with $\lambda > 0$ to be defined later, can be considered; it is negative-definite if

$$\begin{aligned} &\begin{bmatrix} \lambda X^{-1} & 0 \\ 0 & P \end{bmatrix} \begin{bmatrix} A_h + B_h F_h & -B_h F_h \\ 0 & A_h - K_h C_h \end{bmatrix} + \begin{bmatrix} A_h + B_h F_h & -B_h F_h \\ 0 & A_h - K_h C_h \end{bmatrix}^T \begin{bmatrix} \lambda X^{-1} & 0 \\ 0 & P \end{bmatrix} \\ &= \begin{bmatrix} \lambda X^{-1}(A_h + B_h F_h) + (*) & -\lambda X^{-1} B_h F_h \\ -\lambda F_h^T B_h^T X^{-1} & P(A_h - K_h C_h) + (*) \end{bmatrix} < 0, \end{aligned}$$

which, by Schur complement in Lemma A.1 yields

$$\lambda R(\cdot) - S(\cdot) > 0, \tag{6.48}$$

with

$$\begin{aligned} R(\cdot) &= X^{-1} B_h F_h \left(P(A_h - K_h C_h) + (*)\right)^{-1} F_h^T B_h^T X^{-1}, \\ S(\cdot) &= X^{-1}(A_h + B_h F_h) + (A_h + B_h F_h)^T X^{-1}. \end{aligned}$$

But, conditions (6.46) guarantee that $\lambda_{\min}(R(\cdot)) < 0$ and $\lambda_{\max}(S(\cdot)) < 0$, which means λ can be chosen as to hold $\lambda\,(\lambda_{\min}(R(\cdot))) > \lambda_{\max}(S(\cdot))$, which guarantees (6.48). Thus, $V(\boldsymbol{x}, \boldsymbol{e})$ is a Lyapunov function proving $[\boldsymbol{x}^T\ \boldsymbol{e}^T]^T = 0$ is asymptotically stable.

Naturally, since the sign of the different expressions in the previous developments does not change as long as the MFs $h_i(z)$ hold the convex sum property, it is guaranteed that any trajectory starting within the outermost Lyapunov level $V(\boldsymbol{x}, \boldsymbol{e}) = [\boldsymbol{x}^T\ \boldsymbol{e}^T]\text{block-diag}(\lambda X^{-1}, P)[\boldsymbol{x}^T\ \boldsymbol{e}^T]^T \leq k, k > 0$, in a compact set of the error/state space $\mathcal{X} \times \mathcal{E}$ holding Assumptions 4.1, 4.2, 6.1, and 6.2, goes asymptotically to the origin, as these assumptions guarantee $z \in \mathcal{C}$, where convexity of the MFs $h_i(z)$ is ensured. □

Remark 6.13 The theorem above means that it is possible to design separately the observer and the controller, and therefore one can make use of the results and relaxations previously discussed.

The following example illustrates observer-based control design:

Example 6.10 Consider a continuous-time TS model with two vertices with the local matrices expressed as

$$A_1 = \begin{bmatrix} -7 & 1.25 \\ 0.85 & -1.6 \end{bmatrix}, \quad B_1 = \begin{bmatrix} 1 \\ 0 \end{bmatrix}, \quad C_1 = \begin{bmatrix} 1 & 0 \end{bmatrix},$$

$$A_2 = \begin{bmatrix} -1.45 & -0.4 \\ 0.57 & 0.7 \end{bmatrix}, \quad B_2 = \begin{bmatrix} 1 \\ 0 \end{bmatrix}, \quad C_2 = \begin{bmatrix} 0 & 1 \end{bmatrix},$$

and membership functions $h_1(z) = (1 - \sin z)/2$, $h_2(z) = (1 + \sin z)/2$. Note that the scheduling variable is an exogenous measured variable and thus can be used in the observer.

Conditions (6.46) are feasible; they yield the solutions:

$$X = \begin{bmatrix} 0.06 & 0.19 \\ 0.19 & 1.53 \end{bmatrix}, \qquad P = \begin{bmatrix} 0.10 & -0.05 \\ -0.05 & 0.46 \end{bmatrix},$$

$$K_1 = \begin{bmatrix} -2.77 \\ 1.75 \end{bmatrix}, \qquad K_2 = \begin{bmatrix} 2.35 \\ 2.65 \end{bmatrix},$$

$$F_1 = \begin{bmatrix} -17.35 & 2.65 \end{bmatrix}, \qquad F_2 = \begin{bmatrix} 16.08 & -3.93 \end{bmatrix}.$$

A trajectory of the estimation error and of the states of the closed-loop system using the observer and controller gains above is presented in Fig. 6.8a,b. The first membership function is represented in Fig. 6.8c. For this particular trajectory, the true initial states were $\boldsymbol{x}(0) = [1\ \ -1]^T$, and the estimated initial states were $\hat{\boldsymbol{x}}(0) = [0\ 0]^T$. As can be seen, all the variables converge to zero. □

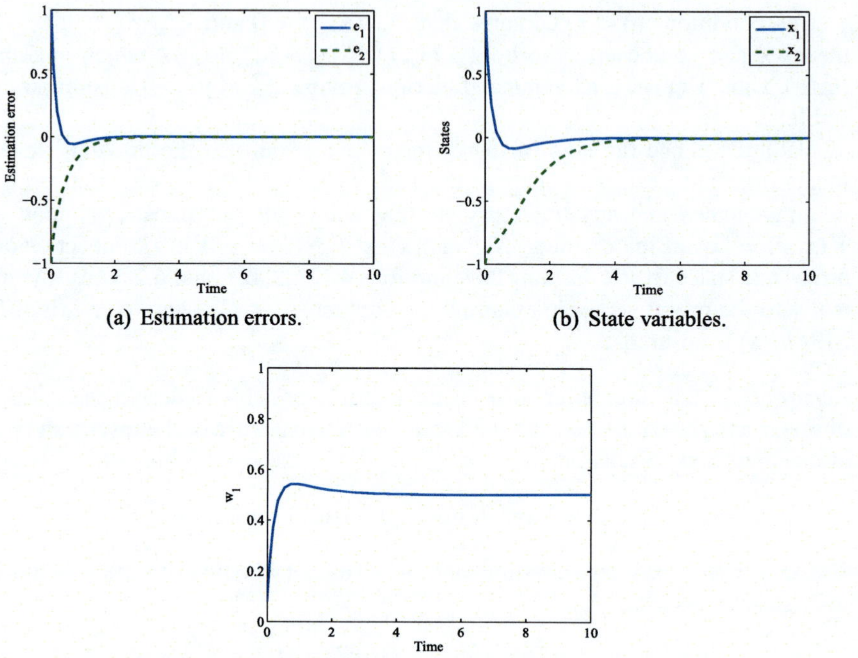

(a) Estimation errors.

(b) State variables.

(c) First membership function.

Fig. 6.8 Simulation results for Example 6.10

Discrete-Time Case

Let us consider now the discrete-time case. For this, we use a TS model in shorthand notation:

$$\begin{aligned} \boldsymbol{x}(t+1) &= A_h \boldsymbol{x}(t) + B_h \boldsymbol{u}(t), \\ \boldsymbol{y}(t) &= C_h \boldsymbol{x}(t), \end{aligned} \tag{6.49}$$

under the usual definitions and hypothesis about z depending only on known (measured) variables, except $\boldsymbol{u}$.

The observer considered is of the form

$$\begin{aligned} \widehat{\boldsymbol{x}}(t+1) &= A_h \widehat{\boldsymbol{x}}(t) + B_h \boldsymbol{u}(t) + K_h(\boldsymbol{y}(t) - \widehat{\boldsymbol{y}}(t)), \\ \widehat{\boldsymbol{y}}(t) &= C_h \widehat{\boldsymbol{x}}(t), \end{aligned} \tag{6.50}$$

and the controller is

$$\boldsymbol{u}(t) = F_h H^{-1} \widehat{\boldsymbol{x}}(t). \tag{6.51}$$

The estimation error is

$$e(t+1) = (A_h - K_h C_h)e(t),$$

and the closed-loop dynamics using the estimate-based control law is

$$x(t+1) = (A_h + B_h F_h H^{-1})x(t) - B_h F_h e(t), \tag{6.52}$$

which means the following combined dynamics are obtained:

$$\begin{bmatrix} x(t+1) \\ e(t+1) \end{bmatrix} = \begin{bmatrix} A_h + B_h F_h H^{-1} & -B_h F_h H^{-1} \\ 0 & A_h - K_h C_h \end{bmatrix} \begin{bmatrix} e(t) \\ x(t) \end{bmatrix}, \tag{6.53}$$

which has the same structure as in the continuous-time case.

Since the separation principle holds, it is possible to design separately the observer and the controller, and make use of the results and relaxations presented. For instance, using a common quadratic Lyapunov function for observer design and another one for controller design, the following result can be established:

Theorem 6.12 *The origin* $[x^T \ e^T]^T = 0$ *of the closed-loop dynamics (6.53), resulting from the PDC control law (6.51) which employs the states of the observer (6.50), is asymptotically stable if there exist* $X = X^T > 0$, $P = P^T > 0$, H, F_i, N_i, $i \in \{1, 2, \ldots, r\}$, *such that*

$$\begin{gathered} \begin{bmatrix} -H - H^T + X & H^T A_h^T + F_h^T B_h^T \\ A_h H + B_h F_h & -X \end{bmatrix} < 0, \\ \begin{bmatrix} -P & A_h^T P - C_h^T N_h^T \\ P A_h - N_h C_h & -P \end{bmatrix} \leq 0 \end{gathered} \tag{6.54}$$

holds. In that case, the observer gains are $K_i = P^{-1} N_i$, $i \in \{1, 2, \ldots, r\}$, *and there exists* $\lambda > 0$ *such that any trajectory starting within the outermost Lyapunov level*

$$V(x, e) = \begin{bmatrix} x \\ e \end{bmatrix}^T \begin{bmatrix} \lambda X^{-1} & 0 \\ 0 & P \end{bmatrix} \begin{bmatrix} x \\ e \end{bmatrix} \leq k, \quad k > 0,$$

in a compact set $\mathcal{X} \times \mathcal{E}$ *of the error state space guaranteeing* $z \in \mathcal{C} \subseteq \mathbb{R}^p$ *goes asymptotically to* $[x^T \ e^T]^T = 0$.

Proof Conditions (6.54) are based on Theorems 5.3 and 6.2 for discrete-time stabilisation and discrete-time observation. Using these facts along with the same outline of Theorem 6.11 the desired result follows.

The following example illustrates observer-based controller design in the discrete-time case:

Example 6.11 Consider a discrete-time TS model with two vertex models described by matrices

$$A_1 = \begin{bmatrix} -1.2 & 0.8 \\ 0.26 & 0.9 \end{bmatrix}, \; B_1 = \begin{bmatrix} 0 \\ 1 \end{bmatrix}, \; C_1 = \begin{bmatrix} 1 & 0 \end{bmatrix},$$

$$A_2 = \begin{bmatrix} 2.34 & -1 \\ 1.21 & -1.12 \end{bmatrix}, \; B_2 = \begin{bmatrix} 0 \\ -1 \end{bmatrix}, \; C_2 = \begin{bmatrix} 0 & 1 \end{bmatrix},$$

and membership functions $h_1 = (1 - \sin z)/2$, $h_2 = (1 + \sin z)/2$, where z is an exogenous measurable variable.

Solving the LMIs, one obtains

$$P = \begin{bmatrix} 0.18 & 0.30 \\ 0.30 & 0.57 \end{bmatrix}, \quad X = \begin{bmatrix} 0.37 & -0.47 \\ -0.47 & 0.70 \end{bmatrix}, \quad H = \begin{bmatrix} 0.13 & 0.17 \\ 0.22 & 0.40 \end{bmatrix},$$

$$K_1 = \begin{bmatrix} -1.07 \\ 0.79 \end{bmatrix}, \quad K_2 = \begin{bmatrix} 2.67 \\ 1.18 \end{bmatrix}, \quad F_1 = \begin{bmatrix} -0.20 & -0.20 \end{bmatrix}, \; F_2 = \begin{bmatrix} -0.23 & -0.23 \end{bmatrix}.$$

A trajectory of the estimation error and of the states of the closed-loop system using the observer and controller gains above is presented in Fig. 6.9a,b. The first membership function is illustrated in Fig. 6.9c. For this particular trajectory, the true initial states were $\boldsymbol{x}(0) = [1\ 1]^T$, and the estimated initial states were $\hat{\boldsymbol{x}}(0) = [0\ 0]^T$. As can be seen, all the variables converge to zero. □

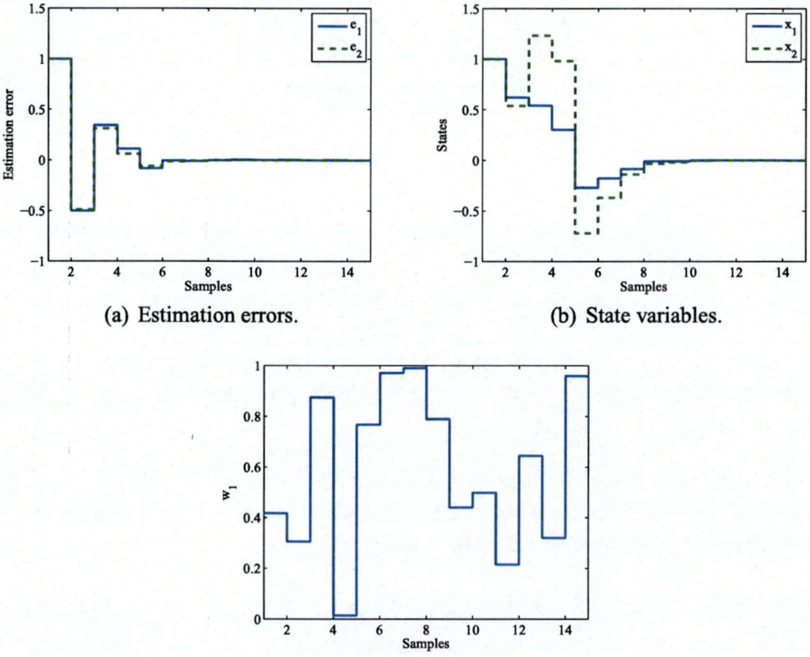

(a) Estimation errors.

(b) State variables.

(c) First membership function.

Fig. 6.9 Simulation results for Example 6.11

6.4 Output Feedback for LFT Models

Once we are in an LFT framework, most linear controller designs can be cast as static output feedback problems (both state feedback, dynamic output feedback and—of course!—actual static output feedback).

Indeed, any dynamic state-space controller representation can be "disguised" as a static one if, in the equations of the plant, an extra state equation $\dot{\boldsymbol{x}}_c = \boldsymbol{u}_{c1}$ (pure integrator, $\boldsymbol{u}_{c1}$ being an artificial dummy input), and an extra output equation $\boldsymbol{y}_{c1} = \boldsymbol{x}_c$ are artificially inserted, so the apparently *static* 2-input 2-output output-feedback controller:

$$\begin{bmatrix} \boldsymbol{u}_{c1} \\ \boldsymbol{u} \end{bmatrix} = \begin{bmatrix} A_c & B_c \\ C_c & D_c \end{bmatrix} \begin{bmatrix} \boldsymbol{y}_{c1} \\ \boldsymbol{y} \end{bmatrix} \tag{6.55}$$

would actually mean a dynamic output-feedback controller, as usual:

$$\begin{bmatrix} \dot{\boldsymbol{x}}_c \\ \boldsymbol{u} \end{bmatrix} = \begin{bmatrix} A_c & B_c \\ C_c & D_c \end{bmatrix} \begin{bmatrix} \boldsymbol{x}_c \\ \boldsymbol{y} \end{bmatrix}. \tag{6.56}$$

So, let us concentrate on the static output-feedback setting, and design robust controllers (i.e. constant matrices in (6.55)) with $\mathcal{H}_\infty$ bound performance criteria or, alternatively, gain-scheduled ones (with matrices in (6.55) now depending on δ, assumed measurable).

Remark 6.14 Even if many problems can be *written* as a static output feedback one, some rank and inverse-matrix issues will *not* allow some cases to be *solved* in an efficient convex way, as later discussed.

Consider the system depicted in Fig. 6.10, with state ψ, where we have three inputs:

- $\boldsymbol{w}_\delta$ from the uncertain block (related to robustness or gain scheduling);
- $\boldsymbol{w}_1$, disturbances to be rejected (related to disturbance attenuation performance);
- $\boldsymbol{u}$: controller output

and three outputs:

- $\boldsymbol{z}_\delta$, the input to the uncertainty;
- $\boldsymbol{z}_1$, the performance-related output;
- $\boldsymbol{y}$, information to the controller.

In the simplest case, we might consider, additionally to the above plant, an LFT-connected uncertainty $\boldsymbol{w}_\delta = \Delta \boldsymbol{z}_\delta$ or, in a more general setup, as earlier done in (4.88) for stability analysis, we can consider a subspace $\mathcal{S}(\Delta)$ where the joint vector $[\boldsymbol{w}_\delta{}^T, \boldsymbol{z}_\delta{}^T]^T$ should belong.

We will have the performance goal of proving that, assuming initial conditions at zero state, the following holds:

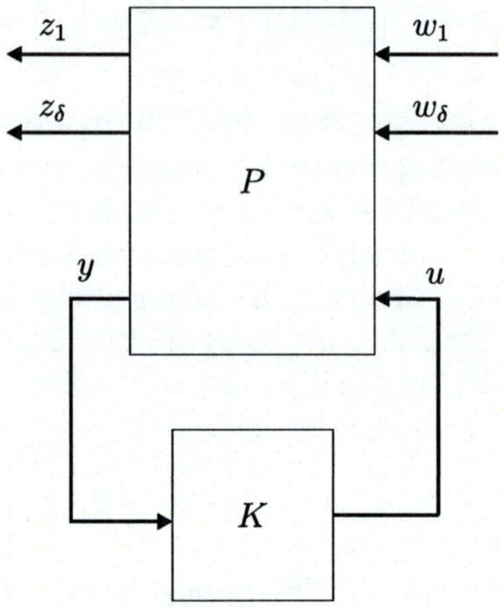

Fig. 6.10 A 3-input, 3-output representation of the problem in consideration

$$\int_0^t \begin{bmatrix} \boldsymbol{w}_1 \\ z_1 \end{bmatrix}^T \begin{bmatrix} Q & S \\ S^T & R \end{bmatrix} \begin{bmatrix} \boldsymbol{w}_1 \\ z_1 \end{bmatrix} dt < 0 \quad \forall t \geq 0. \tag{6.57}$$

For instance, a particular case of (6.57) is $Q = -\gamma I$, $S = 0$, $R = \gamma^{-1} I$, leading to the standard $\mathcal{H}_\infty$ bound norm $\|z_1\|_2 \leq \gamma \|\boldsymbol{w}_1\|_2$.

The generic state-space representation of the process in Figure 6.10 will be represented by

$$\begin{bmatrix} \dot{\boldsymbol{x}} \\ z_\delta \\ z_1 \\ \boldsymbol{y} \end{bmatrix} = \begin{bmatrix} A & B_\delta & B_1 & B_2 \\ C_\delta & D_{\delta\delta} & D_{\delta 1} & D_{\delta 2} \\ C_1 & D_{1\delta} & D_{11} & D_{12} \\ C_2 & D_{2\delta} & D_{21} & 0 \end{bmatrix} \begin{bmatrix} \boldsymbol{x} \\ \boldsymbol{w}_\delta \\ \boldsymbol{w}_1 \\ \boldsymbol{u} \end{bmatrix}, \tag{6.58}$$

which, jointly with some specification of a subspace $\mathcal{S}(\Delta)$ where the joint vector $[\boldsymbol{w}_\delta{}^T, \boldsymbol{z}_\delta{}^T]^T$ should belong, completes the model, as earlier done in (4.88) for stability analysis.

Controller Structure

We can now envision several control options for the above LTV/LPV plant. First, a robust LTI controller $\boldsymbol{u} = X\boldsymbol{y}$, with constant X can be considered (as already discussed, static output-feedback is a "tweak" to actually design dynamic linear controllers). As an alternative, we could have a time-varying gain-scheduled controller, $\boldsymbol{u} = \hat{X}(\Delta)\boldsymbol{y}$.

For brevity, let us discuss here the gain-scheduled controllers, as the robust controllers may be considered a particular case, and they are well studied in the linear robust control community, in standard linear $\mathcal{H}_\infty$ (full unstructured Δ) and μ-

synthesis literature [37]. For instance, it is well known that most structured uncertainty and reduced-order feedback robust control problems are not convex.

Such gain-scheduled controller will be sought in the class of controllers which can be expressed in LFT form:

$$\begin{bmatrix} \boldsymbol{u}_{c1} \\ \boldsymbol{\eta}_\delta \\ \boldsymbol{u} \end{bmatrix} = \Sigma \cdot \begin{bmatrix} \boldsymbol{x}_c \\ \boldsymbol{y}_\delta \\ \boldsymbol{y} \end{bmatrix} = \Sigma \cdot \begin{bmatrix} \boldsymbol{x}_c \\ \boldsymbol{y}_\delta \\ C_2\boldsymbol{x} + D_{2\delta}\boldsymbol{w}_\delta + D_{21}\boldsymbol{w}_1 \end{bmatrix}, \tag{6.59}$$

where Σ is a to-be-computed freely assignable constant matrix and $[\boldsymbol{y}_\delta{}^T, \boldsymbol{\eta}_\delta{}^T]^T$ must belong to a subspace $\mathcal{S}_c(\Delta)$ which can be either:

- Coincident with the plant one $\mathcal{S}_c(\Delta) \equiv \mathcal{S}(\Delta)$, or
- General scheduling subspace functions, to be computed, jointly with Σ, as a result of the Gain-Scheduling LMIs (not necessarily related to the plant's uncertainty subspace $\mathcal{S}$).

The first case $\mathcal{S}_c(\Delta) \equiv \mathcal{S}(\Delta)$, for the uncertainty structure $\boldsymbol{w}_\delta = \Delta \boldsymbol{z}_\delta$, was studied in earlier gain-scheduling literature [38], the second one appears in, for instance, [39], and it is, of course, more powerful. An "intermediate" case in which the general controller scheduling function is restricted to be a convex sum with the same number of vertices as a polytopic Δ appears in [40].

6.4.1 Closed-Loop Equation

Let us consider only the full-order gain-scheduled feedback with Δ fully measurable, which yields a convex problem; most other cases are non-convex. In this case, the state actually contains the plant state and the controller state, both of equal size, so we will expand the state as $\psi := [\boldsymbol{x}^T\ \boldsymbol{x}_c^T]^T$.

Operating with (6.58) and (6.59), replacing $\boldsymbol{x}$ by ψ, and adding the integrators $\dot{\boldsymbol{x}}_c = \boldsymbol{u}_{c1}$ we can write:

$$\begin{bmatrix} \dot{\boldsymbol{x}} \\ \dot{\boldsymbol{x}}_c \\ \boldsymbol{z}_\delta \\ \boldsymbol{\eta}_\delta \\ \boldsymbol{z}_1 \end{bmatrix} = \begin{bmatrix} A & 0 & B_\delta & 0 & B_1 \\ 0 & 0 & 0 & 0 & 0 \\ C_\delta & 0 & D_{\delta\delta} & 0 & D_{\delta 1} \\ 0 & 0 & 0 & 0 & 0 \\ C_1 & 0 & D_{1\delta} & 0 & D_{11} \end{bmatrix} \begin{bmatrix} \boldsymbol{x} \\ \boldsymbol{x}_c \\ \boldsymbol{w}_\delta \\ \boldsymbol{y}_\delta \\ \boldsymbol{w}_1 \end{bmatrix} + \begin{bmatrix} 0 & 0 & B_2 \\ I & 0 & 0 \\ 0 & 0 & D_{\delta 2} \\ 0 & I & 0 \\ 0 & 0 & D_{12} \end{bmatrix} \begin{bmatrix} \boldsymbol{u}_{c1} \\ \boldsymbol{\eta}_\delta \\ \boldsymbol{u} \end{bmatrix}$$

$$= \left[\begin{bmatrix} A & 0 & B_\delta & 0 & B_1 \\ 0 & 0 & 0 & 0 & 0 \\ C_\delta & 0 & D_{\delta\delta} & 0 & D_{\delta 1} \\ 0 & 0 & 0 & 0 & 0 \\ C_1 & 0 & D_{1\delta} & 0 & D_{11} \end{bmatrix} + \begin{bmatrix} 0 & 0 & B_2 \\ I & 0 & 0 \\ 0 & 0 & D_{\delta 2} \\ 0 & I & 0 \\ 0 & 0 & D_{12} \end{bmatrix} \cdot \Sigma \cdot \begin{bmatrix} 0 & I & 0 & 0 & 0 \\ 0 & 0 & 0 & I & 0 \\ C_2 & 0 & D_{2\delta} & 0 & D_{21} \end{bmatrix} \right] \cdot \begin{bmatrix} \boldsymbol{x} \\ \boldsymbol{x}_c \\ \boldsymbol{w}_\delta \\ \boldsymbol{y}_\delta \\ \boldsymbol{w}_1 \end{bmatrix}. \tag{6.60}$$

Now, considering the dissipativity Lyapunov equation in the augmented state space, for some $V(\psi) \geq 0$:

$$\dot{V} + \begin{bmatrix} w_1 \\ z_1 \end{bmatrix}^T \begin{bmatrix} Q & S \\ S^T & R \end{bmatrix} \begin{bmatrix} w_1 \\ z_1 \end{bmatrix} \leq 0 \tag{6.61}$$

and taking integrals, the performance goal (6.57) holds if (6.61) holds. So, we need to express the latter as a matrix inequality.

Let us, then, write the above Lyapunov equation with $\dot{V} = \psi^T P \dot{\psi} + \dot{\psi}^T P \psi$, and introduce a multiplier, so using full-block S-procedure in Property A.2, the inequality (6.61) holds for all

$$[w_\delta^T, y_\delta^T, z_\delta^T, \eta_\delta^T]^T \in \mathcal{S}(\Delta) \times \mathcal{S}_c(\Delta)$$

if and only if there exists a multiplier M which fulfils, suitably partitioning M in 4×4 blocks:

$$(*)^T \underbrace{\begin{bmatrix} Q_{\delta,11} & Q_{\delta,12} & S_{\delta,11} & S_{\delta,12} \\ Q_{\delta,12}^T & Q_{\delta,22} & S_{\delta,21} & S_{\delta,22} \\ S_{\delta,11}^T & S_{\delta,21}^T & R_{\delta,11} & R_{\delta,12} \\ S_{\delta,12}^T & S_{\delta,22}^T & R_{\delta,12}^T & R_{\delta,22} \end{bmatrix}}_{M} \begin{bmatrix} w_\delta \\ y_\delta \\ z_\delta \\ \eta_\delta \end{bmatrix} > 0, \quad \forall \begin{bmatrix} w_\delta \\ y_\delta \\ z_\delta \\ \eta_\delta \end{bmatrix} \in \mathcal{S}(\Delta) \times \mathcal{S}_c(\Delta), \tag{6.62}$$

and there exists a matrix $P > 0$ of size $2n \times 2n$ such that

$$P := \begin{bmatrix} P_{11} & P_{12} \\ P_{12}^T & P_{22} \end{bmatrix},$$

$$(*) \underbrace{\begin{bmatrix} 0 & 0 & 0 & 0 & 0 & P_{11} & P_{12} & 0 & 0 & 0 \\ 0 & 0 & 0 & 0 & 0 & {P_{12}}^T & P_{22} & 0 & 0 & 0 \\ 0 & 0 & Q_{\delta,11} & Q_{\delta,12} & 0 & 0 & 0 & S_{\delta,11} & S_{\delta,12} & 0 \\ 0 & 0 & Q_{\delta,12}^T & Q_{\delta,22} & 0 & 0 & 0 & S_{\delta,21} & S_{\delta,22} & 0 \\ 0 & 0 & 0 & 0 & Q & 0 & 0 & 0 & 0 & S \\ P_{11} & P_{12} & 0 & 0 & 0 & 0 & 0 & 0 & 0 & 0 \\ {P_{12}}^T & P_{22} & 0 & 0 & 0 & 0 & 0 & 0 & 0 & 0 \\ 0 & 0 & S_{\delta,11}^T & S_{\delta,21}^T & 0 & 0 & 0 & R_{\delta,11} & R_{\delta,12} & 0 \\ 0 & 0 & S_{\delta,12}^T & S_{\delta,22}^T & 0 & 0 & 0 & R_{\delta,12}^T & R_{\delta,22} & 0 \\ 0 & 0 & 0 & 0 & S^T & 0 & 0 & 0 & 0 & R \end{bmatrix}}_{\Xi} \begin{bmatrix} x \\ x_c \\ w_\delta \\ y_\delta \\ w_1 \\ \dot{x} \\ \dot{x}_c \\ z_\delta \\ \eta_\delta \\ z_1 \end{bmatrix} < 0. \tag{6.63}$$

In the sequel, let us denote as Ξ the 10×10 block matrix in (6.63) above.

Now, replacing the 10-element righthand-side column vector, denoting it by ξ for brevity, we can express it as

$$\xi = \begin{bmatrix} I \\ \mathbb{C} + \mathbb{A}^T \Sigma \, \mathbb{B} \end{bmatrix} \begin{bmatrix} \boldsymbol{x} \\ \boldsymbol{x}_c \\ \boldsymbol{w}_\delta \\ \boldsymbol{y}_\delta \\ \boldsymbol{w}_1 \end{bmatrix}, \tag{6.64}$$

where matrices $\mathbb{A}$, $\mathbb{B}$ and $\mathbb{C}$ are readily obtained from (6.60), as follows:

$$\mathbb{A} = \begin{bmatrix} 0 & I & 0 & 0 & 0 \\ 0 & 0 & 0 & I & 0 \\ B_2^T & 0 & D_{\delta 2}^T & 0 & D_{12}^T \end{bmatrix}, \tag{6.65}$$

$$\mathbb{B} = \begin{bmatrix} 0 & I & 0 & 0 & 0 \\ 0 & 0 & 0 & I & 0 \\ C_2 & 0 & D_{2\delta} & 0 & D_{21} \end{bmatrix}, \tag{6.66}$$

$$\mathbb{C} = \begin{bmatrix} A & 0 & B_\delta & 0 & B_1 \\ 0 & 0 & 0 & 0 & 0 \\ C_\delta & 0 & D_{\delta\delta} & 0 & D_{\delta 1} \\ 0 & 0 & 0 & 0 & 0 \\ C_1 & 0 & D_{1\delta} & 0 & D_{11} \end{bmatrix}. \tag{6.67}$$

The Elimination Lemma A.4 can now be readily used. However, such a lemma involves both Ξ in (6.63) and its *inverse*. Hence, even if Σ can be eliminated via the referred lemma, the result is a pair of expressions, one on that matrix and other on the inverse which, in a general case is *not* an LMI problem. However, given the particular structure of the matrices in (6.60), this particular case can be expressed as a convex problem. The main idea underlying the convexifications is the fact that the null spaces in Lemma A.4 will make the LMIs depend only on the upper-left block P_{11} and a similarly located block of P^{-1}, so Lemma A.5 will be applicable to build the full P; indeed, the next section will explain these developments in further detail.

6.4.2 *Convexification of the Synthesis Conditions*

Recalling (6.66), note that null($\mathbb{B}$) $= [\xi_1^T \; 0 \; \xi_2^T \; 0 \; \xi_3^T]^T$ with ξ_1, ξ_2, ξ_3 such that

$$C_2\xi_1 + D_{2\delta}\xi_2 + D_{21}\xi_3 = 0, \tag{6.68}$$

so it depends only on the "output" equation (how y is obtained) and will give the so-called "observer" LMI in Ξ. Denoting by ψ_B a basis of null($\mathbb{B}$), constraining (6.63) to such null space, we have the expression of (A.22) for the problem in consideration here:

$$\psi_B^T \begin{bmatrix} I \\ \mathbb{C} \end{bmatrix}^T \Xi \begin{bmatrix} I \\ \mathbb{C} \end{bmatrix} \psi_B < 0. \tag{6.69}$$

Analogously, recalling (6.65), we can write null($\mathbb{A}$) $= [\psi_1^T\ 0\ \psi_2^T\ 0\ \psi_3^T]^T$, such that $B_2^T\psi_1 + D_{\delta 2}^T\psi_2 + D_{12}^T\psi_3 = 0$, so it depends on the input side (effect of u); it will yield the so-called controller LMI in Ξ^{-1}. Denoting by ψ_A a basis of null($\mathbb{A}$) we have condition (A.23):

$$\psi_A^T \begin{bmatrix} -\mathbb{C}^T \\ I \end{bmatrix}^T \Xi^{-1} \begin{bmatrix} -\mathbb{C}^T \\ I \end{bmatrix} \psi_A > 0. \tag{6.70}$$

Let us now study in more detail both LMIs to check their structure, given that $\mathbb{C}$ has many zero blocks that might make them amenable to useful manipulations.

Observer LMI

As $\mathbb{C}$ has (block) rows 2 and 4 equal to zero[2], when considering $\begin{bmatrix} I\ \mathbb{C}^T \end{bmatrix}^T$, it will have rows 7 and 9 equal to zero, as the identity here has a 5-block size to fit the 10-block structure in (6.63). Now, as ψ_B has null second and fourth component, when evaluating $\begin{bmatrix} I\ \mathbb{C}^T \end{bmatrix}^T \psi_B$ that means that columns 2 and 4 of $\begin{bmatrix} I\ \mathbb{C}^T \end{bmatrix}^T$ will be multiplied by zero. That will make, too, that rows 2 and 4 of $\begin{bmatrix} I\ \mathbb{C}^T \end{bmatrix}^T \psi_B$ are zero.

So, the four zero rows will eliminate the effect of columns 2,4,7, and 9 of Ξ. By symmetry, analysing the factor $\begin{bmatrix} I\ \mathbb{C}^T \end{bmatrix}$ at the left-hand side of (6.69), rows 2,4,7, and 9 of Ξ will be,too, irrelevant. Hence, deleting everything which ends up multiplied by zero, the observer LMI (6.69) renders in the form:

$$(*)^T \begin{bmatrix} 0 & 0 & 0 & P_{11} & 0 & 0 \\ 0 & Q_{\delta,11} & 0 & 0 & S_{\delta,11} & 0 \\ 0 & 0 & Q & 0 & 0 & S \\ P_{11} & 0 & 0 & 0 & 0 & 0 \\ 0 & S_{\delta,11}^T & 0 & 0 & R_{\delta,11} & 0 \\ 0 & 0 & S^T & 0 & 0 & R \end{bmatrix} \begin{bmatrix} I & 0 & 0 \\ 0 & I & 0 \\ 0 & 0 & I \\ A & B_\delta & B_1 \\ C_\delta & D_{\delta\delta} & D_{\delta 1} \\ C_1 & D_{1\delta} & D_{11} \end{bmatrix} \psi_B^* > 0, \tag{6.71}$$

the columns of ψ_B^* being a basis of null$[C_2\ D_{2\delta}\ D_{21}]$, from (6.68).

The above condition must be complemented, parallel to the developments in the stability analysis conditions (4.99), with constraints on the multiplier, i.e. the multiplier must verify:

$$\eta^T \begin{bmatrix} Q_{\delta,11} & S_{\delta,11} \\ S_{\delta,11}{}^T & R_{\delta,11} \end{bmatrix} \eta > 0 \qquad \forall \eta \in \mathcal{S}(\Delta) \tag{6.72}$$

---

[2] Note that, for convenience, numbers refer to the block structure (6.67), not actual row numbers.

for the full-block S-procedure to apply. These conditions are the particularisation to the relevant subspace of (6.62).

Controller LMI

First note that since Ξ in (6.63) can be made a block-diagonal matrix (after some permutations of rows and columns), Ξ^{-1} can be expressed from the corresponding block-by-block inverse, permuting back to the original layout. One block arises from the Lyapunov function, whose inverse will be denoted as $X = P^{-1}$:

$$\begin{bmatrix} 0 & P \\ P & 0 \end{bmatrix}^{-1} = \begin{bmatrix} 0 & P^{-1} \\ P^{-1} & 0 \end{bmatrix} := \begin{bmatrix} 0 & X \\ X & 0 \end{bmatrix}; \tag{6.73}$$

another block is the uncertainty multiplier M, so the corresponding elements of M^{-1} will be denoted with tilde, i.e. $\tilde{Q}_{\delta,11}$, etc... and another block is the inverse of the quadratic performance weight, which is not a decision variable but, nevertheless, will be denoted also with $\tilde{Q}$, $\tilde{S}$, $\tilde{R}$ (known matrices).

Note that X is of dimensions $2n \times 2n$, so it will be assumed partitioned as

$$X := \begin{bmatrix} X_{11} & X_{12} \\ X_{12}^T & X_{22} \end{bmatrix}.$$

Following a similar argumentation with (6.70) and $\begin{bmatrix} -\mathbb{C} & I \end{bmatrix}^T \psi_A$, it can be checked that columns 2 and 4 (from null columns 2 and 4 of $\mathbb{C}$) and 7, 9 from the zeros in the basis of null($\mathbb{A}$) will yield an LMI in the form:

$$(*)^T \begin{bmatrix} 0 & 0 & 0 & X_{11} & 0 & 0 \\ 0 & \tilde{Q}_{\delta,11} & 0 & 0 & \tilde{S}_{\delta,11} & 0 \\ 0 & 0 & \tilde{Q} & 0 & 0 & \tilde{S} \\ X_{11} & 0 & 0 & 0 & 0 & 0 \\ 0 & \tilde{S}_{\delta,11}^T & 0 & 0 & \tilde{R}_{\delta,11} & 0 \\ 0 & 0 & \tilde{S}^T & 0 & 0 & \tilde{R} \end{bmatrix} \begin{bmatrix} -A^T & -C_\delta^T & -C_1{}^T \\ -B_\delta^T & -D_{\delta\delta}^T & -D_{1\delta}^T \\ -B_1^T & -D_{\delta 1}^T & -D_{11}^T \\ I & 0 & 0 \\ 0 & I & 0 \\ 0 & 0 & I \end{bmatrix} \psi_A^* > 0, \tag{6.74}$$

the columns of ψ_A^* being a basis of null$[(B_2^T\ D_{\delta 2}{}^T\ D_{12}^T)]$.

For the full-block S-procedure to apply, as signs have changed and the explored subspaces are orthogonal to those in the observer LMIs, the *inverse* multiplier must verify:

$$\eta^T \begin{bmatrix} \tilde{Q}_{\delta,11} & \tilde{S}_{\delta,11} \\ \tilde{S}_{\delta,11}{}^T & \tilde{R}_{\delta,11} \end{bmatrix} \eta < 0 \ \forall \eta \in \mathcal{S}(\Delta)^\perp \tag{6.75}$$

from (6.62) and Lemma A.2. Say, in the particular case of $\boldsymbol{w}_\delta = \Delta \boldsymbol{z}_\delta$,

$$S(\Delta) = \text{Im}\begin{bmatrix} \Delta \\ I \end{bmatrix}, \quad S(\Delta)^{\perp} = \begin{bmatrix} I \\ -\Delta^T \end{bmatrix}, \text{ as, indeed } \begin{bmatrix} \Delta \\ I \end{bmatrix}^T \cdot \begin{bmatrix} I \\ -\Delta^T \end{bmatrix} = 0.$$

The above observer and controller conditions, and the two multiplier ones (6.72) and (6.75) must be completed by positive-definiteness of the Lyapunov function and its inverse, as well as full rank of $P_{11} - X_{11}^{-1}$, needed for latter application of completion Lemma A.5, which is guaranteed by

$$\begin{bmatrix} P_{11} & I \\ I & X_{11} \end{bmatrix} > 0. \tag{6.76}$$

For brevity, the reader is referred to the detailed derivations in the seminal contribution [39] for ampler detail.

6.4.3 Controller Reconstruction

From the above conditions and particular suitable descriptions of Δ in each problem to be solved, the values for LMI decision variables $P_{11}, X_{11}, Q_{\delta,11}, \tilde{Q}_{\delta,11}, S_{\delta,11}, \tilde{S}_{\delta,11}, R_{\delta,11}, \tilde{R}_{\delta,11}$ can be readily obtained by convex optimisation in quite a few cases.

The next step in order to obtain the full matrix Ξ in (6.63) is resorting to the matrix completion lemma, i.e. Lemma A.5. Indeed, for the joint Lyapunov function P, Ξ can be obtained from the reduced-size "observer" one P_{11} and the "controller" one X_{11} directly. Regarding the multiplier, it is necessary to apply a row and column permutation (2nd and third) so the known variables appear in the top left of the multiplier M before the direct use of the completion lemma. Then, there are two options:

1. With *unrestricted* multipliers, apply the matrix completion Lemma A.5, to obtain (after reversing the permutation) the full multiplier M in (6.62).
2. Searching for multipliers in *restricted* families (block diagonal, for instance).

Let us give some comments on such choices.

Full-Block Multipliers

The advantage of the first approach (general "full-block" multipliers) is that, of course, they are less conservative than the second option. However, Lemma A.5 does not take into account the existence of $\mathcal{S}_c(\Delta)$, i.e. the controller scheduling function. Hence, $\mathcal{S}_c(\Delta)$ must be built *a posteriori* from the positive eigenvectors of a matrix which depends on Δ in order to fulfil (6.62). Details are omitted for brevity, and the reader is referred to [39]. The basic idea of this very elegant solution is that there is no need to *a priori* prescribe which is the relation between η_δ

and $\boldsymbol{y}_\delta$, thus providing superior controllers to earlier literature which (conservatively) assumed $\boldsymbol{\eta}_\delta = \Delta \boldsymbol{y}_\delta$, i.e. the "uncertainty" in controller enters exactly like the plant. However, for implementation, it requires solving a matrix inversion and an eigenvalue-eigenvector problem at every sample; there might not be a simple $\mathcal{S}_c(\Delta)$ with an analytical expression.

Structured Multipliers

On the other hand, the search for multipliers in a restricted family is less powerful but, in exchange, it allows for "intuitive" solutions such as $\mathcal{S}_C(\Delta) \equiv \mathcal{S}(\Delta)$, i.e. scheduling functions of the controller analogous to the plant's. This is the gain-scheduling equivalent of the PDC idea in TS systems, proposed in, e.g. [38]. In this particular case, with $\boldsymbol{w}_\delta = \Delta \boldsymbol{z}_\delta$, $\Delta^T \Delta < I$, the multipliers can be conservatively chosen to be in the form $Q_{\delta,11} = -\Sigma_p$, $S_{\delta,11} = 0$, $R_{\delta,11} = \Sigma_p$. The inverse multiplier must be $\tilde{Q}_{\delta,11} = -\tilde{\Sigma}_p$, $\tilde{S}_{\delta,11} = 0$, $\tilde{R}_{\delta,11} = \tilde{\Sigma}_p$.

Block-Diagonal Scalings, Norm-Bounded Δ

By the matrix reconstruction lemma, choosing any $T = \text{block-diag}(-T_{pc}, T_{pc})$ will revert to a *repeated* multiplier (doing and undoing row and column $2 \leftrightarrow 3$ permutations):

$$M = \begin{bmatrix} -\Sigma_p & -T_{pc} & 0 & 0 \\ -T_{pc}{}^T & -\Sigma_c & 0 & 0 \\ 0 & 0 & \Sigma_p & T_{pc} \\ 0 & 0 & T_{pc}{}^T & \Sigma_c \end{bmatrix} \tag{6.77}$$

with $\Sigma_c = T_{pc}^T(\Sigma_p - \tilde{\Sigma}_p^{-1})^{-1} T_{pc}$. So, if Σ and T_{pc} commute with Δ, having a suitable block-diagonal structure (see Section 4.7.3), it is straightforward to prove that

$$\begin{bmatrix} \Sigma_p & T_{pc} \\ T_{pc}{}^T & -\Sigma_c \end{bmatrix} \begin{bmatrix} \Delta & 0 \\ 0 & \Delta \end{bmatrix} = \begin{bmatrix} \Delta & 0 \\ 0 & \Delta \end{bmatrix} \begin{bmatrix} \Sigma_p & T_{pc} \\ T_{pc}{}^T & -\Sigma_c \end{bmatrix}, \tag{6.78}$$

so M and M^{-1} verify the required conditions on the "repeated Δ" arising from identical $\mathcal{S}_C(\Delta)$, and $\mathcal{S}(\Delta)$ given from the plant's description $\boldsymbol{w}_\delta = \Delta \boldsymbol{z}_\delta$ and the conservative but *intentionally chosen* $\boldsymbol{y}_\delta = \Delta \boldsymbol{\eta}_\delta$. This result shows that the results in [38] are a particular case of those in [39], as discussed in the latter work.

Convex Δ^p (Plant), Convex Controller Scheduling $\Delta^{[C]} \neq \Delta^p$

An alternative approach allowing for a different controller scheduling, but still convex and not needing eigenvalue computations at each sample, appears in [40]. The basic idea is described below. As there will be two different Δ's, contrarily to the above approach, Δ^p will refer to the "plant" one, and $\Delta^{[C]}$ to the to-be-computed controller one. Consider an uncertainty structure:

$$\boldsymbol{w}_\delta = \Delta_h^p \boldsymbol{z}_\delta = \left(\sum_{i=1}^r h_i(z)\Delta_i\right) \boldsymbol{z}_\delta \tag{6.79}$$

and a controller structure in LFT form (6.59) with

$$\boldsymbol{\eta}_\delta = \Delta_h^{[C]} \boldsymbol{y}_\delta = \left(\sum_{i=1}^r h_i(z)\Delta_i^{[C]}\right) \boldsymbol{y}_\delta, \tag{6.80}$$

where the arguments to h_i are assumed known or measurable so the scheduling (6.80) can be carried out in real-time implementation. The new vertices $\Delta_i^{[C]}$ should be obtained as a result of the multiplier manipulation.

Let us, for brevity, consider in this section $\Delta :=$ block-diag$(\Delta_h^p, \Delta_h^{[C]})$, and denote its inputs and outputs as $\boldsymbol{w}_{\delta,pc} := [\boldsymbol{w}_\delta^T, \boldsymbol{\eta}_\delta^T]^T$, $\boldsymbol{z}_{\delta,pc} := [\boldsymbol{z}_\delta^T, \boldsymbol{y}_\delta^T]^T$. Hence, $\boldsymbol{w}_{\delta,pc} = \Delta \boldsymbol{z}_{\delta,pc}$. Consider now a multiplier structure

$$M := \begin{bmatrix} Q_M & 0 \\ 0 & R_M \end{bmatrix}, \quad Q_M < 0,\ R_M > 0, \tag{6.81}$$

where Q_M and R_M are actually free from any "commute with Δ" condition, and, Δ does not need to have a unity norm bound.

The multiplier conditions which should hold in the suitable subspace are

$$\boldsymbol{w}_{\delta,pc}^T Q_M \boldsymbol{w}_{\delta,pc} + \boldsymbol{z}_{\delta,pc}^T R_M \boldsymbol{z}_{\delta,pc} > 0, \tag{6.82}$$

which, replacing the expression of $\boldsymbol{w}_{\delta,pc}$ results in

$$\boldsymbol{z}_{\delta,pc}^T (\Delta^T Q_M \Delta + R_M) \boldsymbol{z}_{\delta,pc} > 0, \tag{6.83}$$

so, from a Schur complement in Lemma A.1 and the definiteness conditions in (6.81) above, we get the equivalent condition:

$$\begin{bmatrix} R_M & \Delta^T \\ \Delta & -Q_M^{-1} \end{bmatrix} > 0. \tag{6.84}$$

As a by-product, reverting the Schur complement inverting R_M, we can write that the above expression is equivalent to

$$\psi^T(Q_M^{-1} + \Delta R_M^{-1}\Delta^T)\psi < 0\ \forall\psi \neq 0, \tag{6.85}$$

which amounts to the complementary condition of the inverse multiplier in the orthogonal subspace:

$$(*)\begin{bmatrix} Q_M^{-1} & 0 \\ 0 & R_M^{-1} \end{bmatrix}\begin{bmatrix} I \\ -\Delta^T \end{bmatrix} < 0. \tag{6.86}$$

Hence, with the multiplier definition in (6.81), condition (6.84) ensures that sign conditions on both $\mathcal{S}(\Delta)$ and its orthogonal hold.

Hence, the multiplier is plugged into the general conditions (6.63) and the resulting (6.71) and (6.74) are obtained. Note that (6.84) is sufficient for (6.72) and (6.75) so they do not need to be present.

As a result, we have a condition on the full Q_M^{-1} and R_M which is (6.84) as well as conditions (6.71) and (6.74), which only affect a submatrix of Q_M, R_M and their inverses. The simultaneous appearance of matrices, their inverses, and submatrices seem to indicate that the problem is not convex. However, the elegant developments in [40] allow, after some changes of variable, to express the problem in a convex way. Basically, what is needed is to write (6.84) only in terms of the (1, 1) blocks of Q_M, R_M, Q_M^{-1} and R_M^{-1}. The key step is defining:

$$R_M = \begin{bmatrix} R_{11} & R_{12} \\ R_{12}{}^T & R_{22} \end{bmatrix}, \quad R_M^{-1} = \begin{bmatrix} \tilde{R}_{11} & \tilde{R}_{12} \\ \tilde{R}_{12}{}^T & \tilde{R}_{22} \end{bmatrix}, \tag{6.87}$$

$$Q_M = \begin{bmatrix} Q_{11} & Q_{12} \\ Q_{12}{}^T & Q_{22} \end{bmatrix}, \quad Q_M^{-1} = \begin{bmatrix} \tilde{Q}_{11} & \tilde{Q}_{12} \\ \tilde{Q}_{12}{}^T & \tilde{Q}_{22} \end{bmatrix} \tag{6.88}$$

and, using the fact that

$$R_M \cdot R_M^{-1} = \begin{bmatrix} I & 0 \\ 0 & I \end{bmatrix} = \begin{bmatrix} R_{11}\tilde{R}_{11} + R_{12}\tilde{R}_{12}^T & R_{11}\tilde{R}_{12} + R_{12}\tilde{R}_{22} \\ (*) & R_{12}^T\tilde{R}_{12} + R_{22}\tilde{R}_{22} \end{bmatrix} \tag{6.89}$$

and a similar equality with Q and Q^{-1}. Defining

$$\Pi_H = \begin{bmatrix} I & \tilde{R}_{11} \\ 0 & \tilde{R}_{12}^T \end{bmatrix}, \quad \Pi_Q = \begin{bmatrix} Q_{11} & I \\ Q_{12}^T & 0 \end{bmatrix} \tag{6.90}$$

after some manipulations and repeated use of (6.89), it is shown in [40] that

$$\begin{bmatrix} \Pi_H^T & 0 \\ 0 & \Pi_Q{}^T \end{bmatrix} \begin{bmatrix} R & \Delta^T \\ \Delta & -Q^{-1} \end{bmatrix} \begin{bmatrix} \Pi_H & 0 \\ 0 & \Pi_Q \end{bmatrix} = \begin{bmatrix} R_{11} & I & \Delta_h^{pT} Q_{11} & \Delta_h^{pT} \\ I & \tilde{R}_{11} & \Delta_{K,z}^T & \tilde{R}_{11}\Delta_h^{pT} \\ Q_{11}\Delta_h^p & \Delta_{K,z} & -Q_{11} & I \\ \Delta_h^p & \Delta_h^p \tilde{R}_{11} & -I & -\tilde{Q}_{11} \end{bmatrix}, \quad (6.91)$$

where the change of variable

$$\Delta_{K,z} = Q_{11}\Delta_h^p \tilde{R}_{11} + Q_{12}\Delta_h^{[C]}\tilde{R}_{12}^T$$

has been made, replacing the original controller vertices $\Delta_i^{[C]}$, $i = 1, \ldots, r$ by a transformed set of decision variables $\Delta_{K,i}$ in order to linearise the product between Q_{11} and $\tilde{R}_{11}$.

The resulting (6.91) is a convex sum so all the "11" variables and $\Delta_{K,i}$ can be obtained via LMIs.

Once they are solved, the matrix reconstruction lemma for Q, R, P can be invoked and the actual controller vertices in (6.80) can be obtained by inverting the change of variable above, i.e.

$$\Delta_i^{[C]} = Q_{12}^{-1}\left(\Delta_{K,i} - Q_{11}\Delta_i^p \tilde{R}_{11}\right)\tilde{R}_{12}^{-T}. \quad (6.92)$$

Details are omitted for brevity. The reader is referred to [40] and references therein for ampler discussion.

A generalisation of the above discussion to the non-quadratic Lyapunov function presented in Section 4.8 appears in [41]. For brevity, it is not discussed here and the interested reader is referred to the original work for details.

6.5 Application Case Study: Sitting Control for People with Spinal Injury

People living with a complete spinal cord injury (SCI) lose all sensibility and mobility below their injury level. After an SCI, sitting control in everyday life becomes one of the crucial goals of rehabilitation [42]. The control of the sitting position can be realised with the upper limbs and head, instead of the lumbar muscle, as a compensatory strategy to maintain equilibrium [43]. The goal of most studies is to distinguish the voluntary forces generated at the upper limbs in order to stabilise the position from the passive torques produced at the lumbar level by the stretching of soft tissues [44] and internal organs compression. Several models have been derived, from the simplest one called *head-2-arm-and-trunk* (H2AT) model [45] and its variation presented thereafter [46] to the *seated-3-segment* (S3S) version [47], which are out of the scope of this book. Therefore, the goal is to be able to recover the unknown and unmeasurable torques produced, especially the passive one at the lumbar level.

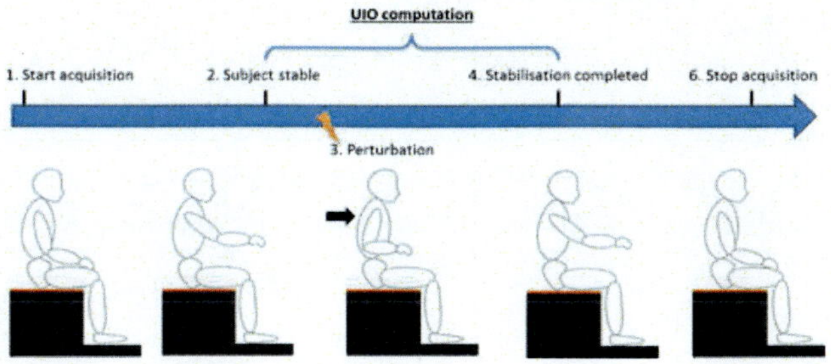

Fig. 6.11 Experimental procedure

A discrete-time convex descriptor model for a PI-observer, including the treatment of unmeasured premises, has been derived. Experiments have been conducted at the Pathokinesiology Laboratory of Montreal: they have validated the approach[3].

Experimental Protocol

Full description of the manipulations and results can be found in [45] and [46]. Figure 6.11 summarises the experimental protocol.

Four subjects who sustained a complete Spinal Cord Injury (SCI) were asked to perform 9 manipulations. The goal is to maintain a sitting position on a height-adjustable table without back support with hip and knees flexed to 90°, feet resting on the floor and upper limbs flexed to 90° at the elbow level. When stability of the position is reached, a light destabilising force is applied in the back via a foam-coated wooden pole. Then, the PI-observer is used until the subject recovers stability. Kinematic is recorded via 39 skin-fixed infrared light-emitting diodes.

Modelling

The H2AT (see Fig. 6.12) is composed of two rods, the first one represents the trunk as a classical inverted pendulum rotating at the lumbar joint (point T) while the second rod, representing the head and upper limbs, slides at the top of the first one with a constant angle a and crosses the first rod at point E. The output of the system is the position $x_H(t)$ of the COM_{H2A}, i.e. the head and upper limbs barycenter of mass,

[3] Ethical approval has been obtained from the Research Ethics Committee of the Center for Interdisciplinary Research in Rehabilitation of Greater Montreal (CRIR-1083-0515R). The participants read and signed the informed consent form prior to initiating the measurements. We wish to thank Pr. Dany H. Gagnon from the Pathokinesiology Laboratory of Montreal, Canada, for giving us the opportunity to use the technical equipment and apply our theoretical tools to subjects living with a SCI.

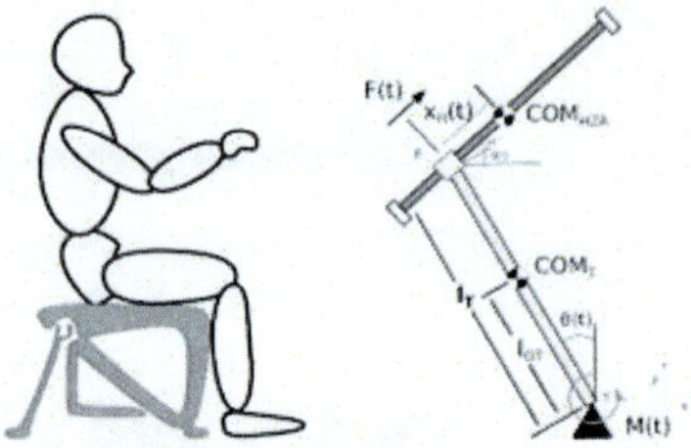

Fig. 6.12 H2AT Model

and the angular position of the trunk $\theta(t)$. Inputs of the system are the controlling force $F(t)$ which makes the upper rod slide over the lower one and the passive lumbar torque $M(t)$ which represents the passive resistance in the lumbar region. This model is generic and requires a minimum of biomechanical parameters.

For simulation, we consider a 80kg male subject. The notation $m_H = 16.1$kg stands for the mass of the upper segment, corresponding to the head, neck, and upper limbs; $m_T = 26.64$kg for the mass of the trunk; $l_T = 477$mm for the length of the trunk; and $l_{G_T} = 276.66$mm for the length to the center of mass of the trunk. A full neck flexion with both arms stretched gives a value of $x_H = 105.27$mm, whereas an extension of the neck and upper limbs gives $x_H = -75.18$mm. Experimental data of people living with SCI when exposed to perturbation [48] allows us to define the following compact set which is used for convex modelling:

$$\mathcal{X} = \left\{(x_H, \theta, \dot{\theta}) : -0.075 \leq x_H \leq 0.105,\ -0.087 \leq \theta \leq 0.122,\ -0.209 \leq \dot{\theta} \leq 0.209\right\},$$

where x_H, θ, and $\dot{\theta}$ have been expressed in meters, radian, and radian/s, respectively.

The modeling part using Euler-Lagrange formalism can be found in [46]; it results in the following nonlinear descriptor model with state vector $\boldsymbol{x}(t) = [x_H\ \theta\ \dot{x}_H\ \dot{\theta}]^T$ and $t \in \mathbb{R}$:

$$\begin{cases} E(\boldsymbol{x})\dot{\boldsymbol{x}}(t) = A(\boldsymbol{x})\boldsymbol{x}(t) + \boldsymbol{B}\boldsymbol{u}(t), \\ \boldsymbol{y}(t) = C\boldsymbol{x}(t), \end{cases} \tag{6.93}$$

the input vector redefined as

$$\boldsymbol{u}(t) = \begin{bmatrix} F(t) - m_H g \sin a \cos\theta \\ M(t) \end{bmatrix},$$

and system matrices

$$E(\boldsymbol{x}) = \begin{bmatrix} 1\,0 & 0 & 0 \\ 0\,1 & 0 & 0 \\ 0\,0 & m_H & -m_H l_T \cos a \\ 0\,0 & -m_H l_T \cos a & J(x) \end{bmatrix}, \quad A(\boldsymbol{x}) = \begin{bmatrix} 0 & 0 & 1 & 0 \\ 0 & 0 & 0 & 1 \\ 0 & a_{32} & 0 & a_{34} \\ a_{41} & a_{42} & a_{43} & 0 \end{bmatrix},$$

$$B = \begin{bmatrix} 0\,0\,0\,1 \\ 0\,0\,1\,0 \end{bmatrix}^T, \quad C = \begin{bmatrix} 1\,0\,0\,0 \\ 0\,1\,0\,0 \end{bmatrix},$$

where

$$J(x) = I_{G_T} + m_T I_{G_T}^2 + I_{G_H} + m_H I_T^2 + m_H x_H^2 + 2m_H x_H l_T \sin a,$$
$$a_{32} = -m_H g \cos a \frac{\sin\theta}{\theta}, \quad a_{34} = m_H \dot{\theta}(x_H + l_T \sin a),$$
$$a_{41} = -m_H g \cos(\theta + a), \quad a_{42} = (m_H l_T + m_T l_{G_T}) g \frac{\sin\theta}{\theta},$$
$$a_{43} = -2m_H \dot{\theta}(x_H + l_T \sin a).$$

In order to express (6.93) in discrete-time, the classical Euler's approximation is used so that $\dot{x} \approx (x(t+1) - x(t))/T_s$, where T_s is the sampling period and $t \in \mathbb{Z}$:

$$\begin{aligned} E(x)x(t+1) &= A_d(x)x(t) + T_s B u(t), \\ y(t) &= Cx(t), \end{aligned} \tag{6.94}$$

with $T_s = 0.01$s and $A_d(x) = T_s A(x) + E(x)$.

Four nonlinear bounded terms (due to the anatomical constraints) are present in (6.94):

$$z_1 = \frac{\sin\theta}{\theta} \in [0.9975, 1], \quad z_2 = \cos(\theta_k + a) \in [0.9782, 1],$$
$$z_3 = x_H\dot{\theta} + \dot{\theta} l_T \sin a \in [-0.0568, 0.0568],$$
$$z_4 = x_H(x_H + 2l_T \sin\alpha) \in [0, 0.0134],$$

based on which the scheduling vector $z = [z_1\, z_2\, z_3\, z_4]^T$ is defined, allowing in turn to define weights $w_j^i(z)$, $i \in \{1, 2, 3, 4\}$, $j \in \{0, 1\}$, and MFs $h_i(z)$ computed as all the different product combinations $w_{j_1}^1 w_{j_2}^2 w_{j_3}^3 w_{j_4}^4$.

Thus, sector nonlinearity allows finding a convex descriptor model with $2^4 = 16$ vertices, namely

$$\begin{aligned} E_h x(t+1) &= A_{dh} x(t) + T_s B u(t), \\ y(t) &= Cx(t), \end{aligned} \tag{6.95}$$

where

$$E_h = \sum_{i=1}^{16} h_i(z) E_i, \quad A_{dh} = \sum_{i=1}^{16} h_i(z) A_{di},$$
$$E_i = E(x)|_{h_i(z)=1}, \quad A_{di} = A_d(x)|_{h_i(z)=1}.$$

Obviously, the H2AT model being open-loop unstable, a control law has to be designed. Since we are interested in illustrating observer design, this part is out of the scope; the interested reader is referred to [46].

PI-Observer Design with Unmeasurable Premises

Classical schemes for observer design are not directly applicable as some membership functions are not available. Due to the discrete-time descriptor nature of the observed system, the proposed scheme will also adopt a descriptor convex form. In addition, it will be a proportional-integral (PI) one as a 2nd order discrete integrator is used to capture the dynamics of the unknown input d; in the z-domain (not to be confused with scheduling variables z!) this integrator is $(z-1)^2 d(t) \approx 0$. This choice represents a good compromise between accuracy of results and complexity of the observer. Thus, with

$$D_M = \begin{bmatrix} d_M(t) \\ d_{M(t+1)} \end{bmatrix}, \quad D_F = \begin{bmatrix} d_F(t) \\ d_F(t+1) \end{bmatrix},$$

it yields the following 8th-order extended system:

$$\begin{aligned} E_h^e \boldsymbol{x}^e(t+1) &= A_{dh}^e \boldsymbol{x}^e(t), \\ \boldsymbol{y}(t) &= C^e \boldsymbol{x}^e(t), \end{aligned} \tag{6.96}$$

where

$$\boldsymbol{x}^e = \begin{bmatrix} \boldsymbol{x} \\ D_F \\ D_M \end{bmatrix} \in \mathbb{R}^8, \; E_h^e = \begin{bmatrix} E_h & 0_{4\times4} \\ 0_{4\times4} & I_{4\times4} \end{bmatrix}, \; B = \begin{bmatrix} B_F & B_M \end{bmatrix}, \; C^e = \begin{bmatrix} C & 0_{2\times4} \end{bmatrix},$$

$$A_{d_h}^e = \begin{bmatrix} A_{d_h} & \begin{bmatrix} T_s B_F & 0_{4\times1} \end{bmatrix} & \begin{bmatrix} T_s B_M & 0_{4\times1} \end{bmatrix} \\ 0_{2\times4} & \Gamma & 0_{2\times2} \\ 0_{2\times4} & 0_{2\times2} & \Gamma \end{bmatrix}, \; \Gamma = \begin{bmatrix} 0 & 1 \\ 2 & -1 \end{bmatrix}.$$

Note that E_h^e is invertible for all $\boldsymbol{x} \in \mathcal{X}$. Recall that $\dot{\theta}$ is not measured and must be estimated to be used in the MFs computation.

Using the results in [20], the following observer is proposed:

$$\begin{aligned} E_\alpha^e \hat{\boldsymbol{x}}^e(t+1) &= A_{\alpha\hat{\beta}}^e \hat{\boldsymbol{x}}^e(t) + G_{\hat{\beta}}^{-1} K_{\alpha\hat{\beta}}(\boldsymbol{y}(t) - \hat{\boldsymbol{y}}(t)), \\ \hat{\boldsymbol{y}}(t) &= C^e \hat{\boldsymbol{x}}^e(t), \end{aligned} \tag{6.97}$$

where $\hat{\boldsymbol{x}}^e \in \mathbb{R}^8$ is the extended estimation vector, $\alpha \in \{\theta, x_H\}$ stands for the set of measurable variables, and $\beta = x_H\dot{\theta}$ is the only unmeasured variable. Matrices $G_{\hat{\beta}} \in \mathbb{R}^{8\times8}$ and $K_{\alpha\hat{\beta}} \in \mathbb{R}^{8\times2}$ are nonlinear gains to be designed. To this end, let us define the estimation error $\boldsymbol{e}(t) = \boldsymbol{x}^e(t) - \hat{\boldsymbol{x}}^e(t)$, and its dynamics via the descriptor form $E_\alpha^e \boldsymbol{e}(t+1) = E_\alpha^e \boldsymbol{x}^e(t+1) - E_\alpha^e \hat{\boldsymbol{x}}^e(t+1)$, from which $E_\alpha^e \boldsymbol{e}(t+1) = A_{\alpha\beta}^e \boldsymbol{x}^e(t) - A_{\alpha\hat{\beta}}^e \hat{\boldsymbol{x}}^e(t) - G_{\hat{\beta}}^{-1} K_{\alpha\hat{\beta}} \hat{\boldsymbol{x}}^e(t)(\boldsymbol{y}(t) - \hat{\boldsymbol{y}}(t))$, or equivalently:

$$E_\alpha^e \boldsymbol{e}(t+1) = \left(A_{\alpha\beta}^e - G_{\hat{\beta}}^{-1} K_{\alpha\hat{\beta}} C^e\right) \boldsymbol{e}(t) + \left(A_{\alpha\beta}^e - A_{\alpha\hat{\beta}^e}\right) \hat{\boldsymbol{x}}^e(t). \tag{6.98}$$

Notice that

$$\left(A^e_{\alpha\beta} - A^e_{\alpha\hat{\beta}}\right)\hat{x}^e(t) = \begin{bmatrix} A_{\alpha\beta} - A_{\alpha\hat{\beta}} & 0_{4\times4} \\ 0_{4\times4} & 0_{4\times4} \end{bmatrix}\hat{x}^e(t),$$

with $A_{\alpha\beta} - A_{\alpha\hat{\beta}}$ having only 2 non-zero terms in entries (3, 4) and (4, 2). Therefore, after some straightforward manipulations (6.98) can be written as

$$\begin{aligned} E^e_\alpha \boldsymbol{e}(t+1) &= \left(A^e_{\alpha\beta} - G^{-1}_{\hat{\beta}} K_{\alpha\hat{\beta}} C^e\right)\boldsymbol{e}(t) \\ &\quad + T_s m_H \left(\dot{\theta}(t) - \hat{\dot{\theta}}(t)\right)\begin{bmatrix} 0_{2\times1} \\ x_H(t)\hat{\dot{\theta}}(t) + l_T\sin(a)\hat{\dot{\theta}}(t) \\ -2x_H(t)\hat{\dot{x}}_H(t) - 2l_T\sin(a)\hat{\dot{x}}_H(t) \\ 0_{4\times1} \end{bmatrix} \end{aligned} \tag{6.99}$$

and as $\dot{\theta} - \hat{\dot{\theta}}$ is the fourth term of $\boldsymbol{e}$, (6.99) yields:

$$E^e_\alpha \boldsymbol{e}(t+1) = \left(A^e_{\alpha\beta} - G^{-1}_{\hat{\beta}} K_{\alpha\hat{\beta}} C^e + T_s m_H \Delta(t)\right)\boldsymbol{e}(t), \tag{6.100}$$

where $\Delta(t) \in \mathbb{R}^{8\times8}$ is a null matrix except for the entries

$$\begin{aligned} \Delta_{(3,4)}(t) &= x_H(t)\hat{\dot{\theta}}(t) + l_T\sin(a)\hat{\dot{\theta}}(t), \\ \Delta_{(4,4)}(t) &= -2x_H\hat{\dot{x}}_H(t) - 2l_T\sin(a)\hat{\dot{x}}_H(t). \end{aligned}$$

As with many other robustness arguments, consider any bounded uncertainty $\mu(t) \in [\underline{\mu}, \overline{\mu}]$ to be expressed as $\mu(t) = c + r \times \delta(t)$ with $\delta(t) \in [-1, 1]$, the center: $c = 0.5(\overline{\mu} - \underline{\mu})$, and the radius $r = 0.5(\overline{\mu} - \underline{\mu})$. Therefore defining accordingly $c_{(3,4)}$, $r_{(3,4)}$, $c_{(4,4)}$, and $r_{(4,4)}$, (6.100) becomes:

$$E^e_\alpha \boldsymbol{e}(t+1) = \left(A^e_{\alpha\beta} - G^{-1}_{\hat{\beta}} K_{\alpha\hat{\beta}} C^e + T_s m_H A_c + H\delta(t)F\right)\boldsymbol{e}(t), \tag{6.101}$$

with

$$H = \begin{bmatrix} 0_{2\times2} \\ I_2 \\ 0_{4\times2} \end{bmatrix}, \quad F = \begin{bmatrix} 0_{2\times3} & T_s m_H \begin{bmatrix} r_{(3,4)} \\ -2r_{(4,4)} \end{bmatrix} & 0_{2\times4} \end{bmatrix},$$

and A_c being a null matrix except for $A_c(3,4) = c_{(3,4)}$, $A_c(4,4) = c_{(4,4)}$.

Let us express $A'_{\alpha\beta} = A_{\alpha\beta} + T_s m_H A_c$ in a more convenient way:

$$A'_{\alpha\beta} = A_{\alpha\beta} + \begin{bmatrix} 0_{2\times2} & 0_{2\times2} \\ 0_{2\times2} & \begin{bmatrix} 0 & T_s m_H c_{(3,4)} \\ 0 & T_s m_H c_{(4,4)} \end{bmatrix} \end{bmatrix}$$

from which it follows that

$$A'^{e}_{\alpha\beta} = \begin{bmatrix} A'_{\alpha\beta} & [T_s B_F \; 0_{4\times1}] & [T_s B_M \; 0_{4\times1}] \\ 0_{2\times4} & \Gamma & 0_{2\times2} \\ 0_{2\times4} & 0_{2\times2} & \Gamma \end{bmatrix}.$$

Therefore, we transformed the non-measurable premise problem into a problem of asymptotic convergence of the following uncertain system:

$$E^e_\alpha e(t+1) = \left(A'^{e}_{\alpha\beta} - G^{-1}_{\hat\beta} K_{\alpha\hat\beta} C^e + T_s m_H A_c + H\delta(t)F\right) e(t). \tag{6.102}$$

Using a non-quadratic Lyapunov function candidate

$$V(e) = e^T P_{\alpha(t-1)} e, \quad P_{\alpha(t-1)} = P^T_{\alpha(t-1)} > 0, \tag{6.103}$$

the origin $e = 0$ of the error dynamics (6.98) is asymptotically stable if there exists matrices $P_{\alpha(t-1)} = P^T_{\alpha(t-1)} > 0$, $G_{\hat\beta} \in \mathbb{R}^{8\times8}$, and $K_{\alpha\hat\beta} \in \mathbb{R}^{8\times2}$, and scalars $\eta_{\alpha\beta\hat\beta} > 0$, $0 < \lambda < 1$, such that the following inequality holds:

$$\begin{bmatrix} -\lambda P_{\alpha(t-1)} + \eta_{\alpha\beta\hat\beta} F^T F & (*) & 0 \\ G_{\hat\beta} A'^{e}_{\alpha\beta} - K_{\alpha\hat\beta} C^e & P_{\alpha(t)} - G_{\hat\beta} E^e_\alpha - (E^e_\alpha)^T G^T_{\hat\beta} & (*) \\ 0 & H^T G^T_{\hat\beta} & -\eta_{\alpha\beta\hat\beta} I_{2\times2} \end{bmatrix} < 0.$$

The interested reader is referred to [46] for details. The results have been validated both in simulation and real-time experiments, which will be presented next.

Real-Time Experimental Results

The estimation error between measured and estimated data has been computed. Figure 6.13 presents an example of angle $\theta(t)$ and displacement $x_H(t)$ and relative estimation error during one acquisition.

The root mean square error (RSME) for both measured variables during all acquisitions has been computed as 1.8×10^{-4}m for $x_H(t)$ and 5.1×10^{-4}rad for $\theta(t)$, with maximum value at 4.2×10^{-4}m for $x_H(t)$ and 8.2×10^{-4}rad for $\theta(t)$, which exhibits a good behaviour of the observer. Figures 6.14 and 6.15 present the estimated internal force and torque, respectively, for all acquisitions of the four subjects. In order to compare internal efforts between subjects, the data are normalised by the mass of each subjects. Figure 6.14 shows that for subjects S1 and S2 the normalised amplitude is greater than that for subjects S3 and S4 with a longer time of stabilisation after perturbation. Many details are omitted for brevity, but the results shown in this section illustrate the applicability of the convex modelling and optimisation techniques. For an interpretation of conclusions in a biomechanical sense, the inter-

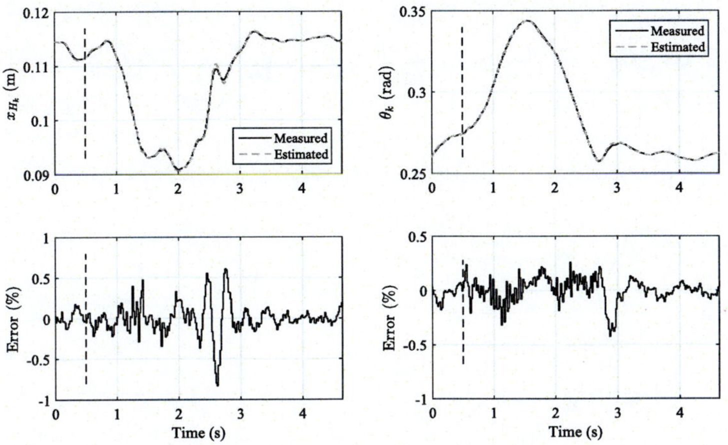

Fig. 6.13 (Top) Experimental input measurements (black) and estimation (dashed-grey). (Bottom) Estimation error. The vertical dashed line represents the perturbation instant

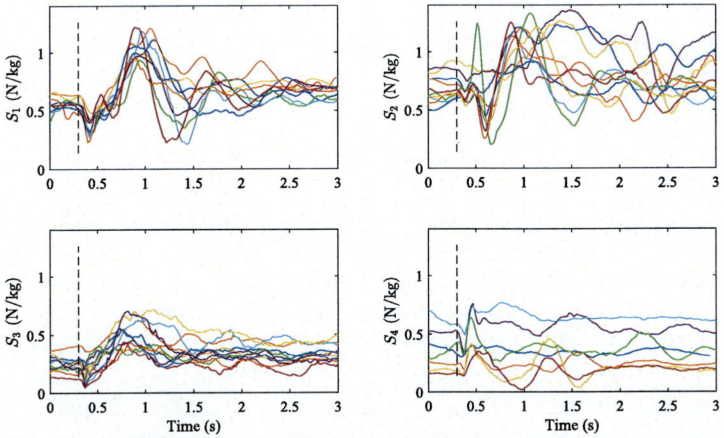

Fig. 6.14 Estimated internal force $\hat{F}(t)$ (N/kg) from (6) for all acquisitions, subjects 1, 2 (top) 3, 4 (bottom)

ested reader is referred to the works this section is based on, as they are out of the scope of this book.

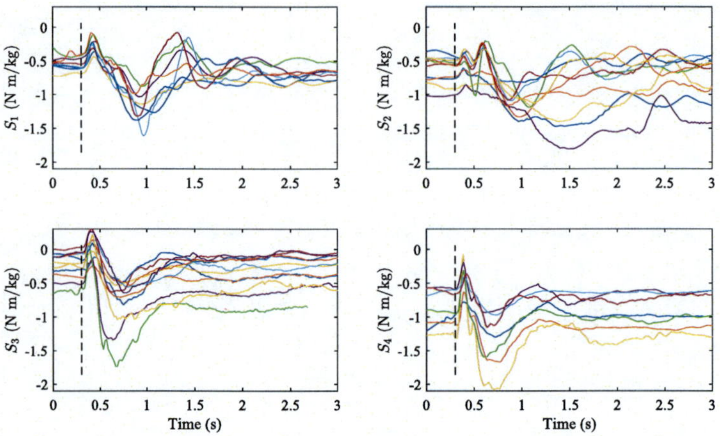

Fig. 6.15 Estimated internal torque $\hat{M}(t)$ $(N \cdot m/kg)$ from (6) for all acquisitions, subjects 1, 2 (top) 3, 4 (bottom)

6.6 Application Case Study: Observer-Based Control for Wheelchairs

This work[4] is concerned with a second motorisation kit, a power-assisted wheelchair (PAW) that has no joystick and whose goal is to adapt the electrical assistance to any person with reduced mobility (PRM) in any circumstances. One particularity of this assistance kit, designed by the Autonomad Mobility company, is that it does not require sensors to measure the human torque, reducing the cost of the kit in a significant way. They have been replaced with an unknown-input observer in the form of a PI-observer [49, 50]. The challenge faced in this work is how to go from "one-user, fixed-conditions" to a solution that is "wheelchair- and PRM-independent". Importantly also, the strategy has to "understand" and include the user's intention. These objectives have to be transformed into a control framework. Three main steps are necessary, presented Fig. 6.16 [51]:

1. Robust human torque estimation, whatever the pushing features are; it delivers the estimated speed and rotation of the wheelchair $\left[\hat{\omega}(t)\ \hat{\phi}(t)\right]^T$ [50].
2. A reference trajectory generation $\left[\omega_r\ \phi_r\right]^T$, in order to transform the perceived intentions of the user into signals usable by the control.
3. A robust observer-based tracking controller design. The design has to integrate the uncertainties coming from: the human torque estimation, the measurements

[4] This research was conducted with support of the European Union through the European Regional Development Fund (ERDF), the French Ministry of Higher Education and Research, the National Center for Scientific Research (CNRS), and the Hauts-de-France Region. It is a collaboration between the LAMIH UMR CNRS 8201, Polytechnic University Hauts-de-France in Valenciennes and the company Autonomad Mobility.

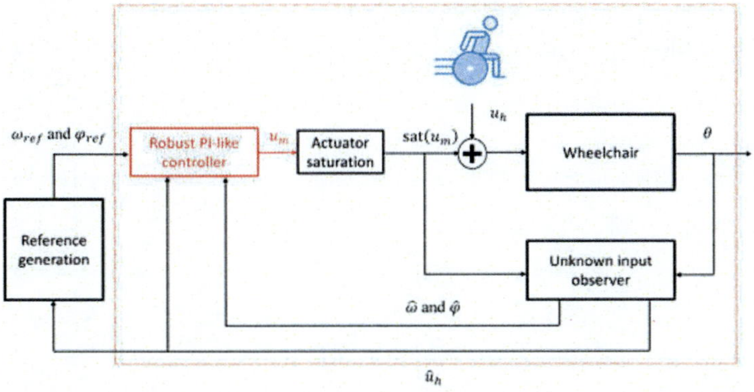

Fig. 6.16 Assisting control loop. Reproduced from [52] by Guoxi Feng with permission from Guoxi Feng

(environment conditions), the unknown model parameters (PRM and wheelchair masses, friction, etc.), and the non-modelled dynamics (casters wheels).

Finally, the overall strategy, not only must be safe and secure, but also friendly to convince the potential end-users. The idea of the observer design is not to use the amplitude of the signals estimated, because, due to the uncertainties, they cannot be accurate enough, but to use the estimation of their frequency. This estimation can be used to detect the user's intentions, as shown in [50]. Nevertheless, the full problem has not a key-in-hand solution, therefore a 2-step procedure has been used: first, design a robust control, and second, design the observer while ensuring global performance and stability of the loop. In the sequel, the subscript h will not have the usual meaning as it will stand only for "human signal"; similarly, z is not an scheduling vector as in the rest of the book, but just an auxiliary state.

Modelling

A precise modelling including any PRM (weight, force/torque available, disease, state of fatigue) any wheelchair (as the kit is adaptable on any kind of manual wheelchair), the ground and weather conditions is unrealistic; therefore, the wheelchair is just considered as a two-wheeled transporter, see Fig. 6.17. The physical parameters of the wheelchair used for the experiments are given in Table 6.3. The two-wheeled PAW is described by the dynamics [53]:

$$\alpha\dot{\theta}_R + \beta\dot{\theta}_L = T_{mr} + T_{hr} - \mathcal{K}\theta_R, \quad \alpha = \frac{mr^2}{4} + \frac{\left(I_a + md^2\right)\zeta^2}{b^2} + I_0,$$
$$\alpha\dot{\theta}_L + \beta\dot{\theta}_R = T_{ml} + T_{hl} - \mathcal{K}\theta_L, \quad \beta = \frac{mr^2}{4} - \frac{\left(I_a + md^2\right)\zeta^2}{b^2},$$

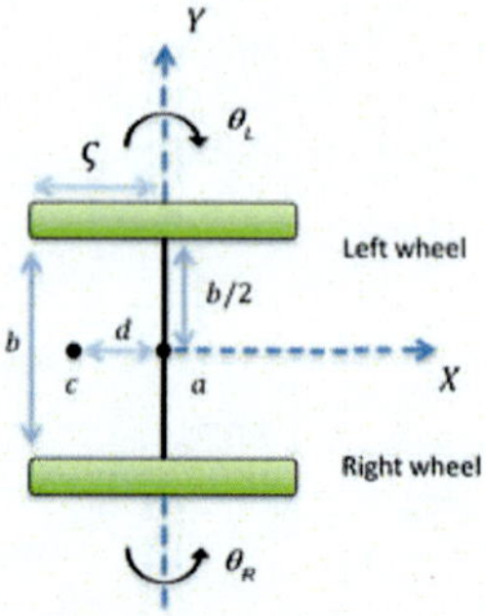

Fig. 6.17 Simplified top view of the wheelchair

where the propulsion consists in the human torques T_{hr} and T_{hl} as well as the motors torques T_{mr} and T_{ml}; the left angular velocity and the right angular velocity are θ_L and θ_R, respectively; ω is the velocity of the mass center, and ϕ the yaw velocity: they are the variables naturally and implicitly controlled by the user for her/his displacements, i.e.:

$$\begin{bmatrix} \omega \\ \phi \end{bmatrix} = \begin{bmatrix} \zeta/2 & \zeta/2 \\ \zeta/b & -\zeta/b \end{bmatrix} \begin{bmatrix} \theta_R \\ \theta_L \end{bmatrix}. \tag{6.104}$$

Considering the sampling time $T_s = 500\text{ms}$, imposed by the company hardware, $t \in \mathbb{Z}$, and $\boldsymbol{x} = \begin{bmatrix} \omega \ \phi \end{bmatrix}^T$ the state vector, using Euler's approximation $\dot{\boldsymbol{x}} \approx (\boldsymbol{x}(t+1) - \boldsymbol{x}(t))/T_s$, a discrete-time descriptor form is obtained:

$$E(m)\boldsymbol{x}(t+1) = A(m, \mathcal{K})\boldsymbol{x}(t) + B(\boldsymbol{u}_h(t) + \boldsymbol{u}_m(t)), \quad \boldsymbol{y} = C\boldsymbol{x}, \tag{6.105}$$

where

$$\boldsymbol{y} = \begin{bmatrix} \theta_R \\ \theta_L \end{bmatrix}, \ \boldsymbol{u}_h = \begin{bmatrix} T_{hr} \\ T_{hl} \end{bmatrix}, \ \boldsymbol{u}_m = \begin{bmatrix} T_{mr} \\ T_{ml} \end{bmatrix},$$

stand for the output, the human torque, and the motor torque, respectively. Both the mass m and the viscous friction coefficient $\mathcal{K}$ are uncertain; they are assumed to lie in a fixed interval:

$$\underline{m} \leq m \leq \overline{m}, \quad \underline{\mathcal{K}} \leq \mathcal{K} \leq \overline{\mathcal{K}}. \tag{6.106}$$

The corresponding matrices of (6.105) are

$$A(m, \mathcal{K}) = \begin{bmatrix} \alpha - T_e\mathcal{K} & \beta \\ \beta & \alpha - T_e\mathcal{K} \end{bmatrix} \begin{bmatrix} \zeta/2 & \zeta/2 \\ \zeta/b & -\zeta/b \end{bmatrix}, \ B = T_s I,$$

$$E(m) = \begin{bmatrix} \alpha & \beta \\ \beta & \alpha \end{bmatrix} \begin{bmatrix} \zeta/2 & \zeta/2 \\ \zeta/b & -\zeta/b \end{bmatrix}, \ C = \begin{bmatrix} 1/\zeta & b/(2\zeta) \\ 1/\zeta & -b/(2\zeta) \end{bmatrix},$$

Table 6.3 Model parameters

Symbol	Description	Value
ζ	Wheel radius [m]	0.33
b	Distance between two wheels [m]	-
d	Distance between the point a and the point c [m]	0.6
c	Centre of gravity of the wheelchair with the human	-
a	Middle point between two wheels	-
m_0	Nominal mass of the wheelchair including the human [kg]	100
$\mathcal{K}_0$	Nominal viscous friction coefficient [Nms]	5
I_a	Inertia of the wheelchair with respect to the vertical axis through the point a [kgm2]	16
I_0	Inertia of each wheel around the wheel axis [kgm2]	0.25
T_s	Sampling time [s]	0.05

based on which a tensor product (TP) model is obtained via sector nonlinearity approach considering the intervals above for the weighting functions w_i^j, $i \in \{0, 1\}$, $j \in \{1, 2\}$:

$$\sum_{i=0}^{1} w_i^1(m) E_i \boldsymbol{x}(t+1) = \sum_{i=0}^{1}\sum_{j=0}^{1} w_i^1(m) w_j^2(\mathcal{K}) A_{ij} \boldsymbol{x}(t) + B(\boldsymbol{u}_h(t) + \boldsymbol{u}_m(t)), \tag{6.107}$$

with: $E_0 = E(\overline{m})$, $E_1 = E(\underline{m})$, $A_{00} = E(\overline{m}, \overline{\mathcal{K}})$, $A_{01} = E(\underline{m}, \overline{\mathcal{K}})$, $A_{10} = E(\overline{m}, \underline{\mathcal{K}})$, and $A_{11} = E(\underline{m}, \underline{\mathcal{K}})$.

Robust Control

As none of the premise variables are measurable, the control law is a linear PI-like controller with $\boldsymbol{\chi}$ the integrator state, i.e.

$$\begin{cases} \boldsymbol{u}_m(t) = LM^{-1}\begin{bmatrix} \boldsymbol{x}(t) \\ \boldsymbol{\chi} \end{bmatrix} - u_h(t), \\ \boldsymbol{\chi}(t+1) = \boldsymbol{\chi}(t) - C\boldsymbol{x}(t). \end{cases} \tag{6.108}$$

Thus, an extended model is readily obtained using the extended state $\overline{\boldsymbol{x}}^T = \left[\boldsymbol{x}^T\ \boldsymbol{\chi}^T\right]$:

$$\overline{E}(m)\overline{\boldsymbol{x}}(t+1) = \overline{A}(m, \mathcal{K})\boldsymbol{x}(t) + \overline{B}(\boldsymbol{u}_h(t) + \boldsymbol{u}_m(t)), \tag{6.109}$$

with

$$\overline{E}(m) = \begin{bmatrix} E(m) & 0 \\ 0 & I_2 \end{bmatrix}, \ \overline{A}(m, \mathcal{K}) = \begin{bmatrix} A(m, \mathcal{K}) & 0 \\ -C & I_2 \end{bmatrix}, \ \overline{B} = \begin{bmatrix} B \\ 0 \end{bmatrix}.$$

Robust control design can be applied with the non-quadratic Lyapunov function

$$V(\overline{\boldsymbol{x}}) = \overline{\boldsymbol{x}}^T \sum_{i=0}^{1} \sum_{j=0}^{1} w_i^1(m) w_j^2(\mathcal{K}) P_{ij} \overline{\boldsymbol{x}}. \tag{6.110}$$

Notice that only the friction is time varying, the mass, even if unknown, is constant for a given trial. Thus, amongst the possible LMI conditions to hold, the choice is not to have crossed-terms, i.e. find X_{ij}, M and L such that

$$\begin{bmatrix} -X_{kj} & M^T \overline{A}_{ij}^T + L^T \overline{B}^T \\ \overline{A}_{ij} M + \overline{B} L & X_{ij} - \overline{E}_j M - M^T \overline{E}_j^T \end{bmatrix} < 0, \quad i, j, k \in \{0, 1\}. \tag{6.111}$$

Estimation and Closed-Loop Stability/Performance

The goal of the output control is to integrate an observer to estimate the human torques and to guarantee a good closed-loop performance. The observer design uses LMI constraints such that the uncertain system (6.105) with the proposed observer-based tracking controller satisfies the following requirements:

- When the reference signal $\boldsymbol{x}_r = 0$ the state of the uncertain system (6.105) and the estimation error $\boldsymbol{e}_{obs}$ converge asymptotically to the origin.
- When the reference signal $\boldsymbol{x}_r \neq 0$ under null initial conditions (state and estimation error), the $\mathcal{L}_2$-norm of the estimation error is bounded as

$$\sum_{k=0}^{\infty} \boldsymbol{e}_o^T \boldsymbol{e}_o \leq \gamma \sum_{k=0}^{\infty} \boldsymbol{x}_r^T \boldsymbol{x}_r,$$

where $\boldsymbol{e}_o$ is to be defined later.

The uncertain parameters m and $\mathcal{K}$ are not measured; therefore, the state estimation is based solely on the nominal parameters m_0 and $\mathcal{K}_0$ with their corresponding matrices $E = E(m_0)$ and $A = A(m_0, \mathcal{K}_0)$ (no arguments as they are constant). Moreover, to estimate the human torques $\hat{\boldsymbol{u}}_h$ a PI-observer is used with a double integrator structure for each torque, i.e. the system (6.107) is rewritten as

$$\begin{bmatrix} E(m) & 0 \\ 0 & I_4 \end{bmatrix} z(t+1) = \begin{bmatrix} A(m,\mathcal{K}) & B_u \\ 0 & \Gamma \end{bmatrix} z(t) + \begin{bmatrix} B \\ 0 \end{bmatrix} \boldsymbol{u}_m(t), \quad y(t) = C\boldsymbol{x}(t), \tag{6.112}$$

where

$$z = \begin{bmatrix} \boldsymbol{x} \\ \bar{\boldsymbol{u}}_h \end{bmatrix}, \; \bar{\boldsymbol{u}}_h = \begin{bmatrix} T_{hr} \\ \dot{T}_{hr} \\ T_{hl} \\ \dot{T}_{hl} \end{bmatrix}, \; \Gamma = \begin{bmatrix} 2 & -1 \\ 1 & 0 \end{bmatrix} \bigotimes I_2, \; B_u = \begin{bmatrix} T_s & 0 & 0 & 0 \\ 0 & 0 & T_s & 0 \end{bmatrix}.$$

Thus, an observer corresponds to

$$\begin{bmatrix} E & 0 \\ 0 & I_4 \end{bmatrix} \hat{z}(t+1) = \begin{bmatrix} A & B_u \\ 0 & \Gamma \end{bmatrix} \hat{z}(t) + \begin{bmatrix} B \\ 0 \end{bmatrix} \boldsymbol{u}_m(t) + G^{-1}W(y(t) - \hat{y}(t)), \tag{6.113}$$

from where we can write the dynamics of the estimation error $\boldsymbol{e}_o = z - \hat{z}$

$$\begin{aligned} &\begin{bmatrix} E(m) - E & 0 \\ 0 & 0 \end{bmatrix} z(t+1) + \begin{bmatrix} E & 0 \\ 0 & I_4 \end{bmatrix} \boldsymbol{e}_o(t+1) \\ &= \begin{bmatrix} A(m,\mathcal{K}) & B_u \\ 0 & \Gamma \end{bmatrix} z(t) - \begin{bmatrix} A & B_u \\ 0 & \Gamma \end{bmatrix} \hat{z}(t) - G^{-1}W(y(t) - \hat{y}(t)), \end{aligned}$$

or, equivalently:

$$\begin{aligned} &\begin{bmatrix} E(m) - E \\ 0 \end{bmatrix} \boldsymbol{x}(t+1) + \begin{bmatrix} E & 0 \\ 0 & I_4 \end{bmatrix} \boldsymbol{e}_o(t+1) \\ &= \left(\begin{bmatrix} A & B_u \\ 0 & \Gamma \end{bmatrix} - G^{-1}W\overline{C} \right) \boldsymbol{e}_o + \begin{bmatrix} A(m,\mathcal{K}) - A \\ 0 \end{bmatrix} \boldsymbol{x}(t). \end{aligned}$$

For the full closed-loop in Fig. 6.16, the control law (6.108) corresponds to

$$\begin{aligned} \boldsymbol{u}_m(t) &= LM^{-1} \begin{bmatrix} \boldsymbol{x}(t) \\ \chi(t) \end{bmatrix} - \hat{\boldsymbol{u}}_h(t), \\ \chi(t+1) &= \chi(t) - C\boldsymbol{x} + \boldsymbol{x}_r, \end{aligned} \tag{6.114}$$

where $\hat{\boldsymbol{u}}_h$ is the torque estimation and $\boldsymbol{x}_r$ is the reference velocity. Thus, the closed-loop dynamics considering (6.109) together with the observer-based controller (6.113) and (6.114) is

$$\begin{bmatrix} \tilde{A}(m,\mathcal{K}) & -\tilde{E}(m) & -\tilde{B} \end{bmatrix} \begin{bmatrix} \tilde{\boldsymbol{x}}(t-1) \\ \tilde{\boldsymbol{x}}(t) \\ \boldsymbol{x}_r(t) \end{bmatrix} = 0, \quad \tilde{\boldsymbol{x}} = \begin{bmatrix} \boldsymbol{x} \\ \boldsymbol{e}_o \end{bmatrix}, \tag{6.115}$$

with

$$\tilde{A}(m,\mathcal{K}) = \begin{bmatrix} \overline{A}(m,\mathcal{K}) + \overline{B}LM^{-1} & \begin{bmatrix} 0 & B_u \\ 0 & 0 \end{bmatrix} \\ \begin{bmatrix} A(m,\mathcal{K}) - A & 0 \\ 0 & 0 \end{bmatrix} & \begin{bmatrix} A & B_u \\ 0 & \Gamma \end{bmatrix} - G^{-1}W\overline{C} \end{bmatrix},$$

$$\tilde{E}(m) = \begin{bmatrix} \overline{E}(m) & 0 \\ \begin{bmatrix} E(m) - E & 0 \\ 0 & 0 \end{bmatrix} & \begin{bmatrix} E & 0 \\ 0 & I_4 \end{bmatrix} \end{bmatrix}, \ \tilde{B} = \begin{bmatrix} 0 \\ I \\ 0 \end{bmatrix}.$$

As for the control (6.110), we consider a Lyapunov function including the uncertainties:

$$V(\tilde{x}) = \tilde{x}^T \sum_{i=0}^{1}\sum_{j=0}^{1} w_i^1(m) w_j^2(\mathcal{K}) \tilde{P}_{ij} \tilde{x},$$

with the same definitions for MFs. If a solution exists to the following ε parameter-dependent LMI problem, it solves the problem of estimation and closed-loop stability, the matrices M and L are fixed by the first step (6.111):

$$\text{find } \tilde{P}_{ij} > 0,\ W,\ \tilde{G} = \begin{bmatrix} G_1 & 0 \\ G_1 & G \end{bmatrix} \text{ such that:}$$

$$\begin{bmatrix} -\tilde{P}_{jk} + \tilde{C}^T\tilde{C} + \varepsilon\tilde{G}\tilde{A}_{ij} & (*) & (*) \\ \tilde{G}\tilde{A}_{ij} - \varepsilon\left(\tilde{G}\tilde{E}_j\right)^T & \tilde{P}_{ij} - \tilde{G}\tilde{E}_j - \left(\tilde{G}\tilde{E}_j\right)^T & (*) \\ -\varepsilon\left(\tilde{G}\tilde{B}\right)^T & -\left(\tilde{G}\tilde{B}\right)^T & -\gamma I \end{bmatrix} < 0, \quad i, j, k \in \{0, 1\}, \tag{6.116}$$

with

$$\tilde{G}\tilde{A}_{ij} = \begin{bmatrix} G_1 & 0 \\ G_1 & G \end{bmatrix} \begin{bmatrix} \overline{A}_{ij} + \overline{B}LM^{-1} & \begin{bmatrix} 0 & B_u \\ 0 & 0 \end{bmatrix} \\ G\begin{bmatrix} A_{ij} - A & 0 \\ 0 & 0 \end{bmatrix} & G\begin{bmatrix} A & B_u \\ 0 & \Gamma \end{bmatrix} - W\overline{C} \end{bmatrix}.$$

The inequalities (6.116) are LMI conditions for a given scalar ε. A numerical search for ε is carried out in a prescribed interval [52].

Experimental Validation

Experimental validation is carried out with a fully equipped wheelchair from Autonomad Mobility (see Fig. 6.18), which is subject to different soils with passages from one to another.

The wheelchair in Fig. 6.18 is equipped with two brushless DC motors powered by a DC battery (autonomy range of approximately 15km). The DC motor's maximum torque is around 40Nm. The motors receive the control signals (voltage or current) via a Texas Instruments C2000 real-time micro-controller. The sensors are two incremental encoders to measure the angular velocities with pulse signals as outputs. Two torque sensors with wireless transmissions supplied by CapInstrumentation are also installed; they are only available on the prototype for the observer validation purpose.

In order to experiment the maneuverability of the wheelchair, specific predefined tasks have to be achieved by the users; we reproduce one of them. The user is asked to perform eight-shaped and oval trajectories (Fig. 6.19, dotted red line) in the parking of Autonomad Mobility. This trial includes obstacle avoidance, different ground adhesions (Fig. 6.19, clear areas), and different (reasonable) slopes. Of course, the red trajectory does not have to be perfectly followed, it gives a global path to achieve. For the user, this path represents sequentially performing eight turns, numbered at the right side of Fig. 6.19, corresponding to 5 left turns followed by 1 right turn followed by 2 left turns.

Figure 6.20 shows the human torque measured by the sensors (blue line), the estimated torque (orange line), and mode, 0 for manual and 1 for assistance (red line). These experimental results show that the PI-observer estimation is sufficient to produce smooth references for pushing frequency and direction [50]. This capability is mainly due to the fact that the strategy is based on the frequency of user's propulsion, thus it is not the amplitude of the estimated torque signals $\hat{\boldsymbol{u}}_h$ that is important, but the detection of its frequency. The outputs of the reference trajectory are $\boldsymbol{x}_r = \left[\omega_r \ \phi_r\right]^T$. Note that higher (resp. lower) frequency detected corresponds to an increase (resp. a decrease) of the speed reference ω_r. To turn the wheelchair, the user just brakes the wheel corresponding to the desired direction. When detected,

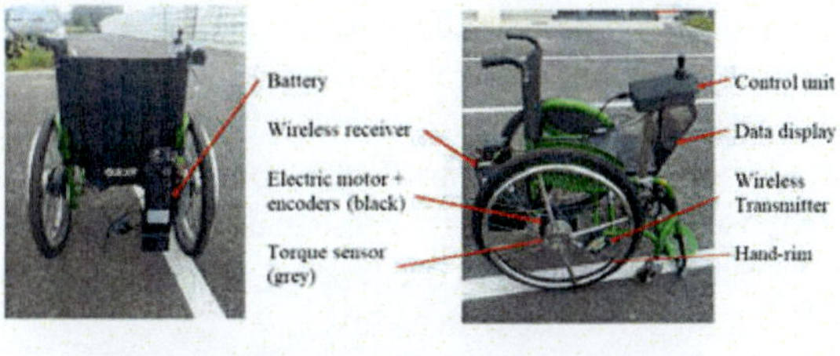

Fig. 6.18 Wheelchair prototype and its components

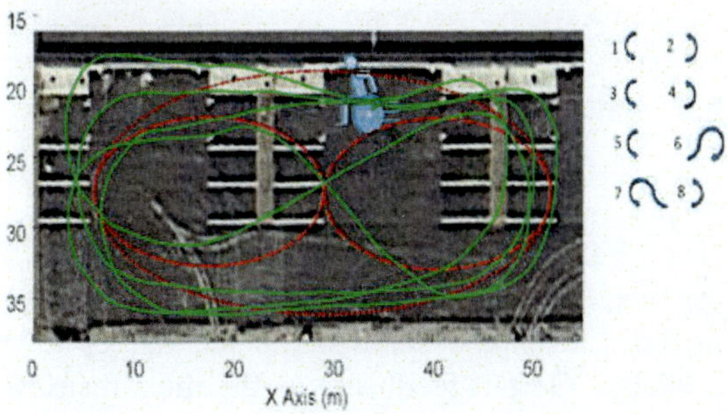

Fig. 6.19 In red: proposed path; in green: 2 oval-shaped trajectories and one eight-shaped trajectory performed by user A under assistive control

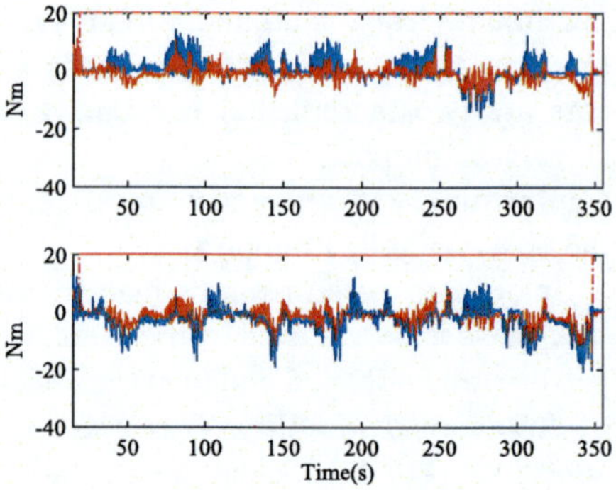

Fig. 6.20 User's torque (blue), estimated torque (orange), mode 0 for manual, 1 for assistance (red dashed line). At the top for the right torque, in the bottom for the left torque

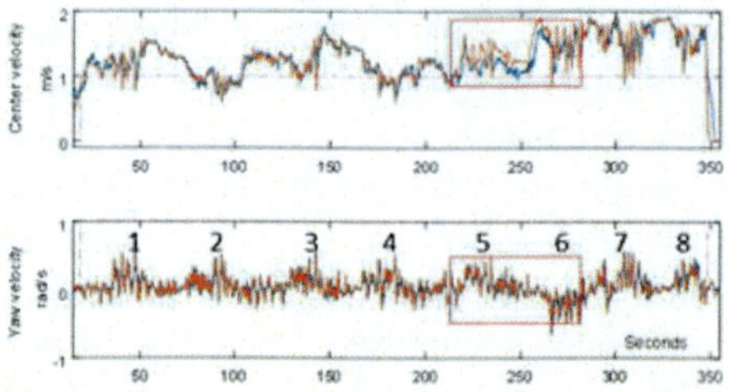

Fig. 6.21 Centre velocity (top), yaw velocity (bottom). In blue the reference generated via the strategy, in red the real outputs. The 8 actions are marked

the strategy reduces the reference center velocity ω_r and generates the rotation speed reference ϕ_r assisting the user to achieve her/his goal. To brake or stop the wheelchair (except an emergency stop, device available via a red stop button), the user brakes both wheels.

Based on the reference signals generated by the strategy in Fig. 6.21 (blue colour), the motor assistance torques are presented in Fig. 6.22. User A was able to accomplish the requested task aided by the proposed assisting algorithm. Importantly, the strategy provided no contra-intuitive behaviours (inappropriate detection of direction, misunderstanding the "intentions") and the user's feedback was a friendly assistance behaviour. Many trials have been conducted to test the manoeuverability of the proposed strategy and its robustness with respect to the mass and road conditions. The reader can refer to the link http://rocon.utcluj.ro/files/pwa_demo.mp4 for more details on some recorded trials.

6.7 Summary and Conclusions

A variety of results modifying controller schemes have been discussed in this chapter. Some of them are minor performance specifications such as constraints on the input/output or disturbance rejection; some others require designing observers in order to extend stabilisation techniques to the case where the states are not available. Output feedback, as the most restricted of the problems, has been pre-

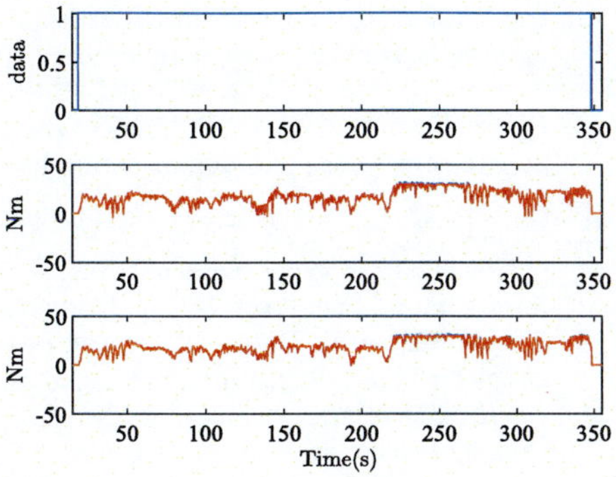

Fig. 6.22 Manual mode 0 and assistance mode 1 (top), motor torque (blue), and saturated motor torque (orange). Right motor (middle), and left motor (bottom)

sented in the last section for LFT models as their structure is better prepared for such a demanding task. The presented material only covers the basics as, naturally, much more up-to-date results on these subjects can be found in the given references.

References

1. Wang, H., Tanaka, K., Griffin, M.: An approach to fuzzy control of nonlinear systems: stability and design issues. IEEE Trans. Fuzzy Syst. **4**(1), 14–23 (1996)
2. Kim, E., Lee, H.: New approaches to relaxed quadratic stability condition of fuzzy control systems. IEEE Trans. Fuzzy Syst. **8**(5), 523–534 (2000)
3. Bergsten, P.: Observers and Controllers for Takagi–Sugeno Fuzzy Systems. Ph.D. thesis, Örebro University (2001)
4. Liu, X., Zhang, Q.: New approaches to H_∞ controller designs based on fuzzy observers for T–S fuzzy systems via LMI. Automatica **39**(9), 1571–1582 (2003)
5. Liu, X., Zhang, Q.: Approaches to quadratic stability conditions and H_∞; control designs for T–S fuzzy systems. IEEE Trans. Fuzzy Syst. **11**(6), 830–839 (2003)
6. Tanaka, K., Ikeda, T., Wang, H.: Fuzzy regulators and fuzzy observers: relaxed stability conditions and LMI-based designs. IEEE Trans. Fuzzy Syst. **6**(2), 250–265 (1998)
7. Lendek, Z., Guerra, T., Babuška, R., De Schutter, B.: Stability Analysis and Nonlinear Observer Design Using Takagi–Sugeno Fuzzy Models: volume 262 of Studies in Fuzziness and Soft Computing. Springer, Berlin (2010)
8. Chadli, M., Guerra, T.: LMI solution for robust static output feedback control of discrete Takagi–Sugeno fuzzy models. IEEE Trans. Fuzzy Syst. **20**(6), 1160–1165 (2012)
9. Tanaka, K., Wang, H.O.: Fuzzy Control System Design and Analysis: A Linear Matrix Inequality Approach. Wiley, New York (2001)
10. Mozelli, L.A., Palhares, R.M., Souza, F.O., Mendes, E.M.A.M.: Reducing conservativeness in recent stability conditions of TS fuzzy systems. Automatica **45**(6), 1580–1583 (2009)
11. Aguiar, B., Márquez, R., Bernal, M.: An LMI-based global non-quadratic observer design via Takagi–Sugeno models and Levant's robust differentiators. In: American Control Conference (ACC), pp. 119–124. IEEE, New York (2016)
12. Kerkeni, H., Guerra, T.M., Lauber, J.: Controller and observer designs for discrete TS models using an efficient Lyapunov function. In: Proceedings of the IEEE International Conference on Fuzzy Systems, pp. 1–7. IEEE, New York (2010)

13. Estrada-Manzo, V., Guerra, T.M., Lendek, Z., Pudlo, P.: Discrete-time Takagi–Sugeno descriptor mdoels: controller design. In: Proceedings of the IEEE International Conference on Fuzzy Systems, pp. 1–5. IEEE, New York (2014b)
14. Guerra, T.M., Guelton, K., Delprat, S.: A class of nonlinear observers in descriptor form: LMI based design with applications in biomechanics. In: Proceedings of the Workshop IFAC/AFNC (2004)
15. Guelton, K., Delprat, S., Guerra, T.-M.: An alternative to inverse dynamics joint torques estimation in human stance based on a Takagi–Sugeno unknown-inputs observer in the descriptor form. Control. Eng. Pract. **16**(12), 1414–1426 (2008)
16. Estrada-Manzo, V., Guerra, T. M., and Lendek, Zs. (2014a). An LMI approach for observer design for Takagi–Sugeno descriptor models. In *Proc. International Conference on Automation, Quality and Testing, Robotics*, pages 1–5, Cluj-Napoca, Rumania
17. Guerra, T., Estrada-Manzo, V., Lendek, Z.: Observer design for Takagi–Sugeno descriptor models: An LMI approach. Automatica **52**, 154–159 (2015)
18. Guerra, T. M., Bernal, M., Kruszewski, A., and Afroun, M. (2007). A way to improve results for the stabilization of continuous-time fuzzy descriptor models. In *46th IEEE Conference on Decision and Control*, pages 5960–5964
19. Estrada-Manzo, V., Lendek, Zs., and Guerra, T. M. (2014c). Discrete-time Takagi–Sugeno descriptor models: observer design. In *Proc. 19th World IFAC Congress*, pages 1–5, Cape Town, South Africa
20. Estrada-Manzo, V., Lendek, Z., Guerra, T.M.: Generalized LMI observer design for discrete-time nonlinear descriptor models. Neurocomputing **182**, 210–220 (2016)
21. Luenberger, D.G.: Observers for multivariable systems. IEEE Trans. Autom. Control **11**(2), 190–197 (1966)
22. Palm, R. and Bergsten, P. (2000). Sliding mode observer for a Takagi–Sugeno fuzzy system. In *Proceedings of the 9th IEEE International Conference on Fuzzy Systems*, volume 2, pages 665–670, San Antonio, TX, USA
23. Oudghiri, M., Chadli, M., and El Hajjaji, A. (2007). Lateral vehicle velocity estimation using fuzzy sliding mode observer. In *Proceedings of the Mediterranean Conference on Control and Automation*, pages 1–6, Athens, Greece
24. Palm, R. and Driankov, D. (1999). Towards a systematic analysis of fuzzy observers. In *Proceedings of the 18th International Conference of the North American Fuzzy Information Processing Society*, pages 179–183, New York, NY, USA
25. Bergsten, P., Palm, R.: Thau-Luenberger observers for TS fuzzy systems. In: The Ninth IEEE International Conference on Fuzzy Systems, pp. 671–676. San Antonio, TX, USA (2000)
26. Ichalal, D., Marx, B., Ragot, J., and Maquin, D. (2008). Design of observers for Takagi–Sugeno systems with immeasurable premise variables: an L_2 approach. In *17th World Congress IFAC*, pages 2768–2773
27. Ichalal, D., Marx, B., Ragot, J., Maquin, D.: State estimation of Takagi–Sugeno systems with unmeasurable premise variables. IET Control Theory Applications **4**(5), 897–908 (2010)
28. Guerra, T., Márquez, R., Kruszewski, A., Bernal, M.: H infinity LMI-based observer design for nonlinear systems via Takagi–Sugeno models with unmeasured premise variables. IEEE Trans. Fuzzy Syst. **26**(3), 1498–1509 (2018)
29. Quintana, D., Estrada-Manzo, V., and Bernal, M. (2020). An exact handling of the gradient for overcoming persistent problems in nonlinear observer design via convex optimization techniques. *Fuzzy Sets and Systems*
30. Uang, H.-J. and Chen, B.-S. (2000). Fuzzy decentralized controller and observer design for nonlinear interconnected systems. In *Proceedings of the Ninth IEEE International Conference on Fuzzy Systems*, volume 2, pages 945–948, San Antonio, TX, USA
31. Tseng, C.-S.: A novel approach to H_∞ decentralized fuzzy-observer-based fuzzy control design for nonlinear interconnected systems. IEEE Trans. Fuzzy Syst. **16**(5), 1337–1350 (2008)
32. Yoneyama, J., Nishikawa, M., Katayama, H., Ichikawa, A.: Design of output feedback controllers for Takagi–Sugeno fuzzy systems. Fuzzy Sets Syst. **121**(1), 127–148 (2001)

33. Guerra, T.M., Kruszewski, A., Vermeiren, L., Tirmant, H.: Conditions of output stabilization for nonlinear models in the Takagi–Sugeno's form. Fuzzy Sets Syst. **159**, 1248–1259 (2006)
34. Ma, X.-J., Sun, Z.-Q., He, Y.-Y.: Analysis and design of fuzzy controller and fuzzy observer. IEEE Trans. Fuzzy Syst. **6**(1), 41–51 (1998)
35. Yoneyama, J., Nishikawa, M., Katayama, H., Ichikawa, A.: Output stabilization of Takagi–Sugeno fuzzy systems. Fuzzy Sets Syst. **111**(2), 253–266 (2000)
36. Boyd, S., El Ghaoui, L., Féron, E., and Balakrishnan, V. (1994). *Linear Matrix Inequalities in System and Control Theory*. Studies in Applied Mathematics. Society for Industrial and Applied Mathematics, Philadelphia, PA, USA
37. Skogestad, S., Postlethwaite, I.: Multivariable feedback control: analysis and design, vol. 2. Wiley, New York (2007)
38. Apkarian, P., Gahinet, P.: A convex characterization of gain-scheduled h_∞ controllers. IEEE Trans. Autom. Control **40**(5), 853–864 (1995)
39. Scherer, C.W.: LPV control and full block multipliers. Automatica **37**(3), 361–375 (2001)
40. Tuan, H.D., Apkarian, P., Narikiyo, T., Kanota, M.: New fuzzy control model and dynamic output feedback parallel distributed compensation. IEEE Trans. Fuzzy Syst. **12**(1), 13–21 (2004)
41. Wu, F., Dong, K.: Gain-scheduling control of LFT systems using parameter-dependent Lyapunov functions. Automatica **42**(1), 39–50 (2006)
42. Potten, Y., Seelen, H., Drukker, J., Reulen, J., Drost, M.: Postural muscle responses in the spinal cord injured persons during forward reaching. Ergonomics **42**(9), 1200–1215 (1999)
43. Grangeon, M., Gagnon, D., Gauthier, C., Jacquemin, G., Masani, K., Popovic, M.: Effects of upper limb positions and weight support roles on quasi-static seated postural stability in individuals with spinal cord injury. Gait & posture **36**(3), 572–579 (2012)
44. Bergmark, A.: Stability of the lumbar spine: a study in mechanical engineering. Acta Orthop. Scand. **60**(sup230), 1–54 (1989)
45. Blandeau, M. (2018). *Modélisation et caractérisation de la stabilité en position assise chez les personnes vivant avec une lésion de la moelle épinière*. PhD thesis
46. Blandeau, M., Guerra, T., and Pudlo, P. (2020). Application of a ts unknown input observer for studying sitting control for people living with spinal cord injury. In *Control Applications for Biomedical Engineering Systems*, pages 169–195. Elsevier
47. Guerra, T., Blandeau, M., Nguyen, A., Srihi, H., Dequidt, A.: Stabilizing unstable biomechanical model to understand sitting stability for persons with spinal cord injury. Berlin, Gemany. Accepted, In IFAC World Congress (2020)
48. Bjerkefors, A., Carpenter, M., Thorstensson, A.: Dynamic trunk stability is improved in paraplegics following kayak ergometer training. Scandinavian journal of medicine & science in sports **17**(6), 672–679 (2007)
49. Mohammad, S., Guerra, T. M., and Pudlo, P. (2018). Method and device assisting with the electric propulsion of a rolling system, wheelchair kit comprising such a device and wheelchair equipped with such a device. US Patent 10,028,871
50. Feng, G., Guerra, T.M., Mohammad, S., Busoniu, L.: Observer-based assistive control design under time-varying sampling for power-assisted wheelchairs. IFAC-PapersOnLine **51**(10), 151–156 (2018)
51. Feng, G., Guerra, T. M., Busoniu, L., and Mohammad, S. (2021). Robust observer-based tracking control under actuator constraints for power-assisted wheelchairs (under review). *IFAC-Control Engineering Practice*
52. Feng, G. (2019). *Mobility aid for the disabled using unknown input observers and reinforcement learning*. PhD thesis
53. Tsai, M. C. and Hsueh, P. W. (2012). Synchronized motion control for 2d joystick-based electric wheelchair driven by two wheel motors. In *2012 IEEE/ASME International Conference on Advanced Intelligent Mechatronics (AIM)*, pages 702–707. IEEE

Chapter 7
Conclusions and Perspectives

This book has been subject to a pair of somehow opposed forces: on the one hand, it acknowledges its fuzzy control origins and its strong relationship with the LPV community; on the other hand, it focuses on model-based nonlinear control via convex optimisation, thus getting rid of most "approximate" perspectives. The direct Lyapunov method and a variety of convex models lie at the core of its developments. Admittedly, it fell short from integrating their driving forces, but its main objective—i.e. drawing conclusions on nonlinear systems via exact convex representations and convex optimisation tools—has been attained in a variety of ways, clearly illustrating the maturity and challenges of this legitimate framework.

Nevertheless, the reader must be warned of a variety of issues. Within the control community, there is not enough understanding about convex optimisation methods: the key linear characteristic of matrix inequalities and its relationship with its solution being decidable in polynomial time is usually dismissed, leading to non-convex solutions being presented as LMIs or as equally valid "alternatives"; the process of rewriting nonlinear expressions as convex sums is often misinterpreted as approximate or LPV; the possibility of systematically calculating gains or Lyapunov functions via numerical tools is disregarded in favour of hardly applicable analytical methods which require good doses of ingenuity on the part of the designer. At the same time, the fuzzy and LPV communities have not made easy legitimating convex modelling and optimisation as a solid framework for nonlinear system analysis and design: the heuristic rule-based setup of fuzzy expert knowledge and the rich family of approximate convex models have often being mixed up with exact rewriting of nonlinear expressions and formal Lyapunov-based developments, paying little attention to the relevance of the specific dependence of membership functions on parameters, time, or states. In this regard, the present book constitutes an effort to "close the gap" between the convex modelling and optimisation framework widely used in fuzzy and LPV communities and rigorous nonlinear control techniques.

As presented along the previous pages, given a continuous- or discrete-time nonlinear model of a plant, there are several options for convex modelling

M. Bernal et al., *Analysis and Synthesis of Nonlinear Control Systems*, Studies in Systems, Decision and Control 408, https://doi.org/10.1007/978-3-030-90773-0_7

(TS, descriptor, polynomial, LFT), a variety of Lyapunov functions to be employed for analysis and design (quadratic, convex, piecewise), many ways of deriving sufficient LMI or SOS conditions with different levels of conservatism and computational burden (matrix properties, relaxations), and a number of performance specialisations addressing robustness, output feedback, etc. The book has left out the increasing amount of applications of exact convex modelling techniques in specific areas of nonlinear control, e.g. singular systems, fractional-order models, sliding modes, output regulation, etc., where their appearance has provided insightful perspectives, not only systematicness of calculations.

What remains ahead for convex optimisation-based analysis and design of nonlinear systems? Evidently, we will continue to witness the "convexification" of different areas of nonlinear control, some of them more amenable to LMI- or SOS-based design than others; it might be harder to find breakthroughs within the "pure" TS community as most of its developments are now only incremental, but of course they cannot be discarded. Any novelty in the numerical methods employed to solve LMI and SOS conditions—like those arising from the use of interior-point methods thirty years ago—might be beneficial for the framework addressed in this book. We hope young researchers to be inspired to go further, well beyond what has been described so far.

Appendix A
Useful Matrix and Norm Results

In this appendix, some definitions and properties regarding matrices and norms (including those for signals and systems) are outlined; most of the material on matrices comes from Lancaster and Tismenetsky [1]; the rest can be found in Boyd et al. [2], and Scherer and Weiland [3]. Some of these concepts are not directly used, but serve as the basis for further used properties and lemmas.

A.1 Column, Row and Null Spaces

Definition A.1 The column space of a matrix $A \in \mathbb{R}^{m\times n}$, denoted as $\mathscr{C}(A)$, is the set of all linear combinations of its columns, i.e. the set of all $\boldsymbol{y} \in \mathbb{R}^m$ expressed as $\boldsymbol{y} = A\boldsymbol{x}$ where $\boldsymbol{x}$ is a vector of n arbitrary coefficients. The column space is a subspace of $\mathbb{R}^m$.

Definition A.2 The row space of a matrix $A \in \mathbb{R}^{m\times n}$, denoted as $\mathscr{R}(A)$, is the set of all linear combinations of the rows; it is a subspace of $\mathbb{R}^n$, and it is isomorph to the column space of A^T, i.e. $\mathscr{C}(A^T)$.

Definition A.3 The null space (also known as the kernel) of a matrix $A \in \mathbb{R}^{m\times n}$, denoted as $\mathscr{N}(A)$ or $\ker A$, is the set of all $\boldsymbol{y} \in \mathbb{R}^n$ such that $A\boldsymbol{y} = 0$.

Note that the null space is the set of vectors orthogonal to all rows of A (columns of A^T): $\mathscr{N} \perp \mathscr{C}(A^T)$. Also, $\mathscr{C}(A) \cup \mathscr{N}(A^T) = \mathbb{R}^m$.

Definition A.4 The dimension of the subspace $\mathscr{C}(A)$ is known as the *rank* of matrix A. It is equal to the dimension of the row space and to the dimension of the biggest non-zero[1] minor (determinant of square submatrix).

[1] This definition is not useful for numeric calculations.

M. Bernal et al., *Analysis and Synthesis of Nonlinear Control Systems*, Studies in Systems, Decision and Control 408, https://doi.org/10.1007/978-3-030-90773-0

Proposition A.1 *A* solution *to the system of equations* $A\boldsymbol{x} = \boldsymbol{b}$, *for* $\boldsymbol{b} \in \mathscr{C}(A)$, *can be expressed as* $\boldsymbol{x}_0 + \boldsymbol{x}_h$, *where* $\boldsymbol{x}_0$ *is a* particular *solution and* $\boldsymbol{x}_h$ *is any vector in* $\mathscr{N}(A)$ *(solution of* $A\boldsymbol{x} = 0$*). If* $\boldsymbol{b} \notin \mathscr{C}(A)$, *the system of equations has no solution.*

MATLAB® : Related commands are `null,orth,rank,det`.

A.2 Matrix Inversion

Definition A.5 The *inverse* of a matrix $A \in \mathbb{R}^{n\times n}$, denoted as A^{-1}, is a square matrix of the same dimensions as A, such that $AA^{-1} = I$ and $A^{-1}A = I$. A square matrix A with non-zero determinant allows the calculation of its inverse, A^{-1} as

$$A^{-1} = \frac{1}{\det A}\text{adj}A \qquad (\text{adj}A)_{ij} = (-1)^{i+j}\det A^{ji}, \tag{A.1}$$

where A^{ji} is a submatrix formed by deleting the row j and the column i from A. The determinant of A^{-1} is $1/\det(A)$. These matrices are called *invertible*.

Definition A.6 A matrix, U, is unitary if it verifies

$$U^* = U^{-1},$$

where superscript * stands for complex conjugate transpose (transpose in the real case). In that case, all rows have norm 1 and are orthogonal to the rest.[2]

Definition A.7 If a matrix $A \in \mathbb{R}^{m\times n}$ with $m \geq n$ has rank n (full column rank), the matrix

$$^{\dagger}A = (A^TA)^{-1}A^T \tag{A.2}$$

fulfils

$$^{\dagger}AA = I_{n\times n}.$$

Matrix $^{\dagger}A$, of dimensions $n \times m$, is denoted as the *left pseudoinverse* of A.

Given a system of equations, $A\boldsymbol{x} = \boldsymbol{y}$, under full column rank assumptions, the value of x minimising the least squares error criteria $J = \|A\boldsymbol{x} - \boldsymbol{y}\|_2$ (see norm definitions in Appendix A.7) with more equations than unknowns is

$$\boldsymbol{x}_{opt} = {}^{\dagger}A\boldsymbol{y}.$$

Definition A.8 If a matrix $A \in \mathbb{R}^{m\times n}$ with $m \leq n$ has rank m (full row rank), matrix

[2] In the following, mostly matrices with real elements will be considered. To extend the results to the complex case, any transposition must be replaced by conjugate transpose.

$$A^\dagger = A^T(AA^T)^{-1} \tag{A.3}$$

fulfils

$$AA^\dagger = I_{m\times m}.$$

So matrix $A^\dagger$, with dimensions $n \times m$, is denoted as *right pseudoinverse*.

In this case, a system, $Ax = y$, has more unknowns than equations, which means there are infinite solutions. The solution with minimum norm is

$$x_{opt} = A^\dagger y.$$

If $A_{m\times n}$ has rank less than $\min(m, n)$, no left or right pseudoinverse exists. A more general definition (Moore–Penrose pseudoinverse) exists.

MATLAB® : Related commands are `inv,pinv`.

A.3 Eigenvalues and Eigenvectors

Definition A.9 A possibly complex number $\lambda \in \mathbb{C}$ is an *eigenvalue* of a square matrix $A \in \mathbb{R}^{n\times n}$ if $Ax = \lambda x$ holds for a vector $x \neq 0$. Such vector is called an *eigenvector* of A.

The eigenvalues of a square matrix A can be found by solving the equation

$$\det(A - \lambda I) = 0,$$

where the left-hand side is a polynomial in λ (assumed to be a complex variable) known as the *characteristic polynomial* of A, and the equation above is named *characteristic equation* of A. There exist n (possibly repeated) eigenvalues. Diagonal and triangular matrices have the diagonal entries as eigenvalues.

Eigenvectors for different eigenvalues are linearly independent. That may not be the case for repeated eigenvalues. If there are n linearly independent eigenvectors, a matrix T can be formed by juxtaposing them in column form so that

$$AT = T\Lambda,$$

where Λ is a diagonal matrix, so $\Lambda = T^{-1}AT$. In that case, it is said that A is diagonalisable. If matrix T cannot be formed with independent eigenvalues, the so-called Jordan upper triangular canonical form can be obtained.

Definition A.10 A matrix A is *positive-definite*, denoted as $A > 0$, if for all $x \neq 0$ it holds $x^T Ax > 0$; this translates into symmetric matrix A having all its eigenvalues real and positive. Similarly, a *positive semidefinite* matrix A, denoted as $A \geq 0$, is such that $x^T Ax \geq 0$ for all $x \neq 0$ which is equivalent to being symmetric and having its eigenvalues greater or equal to 0. Similar definitions can be made for a *negative-definite* or *negative semidefinite* matrix A.

Note that positive-definiteness as well as the other 3 categories imply that matrix A is symmetric. Therefore, negative-definiteness should not be confused with a *Hurwitz matrix*, i.e. a possibly non-symmetric matrix whose eigenvalues have a negative real part.

Proposition A.2 *A real symmetric matrix has all eigenvalues in the real line, and all eigenvectors have real components.*

Indeed, $\boldsymbol{u}^* A^H = \boldsymbol{u}^* \lambda^*$, so $\boldsymbol{u}^* A = \lambda^* \boldsymbol{u}^*$. On the one hand, $\boldsymbol{u}^* A \boldsymbol{u} = \boldsymbol{u}^* \lambda \boldsymbol{u}$. On the other hand, $(\boldsymbol{u}^* A)\boldsymbol{u} = \lambda^* \boldsymbol{u}^* \boldsymbol{u}$. Hence, $\lambda = \lambda^*$, thus it is real and, obviously, as $(\lambda \boldsymbol{v}) = \lambda(\text{conj}\boldsymbol{v}) = \text{conj}(A\boldsymbol{v}) = \text{conj}(A)\text{conj}(\boldsymbol{v}) = A\text{conj}(\boldsymbol{v})$ where if conj denotes the complex conjugate, then if $\boldsymbol{v}$ is an eigenvector then so it is $\text{conj}(\boldsymbol{v})$, and hence so it is $\boldsymbol{v} + \text{conj}(\boldsymbol{v})$ so there exists a real eigenvector associated to λ.

Proposition A.3 *Eigenvectors corresponding to distinct eigenvalues are orthogonal. All real symmetric matrices are diagonalisable via an orthogonal matrix.*

Indeed, $A\boldsymbol{u}_1 = \lambda_1 \boldsymbol{u}_1$, $A\boldsymbol{u}_2 = \lambda_2 \boldsymbol{u}_2$; then,

$$\boldsymbol{u}_2^T A \boldsymbol{u}_1 = \boldsymbol{u}_2^T (\lambda_1 \boldsymbol{u}_1) = (A\boldsymbol{u}_2)^T \boldsymbol{u}_1 = \lambda_2 \boldsymbol{u}_2^T \boldsymbol{u}_1.$$

If $\lambda_1 \neq \lambda_2$ then forcefully $\boldsymbol{u}_2^T \boldsymbol{u}_1 = 0$. If an eigenvalue has a multiplicity greater than one, we can always find an orthonormal basis of the null space of $(\lambda A - I)$.

Proposition A.4 $B = T^{-1} A T$ *has the same eigenvalues as A, for any invertible* T.

Indeed, $\det(\lambda I - T^{-1} A T) = \det(T^{-1}(\lambda I - A)T) = \det(\lambda I - A)$.

Proposition A.5 *The product of the eigenvalues is equal to the determinant.*

Proposition A.6 *The sum of the eigenvalues is the trace of A (sum of diagonal elements).*

Proposition A.7 *The matrix $A + cI$, c being a real or complex scalar, has as eigenvalues $\lambda_i + c$, and the same eigenvectors.*

Definition A.11 As all symmetric matrices have real eigenvalues, the inertia of a matrix P, denoted as $\text{in}(P)$, is defined as the triplet $(\text{in}_-(P), \text{in}_0(P), \text{in}_+(P))$ with the number of negative, zero, and positive eigenvalues. If columns of matrix S form a basis of a subspace $\mathcal{S}$, $\text{in}(P|_{\mathcal{S}}) := \text{in}(S^T P S)$.

Proposition A.8 *If $\boldsymbol{v}$ is an eigenvector of a symmetric matrix P associated with eigenvalue λ, then $\boldsymbol{v}$ is an eigenvector of P^{-1} associated with λ^{-1}.*

Indeed, $P = V^{-1}\Lambda V$, $P^{-1} = V^{-1}\Lambda^{-1}V$, with Λ diagonal and V being the matrix of eigenvectors (which is, actually, orthogonal).

From the above propositions and definition, the following is evident.

Proposition A.9 *If $in_0(P) = 0$, then $in(P) = in(P^{-1})$.*

Proposition A.10 *From an expression $\mathbf{x}^T P\mathbf{x}$ and the change of variable $\mathbf{x} = T\mathbf{z}$, we have the following:*

1. *If T is non-singular, $in(T^T PT) = in(P)$.*
2. *If $P > 0$ then $T^T PT \geq 0$ for any arbitrary matrix T. In particular, if T is full column rank, $T^T PT > 0$.*

Let us consider a system of ordinary linear differential equations:

$$\frac{d\mathbf{x}}{dt} = A\mathbf{x}. \tag{A.4}$$

If we look for an exponential component in the response, as in scalar ordinary equations, $\mathbf{x}(t) = e^{\lambda t}\mathbf{v}$, then $\dot{\mathbf{x}} = e^{\lambda t}\lambda\mathbf{v}$; hence, as $e^{\lambda t}$ is a non-zero scalar for every t:

$$e^{\lambda t}\lambda\mathbf{v} = Ae^{\lambda t}\mathbf{v} \quad \Rightarrow \quad \lambda\mathbf{v} = A\mathbf{v}$$

so λ must be an eigenvalue of A and $\mathbf{v}$ must be an eigenvector[3]: *the exponential terms in solutions of* (A.4) *are the eigenvalues of* A.

A.4 Singular Values and Matrix Gains

The problem of determining the maximum and minimum "gains" of a matrix is related to an eigenvalue problem, when working with Euclidean vector norms (see Appendix A.7).

Given $\mathbf{y} = A\mathbf{x}$, to determine the maximum and minimum gains, and the directions in which they occur, $\|\mathbf{x}\| = 1$ will be assumed (as the effect of the matrix for any $\mathbf{x}$ will be proportional to that on its unit vector $\mathbf{x}/\|\mathbf{x}\|$). So, the problem is stated as follows:

$$\text{maximise } \|\mathbf{y}\|^2 = \|A\mathbf{x}\|^2 = \mathbf{x}^T A^T A\mathbf{x},\ \mathbf{x} \in \mathbb{R}^n,\ \text{subject to } \mathbf{x}^T\mathbf{x} = 1.$$

The constrained optimisation problem can be transformed into an unconstrained one by using one Lagrange multiplier λ:

$$J = \mathbf{x}^T A^T A\mathbf{x} + \lambda(1 - \mathbf{x}^T\mathbf{x}). \tag{A.5}$$

[3] The same happens for a DT system, $\mathbf{x}_{k+1} = A\mathbf{x}_k$, and a solution in the form $\mathbf{x}_k = \lambda^k\mathbf{v}$.

So, differentiating with respect to $\boldsymbol{x}$, the result is the eigenvalue problem:

$$A^T A\boldsymbol{x} = \lambda \boldsymbol{x}. \tag{A.6}$$

So the maximum and minimum gains occur at some of the *eigenvectors* of $A^T A$, and the eigenvalues are the associated Lagrange multipliers. As $A^T A$ is symmetric, positive semidefinite ($\|A\boldsymbol{x}\| = \boldsymbol{x}^T A^T A\boldsymbol{x} \geq 0$), all its eigenvalues are *real*, nonnegative, and the eigenvectors can be chosen to be an *orthonormal* basis.[4]

It can be shown that for any eigenvector $\boldsymbol{x}_i$, $i = 1, \ldots, n$ of $A^T A$, the output norm $\|\boldsymbol{y}_i\| = \|A\boldsymbol{x}_i\|$ attained is $\sqrt{\boldsymbol{x}_i^T \lambda_i \boldsymbol{x}_i}$=$\sqrt{\lambda_i}\|\boldsymbol{x}_i\|$. So, the maximum gain corresponds to the eigenvector associated with the maximum eigenvalue, and the analogue for the minimum one. Let us denote as $\lambda_1 \geq \lambda_2 \geq \cdots \geq \lambda_n$ the eigenvalues in decreasing sequence, so $\boldsymbol{x}_1$ is the most amplified input direction and $\boldsymbol{x}_n$ is the least.

The intermediate vectors are *saddle* points of the optimisation problem. Furthermore, λ_2 maximises the gain of A subject to $\boldsymbol{x}^T\boldsymbol{x} = 1$ and $\boldsymbol{x}^T\boldsymbol{x}_1 = 0$ (the maximum gain on directions orthogonal to the $\boldsymbol{x}_1$). Carrying on, λ_i maximises the gain of the matrix in directions orthogonal to all $\boldsymbol{x}_j$ for $j < i$.

As the eigenvectors form an orthonormal basis, also the outputs $\boldsymbol{y}_i = A\boldsymbol{x}_i$ and $\boldsymbol{y}_j = A\boldsymbol{x}_j$ are orthogonal as $\boldsymbol{y}_i^T\boldsymbol{y}_j = \boldsymbol{x}_i^T A^T A\boldsymbol{x}_j = \boldsymbol{x}_i\lambda_j\boldsymbol{x}_j = 0$.

So, the action of A on an arbitrary $\boldsymbol{x}$ can be decomposed as follows:

1. Decompose (project) $\boldsymbol{x}$ on the basis formed by the eigenvectors $\boldsymbol{x}_i$. Let us denote as V the matrix whose rows are $\boldsymbol{x}_i$.
2. Scale each component by $\sqrt{\lambda_i}$. If $\lambda_i = 0$, the component is irrelevant.
3. Rotate the result to be aligned with $\boldsymbol{y}_i$ and add up all components.

The result is that matrix A can be written as

$$A = U\Sigma V^T, \tag{A.7}$$

where U and V are orthonormal rotation matrices and Σ is diagonal. Furthermore, if A is $m \times n$, the columns of V $(n \times n)$ are the eigenvectors of $A^T A$, Σ $(m \times n)$ contains the (positive) square root of the eigenvalues ($\sigma_i = \sqrt{\lambda_i}$ are denoted as *singular values*) in the diagonal, and columns of U $(m \times m)$ are the eigenvectors of AA^T (by a similar argument with $A^T = V\Sigma U^T$), as the determinants of U and V are 1, $\det(A) = \det(\Sigma)$.

Equation (A.7) is denoted as the singular-value decomposition (SVD). Usually, the maximum and minimum singular values are written as

$$\overline{\sigma}(A) = \sqrt{\lambda_1}; \qquad \underline{\sigma}(A) = \sqrt{\lambda_n}$$

so the following inequalities hold:

$$\underline{\sigma}(A)\|\boldsymbol{x}\| \leq \|A\boldsymbol{x}\| \leq \overline{\sigma}(A)\|\boldsymbol{x}\|. \tag{A.8}$$

[4] Multiplication by an orthonormal matrix can be considered as a change of basis by rotation.

Since an orthogonal (rotation) matrix has its transpose equal to its inverse, the inverse of a matrix A can be expressed as

$$A^{-1} = V\Sigma^{-1}U^T \tag{A.9}$$

so its singular values are the inverse of those of A. In particular,

$$\overline{\sigma}(A^{-1}) = 1/\underline{\sigma}(A), \tag{A.10}$$

so a matrix is invertible if all its singular values are non-zero. Left and right pseudoinverses can also be expressed in terms of the SVD of A.

The maximum singular value is a *matrix norm*, in the sense of the definitions of Sect. A.7. It satisfies some norm properties, in particular

$$\overline{\sigma}(AB) \le \overline{\sigma}(A)\overline{\sigma}(B); \tag{A.11}$$

so, unless otherwise stated, the following notation will be assumed:

$$\|A\| = \overline{\sigma}(A).$$

The minimum gain, $\underline{\sigma}()$, satisfies $\underline{\sigma}(A)\underline{\sigma}(B) \le \underline{\sigma}(AB)$.

Based on the results above, the *condition number* of a matrix A is defined as

$$\gamma(A) = \frac{\overline{\sigma}(A)}{\underline{\sigma}(A)} \ge 1.$$

From (A.10), it is straightforward to prove that $\gamma(A^{-1}) = \gamma(A)$.

Let us now analyse some important singular-value applications.

Noise Amplification

In a system $A\boldsymbol{x} = \boldsymbol{y}$, where A is a regular matrix and $\boldsymbol{y}$ are noise-corrupted experimental data, an estimate on the relative accuracy of $\boldsymbol{x} = A^{-1}\boldsymbol{y}$ can be obtained.

Let us denote as $\delta\boldsymbol{y}$ the "error" (noise) present on the experimental data. Then, $\delta\boldsymbol{x} = A^{-1}\delta\boldsymbol{y}$ and, based on (A.8),

$$\|\delta\boldsymbol{x}\| \le \bar{\sigma}(A^{-1})\|\delta\boldsymbol{y}\|; \qquad \|\boldsymbol{x}\| \ge \underline{\sigma}(A^{-1})\|\boldsymbol{y}\|$$

so dividing the first inequality by $\|\boldsymbol{x}\|$:

$$\frac{\|\delta\boldsymbol{x}\|}{\|\boldsymbol{x}\|} \le \frac{\bar{\sigma}(A^{-1})\|\delta\boldsymbol{y}\|}{\|\boldsymbol{x}\|} \le \frac{\overline{\sigma}(A^{-1})}{\underline{\sigma}(A^{-1})}\frac{\|\delta\boldsymbol{y}\|}{\|\boldsymbol{y}\|} = \gamma(A^{-1})\frac{\|\delta\boldsymbol{y}\|}{\|\boldsymbol{y}\|}. \tag{A.12}$$

The interpretation of this formula is that the relative error on the experimental data may have a worst-case amplification given by the condition number.[5] A similar argument would prove that relative error in $\boldsymbol{x}$ may appear in $\boldsymbol{y}$ amplified by the factor $\gamma(A)$.

Modelling Error

In this case, the issue to be considered is the difference in solutions for $\boldsymbol{u}$ from a nominal model, $\boldsymbol{y} = G\boldsymbol{u}$, and a perturbed one, $\boldsymbol{y} = (G + \delta G)\boldsymbol{u}^*$, denoting as $\delta\boldsymbol{u}$ the difference between solutions, $\boldsymbol{u}^* = (\boldsymbol{u} + \delta\boldsymbol{u})$. Subtracting these expressions, $0 = G\boldsymbol{u} - (G + \delta G)(\boldsymbol{u} + \delta\boldsymbol{u})$, results in

$$\delta\boldsymbol{u} = -G^{-1}\delta G(\boldsymbol{u} + \delta\boldsymbol{u});$$

so, using (A.11) and (A.10),

$$\|\delta\boldsymbol{u}\| \leq \overline{\sigma}(G^{-1})\overline{\sigma}(\delta G)\|\boldsymbol{u} + \delta\boldsymbol{u}\| = \frac{\overline{\sigma}(\delta G)}{\underline{\sigma}(G)}\|\boldsymbol{u} + \delta\boldsymbol{u}\| = \frac{\overline{\sigma}(G)}{\underline{\sigma}(G)}\frac{\overline{\sigma}(\delta G)}{\overline{\sigma}(G)}\|\boldsymbol{u} + \delta\boldsymbol{u}\|$$

so relative modelling error also may get amplified in the results by $\gamma(G)$:

$$\frac{\delta\boldsymbol{u}}{\|\boldsymbol{u} + \delta\boldsymbol{u}\|} \leq \gamma(G)\frac{\|\delta G\|}{\|G\|}. \tag{A.13}$$

Rank Determination

One application of the above result is that, in matrices obtained from noisy or unaccurate experimental data (even round-off errors), no matrix has “exactly” zero determinant or zero singular values.

To determine in practice the reliability of a matrix inversion, the condition number is calculated. If it is very large, it implies that the matrix inversion is unreliable (very high error figures may be expected, from (A.12) and (A.13)).

It can be easily shown that the rank of a matrix is the number of non-zero singular values. This is an operational definition where a *tolerance* can be specified, on the minimum singular value or on the condition number.

Example A.1 This is an example to show that the value of the determinant, by itself, is not a good indication of how close the matrix is to being singular:

[5] The condition number, as defined, implies that the Euclidean vector norm is used in the above calculations. For comparable results with other norms, the condition number is defined as $\gamma(A) = \|A\| \cdot \|A^{-1}\|$ using the suitable induced matrix norm.

```
A = -38.8843    44.4703    92.1813
      1.8504    61.5432    73.8207
     82.1407    79.1937    17.6266
  det(A)  = 857.5811
  svd(A)  =   153.2032   105.2887   0.0532.
```

Condition number is around 3000, i.e. in some directions, the gain is almost three thousand times larger than in other ones so, depending on the reliability of the data, the "practical" rank might be 2. Any control design philosophy relying on that matrix having rank 3 will be quite sensitive to modelling errors. If element $A(1, 1)$ were -38.7042, the matrix would become singular: the allowable relative error for singularity is 0.003 in some elements. This matrix is "singular" (rank 2) if data do not have more than three precise significant figures.

MATLAB® : Related commands are `svd,cond,condeig`.
The reader is referred to Strang [4]; Golub and van Loan [5]; Meyer [6] for further detail on these topics.

Column and Null Space Determination

If all singular values below a tolerance limit are assumed to be zero, then the matrix SVD can be understood as

$$[U_1 U_2]\begin{bmatrix} \Sigma_1 & 0 \\ 0 & 0 \end{bmatrix}[V_1 V_2]^T,$$

where Σ_1 contains the non-zero singular values. In this case, V_2 is a basis of the *null* space, and U_1 is a basis of the *column* space. This is a numerically reliable way of calculating such spaces, used in MATLAB®.

A.5 Properties Usually Employed to Derive LMI Conditions

The Schur Complement

Let us consider a symmetric $m \times m$ matrix

$$P = \begin{bmatrix} P_{11} & P_{12} \\ P_{12}^T & P_{22} \end{bmatrix}, \tag{A.14}$$

where P_{11} is square, size $m_1 \times m_1$, and P_{22} is square, size $m_2 \times m_2$, $m = m_1 + m_2$. If P_{11} is non-singular, then it is easy to see that

$$P = \begin{bmatrix} I_{m_1} & 0 \\ P_{12}^T P_{11}^{-1} & I_{m_2} \end{bmatrix} \begin{bmatrix} P_{11} & 0 \\ 0 & P_{22} - P_{12}^T P_{11}^{-1} P_{12} \end{bmatrix} \begin{bmatrix} I_{m_1} & P_{11}^{-1} P_{12} \\ 0 & I_{m_2} \end{bmatrix} \tag{A.15}$$

by simply carrying out the blockwise operations.

Note also that the leftmost and rightmost matrices are actually transpose of one another, so expression (A.15) is a congruence transformation. Hence, we can state, based on Proposition A.10:

Lemma A.1 *$P > 0$ if and only if $P_{11} > 0$ and $Sch_1(P) = P_{22} - P_{12}^T P_{11}^{-1} P_{12} > 0$. Equivalently, by elementary row and column permutations, $P > 0$ if and only if $P_{22} > 0$ and $Sch_2(P) = P_{11} - P_{12} P_{22}^{-1} P_{12}^T > 0$.*

The above result is helpful in expressing some quadratic nonlinear matrix inequalities as LMIs on larger matrices.

Dualisation

The congruence transformation in (A.15) allows, using the inertia statement in Proposition A.9, proving the following result, which generalises (A.1) for non- positive-definite cases.

Proposition A.11 *For any partitioned matrix P, as in (A.14):*

$$in(P) = in(P_{11}) + in(Sch_1(P)).$$

Also, from (A.15), and the fact that

$$\begin{bmatrix} I & P_{11}^{-1} P_{12} \\ 0 & I \end{bmatrix}^{-1} = \begin{bmatrix} I & -P_{11}^{-1} P_{12} \\ 0 & I \end{bmatrix},$$

we can derive a partitioned-inverse formula based on the Schur complement's inverse:

$$P^{-1} = \begin{bmatrix} I_{m_1} & -P_{11}^{-1} P_{12} \\ 0 & I_{m_2} \end{bmatrix} \begin{bmatrix} P_{11}^{-1} & 0 \\ 0 & (P_{22} - P_{12}^T P_{11}^{-1} P_{12})^{-1} \end{bmatrix} \begin{bmatrix} I_{m_1} & 0 \\ -P_{12}^T P_{11}^{-1} & I_{m_2} \end{bmatrix}, \tag{A.16}$$

which results, denoting $H = \text{Sch}_1(P)^{-1}$, in

$$P^{-1} = \begin{bmatrix} P_{11}^{-1}(I + P_{12} H P_{12}^T P_{11}^{-1}) & -P_{11}^{-1} P_{12} H \\ -H P_{12}^T P_{11}^{-1} & H \end{bmatrix} \tag{A.17}$$

and carrying out similar operations with $H_2 = \text{Sch}_2(P)^{-1}$:

$$P^{-1} = \begin{bmatrix} H_2 & -H_2 P_{12} P_{22}^{-1} \\ (*) & P_{22}^{-1}(I + P_{12}^T H_2 P_{12} P_{22}^{-1}) \end{bmatrix}. \tag{A.18}$$

Hence, using Proposition A.9, we can rewrite Proposition A.11 as $\text{in}(P) = \text{in}(P_{11}) + \text{in}(H)$. Denote now as S the first m_1 columns of the identity matrix of size m, and denote as $S^\perp$ the last m_2 columns (indeed, S and $S^\perp$ are orthogonal). Then $\text{in}(P_{11}) = in(S^T P S)$, and $\text{in}(H) = \text{in}((S^\perp)^T P^{-1} S^\perp)$.

As, given any subspace $\mathcal{S}$ and its orthogonal complement $\mathcal{S}^\perp$, we can carry out a coordinate transformation so that we transform it to the canonical basis in the above paragraph, we can state a key result known as dualisation lemma Scherer [7].

Lemma A.2 (Dualisation) *For any matrix P and any subspace $\mathcal{S}$:*

$$in(P) = in(P|_{\mathcal{S}}) + in(P^{-1}|_{\mathcal{S}^\perp}) \tag{A.19}$$

and, hence, if $in(P) = (m_-, 0, m_+)$ and S is a (column) rank m_- matrix,

$$S^T P S < 0 \Leftrightarrow (S^\perp)^T P^{-1} S^\perp > 0. \tag{A.20}$$

This result allows transforming an LMI in an unknown P which should hold in a reduced-dimension subspace into an LMI in its inverse and the orthogonal subspace where convenient.

Elimination Lemmas

For variables which appear *linearly* in an LMI, we can use the below lemma:

Lemma A.3 (Finsler's lemma) *[8] (Finsler's Lemma) Consider a vector $\boldsymbol{x} \in \mathbb{R}^{n_x}$ and two matrices $Q = Q^T \in \mathbb{R}^{n_x \times n_x}$ and $R \in \mathbb{R}^{m \times n_x}$ such that rank$(R) < n_x$. The two following expressions are equivalent:*

1. $\boldsymbol{x}^T Q \boldsymbol{x} < 0,\ \boldsymbol{x} \in \{\boldsymbol{x} \in \mathbb{R}^{n_x}, \boldsymbol{x} \neq 0, R\boldsymbol{x} = 0\}$.
2. $\exists M \in \mathbb{R}^{m \times n_x}$ *such that* $Q + MR + R^T M^T < 0$.

The following lemma applies to some *quadratic* expressions in a variable X, and is a version of a key result in $\mathcal{H}_\infty$ output feedback synthesis.

Lemma A.4 (Scherer's elimination lemma) *Let P : $in(P) = (m_-, 0, m_+)$; then, the quadratic inequality in X, P*

$$\begin{bmatrix} I_{m_-} & C^T + B^T X^T A \end{bmatrix} P \begin{bmatrix} I_{m_-} \\ A^T X B + C \end{bmatrix} < 0 \tag{A.21}$$

with C of dimensions $m_+ \times m_-$ has a solution if and only if *both of the following conditions hold:*

$$\Psi_B^T [I_{m_-} \ C^T] P \begin{bmatrix} I_{m_-} \\ C \end{bmatrix} \Psi_B < 0 \tag{A.22}$$

$$\Psi_A^T [-C \ I_{m_+}] P^{-1} \begin{bmatrix} -C^T \\ I_{m_+} \end{bmatrix} \Psi_A > 0, \tag{A.23}$$

Ψ_B being a matrix whose columns form a basis of the null space of B and Ψ_A a basis of the null space of A.

The proof that (A.21) implies (A.22) and (A.23) is easy. Indeed, (A.22) results from (A.21) from the fact that $B\Psi_B = 0$. Now, note that

$$\begin{bmatrix} -B^T X^T A - C^T \\ I_{m_+} \end{bmatrix}^{\perp} = \begin{bmatrix} I_{m_-} \\ A^T X B + C \end{bmatrix}. \tag{A.24}$$

Hence, by using Lemma A.2, the original expression (A.21) is then equivalent to

$$\begin{bmatrix} -B^T X^T A - C^T \\ I_{m_+} \end{bmatrix}^T P^{-1} \begin{bmatrix} -B^T X^T A - C^T \\ I_{m_+} \end{bmatrix} > 0 \tag{A.25}$$

so, as $A\Psi_A = 0$ then (A.23) ensues.

The converse (i.e. proving that if (A.22) and (A.23) hold, then there exists X such that (A.21) holds) is substantially more involved, and the reader is referred to Scherer [7] for details. Note that the proof in the cited reference is *constructive* so X can be computed from P and the SVD decomposition of A and B, by solving an eigenvalue–eigenvector problem; details are omitted for brevity.

Matrix Completion

Let us consider the problem of having

$$P = \begin{bmatrix} P_{11} & P_{12} \\ P_{12}^T & P_{22} \end{bmatrix}, \quad P^{-1} = X = \begin{bmatrix} X_{11} & X_{12} \\ X_{12}^T & X_{22} \end{bmatrix}, \tag{A.26}$$

where P is of size $2n \times 2n$ and P_{11} is of size $n \times n$.

The following result is key in gain scheduling and $\mathcal{H}_\infty$ settings in order to build "full" Lyapunov functions and multipliers from some small known "pieces" of them, in particular, assuming P_{11} and X_{11} are known.

Lemma A.5 *Let T be an arbitrarily chosen non-singular matrix, and let $P_{11} - X_{11}^{-1}$ be full rank.*[6] *Then, the inverse of the suitably partitioned matrix:*

[6] if $P_{11} > 0$ and $X_{11} > 0$ come from LMI constraints, the additional constraint below, based on Schur complement, ensures $P_{11} - X_{11}^{-1}$ being full-rank:

$$P = \begin{bmatrix} P_{11} & T \\ T^T & T^T(P_{11} - X_{11}^{-1})^{-1}T \end{bmatrix} \tag{A.28}$$

has X_{11} *as its* $(1, 1)$ *block.*

Proof As $\mathrm{Sch}_2(P) = P_{11} - TT^{-1}(P_{11} - X_{11}^{-1})T^{-T}T^T = X_{11}^{-1}$, applying (A.18) the result is immediate.

Other references such as Apkarian and Gahinet [9] and Tuan et al. [10] propose the matrix completion starting by solving $P_{11}X_{11} + P_{12}X_{12}^T = I$, which arises directly from matrix multiplications given the blocks in (A.26). Later, the matrix is completed from $P_{12}^T X_{11} + P_{22}X_{12}^T = 0$, also arising from (A.26). It is easily seen that, setting $P_{12} = T$, the result is (A.28). Other authors propose solving for $P_{12}X_{12}^T$ from an SVD of $I - P_{11}X_{11} = USV^T = U\sqrt{S} \cdot \sqrt{S}V^T$; this amounts to a particular choice of $T = U\sqrt{S}$.

Other Properties

Property A.1 (Congruence) *Given a matrix* $P = P^T$ *and a full column rank matrix* Q*, it holds that*

$$P > 0 \Rightarrow QPQ^T > 0.$$

Property A.2 (S-procedure) *Consider matrices* $F_i = F_i^T \in \mathbb{R}^{n\times n}$, $\boldsymbol{x} \in \mathbb{R}^n$, *such that* $\boldsymbol{x}^T F_i \boldsymbol{x} \geq 0$, $\forall \boldsymbol{x}$, $i = 1, \ldots, p$, *and the quadratic inequality condition*

$$\boldsymbol{x}^T F_0 \boldsymbol{x} > 0 \tag{A.29}$$

$\boldsymbol{x} \neq 0$. *A sufficient condition for* (A.29) *to hold is there exist* $\tau_i \geq 0$, $i = 1, \ldots, p$, *such that* $F_0 - \sum_{i=1}^{p} \tau_i F_i > 0$.

Property A.3 (Completion of squares) *Given two matrices* X *and* Y *of appropriate sizes and* $Q = Q^T > 0$, *the following inequality holds:*

$$X^T Y + Y^T X \leq X^T Q X + Y^T Q^{-1} Y.$$

Property A.4 *Let* A *and* B *be matrices of appropriate dimensions and ranks, with* $B = B^T > 0$. *Then*

$$(A - B)^T B^{-1}(A - B) \geq 0. \iff A^T B^{-1} A \geq A + A^T - B.$$

$$\begin{bmatrix} P_{11} & I \\ I & X_{11} \end{bmatrix} > 0. \tag{A.27}$$

Also, the full-rank condition can be relieved with some modifications Scherer [7].

A.6 Matrix Exponential

The definition of a matrix exponential is the same as in the exponential of a scalar:

$$\exp(A) = e^A := \sum_{n=0}^{\infty} \frac{A^n}{n!} = I + A + \frac{A^2}{2} + \frac{A^3}{6} + \cdots \tag{A.30}$$

In solving linear differential equations, matrix exponentials such as

$$\exp(At) = e^{At} = \sum_{n=0}^{\infty} \frac{A^n t^n}{n!} = I + At + \frac{A^2 t^2}{2} + \frac{A^3 t^3}{6} + \cdots \tag{A.31}$$

appear, where $t \in \mathbb{R}$ stands for time. The matrix exponential has the following properties:

1. $e^{0_{n\times n}} = I_{n\times n}$.
2. $e^{At_1} e^{At_2} = e^{A(t_1+t_2)}$.
3. $\left(e^{At}\right)^{-1} = e^{-At}$.
4. $\frac{d}{dt}\left(e^{At}\right) = Ae^{At} = e^{At}A$. Indeed,

$$\frac{d}{dt}\left(e^{At}\right) = \sum_{n=1}^{\infty} \frac{A^n\, n t^{n-1}}{n!} = A + \frac{A^2\, 2t}{2} + \cdots = A(I + \frac{A^2\, t}{1} + \frac{A^3\, t^2}{2} + \cdots)$$
$$= \sum_{n=1}^{\infty} \frac{A^n\, t^{n-1}}{(n-1)!} = A \sum_{n=0}^{\infty} \frac{A^n t^n}{n!} = (\sum_{n=0}^{\infty} \frac{A^n t^n}{n!})A = Ae^{At} = e^{At}A. \tag{A.32}$$

5. Based on the previous property, if A is non-singular, $\int_0^t e^{A\tau}\, d\tau = A^{-1}(e^{At} - I)$. The integral also exists for singular A, but it must be carried out with the power series.

Calculation

If the matrix is diagonal, the exponential is just the exponential of the diagonal elements:

$$\exp\left(\begin{bmatrix} c_1 & 0 \\ 0 & c_2 \end{bmatrix}\right) = \begin{bmatrix} e^{c_1} & 0 \\ 0 & e^{c2} \end{bmatrix}. \tag{A.33}$$

If the matrix is diagonalisable, $A = V^{-1}\Lambda V$, as

$$A^2 = V^{-1}\Lambda V V^{-1}\Lambda V = V^{-1}\Lambda^2 V; \quad \ldots \quad A^n = V^{-1}\Lambda^n V, \tag{A.34}$$

it can be shown that

$$e^{At} = V^{-1} e^{\Lambda t} V. \tag{A.35}$$

If the matrix is not diagonalisable, it admits a canonical Jordan form. It can be shown that a Jordan block of dimension m has a matrix exponential given by

$$\exp\left(\begin{bmatrix} \lambda & 1 & 0 \ldots \\ 0 & \lambda & 1 \ldots \\ 0 \ldots & \lambda & 1 \\ 0 \ldots & 0 & \lambda \end{bmatrix}\right) = e^{\lambda t} \times \begin{bmatrix} 1 & t & \ldots & \frac{t^{m-1}}{(m-1)!} \\ 0 & 1 & \ldots & \frac{t^{m-2}}{(m-2)!} \\ \vdots & \vdots & \ddots & \vdots \\ 0 & 0 & \cdots & 1 \end{bmatrix}. \tag{A.36}$$

Calculation of matrix exponentials via a computer must be made by taking a few terms of the power series defining the matrix exponential. However, the *Padé* approximation, $e^{At} = e^{At/2}\left(e^{-At/2}\right)^{-1}$, gives better results, evaluating $e^{At/2}$ and $e^{-At/2}$ by a truncated Taylor series expansion.

MATLAB® : Related commands are `expm`, `expm1`, `expm2`, `expm3`, `funm`.

Differential Equations

The solution of the equation:

$$\dot{\boldsymbol{x}}(t) = A\boldsymbol{x}(t) + B\boldsymbol{u}(t) \tag{A.37}$$

can be expressed in terms of a matrix exponential, as

$$\frac{d}{dt}\left(e^{-At}\boldsymbol{x}\right) = e^{-At}\dot{\boldsymbol{x}} - e^{-At}A\boldsymbol{x} = e^{-At}B\boldsymbol{u}$$

$$\Longleftrightarrow\; e^{-At}\boldsymbol{x}(t) - e^{-A\cdot 0}\boldsymbol{x}(0) = \int_0^t e^{-A\tau}B\boldsymbol{u}\,d\tau$$

so, as $e^0 = I$, multiplying by e^{At} the result is

$$\boldsymbol{x}(t) = e^{At}\boldsymbol{x}(0) + e^{At}\int_0^t e^{-A\tau}B\boldsymbol{u}\,d\tau = e^{At}\boldsymbol{x}(0) + \int_0^t e^{A(t-\tau)}B\boldsymbol{u}\,d\tau.$$

A.7 Signal and System Norms

Definition A.12 A norm is a real function $\|\cdot\| : \mathcal{V} \to \mathbb{R}_+$ where $\mathcal{V}$ is a vector space, satisfying the following conditions:

1. $\|x\| \geq 0 \quad \forall x \in \mathcal{V}$.
2. $\|x\| = 0 \Leftrightarrow x$ is the null vector ($x = 0$).
3. $\|\alpha x\| = |\alpha| \cdot \|x\| \quad \forall x \in \mathcal{V}$ and for any scalar α (real or complex, $|\alpha|$ being the absolute value or modulus).
4. $\|x + y\| \leq \|x\| + \|y\| \quad \forall x, y \in \mathcal{V}$ (triangle inequality).

Examples of those norms are, in $\mathbb{R}^n$, the p-norms:

$$\|(a_1, \ldots, a_n)\|_p = \left(\sum_{i=1}^{n} |a_i|^p\right)^{\frac{1}{p}} \quad p \geq 1.$$

The 2-norm, $\|a\|_2 = \sqrt{a^T a}$, is usually named the *Euclidean* norm.[7]

Another example is the *maximum* norm:

$$\|(a_1, \ldots, a_n)\|_{max} = \max_i |a_i|.$$

This norm is also usually denoted as the "infinity" norm, $\|a\|_\infty$, as p-norms tend to the maximum norm for $p \to \infty$.

Function Spaces

Considering two normed vector spaces $(\mathcal{V}_a, \|\cdot\|_a)$ and $(\mathcal{V}_b, \|\cdot\|_b)$, the set of applications (functions) between $\mathcal{V}_a$ and $\mathcal{V}_b$ is also a vector space, defining the sum of functions and the product by a scalar α as $(f_1 + f_2)(x) = f_1(x) + f_2(x)$ and $(\alpha f)(x) = \alpha f(x)$, respectively.

Definition A.13 In a function vector space, the norm of a continuous function $f : \mathcal{V}_a \to \mathcal{V}_b$ such that $f(0) = 0$ can be defined as

$$\|f\|_{ab} = \sup_{\|v\|_a \neq 0} \frac{\|f(v)\|_b}{\|v\|_a}. \tag{A.38}$$

Remark A.1 The norm of a function is a measure of its maximum amplification (worst-case gain), as

$$\|f(v)\|_b \leq \|f\|_{ab} \|v\|_a \tag{A.39}$$

and the norm can be interpreted as the size of the smallest circular "cone" containing $f(v)$ for all possible v. Nonlinearities with finite norm are called *sector-bounded* nonlinearities.

The above-defined norms are also denoted as *induced norms*.

---

[7] Vector a is assumed to be in column form.

If f is *linear*, the norm definition is equivalent to

$$\|f\|_{ab} = \sup_{\|v\|_a=1} \|f(v)\|_b \tag{A.40}$$

as, indeed, the results of scaling v will be proportional to those with $\|v\| = 1$.

Example A.2 Any norm of function $y = u + \sin u$ is 2, as its graph is tangent to the cone $y = \pm 2u$. $\sqrt{|u|}$ does not have a finite norm.

Example A.3 Consider the set of linear functions between $\mathbb{R}^m$ and $\mathbb{R}^n$, both with the Euclidean norm, $\|\cdot\|_2$. Note that each linear function can be uniquely represented by a matrix, A, of dimensions $m \times n$. The norm of the function $y = Ax$ (termed the norm of matrix A, with a slight abuse of notation) is the maximum singular value defined in Sect. A.4:

$$\|A\| = \sup_{\|x\|=1} \|Ax\| = \overline{\sigma}(A). \tag{A.41}$$

Submultiplicative property.
One important property of these norms in function spaces is the *submultiplicative* property regarding *composition*, $\circ$, of functions $f_1 : \mathcal{V}_b \to \mathcal{V}_c$, $f_2 : \mathcal{V}_a \to \mathcal{V}_b$:

$$\|f_1 \circ f_2\|_{ac} = \sup_{\|v\|\neq 0} \frac{\|f_1(f_2(v))\|_c}{\|v\|_a} \leq \sup_{\|v\|\neq 0} \frac{\|f_1\|_{bc}\|f_2(v)\|_b}{\|v\|_a} = \|f_1\|_{bc} \cdot \|f_2\|_{ab}, \tag{A.42}$$

where the inequality comes from (A.39).

The composition of linear functions is equivalent to *matrix multiplication*. The above property is useful for evaluating bounds on the maximum gain of series-connected systems.

MATLAB® : Related commands are `norm, lti/norm`.

Signal Norms

A particular application of some of the above definitions to signal spaces (functions of time) and dynamical systems (interpreted as functions between two signal spaces) allows the following definitions.

Signals are defined as functions of time $f : \mathbb{R} \to \mathbb{R}^n$. If a certain norm $\|\ \|$ on $\mathbb{R}^n$ is assumed, it can be proved that the following expression constitutes a norm in the space of signals[8]:

$$\|f\|_p = \left(\int_{-\infty}^{\infty} \|f(t)\|^p \, dt\right)^{\frac{1}{p}} \quad p \geq 1. \tag{A.43}$$

[8] To verify property 2 in definition A.12, some technical conditions defining (a) equality of two functions if they differ on a zero-measure set, and (b) existence of the defining integral are needed.

For instance, usual signal norms, applied to an error signal vector, $\boldsymbol{e}(t)$, with components $e_i(t)$, $i \in \{1, 2, \ldots, n\}$, are

1-norm (integral of absolute error IAE):

$$\|\boldsymbol{e}\|_1 = \int_{-\infty}^{\infty} \sum_{i=1}^{n} |e_i(t)|\, dt. \tag{A.44}$$

2-norm (integral of squared error ISE):

$$\|\boldsymbol{e}\|_2 = \sqrt{\int_{-\infty}^{\infty} \sum_{i=1}^{n} e_i(t)^2\, dt}. \tag{A.45}$$

Maximum-norm (∞-norm):

$$\|\boldsymbol{e}\|_\infty = \sup_t \max_i |e_i(t)|. \tag{A.46}$$

Unless otherwise stated, it will be understood that *bounded* signals are those with the finite maximum norm and *bounded-energy* ones are those with finite 2-norm.

In most cases, signals are defined only for positive t and the lower time limits in the above expressions may be changed to zero.

System Norms

Systems are functions from one signal space (input space) to another one (output space). So, norms of these can be defined according to Definition A.13.

In the particular case of norms of *linear* systems, these systems can be characterised by the impulse response, $G(t)$, or frequency response, $G(j\omega)$, matrices. Define

$$\|G(s)\|_2 = \sqrt{\frac{1}{2\pi} \int_{-\infty}^{+\infty} \sum_{ij} |G_{ij}(j\omega)|^2\, d\omega}, \tag{A.47}$$

$$\|G(s)\|_\infty = \max_\omega \bar{\sigma} G(j\omega), \tag{A.48}$$

$$\|G(t)\|_2 = \sqrt{\int_0^{+\infty} \sum_{ij} |G_{ij}(t)|^2\, dt}, \tag{A.49}$$

$$\|G(t)\|_1 = \int_0^{+\infty} \max_i \sum_j |G_{ij}(t)|\, dt. \tag{A.50}$$

Table A.1 Induced system norms for specific input and output norms

	$\|u\|_2$	$\|u\|_\infty$
$\|y\|_2$	$\|G(s)\|_\infty$	∞
$\|y\|_\infty$	$\|G(s)\|_2$	$\|G(t)\|_1$

Parseval theorem proves that $\|G(t)\|_2 = \|G(s)\|_2$. The 2-norm of a system can also be interpreted as the variance to a white noise input.

It can be shown that the worst-case gain (i.e. the system norm in definition A.13) using specific input and output norms are those in Table A.1.

MATLAB® : Related commands are `norm (lti/norm)`.

A.8 BIBO Stability and the Small-Gain Theorem

Given two signal spaces, $\mathcal{U}$ and $\mathcal{Y}$, with a particular norm on them, and a dynamical system, Σ, relating signals in those spaces by $y = \Sigma(u)$, the system is said to be bounded input bounded output (BIBO)-stable if Σ has a finite induced norm, i.e. if a constant c exists so that $\|y\| \leq c\|u\|$, for zero initial conditions (equilibrium point at the origin is implicitly assumed, i.e. $0 = \sigma 0$). It can be shown that the infimum of those constants is the norm of $\|\Sigma\|$ defined in the sense of Definition A.13.

As systems are interconnected to form more complex ones, it is interesting to investigate the stability properties of a system from the stability properties of its subsystems and its structure.

Series and Parallel Connections

Two systems Σ_1 and Σ_2 are series-connected if

$$y_2 = \Sigma_2(y_1); \qquad y_1 = \Sigma_1(u).$$

Equation (A.42) shows that if two systems are BIBO-stable, its series connection also has that property. The result is also straightforward in parallel connections, $y = \Sigma_1(u) + \Sigma_2(u)$.

Feedback Connection (Small-Gain Theorem)

Let us analyse the feedback loop in Fig. A.1 in order to derive a sufficient BIBO-stability condition.

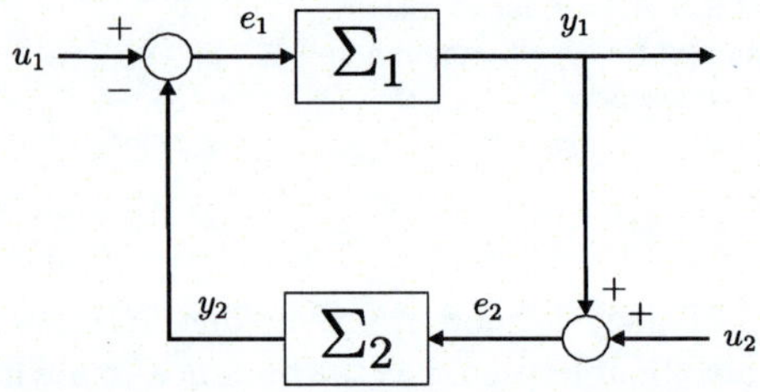

Fig. A.1 Feedback connection

Let us assume that Σ_1 and Σ_2 are BIBO-stable (i.e. that a finite norm can be calculated so that $\|y_2\| \leq \|\Sigma_1\| \cdot \|e_1\|$ and $\|y_1\| \leq \|\Sigma_2\| \cdot \|e_2\|$). By successively applying the definition of a norm (maximum gain) and the triangle inequality,

$$\|y_2\| \leq \|\Sigma_1\| \cdot \|e_1\| \leq \|\Sigma_1\|(\|u_1\| + \|y_1\|) \leq \\ \|\Sigma_1\|(\|u_1\| + \|\Sigma_2\| \cdot \|e_2\|) \leq \|\Sigma_1\|(\|u_1\| + \|\Sigma_2\|(\|u_2\| + \|y_2\|)). \quad \text{(A.51)}$$

Solving for $\|y_2\|$ gives

$$(1 - \|\Sigma_1\| \cdot \|\Sigma_2\|)\|y_2\| \leq \|\Sigma_1\| \cdot \|u_1\| + \|\Sigma_1\| \cdot \|\Sigma_2\| \cdot \|u_2\|; \quad \text{(A.52)}$$

so, if $\|\Sigma_1\| \cdot \|\Sigma_2\| < 1$, any solution y_2 verifies that gains from inputs to y_2 are bounded (BIBO stability). Similar expressions can be obtained analogously for the rest of the signals, so the following result can be stated.

Theorem A.1 (Small-gain Theorem) *For bounded inputs, all signals in the closed loop in Fig. A.1 are bounded if Σ_1 and Σ_2 are stable and*

$$\|\Sigma_1\| \cdot \|\Sigma_2\| \leq 1.$$

The results holds for Σ_1, Σ_2 being any dynamical system (linear or nonlinear) as long as stability and the system norm can be suitably defined. In the application of this theorem to robust control, one of the blocks will be the linear closed-loop system and the other will be the unknown, possibly nonlinear, modelling errors.

Example A.4 In the case of one of the systems (say, Σ_2) being a stable LTI one, and that the signal norm $\|y\|$, $\|u\|$ to be considered is the Euclidean 2-norm (inducing the ∞-norm for systems; see Table A.1), a sufficient condition for stability of the closed-loop system is

$$\|\Sigma_1\| \sup_{w \in \mathbb{R}} \bar{\sigma}(\Sigma_2(j\omega)) \leq 1. \quad \text{(A.53)}$$

Appendix B
Lyapunov Stability

The definitions and theorems in Sects. B.1 and B.2 of this Appendix can be found in Khalil [11, 12]; those in Sects. B.3 and B.4 are adaptations of those in Vidyasagar [13].

B.1 Time-Invariant Continuous-Time Systems

Consider an autonomous time-invariant nonlinear system

$$\dot{\boldsymbol{x}}(t) = \boldsymbol{f}(\boldsymbol{x}) \qquad \boldsymbol{x} \in \mathcal{D}, \tag{B.1}$$

where $t \in \mathbb{R}$, $\boldsymbol{x} \in \mathbb{R}^n$, $\boldsymbol{x}(t)$ denotes the solution of this system from initial condition $\boldsymbol{x}(0)$, $\boldsymbol{x} = 0$ is an equilibrium point, and $\boldsymbol{f}(\boldsymbol{x}) : \mathcal{D} \mapsto \mathbb{R}^n$ is a locally Lipschitz function defined over a domain $\mathcal{D} \subset \mathbb{R}^n$, where $0 \in \mathcal{D}$.

Definition B.1 The equilibrium point $\boldsymbol{x} = 0$ is *asymptotically stable* if $\forall \epsilon > 0$ $\exists \delta(\epsilon) > 0$ such that

$$\|\boldsymbol{x}(0)\| < \delta(\epsilon) \Rightarrow \|\boldsymbol{x}(t)\| < \epsilon \text{ and } \lim_{t \to \infty} \boldsymbol{x}(t) = 0.$$

Definition B.2 The equilibrium point $\boldsymbol{x} = 0$ is *globally asymptotically stable* if $\forall \boldsymbol{x}(0)$ it holds that $\lim_{t \to \infty} \boldsymbol{x}(t) = 0$.

Definition B.3 The equilibrium point $\boldsymbol{x} = 0$ is *exponentially stable* if $\exists\, c > 0, k > 0, \lambda > 0$ such that

$$\|\boldsymbol{x}(t)\| \leq k\|\boldsymbol{x}(0)\|e^{-\lambda t}, \ \forall t \geq 0, \ \|\boldsymbol{x}(0)\| < c.$$

M. Bernal et al., *Analysis and Synthesis of Nonlinear Control Systems*, Studies in Systems, Decision and Control 408, https://doi.org/10.1007/978-3-030-90773-0

A solution $x(t)$ that holds the previous inequality for a given $x(0)$ is said to have a *decay rate* of λ.

Definition B.4 The equilibrium point $\boldsymbol{x} = 0$ is *globally exponentially stable* if it is exponentially stable for every initial condition $\boldsymbol{x}(0)$.

Based on the previous definitions and assumptions, the following theorems can be stated.

Theorem B.1 *If* $\exists V(\boldsymbol{x}) : \mathcal{D} \to \mathbb{R}$ *continuously differentiable such that*

$$\begin{aligned} &V(0) = 0, \\ &V(\boldsymbol{x}) > 0, \quad \forall \boldsymbol{x} \in \mathcal{D} \setminus \{0\}, \\ &\dot{V}(\boldsymbol{x}) < 0, \quad \forall \boldsymbol{x} \in \mathcal{D} \setminus \{0\}, \end{aligned} \tag{B.2}$$

then $\boldsymbol{x} = 0$ *is asymptotically stable; if, in addition,* $\mathcal{D} = \mathbb{R}^n$ *and* $V(\boldsymbol{x})$ *is radially unbounded, then* $\boldsymbol{x} = 0$ *is globally asymptotically stable.*

Theorem B.2 *If* $\exists\, k_1 > 0, k_2 > 0, k_3 > 0, a > 0,$ *and* $V(\boldsymbol{x}) : \mathcal{D} \to \mathbb{R}$ *continuously differentiable such that*

$$\begin{aligned} k_1\|\boldsymbol{x}\|^a \le V(\boldsymbol{x}) \le k_2\|\boldsymbol{x}\|^a, \quad \forall \boldsymbol{x} \in \mathcal{D}, \\ \dot{V}(\boldsymbol{x}) \le -k_3\|\boldsymbol{x}\|^a, \quad \forall \boldsymbol{x} \in \mathcal{D}, \end{aligned} \tag{B.3}$$

then $\boldsymbol{x} = 0$ *is exponentially stable. If* (B.3) *holds globally, then* $\boldsymbol{x} = 0$ *is globally exponentially stable.*

B.2 Time-Variant Continuous-Time Systems

In what follows, consider an autonomous time-variant nonlinear system

$$\dot{\boldsymbol{x}} = \boldsymbol{f}(t, \boldsymbol{x}) \qquad \boldsymbol{x} \in \mathcal{D}, \tag{B.4}$$

where $t \in \mathbb{R}$, $\boldsymbol{x} \in \mathbb{R}^n$, $\boldsymbol{x}(t)$ denotes the solution of this system from initial condition $\boldsymbol{x}(t_0)$, $\boldsymbol{x} = 0$ is an equilibrium point in the sense that $\boldsymbol{f}(t, 0) = 0$, $\forall t \ge 0$, and $\boldsymbol{f}(t, \boldsymbol{x}) : \mathbb{R} \times \mathcal{D} \mapsto \mathbb{R}^n$ is a piecewise continuous function in t, locally Lipschitz in $\boldsymbol{x}$, $\forall t \ge 0$, defined over a domain $\mathcal{D} \subset \mathbb{R}^n$, where $0 \in \mathcal{D}$.

Asymptotic and exponential stability in the time-variant case is defined in terms of class $\mathcal{K}$, $\mathcal{K}_\infty$, and $\mathcal{KL}$ functions, whose definitions follow.

Definition B.5 A *class* $\mathcal{K}$ function $\alpha(r) : [0, a) \mapsto \mathbb{R}$ is that which holds $\alpha(0) = 0$ and $\dot{\alpha}(r) > 0$.

Definition B.6 A *class* $\mathcal{K}_\infty$ function $\alpha(r) : [0, \infty) \mapsto \mathbb{R}$ is that which holds $\alpha(0) = 0$, $\dot{\alpha}(r) > 0$, and $\lim_{r\to\infty} \alpha(r) \to \infty$.

Definition B.7 A *class* $\mathcal{KL}$ function $\beta(r, s) : [0, a) \times [0, \infty) \mapsto \mathbb{R}$ is that which is class $\mathcal{K}$ in r for every fixed s and $\partial\beta/\partial s < 0$ for every fixed r, with $\lim_{s\to\infty} \beta(r, s) = 0$.

We can now extend the definitions of asymptotic and exponential stability to the time-variant case.

Definition B.8 The equilibrium point $\boldsymbol{x} = 0$ is *uniformly asymptotically stable* if there exist a class $\mathcal{KL}$ function $\beta(\cdot)$ and $c > 0$ (independent of t_0) such that

$$\|\boldsymbol{x}(t)\| \le \beta(\|\boldsymbol{x}(t_0)\|, t - t_0), \ \forall t \ge t_0 \ge 0, \ \forall \|\boldsymbol{x}(t_0)\| < c.$$

Definition B.9 The equilibrium point $\boldsymbol{x} = 0$ is *globally uniformly asymptotically stable* if it is uniformly asymptotically stable $\forall \boldsymbol{x}(t_0)$.

Definition B.10 The equilibrium point $\boldsymbol{x} = 0$ is *exponentially stable* if there exist $c > 0$, $k > 0$, and $\lambda > 0$ such that

$$\|\boldsymbol{x}(t)\| \le k\|\boldsymbol{x}(t_0)\| e^{-\lambda(t-t_0)}, \ \forall \|\boldsymbol{x}(t_0)\| < c.$$

Definition B.11 The equilibrium point $\boldsymbol{x} = 0$ is *globally exponentially stable* if it is exponentially stable $\forall \boldsymbol{x}(t_0)$.

Theorem B.3 *If there exists a continuously differentiable function* $V : [0, \infty) \times \mathcal{D} \to \mathbb{R}$ *and three continuous functions* $W_i(\boldsymbol{x}) : \mathcal{D} \to \mathbb{R}$, $W_i(\boldsymbol{x}) > 0 \ \forall \boldsymbol{x} \ne 0$, $W_i(0) = 0$, $i \in \{1, 2, 3\}$, *such that*

$$\begin{aligned} &W_1(\boldsymbol{x}) \le V(t, \boldsymbol{x}) \le W_2(\boldsymbol{x}), \quad \forall t \ge 0, \forall \boldsymbol{x} \in \mathcal{D}, \\ &\frac{\partial V}{\partial t} + \frac{\partial V}{\partial \boldsymbol{x}} \boldsymbol{f}(t, \boldsymbol{x}) \le -W_3(\boldsymbol{x}), \quad \forall t \ge 0, \forall \boldsymbol{x} \in \mathcal{D}, \end{aligned} \tag{B.5}$$

then $\boldsymbol{x} = 0$ *is uniformly asymptotically stable. If, moreover,* $\mathcal{D} = \mathbb{R}^n$ *and* $W_1(\boldsymbol{x})$ *is radially unbounded, then* $\boldsymbol{x} = 0$ *is globally uniformly asymptotically stable.*

Theorem B.4 *If* $\exists\, k_1 > 0$, $k_2 > 0$, $k_3 > 0$, $a > 0$, *and a continuously differentiable function* $V : [0, \infty) \times \mathcal{D} \to \mathbb{R}$, *such that*

$$\begin{aligned} &k_1\|\boldsymbol{x}(t)\|^a \le V(t, \boldsymbol{x}) \le k_2\|\boldsymbol{x}(t)\|^a, \quad \forall t \ge 0, \forall \boldsymbol{x} \in \mathcal{D}, \\ &\frac{\partial V}{\partial t} + \frac{\partial V}{\partial \boldsymbol{x}} \boldsymbol{f}(t, \boldsymbol{x}) \le -k_3\|\boldsymbol{x}(t)\|^a, \quad \forall t \ge 0, \boldsymbol{x} \in \mathcal{D}, \end{aligned} \tag{B.6}$$

then $\boldsymbol{x} = 0$ *is exponentially stable. If, moreover, these conditions are met globally,* $\boldsymbol{x} = 0$ *is globally exponentially stable.*

B.3 Time-Invariant Discrete-Time Systems

Consider an autonomous time-invariant discrete-time system

$$\boldsymbol{x}(t+1) = \boldsymbol{f}(\boldsymbol{x}(t)) \qquad \boldsymbol{x} \in \mathcal{D}, \tag{B.7}$$

where $t \in \mathbb{Z}$, $\boldsymbol{x} \in \mathbb{R}^n$, $\boldsymbol{x}(t)$ denotes the solution of this system from initial condition $\boldsymbol{x}(0)$, $\boldsymbol{x} = 0$ is an equilibrium point, and $\boldsymbol{f}(\boldsymbol{x}) : \mathcal{D} \mapsto \mathbb{R}^n$ is a locally Lipschitz function defined over a domain $\mathcal{D} \subset \mathbb{R}^n$, where $0 \in \mathcal{D}$.

The definitions for asymptotically stable and globally asymptotically stable equilibrium point $\boldsymbol{x} = 0$ of (B.7) are the same as those in B.1 and B.2, respectively; those for exponentially stable and globally exponentially stable equilibrium points of a discrete-time system differ from the continuous-time case.

Definition B.12 The equilibrium point $\boldsymbol{x} = 0$ is *exponentially stable* if $\exists\, c > 0$, $k > 0$, $0 < \lambda < 1$ such that

$$\|\boldsymbol{x}(t)\| \leq k\|\boldsymbol{x}(0)\|\lambda^t, \quad \forall t \geq 0, \quad \|\boldsymbol{x}(0)\| < c.$$

A solution $x(t)$ that holds the previous inequality for a given $x(0)$ is said to have a *decay rate* of λ.

Definition B.13 The equilibrium point $\boldsymbol{x} = 0$ is *globally exponentially stable* if it is exponentially stable for every initial condition $\boldsymbol{x}(0)$.

Theorem B.5 *If* $\exists V(\boldsymbol{x}) : \mathcal{D} \to \mathbb{R}$ *such that*

$$\begin{aligned} &V(0) = 0, \\ &V(\boldsymbol{x}) > 0, \quad \forall \boldsymbol{x} \in \mathcal{D} \setminus \{0\}, \\ &\Delta V := V(\boldsymbol{x}(t+1)) - V(\boldsymbol{x}(t)) < 0, \quad \forall \boldsymbol{x} \in \mathcal{D} \setminus \{0\}, \end{aligned} \tag{B.8}$$

then $\boldsymbol{x} = 0$ *is asymptotically stable; if, in addition,* $\mathcal{D} = \mathbb{R}^n$ *and* $V(\boldsymbol{x})$ *is radially unbounded, then* $\boldsymbol{x} = 0$ *is globally asymptotically stable.*

Theorem B.6 *If* $\exists\, k_1 > 0$, $k_2 > 0$, $a > 0$, *and* $V(\boldsymbol{x}) : \mathcal{D} \to \mathbb{R}$ *such that*

$$\begin{aligned} &k_1\|\boldsymbol{x}\|^a \leq V(\boldsymbol{x}) \leq k_2\|\boldsymbol{x}\|^a, \quad \forall \boldsymbol{x} \in \mathcal{D}, \\ &\Delta V := V(\boldsymbol{x}(t+1)) - V(\boldsymbol{x}(t)) \leq -k_3\|\boldsymbol{x}\|^a, \quad \forall \boldsymbol{x} \in \mathcal{D}, \end{aligned} \tag{B.9}$$

then $\boldsymbol{x} = 0$ *is exponentially stable. If* (B.9) *holds globally, then* $\boldsymbol{x} = 0$ *is globally exponentially stable.*

B.4 Time-Variant Discrete-Time Systems

This section considers autonomous time-variant discrete-time systems of the form:

$$x(t+1) = f(t, x(t)) \qquad x \in \mathcal{D}, \tag{B.10}$$

where $t \in \mathbb{Z}$, $x \in \mathbb{R}^n$, $x(t)$ denotes the solution of this system from initial condition $x(t_0)$, $x = 0$ is an equilibrium point in the sense that $f(t, 0) = 0$, $\forall t \geq 0$, and $f(t, x) : \mathbb{R} \times \mathcal{D} \mapsto \mathbb{R}^n$ is a piecewise continuous function in t, locally Lipschitz in x, $\forall t \geq 0$, defined over a domain $\mathcal{D} \subset \mathbb{R}^n$, where $0 \in \mathcal{D}$.

Theorem B.7 *If* $V : [0, \infty) \times \mathcal{D} \to \mathbb{R}$ *and class* $\mathcal{K}$ *functions* $\alpha_1(\|x\|)$, $\alpha_2(\|x\|)$ *exist, such that*

$$\begin{aligned}
&V(t, 0) = 0, \quad \forall t \geq 0, \\
&V(t, x) > \alpha_1(\|x\|), \quad \forall t \geq 0, \forall x \in \mathcal{D} \setminus \{0\}, \\
&\Delta V := V(t+1, x(t+1)) - V(t, x(t)) < -\alpha_2(\|x\|), \quad \forall t \geq 0, \forall x \in \mathcal{D} \setminus \{0\},
\end{aligned} \tag{B.11}$$

then $x = 0$ *is uniformly asymptotically stable; if, in addition,* $\mathcal{D} = \mathbb{R}^n$ *and* $V(t, x)$ *is radially unbounded, then* $x = 0$ *is globally uniformly asymptotically stable.*

Theorem B.8 *If* $\exists\, k_1 > 0$, $k_2 > 0$, $a > 0$, *and* $V : [0, \infty) \times \mathcal{D} \to \mathbb{R}$ *such that*

$$\begin{aligned}
&k_1\|x\|^a \leq V(t, x) \leq k_2\|x\|^a, \quad \forall t \geq 0, \forall x \in \mathcal{D}, \\
&\Delta V := V(t+1, x(t+1)) - V(t, x(t)) \leq -k_3\|x\|^a, \forall t \geq 0, \forall x \in \mathcal{D},
\end{aligned} \tag{B.12}$$

then $x = 0$ *is exponentially stable. If* (B.12) *holds globally, then* $x = 0$ *is globally exponentially stable.*

Remark B.1 The above theorems present only sufficient conditions; if a Lyapunov function V satisfying them cannot be found, it only means that no conclusion can be drawn regarding the stability of the equilibrium point on the basis of these theorems.

Appendix C
Convex Optimisation Tools

This appendix is a brief introduction to the main tools in convex optimisation, namely linear matrix inequalities (LMIs) and sum of squares (SOS). Since the appearance of the Lyapunov theory at the end of the XIX century, many control problems have found their formulation as LMIs, which were later found to be numerically tractable in an efficient manner. Solving an LMI amounts to establish positive-definiteness of a matrix expression; similarly, when dealing with polynomial systems (or a convex combination of them), many stability and control results end up in trying to prove positiveness (or negativeness) of a polynomial in several variables. A conservative way of doing the latter is proving that they are sum of squares. The basic aspects of such ideas are outlined next.[9]

C.1 Linear Matrix Inequalities

In a broad sense, an LMI is a set of expressions whose variables are linearly related matrices. A formal definition of an LMI is [2]

$$F(\mathbf{x}) = F_0 + \sum_{i=1}^{m} x_i F_i \geq 0, \tag{C.1}$$

where $\mathbf{x} \in \mathbb{R}^m$ is the vector of decision variables and $F_i = F_i^T \in \mathbb{R}^{n\times n}, i = 0, \ldots, m$, are given constant symmetric matrices. As any matrix variable can be decomposed into a base of symmetric matrices, the definition (C.1) involving scalars is easily extended to matrices.

[9] The authors are grateful to Jose Luis Pitarch for his collaboration in the development of this chapter from materials from his Ph.D. thesis.

M. Bernal et al., *Analysis and Synthesis of Nonlinear Control Systems*, Studies in Systems, Decision and Control 408, https://doi.org/10.1007/978-3-030-90773-0

The set of solutions of the LMI (C.1), or the so-called feasibility set, denoted by $S = \{x | x \in \mathbb{R}^m, F(x) \geq 0\}$ is a convex subset of $\mathbb{R}^m$.

Indeed, if $F(x) \geq 0$ and $F(z) \geq 0$, $F(\alpha x + \beta z) = \alpha F(x) + \beta F(z) > 0$ for all $\alpha \geq 0$, $\beta \geq 0$ such that $\alpha + \beta = 1$.

When the solution set S is empty, the LMI is said to be infeasible.

Note that matrix positive-definiteness $F(x) \geq 0$ is equivalent to finding decision variables x such that they prove nonnegativeness of a quadratic form $\psi^T F(x)\psi \geq 0$ for all ψ; the generalisation to higher degree when the elements of ψ are polynomials will be discussed in Sect. C.2.

Semidefinite programming (SDP) is another name for LMI-constrained optimisation. Indeed, minimising a *convex* function of x, say $f : \mathbb{R}^m \mapsto \mathbb{R}$, subject to constraints (C.1), is a convex optimisation problem, thus having a single global minima (some problems may be unbounded or unfeasible), with efficient polynomial-time algorithms available (see below).

Expression (C.1) is stated as a non-strict inequality; in this way, the optimal solution can be at the frontier of the (closed) feasible set; this is the way solvers expect the problems to be posed. If, in the theoretical developments in the main text of this book, a strict LMI $F(x) > 0$ is written, it should be understood as $F(x) \geq \varepsilon I$ for some small tolerance parameter ϵ unless otherwise stated.

As any arbitrary matrix F can be expressed as the sum of each of its elements F_{ij} multiplied by a matrix $M^{[ij]}$ whose elements are all zero except element (i, j) which is equal to one (canonical basis), in this book LMIs will be written as matrices whose elements are linear in some (matrix) decision variables, say $AP + PA^T < -2I$ P being a decision variable. It is straightforward to rewrite the said inequality as (C.1) the decision variables being x_i identified with the elements of matrix P; the matrix-decision-variable notation is, however, more convenient for the stability and control design problem development.

The following well-known convex or quasi-convex optimisation problems are relevant for the analysis and the synthesis of control systems [2, 3].

1. Finding a solution $x \in \mathbb{R}^m$ to the LMI system (C.1) or determining that there is no solution is called the feasibility problem (FP). This problem can be cast, for instance, as minimising the convex function $f : x \to \lambda_{\min}(F(x))$, where $\lambda_{\min}$ denotes the smallest eigenvalue, and then deciding whether the solution is positive (strictly feasible solution), zero (feasible solution), or negative (unfeasible case).
2. Minimising a linear combination of the decision variables $b^T x$ subject to (C.1) is called the eigenvalue problem (EVP), also known as an LMI optimisation problem.
3. Minimising the eigenvalues of a pair of matrices that depend affinely on a variable, subject to a set of LMI constraints or determining that the problem is infeasible, i.e. solving the problem *minimise* λ *subject to*

$$\lambda B(x) - A(x) \geq 0,$$
$$B(x) \geq 0,$$
$$C(x) \geq 0,$$

where $A(\boldsymbol{x})$, $B(\boldsymbol{x})$, and $C(\boldsymbol{x})$ are symmetric and affine with respect to $\boldsymbol{x}$, is called a generalised eigenvalue problem (GEVP). This is not exactly an LMI convex optimisation problem, because the decision variables λ and $\boldsymbol{x}$ are multiplied; however, it is quasi-convex in the sense that it can be solved by bisection in λ, given that determining feasibility of the LMI constraints for fixed λ is a convex problem; hence, efficient polynomial-time solutions are possible.

It is of interest to remark that Linear Programming (LP) and (some) Quadratic Programming (QP) problems are particular cases of the SDP problem here defined; efficiency of some interior-point algorithms in SDP optimisation makes it possible to solve large-scale LP and QP more efficiently than older (say, simplex, ...) methods.

Numerous solvers handle these various optimisation problems, such as LMILAB, SeDuMi, SDPT3, VSDP, and LMIRank. In this book, unless otherwise stated, to solve LMI problems, the *SeDuMi* solver within the *Yalmip* toolbox [14] is used.

C.2 Sum of Squares

Let us denote as $\mathcal{R}_z$ the set of polynomials $p(x) : \mathbb{R}^n \to \mathbb{R}$ in some variables z with real coefficients. For instance, $p(x_1, x_2) = x_1^2 - 3x_2^3 \in \mathcal{R}_{x_1 x_2}$. The arguments to the polynomials will be omitted in the sequel when they are clear from the context.

Definition C.1 A polynomial $p \in \mathcal{R}_z$ is *sum of squares* (SOS) if there exists a finite set of polynomials $f_1, \ldots, f_n \in \mathcal{R}_z$ such that p can be written as

$$p = \sum_{i=1}^{n} f_i^2. \tag{C.2}$$

The set of SOS polynomials in variables z will be denoted as Σ_z.

Finding whether a given polynomial is SOS is actually a semidefinite programming (non-strict LMI) problem.

Theorem C.1 *([15]) A polynomial p is SOS if and only if there exists a constant matrix M and a vector of polynomials ψ such that $p = \psi^T M \psi$ and M is positive semidefinite. Matrix M is sometimes denoted as the* Gram matrix *[1].*

Proof If a polynomial verifies (C.2), then ψ can be made by stacking f_i in vector form and M equal to the identity matrix. Conversely, if there exists such $M \geq 0$, it can be expressed as $M = Q^T Q$ (the so-called Cholesky factorization, standard and numerically reliable for strictly positive-definite matrices and, too, in many cases, for semidefinite ones [16]). Then, using as f_i the elements of the polynomial vector $f = Q\psi$, $p = f^T f$ so the required decomposition has been obtained.

Example C.1 The polynomial

$$p(z) = z_1^4 - 4z_1^3 z_2 + 2z_1^3 + 4z_1^2 z_2^2 - 12z_1^2 z_2 + z_1^2 + 16z_1 z_2^2 - 8z_1 z_2 + 16z_2^2$$

can be written as

$$p(z) = \begin{bmatrix} z_1 & z_2 & z_1 z_2 & z_1^2 \end{bmatrix} \begin{bmatrix} 1 & -4 & -2 & 1 \\ -4 & 16 & 8 & -4 \\ -2 & 8 & 4 & -2 \\ 1 & -4 & -2 & 1 \end{bmatrix} \begin{bmatrix} z_1 \\ z_2 \\ z_1 z_2 \\ z_1^2 \end{bmatrix},$$

and, as the matrix is positive-definite, the Cholesky factor gives the SOS decomposition $p(z) = (z_1 - 4z_2 - 2z_1 z_2 + z_1^2)^2$, so $p(z) \in \Sigma_z$. Obviously, we can assert $p(z) \geq 0$ for any $z \in \mathbb{R}^2$.

Evidently, all SOS polynomials are nonnegative everywhere, but the converse is not true except for polynomials in at most 2 variables, polynomials in any number of variables with degree 2, and polynomials in 3 variables with degree up to 4 [17, 18].

Example C.2 Polynomial

$$p(z) = z_1^4 - 2\alpha z_1^3 z_2 + 2z_1^3 + 4z_1^2 z_2^2 - 2\beta z_1^2 z_2 + z_1^2 + 16z_1 z_2^2 - 8z_1 z_2 + 16z_2^2$$

can be written as

$$p(z) = \begin{bmatrix} z_1 & z_2 & z_1 z_2 & z_1^2 \end{bmatrix} \begin{pmatrix} 1 & -4 & -\beta+\gamma & 1 \\ -4 & 16 & 8 & -\gamma \\ -\beta+\gamma & 8 & 4 & -\alpha \\ 1 & -\gamma & -\alpha & 1 \end{pmatrix} \begin{bmatrix} z_1 \\ z_2 \\ z_1 z_2 \\ z_1^2 \end{bmatrix}$$

because the vector of monomials $\psi = (z_1, z_2, z_1 z_2, z_1^2)$ verifies $\psi_1 \psi_3 = \psi_2 \psi_4 = z_1^2 z_2$. Indeed, $\alpha = 2$, $\beta = 6$, and $\gamma = 4$ recover the previous example. If $p(z)$ were, say, a Lyapunov function candidate, a sufficient condition for $p(z) \geq 0$ would be forcing the central matrix to be positive semidefinite, which is an LMI in decision variables α, β, and γ.

The above decomposition (generating suitable vectors of monomials, building the matrices with the required decision variables, and calling an LMI solver) is transparently carried out by some free software packages, for instance: SOSTOOLS [19], YALMIP SOS module [20], and SOSOPT Seiler [21]. Such software integrates with MATLAB®and allows using some symbolic syntax for manipulation and display, either integrating with MATLAB®symbolic toolbox or defining other polynomial-only symbolic classes. Basically, once SOS constraints are expressed as convex LMIs, several standard problems can be dealt with efficiently: SOS feasibility, optimisation of a linear cost with SOS constraints, and GEVP-like problems require bisection.

There are two key issues with numerical SOS software: first, that the number of monomials increases heavily with the number of involved variables and the degree of them. The number of monomials of degree d in n variables is $(n+d-1)!/d!(n-1)!$; a whole generic polynomial contains all monomials of degrees zero to d. For instance, polynomials of degree 2 in 2 variables have four monomials (1, z_1,z_2, $z_1 z_2$), polynomials of degree 5 in 6 variables have 462 monomials (so associated gram matrices have more than $2 \cdot 10^5$ elements), and so on.

A second issue with SOS software is that semidefinite solutions lie in the frontier of the convex set containing the feasible solutions of a convex matrix optimisation problem. Matrices with zero eigenvalues are ill-conditioned, so testing for positive semidefiniteness of large matrices gives, frequently, rise to algorithms stopping with "numerical problems" maybe giving a solution which is slightly suboptimal (very small eigenvalue on the "positive" side) or slightly unfeasible (very small one on the "negative" side), when eigenvalues fit below some specified tolerances.

The first case might be acceptable, but not the second one. A posteriori checking of the results or incorporating some tolerance safeguards is advised. For instance, SOS optimisation problems may be reformulated as feasibility ones: if the optimisation results in an optimal cost γ^*, a strictly feasible solution is obtained by posing a feasibility problem (no cost) with the same SOS constraints plus, say, a relaxed one $\gamma < 1.001\gamma^*$ (see Lofberg [20] for details).

In many cases, nonlinear systems modelled with a convex structure are not globally stable, and the problem of determining the positiveness of a polynomial in a region Ω needs to be addressed. If the region can be written as $\Omega = \{x | g_i(x) \geq 0, h_i(x) = 0, g_i, h_i \in \mathcal{R}_x\}$, i.e. its frontier is determined by polynomials, the so-called *Positivstellensatz* theorem, a central result from real algebraic geometry comes handy.

Definition C.2 Given polynomials $\{g_1, g_2, \ldots, g_n\} \in \mathcal{R}_z$, the *multiplicative monoid* generated by them is the finite set of all products involving g_j, $j \in \{1, 2, \ldots, n\}$ including 1 (empty product); it is denoted by $\mathcal{M}(g_1, g_2, \ldots, g_n)$.

Definition C.3 Given polynomials $\{f_1, f_2, \ldots, f_r\} \in \mathcal{R}_z$, the *cone* generated by them is the set

$$\wp(f_1, f_2, \ldots, f_r) := \left\{ s_0 + \sum_{i=1}^{l} s_i b_i : l \in \mathbb{Z}^+, \ s_i \in \Sigma_z, b_i \in \mathcal{M}(f_1, f_2, \ldots, f_r) \right\},$$

i.e. the arbitrary sum of products of f_i by SOS multiplier polynomials s_i.

If the polynomials $s_i \in \Sigma_z$ and $f_i \in \mathcal{R}_z$, then $f_i^2 s_i \in \Sigma_z$. This allows expressing a cone of $\{f_1, ..., f_r\}$ as a finite sum of 2^r terms (for instance, with 3 elements, the only needed terms from the monoid are 1, f_1, f_2, f_3, $f_1 f_2$, $f_1 f_3$, $f_2 f_3$, $f_1 f_2 f_3$).

Definition C.4 Given polynomials $\{h_1, h_2, \ldots, h_u\} \in \mathcal{R}_z$, the *ideal* generated by them is the set

$$\mathcal{I}(h_1, h_2, \ldots, h_u) := \left\{ \sum_{k=1}^{u} h_k p_k | p_k \in \mathcal{R}_z \right\},$$

where multipliers p_i are arbitrary (not necessarily SOS as in the cone).

Theorem C.2 (Positivstellensatz Stengle [22]) *Given polynomials* $\{f_1, f_2, \ldots, f_r\}$, $\{g_1, g_2, \ldots, g_n\}$, *and* $\{h_1, h_2, ..., h_u\} \in \mathcal{R}_z$, *the following statements are equivalent:*

1. *The following set is the empty set*

$$\left\{ z \in \mathbb{R}^n \;\middle|\; \begin{array}{l} f_1(z) \geq 0, ..., f_r(z) \geq 0 \\ g_1(z) \neq 0, ..., g_t(z) \neq 0 \\ h_1(z) = 0, ..., h_u(z) = 0 \end{array} \right\}.$$

2. *There exist polynomials* $f \in \wp(f_1, f_2, \ldots, f_r)$, $g \in \mathcal{M}(g_1, g_2, \ldots, g_n)$, $h \in \mathcal{I}(h_1, h_2, \ldots, h_u)$ *such that* $f + g^2 + h = 0$.

This theorem is a powerful result which generalises many well-known results like the S-procedure in Property A.2 and Finsler's lemma A.3 to the polynomial case, by carefully choosing some SOS and arbitrary polynomial multipliers [23, pp. 4–7]. In this way, many control problems within the polynomial framework can be addressed by combining the SOS optimisation tools with the Positivstellensatz theorem.

Note, however, that the above theorem is not constructive and that the degree of the involved polynomials in the second statement may be arbitrarily high in a general situation. Hence, in actual computations, in order to determine whether the first statement of the theorem holds, the second statement is sought only in a family of finite-degree polynomials obtaining then a sufficient but not necessary condition for the first statement.

For instance, consider a region Ω defined by polynomial boundaries as follows:

$$\Omega = \{z : g_1(z) > 0, \ldots, g_{m_g}(z) > 0, h_1(z) = 0, \ldots, h_{m_h}(z) = 0\}. \tag{C.3}$$

Corollary C.1 *If SOS polynomials* $s_i(z) \in \Sigma_z$ *and arbitrary ones* $r_j(z) \in \mathcal{R}_z$ *can be found fulfilling:*

$$p(z) - \sum_{i=1}^{m_g} s_i(z) g_i(z) + \sum_{j=1}^{m_h} r_j(z) h_j(z) \in \Sigma_z, \tag{C.4}$$

then $p(z)$ *is locally nonnegative in the region* Ω.

Proof Indeed, note that, in the region Ω, the term $\sum_i s_i(z) g_i(z)$ is positive and $\sum_j r_j(z) h_i(z)$ is zero, so $p(z) - \epsilon(z) \geq \sum_i s_i(z) g_i(z) \geq 0$ for all $z \in \Omega$.

The polynomials $(s_i(z), r_j(z))$ are denoted as *Positivstellensatz multipliers*, analogous to Lagrange and KKT ones in constrained optimisation [24].

The above corollary is a simplified sufficient condition. In order to be necessary and sufficient, products in the full cone generated by $\{-p, g_1, \ldots\}$ must be added (a total of 2^{m_g+1} terms) and arbitrary high multiplier degrees should be searched for. As the computational requirements increase considerably with the degree of the polynomials, usually Corollary C.1 is used as the first implementation choice.

Example C.3 The polynomial $p(x) = 1 - x^3$ is nonnegative in the interval region of interest $\Omega = \{x : 0.25 - x^2 > 0\}$. Indeed, suitable software finds that the quadratic multiplier $s(x) = 1.4076x^2$ fulfils

$$\begin{aligned} 1 - x^3 - s(x)(0.25{-}x^2) = (-0.61313x^2 + 1)^2 + \\ (-0.53472x^2 + 0.93508x)^2 + (0.86357x^2)^2 \in \Sigma_x \end{aligned} \quad \text{(C.5)}$$

so we proved $1 - x^3 > 0 \; \forall x \in \Omega$.

As a generalisation of the SOS polynomial problem, the matrix-SOS problem is defined as verifying whether a polynomial matrix $P(x)$ is positive semidefinite for and value of x via SOS programming. This is done, thanks to the proposition below.

Proposition C.1 *([19]) Let $L(x)$ be an $N \times N$ symmetric polynomial matrix of degree $2d$ in $x \in \mathbb{R}^n$. Furthermore, let $z(x)$ be a column vector whose entries are all monomials in x with degree no greater than d, and consider the following conditions:*

1. *$L(x) \geq 0 \; \forall x \in \mathbb{R}^n$.*
2. *The polynomial $v^T L(x) v$ (of degree $2d$ in x and $2N$ in v) is SOS is the joint variables x, v.*
3. *There exists a constant positive semidefinite matrix Q such that*

$$v^T L(x) v = (v \otimes z(x))^T \, Q \, (v \otimes z(x)) \, ,$$

where $\otimes$ denotes the Kronecker product.

Then (1) $\Leftarrow$ (2) and (2) $\Leftrightarrow$ (3).

Basically, (2) is the definition of positive-semidefinite matrices (v being an arbitrary vector), conservatively evaluated via a SOS requirement. Then (3) actually builds the Gram matrix associated with (2). In this way, the classical linear matrix inequality framework (positive-definiteness of matrices with linear expressions as elements [2]) is extended to the polynomial case.

Of course, Positivstellensatz argumentations may be used to build sufficient local positive-definiteness conditions.

Example C.4 According to the above result, testing if there exist coefficients a, b, such that the matrix

$$L(x) = \begin{pmatrix} 1 - x^2 & 0.1x + b \\ 0.1x + b & 3 - 0.1x^3 + ax^2 \end{pmatrix}$$

is positive semidefinite for all x in $\Omega = \{x \in \mathbb{R} : x^2 \leq 0.9\}$, can be done using Corollary C.1 by proving that the polynomial $H(x, v) = v^T L(x) v + s(x, v)(0.9 - x^2)$ is SOS in the (augmented) set of variables (x, v). Actually, $H(x, v)$ is

$$H(x, v) = (1 - x^2)v_1^2 + 2(0.1x + b)v_1 v_2 + (3 - 0.1x^3 + ax^2)v_2^2 + s(x, v)(0.9 - x^2).$$

Setting a multiplier s with arbitrary structure, say

$$s(x, v) = m_1 v_2^2 x^2 + m_2 v_1 v_2 x^2 + m_3 v_1 v_2 x + m_4 v_2^2 + m_5 v_1^2 x + m_6 v_1^2,$$

we must search for feasible a, b, m_1, ..., m_6 such that both $H(x, v)$ and $s(x, v)$ are SOS. A solution for this problem using SOSOPT is

$$a = 0.5819, \quad b = -0.00030273,$$
$$m_1 = 1.1185, \; m_2 = 0, \; m_3 = 0.03929,$$
$$m_4 = 1.1366, \; m_5 = 0, \; m_6 = 1.0524.$$

The above proposition increases the complexity due to the introduction of the auxiliary variables v. There exist better ways to deal with polynomial SOS matrices, from a computational point of view, without extra variables.

Proposition C.2 (*Scherer and Hol [25]*) *Let $F(x)$ be an $N \times N$ symmetric polynomial matrix of degree $2d$ in $x \in \mathbb{R}^n$. $F(x)$ is an SOS polynomial matrix if and only if there exist a constant matrix $Q \succeq 0$ and a vector of monomials $z(x)$ satisfying*

$$F(x) = (I \otimes z(x))^T Q(I \otimes z(x)) \quad \forall x \in \mathbb{R}^n. \tag{C.6}$$

Actually, the third condition in Proposition C.1 is equivalent to Proposition C.2.

Appendix D
Convex Sum Relaxations

D.1 Summations Over the Standard Simplex

Many control and estimation problems can be written as a double sum negativity or positivity problem, i.e.

$$\Gamma_{hh} := \sum_{i=1}^{r} \sum_{j=1}^{r} h_i(z)h_j(z)\Gamma_{ij}(\cdot) > 0, \tag{D.1}$$

with the symmetric matrices $\Gamma_{ij}(\cdot)$ being affinely dependent on their arguments and the membership functions $h_i(z)$ holding the convex sum property, i.e. $\sum_{i=1}^{r} h_i(z) = 1$ and $h_i(z) \geq 0$ for all z in some compact set $\mathcal{C} \subseteq \mathbb{R}^p$. In the case of polynomial problems, matrices $\Gamma_{ij}(\cdot)$ have entries which are polynomials of the state, whereas $h_i(z)$ normally capture non-polynomial nonlinearities (or high-degree polynomials, unknown coefficients, etc.). In TS-based problems, whether they are ODEs or descriptors, Γ_{ij} are quadratic forms or constant matrices.

However, evaluating whether (D.1) holds may be a daunting task, as $h_i(z)$ are nonlinear functions of their arguments. Thus, the goal is to obtain sufficient conditions for (D.1) to hold.

Denoting by $\boldsymbol{h}$ the r-dimensional vector formed by the memberships, the first conservative step is setting up the so-called *shape-independent* conditions:

$$\sum_{i=1}^{r} \sum_{j=1}^{r} h_i h_j \Gamma_{ij}(\cdot) > 0, \quad \forall \boldsymbol{h} = \begin{bmatrix} h_1 \; h_2 \; \dots \; h_r \end{bmatrix}^T \in \Delta^{r-1}, \tag{D.2}$$

where arguments on $\boldsymbol{h}$, h_i were purposely omitted as we are no longer interested in their dependence on z, but on the following reformulation of the convex sum property:

M. Bernal et al., *Analysis and Synthesis of Nonlinear Control Systems*, Studies in Systems, Decision and Control 408, https://doi.org/10.1007/978-3-030-90773-0

$$\Delta^{r-1} := \{\mu \in \mathbb{R}^r \mid \sum_{i=1}^{r} \mu_i = 1, \quad 0 \le \mu_i \le 1\}, \tag{D.3}$$

i.e. Δ^{r-1} is the $(r-1)$-dimensional standard simplex in $\mathbb{R}^r$. The left-hand term of expression (D.2) is known as a *double convex summation* or, also, as a *2-dimensional* summation. As an example of conditions in form (D.1), see, for instance, (5.14).

Note that, importantly, the nonlinear functions $h(z)$ have been replaced by conditions which must hold for all possible values of some unknown *constants* ranging in a convex set. The following example illustrates the conservatism of such a step.

Example D.1 We know that $f(x) = \sin(x)x > 0$ for all $x \in [-1, 1]$. However, if we express the term as

$$\sin(x) = h_1(x)\sin(1) + h_2(x)\sin(-1),$$

with

$$h_1(x) = \frac{\sin(x) - \sin(-1)}{\sin(1) - \sin(-1)}, \ h_2(x) = 1 - h_1(x),$$

we get

$$f(x) = \sin(x)x = h_1(x)\sin(1)x + h_2(x)\sin(-1)x.$$

If we try to prove the shape-independent version

$$f_{si}(h, x) = h_1 \sin(1)x + h_2 \sin(-1)x > 0$$

for all positive h_1, h_2 such that $h_1 + h_2 = 1$, we will fail because a linear expression in x cannot be positive for all $x \in [-1, 1]$.

After that conservative step, however, we are ensured that the remaining (D.2) is a *homogeneous polynomial* (all terms of the same degree) in h and a polynomial in x (of degree 2 or higher): all nonlinearity has "vanished".

Related generalisations of the double convex sum problem concern the following multi-dimensional signed summation:

$$\Gamma_{\underbrace{hh\cdots h}_{p}} = \sum_{i_1=1}^{r}\sum_{i_2=1}^{r}\cdots\sum_{i_p=1}^{r} h_{i_1}(z)h_{i_2}(z)\ldots h_{i_p}(z)\Gamma_{i_1 i_2 \ldots i_p}(\cdot) > 0, \tag{D.4}$$

whose shape-independent version would have the form

$$\Xi(\boldsymbol{h}, \cdot) = \sum_{i_1=1}^{r}\sum_{i_2=1}^{r}\cdots\sum_{i_p=1}^{r} h_{i_1}h_{i_2}\ldots h_{i_p}\Gamma_{i_1 i_2 \ldots i_p}(\cdot) > 0 \quad \forall \boldsymbol{h} \in \Delta^{r-1}. \tag{D.5}$$

Obviously, the case $p = 2$ reduces to (D.1). Conditions requiring $p = 3$ are, for instance, the dynamic controllers in Li et al. [26] and Tanaka and Wang [27], using

$Q_{ijk} = E_{ijk} + E_{ijk}^T$, with

$$E_{ijk} = \begin{bmatrix} A_i Q_{11} + B_i \mathcal{C}_{jk} & A_i + B_i \mathcal{D}_j C_k \\ \mathcal{A}_{ijk} & A_i P_{11} + \mathcal{B}_{ij} C_k \end{bmatrix} < 0, \tag{D.6}$$

the output-feedback ones in Fang et al. [28]; Chen et al. [29]; or others in this book.

Notation

In order to streamline notation in multi-dimensional summations such as (D.5), the following notation will be used:

$$\mathbb{I}_p = \{\mathbf{i} = (i_1, i_2, \ldots, i_p) \in \mathbb{N}^p \mid 1 \le i_j \le r \ \forall\ j = 1, 2, \ldots, p\}, \tag{D.7}$$

$$\textstyle\sum_{\mathbf{i}\in\mathbb{I}_p} \Gamma_{\mathbf{i}} = \sum_{i_1=1}^r \sum_{i_2=1}^r \cdots \sum_{i_p=1}^r h_{i_1} h_{i_2} \cdots h_{i_p} \Gamma_{i_1 i_2 \ldots i_p}, \tag{D.8}$$

for some suitably defined multi-dimensional $\Gamma_{\mathbf{i}} = \Gamma_{i_1 i_2 \ldots i_p}$, which will be either a number or a matrix, i.e. boldface symbol $\mathbf{i}$ will denote a multi-index in a p-dimensional index set $\mathbb{I}_p$ ($\mathbb{I}_p$ has r^p elements). For instance, triple-summations in a four-rule fuzzy system will be spanned by a multi-index $\mathbf{i} \in \mathbb{I}_3$, where $\mathbb{I}_3$ has 4^3 elements: $\{1, 1, 1\}, \{1, 1, 2\}, \ldots, \{1, 1, 4\}, \{1, 2, 1\}, \ldots, \{4, 4, 4\}$. Furthermore, when the sum dimension is clear, the original summation (D.4) will be written as $\Gamma_{\mathbf{h}}$ or $\Gamma_{h \ldots h}$. Also, if no confusion arises $\Gamma_{h \ldots h}$ may be used, with a slight abuse of notation, to indicate the shape-independent conditions (which are the ones that, actually, will be solved for in most cases). The usual explicit dependence on $\boldsymbol{x}$ of $\Gamma_{\mathbf{i}}(\boldsymbol{x})$ will also, in most cases, be omitted for brevity.

D.2 Sufficient Shape-Independent Conditions

Note that (D.5) with $p = 1$ can be replaced by exact *necessary and sufficient* conditions:

$$\Gamma_h = \sum_{i=1}^r h_i \Gamma_i(\boldsymbol{x}) > 0,\ \forall h_i \in \Delta^{r-1} \Leftrightarrow \Gamma_i(\boldsymbol{x}) > 0\ \forall i \in \{1, 2, \ldots, r\} \tag{D.9}$$

because of the linearity (hence convexity) in h. Then proving positiveness of quadratic polynomials in x is an LMI problem in the polynomial coefficients, and it is an SOS problem in higher degrees for $\Gamma_i(x)$. However, further conservatism remains ahead to check for (D.2) or the general (D.5), discussed below.

Let us first consider the case $p = 2$, as it is the most frequent. The goal now is to find conditions on Γ_{ij} such that (D.2) holds. Of course, as all h_i are positive, we have from

$$\Gamma_{hh} = h_1^2\Gamma_{11} + h_1h_2(\Gamma_{12} + \Gamma_{21}) + h_2^2\Gamma_{22} + h_1h_3(\Gamma_{13} + \Gamma_{31}) + \dots$$

that a trivial sufficient condition is [30]

$$\begin{aligned} &\Gamma_{ii} > 0, \\ &\Gamma_{ij} + \Gamma_{ji} > 0, \end{aligned} \tag{D.10}$$

for $i \in \{1, 2, \dots, r\}$, $j \in \{i+1, \dots, r\}$.

A widely used refinement, which for $p = 2$ always improves on the conditions (D.10), is

Lemma D.1 *[31] Equation (D.1) is satisfied if*

$$\begin{aligned} &\Gamma_{ii} < 0 \\ &\frac{2}{r-1}\Gamma_{ii} + \Gamma_{ij} + \Gamma_{ji} < 0 \end{aligned} \tag{D.11}$$

for $i \in \{1, 2, \dots, r\}$, $j \in \{1, 2, \dots, r\}$, $i < j$.

Adding auxiliary variables can also be useful to reduce the conservatism of the conditions.

Lemma D.2 *[32] Condition (D.1) is satisfied if there exist matrices* $Q_{ii} > 0$, $i \in \{1, 2, \dots, r\}$, *and* $Q_{ij} = Q_{ji}^T$, $i \in \{1, 2, \dots, r\}$, $j = i+1, i+2, \dots, m$ *such that*

$$\begin{aligned} &\Gamma_{ii} > Q_{ii}, \\ &\Gamma_{ij} + \Gamma_{ji} > Q_{ij} + Q_{ji}, \\ &\begin{pmatrix} Q_{11} & Q_{12} & \dots & Q_{1m} \\ Q_{21} & Q_{22} & \dots & Q_{2m} \\ \vdots & & \ddots & \vdots \\ Q_{m1} & Q_{m2} & \dots & Q_{mm} \end{pmatrix} > 0 \end{aligned} \tag{D.12}$$

for $i \in \{1, 2, \dots, r\}$, $j = i+1, i+2, \dots, m$.

The conditions of Lemmas D.1 and D.2 are only sufficient. The next section will discuss conditions which apply to summations of higher dimensions and, also, as a side result, reduce the conservatism of the above ones.

D.3 Asymptotically Exact Conditions

The main result comes from Pólya [33], refined in Powers and Reznick [34] and adapted to specific control problems in Scherer [35]; Sala and Ariño [36]. Let us briefly outline the main ideas.

Let h be an r-dimensional vector (of memberships) and denote the repeated products of some of its elements in a particular order as $h_{\mathbf{i}} = h_{i_1} h_{i_2} \dots h_{i_p}$ for any given p-tuple, $\mathbf{i} = (i_1, i_2, \dots, i_p) \in \mathbb{I}_p$. Note that if $\mathbf{i}$ is a permutation of $\mathbf{j}$, then $h_{\mathbf{i}} = h_{\mathbf{j}}$.

Let us discuss how to prove positiveness of the multi-dimensional summation below, assuming the previously defined notation:

$$f(\boldsymbol{h}) = \sum_{\mathbf{i} \in \mathbb{I}_p} h_{\mathbf{I}} \Gamma_{\mathbf{i}}. \tag{D.13}$$

Note that $f(\boldsymbol{h})$ is a homogeneous polynomial in $\boldsymbol{h}$, of degree p.

Theorem D.1 *If $f(\boldsymbol{h}) \geq \gamma_0 > 0$ for all $\boldsymbol{h} \in \Delta^{r-1}$, then there exists a finite d, to be denoted as a complexity parameter, such that all coefficients of the homogeneous polynomial of degree $p + d$ given by*

$$f^{[d]}(\boldsymbol{h}) = \left(\sum_{\mathbf{j} \in \mathbb{I}_d} h_{\mathbf{J}} \right) \sum_{\mathbf{i} \in \mathbb{I}_p} h_{\mathbf{I}} \Gamma_{\mathbf{i}} \tag{D.14}$$

are positive. If the coefficients are positive for a particular d, so they are for any $d' \geq d$.

Note that as $\sum_{i=1}^{r} h_i = 1$ then, trivially,

$$\sum_{\mathbf{j} \in \mathbb{I}_d} h_{\mathbf{J}} = \left(\sum_{i=1}^{r} h_i \right) \sum_{\mathbf{j} \in \mathbb{I}_{D-1}} h_{\mathbf{j}} = \left(\sum_{i=1}^{r} h_i \right)^d = 1.$$

Hence, $f^{[d]}(\boldsymbol{h}) \equiv f(\boldsymbol{h})$.

If $\Gamma_{\mathbf{i}}$ are quadratic functions of some variable $\boldsymbol{x}$, $\Gamma_{\mathbf{i}} = \boldsymbol{x}^T Q_{\mathbf{i}} \boldsymbol{x}$ then a matrix version of the result quickly ensues considering the same proposition on positive-definite matrices instead of plain real coefficients, replacing $\Gamma_{\mathbf{i}}$ by $Q_{\mathbf{i}}$; details omitted for brevity; see Scherer [35]; Sala and Ariño [36].

D.3.1 Without Additional Variables

From the above results, the conditions (D.10) for 2-dimensional summations can be generalised to higher dimensions. Let us denote as $\mathcal{P}(\mathbf{i})$ all possible permutations of a multi-dimensional index variable $\mathbf{i}$.

Lemma D.3 *If* $\sum_{\mathbf{j}\in\mathcal{P}(\mathbf{i})} \Gamma_{\mathbf{j}} > 0$ *for all* $\mathbf{i}$*, then* $f(h)$ *in* (D.13) *is positive.*

Indeed, note that $\sum_{\mathbf{j}\in\mathcal{P}(\mathbf{i})} \Gamma_{\mathbf{j}}$ are, actually, the coefficients of the polynomial taking into account that the product of the same memberships $\mathbf{i}$ in any different order (spanned by $\mathbf{j}$) does appear as a single coefficient. For instance, for $p = 2$, the fact that $h_i h_j = h_j h_i$ gives rise to a term $h_i h_j(\Gamma_{ij} + \Gamma_{ji})$. Conditions (D.10) are, hence, a particular case of (D.3).

In order to apply Theorem D.1 to relax higher dimensional summations, we may define $\Gamma_{i_1,i_2,\ldots,i_p,i_{p+1},\ldots,i_{p+d}} := \Gamma_{i_1,i_2,\ldots,i_p}$. Introducing the notation $\mathbf{i}(1:p) = (i_1, i_2, \ldots, i_p)$ representing the extraction of the first p elements of a larger tuple $\mathbf{i}$, we can directly apply Lemma D.3, to relax a p-dimensional summation condition via a $(p+d)$-dimensional one.

Lemma D.4 *If* $\tilde{\Gamma}_{\mathbf{i}} = \sum_{\mathbf{j}\in\mathcal{P}(\mathbf{i})} \Gamma_{\mathbf{j}(1:p)} > 0$ *for all* $\mathbf{i} \in \mathbb{I}_{P+d}$*, then* $f(h)$ *in* (D.13) *is positive.*

Of course, once we have analysed $\tilde{\Gamma}_{\mathbf{i}}$ for one particular index $\mathbf{i}$, it is not necessary to rewrite the above sum for any other index $\mathbf{i}'$ which can be expressed as a permutation of it as $\tilde{\Gamma}_{\mathbf{i}'}$ would be identical to $\tilde{\Gamma}_{\mathbf{i}}$.

Example D.2 The Lyapunov equation of a closed-loop PDC controller in a 2-vertex TS model may be expressed as $\dot{V} = \sum_{\mathbf{i}\in\mathbb{I}_2} h_{\mathbf{i}} \boldsymbol{x}^T Q_{\mathbf{i}} \boldsymbol{x} < 0$, the 2-dimensional sum elements given by $Q_{\mathbf{i}} = Q_{ij} = A_i P + B_i F_j$.

Setting complexity parameter $d = 2$, $\dot{V}$ is proved negative if

$$
\begin{aligned}
(1111)\,&\tilde{Q}_{1111} = Q_{11} < 0,\\
(1112)\,&\tilde{Q}_{1112} = Q_{1112} + Q_{1121} + Q_{1211} + Q_{2111} = 2Q_{11} + Q_{12} + Q_{21} < 0,\\
(1122)\,&\tilde{Q}_{1122} = Q_{1122} + Q_{1212} + Q_{2112} + Q_{2121} + Q_{2211} + Q_{1221}\\
&\qquad = Q_{11} + 2Q_{12} + 2Q_{21} + Q_{22} < 0,\\
(1222)\,&\tilde{Q}_{1222} = Q_{1222} + Q_{2122} + Q_{2212} + Q_{2221} = 2Q_{22} + Q_{12} + Q_{21} < 0,\\
(2222)\,&\tilde{Q}_{2222} = Q_{22} < 0.
\end{aligned}
$$

Every other 4-tuple formed by combining indices "1" and "2" is a permutation of one of the 5 above, so there is no need for more conditions.

Inspired by the example, similar conditions can be set up for other number of rules or other values of the complexity parameter.

D.3.2 Additional Variables

Note that $\Gamma_{\mathbf{h}}$ in (D.4) is a polynomial in the set of variables $\boldsymbol{h}, \boldsymbol{x}$ if $\Gamma_{\mathbf{i}}(\boldsymbol{x})$ are polynomials in $\boldsymbol{x}$. Hence, sufficient conditions for positiveness can be set up as sum-of-squares problems. In order to better "guide" the SOS solver towards obtaining squares, the information on $h_i \geq 0$ will be introduced by the change of variable $h_i = \sigma_i^2$, instead of Positivstellensatz multipliers which would be equivalent (or better) but would greatly increase the number of decision variables.

Hence, the multiple summations in (D.4) can be equivalently written as

$$\Gamma_{\mathbf{h}}^{\sigma} = \sum_{i_1=1}^{r}\sum_{i_2=1}^{r}\cdots\sum_{i_p=1}^{r}\sigma_{i_1}^2\sigma_{i_2}^2\cdots\sigma_{i_p}^2\,\Gamma_{i_1 i_2\ldots i_p}(\boldsymbol{x}) > 0 \quad \forall x \neq 0, \tag{D.15}$$

so a sufficient condition for (D.4) to hold is the polynomial $\Gamma_{\mathbf{h}}^{\sigma}$ to be sum of squares in variables $\boldsymbol{x}, \sigma$. Internally, SOS solvers would form the Gram matrix (see Sect. 4.5) and set up the relevant conditions for each particular problem.

Remark D.1 Note that from the concept of Gram matrix and the matrix-SOS conditions in Sect. 4.5, it is easy to see that, for instance, conditions D.2 are a particular case of the matrix-SOS approach.

Indeed, let us consider showing it with two rules, for simplicity.

Let $\boldsymbol{h} = [h_1\, h_2]^T = [\sigma_1^2\, \sigma_2^2]^T$ and $\sigma^E = (\sigma_1\sigma_2, \sigma_1^2, \sigma_2^2)^T$; we have

$$\Gamma_{\boldsymbol{h}}^{\sigma} = (*)\begin{bmatrix}\Gamma_{11} & \Gamma_{12}\\ \Gamma_{12}^T & \Gamma_{22}\end{bmatrix}(I \otimes h) = (*)\begin{bmatrix}S & 0 & 0\\ 0 & \Gamma_{11} & \Gamma_{12} - S/2\\ 0 & \Gamma_{12}^T - S/2 & \Gamma_{22}\end{bmatrix}(I \otimes \sigma^E), \tag{D.16}$$

where the identity matrices have the same size as the individual Γ_{ij}. Now, defining $Q_{11} = \Gamma_{11}$, $Q_{22} = \Gamma_{22}$, $Q_{12} + Q_{12}^T = \Gamma_{12} + \Gamma_{12}^T - S$, the conditions

$$\begin{bmatrix}Q_{11} & Q_{12}\\ Q_{21} & Q_{22}\end{bmatrix} > 0, \quad S > 0 \tag{D.17}$$

are those from the Schur complement using the rightmost expression in (D.16), and they are actually equivalent to (D.2). Hence, we needed S as an additional matrix full of decision variables.

Of course, the sketch of proof also holds for a higher number of rules and, too, it can be shown that the conditions with additional variables in [36, Theorem 5] are also such a particular case of the matrix-SOS approach (details omitted for brevity).

D.4 Exploiting Tensor-Product Structure

Within the context of Takagi–Sugeno fuzzy models, the work in Ariño and Sala [37] presented relaxed LMI conditions exploiting the tensor-product (TP) model of a nonlinear system; this work was based on fuzzy rules. More recently, an update of such tensor-product model notation has been adopted by Gonzalez et al. [38]: it is based on exact convex representations which, instead of "flattening" individual weights into the so-called membership functions, leave them as simple partitions in order to exploit their interrelations in full. To give some idea of the advantages of such methodology, we will consider a slight rearrangement of the notation introduced in Sect. 3.2.1 by considering an affine-in-control nonlinear system of the form

$$\delta x(t) = A(x)x(t) + B(x)u(t)), \tag{D.18}$$

where it is assumed that the entries of $A(x) \in \mathbb{R}^{n\times n}$ and $B(x) \in \mathbb{R}^{n\times m}$ can be written as sums or products of p terms which are bounded in a compact set of the state space $x \in \Omega$; denoting the set of p bounded terms as $z_1(x) \in [z_1^0, z_1^1]$, $z_2(x) \in [z_2^0, z_2^1], \ldots$, $z_p(x) \in [z_p^0, z_p^1]$, it is clear that each of them can be written as a convex sum of its bounds:

$$z_i(x) = \underbrace{\frac{z_i^1 - z_i(x)}{z_i^1 - z_i^0}}_{w_0^i(x)} z_i^0 + \underbrace{\frac{z_i(x) - z_i^0}{z_i^1 - z_i^0}}_{w_1^i(x)} z_i^1, \tag{D.19}$$

where $w_0^i(x) + w_1^i(x) = 1, i \in \{1, 2, \ldots, p\}$.

Due to convexity of functions $w_j^i(x)$, (D.18) can be written as (omitting arguments)

$$\delta x(t) = \sum_{i_1=0}^{1}\sum_{i_2=0}^{1}\cdots\sum_{i_p=0}^{1} w_{i_1}^1 w_{i_2}^2 \cdots w_{i_p}^p \left(A_{(i_1,i_2,\ldots,i_p)}x + B_{(i_1,i_2,\ldots,i_p)}u\right), \tag{D.20}$$

where

$$A_{(i_1,i_2,\ldots,i_p)} = A(x)|_{w_{i_1}^1 = w_{i_2}^2 = \cdots = w_{i_p}^p = 1},$$
$$B_{(i_1,i_2,\ldots,i_p)} = B(x)|_{w_{i_1}^1 = w_{i_2}^2 = \cdots = w_{i_p}^p = 1}.$$

In writing this model, elementary properties of convex sums have been used that allow stacking them at the leftmost side of expressions, namely if $w_0^j + w_1^j = 1$, $w_{i_j}^j \in [0, 1]$:

$$\sum_{i_1=0}^{1} w_{i_1}^1 z_1^{i_1} + \sum_{i_2=0}^{1} w_{i_2}^2 z_2^{i_2} = \sum_{i_1=0}^{1} w_{i_1}^1 \sum_{i_2=0}^{1} w_{i_2}^2 \left(z_1^{i_1} + z_2^{i_2}\right),$$
$$\left(\sum_{i_1=0}^{1} w_{i_1}^1 z_1^{i_1}\right)\left(\sum_{i_2=0}^{1} w_{i_2}^2 z_2^{i_2}\right) = \sum_{i_1=0}^{1} w_{i_1}^1 \sum_{i_2=0}^{1} w_{i_2}^2 z_1^{i_1} z_2^{i_2}.$$

If notation $\mathbb{B} \equiv \{0, 1\}$, $\mathbf{i} \equiv (i_1, i_2, \ldots, i_p)$, $\mathbf{w_i}(x) = w_{i_1}^1(x) w_{i_2}^2(x) \cdots w_{i_p}^p(x)$, is introduced, the system (D.20) can be further simplified to

$$\delta x(t) = \sum_{\mathbf{i}\in\mathbb{B}^p} \mathbf{w_i}(x)\left(A_\mathbf{i} x + B_\mathbf{i} u\right). \tag{D.21}$$

Even without introducing a PDC-like control law, the TP model (D.21) can take into account a variety of relationships, e.g. a power $(z_j(\boldsymbol{x}))^q$ of a convex representation $z_j(\boldsymbol{x}) = \sum_{i=0}^{1} w_i^j(\boldsymbol{x}) z_j^i$ is also convex:

$$\left(\sum_{i=0}^{1} w_i^j(\boldsymbol{x}) z_j^i\right)^q = \left(\sum_{i_1=0}^{1} w_{i_1}^j(\boldsymbol{x}) z_j^{i_1}\right)\left(\sum_{i_2=0}^{j} w_{i_2}^j(\boldsymbol{x}) z_j^{i_2}\right)\cdots\left(\sum_{i_q=0}^{j} w_{i_q}^j(\boldsymbol{x}) z_j^{i_q}\right)$$
$$= \sum_{i_1=0}^{1}\sum_{i_2=0}^{1}\cdots\sum_{i_q=0}^{1} w_{i_1}^j(\boldsymbol{x}) w_{i_2}^j(\boldsymbol{x})\cdots w_{i_q}^j(\boldsymbol{x}) z_j^{i_1} z_j^{i_2}\cdots z_j^{i_q},$$

where clearly the latter sum allows for the association of similar terms as the product $w_{i_1}^j(\boldsymbol{x}) w_{i_2}^j(\boldsymbol{x})\cdots w_{i_q}^j(\boldsymbol{x})$ produces powers of the same base functions $w_0^j(\boldsymbol{x})$ and $w_1^j(\boldsymbol{x})$, all of them related with the same nonlinearity z_j. These associations are beneficial for stability analysis as the resulting LMI conditions will relax the requirement of having all the "independent" vertices stable.

Naturally, a PDC-like control law of the form

$$\boldsymbol{u}(t) = \sum_{\mathbf{j}\in\mathbb{B}^p} \mathbf{w_j}(\boldsymbol{x}) F_{\mathbf{j}} \boldsymbol{x}, \tag{D.22}$$

with $F_{\mathbf{j}} \in \mathbb{R}^{n\times m}$, induces a richer set of associations for stabilisation of (D.18) via the TP model (D.21). Traditional stabilisation results may translate into theorems of the following form (the proof is trivial).

Theorem D.2 *The origin of the nonlinear system* (D.18) *with TP model* (D.21) *on the compact set* Ω *such that* $0 \in \Omega$*, under the control law* (D.22)*, is asymptotically stable if there exist* $X = X^T > 0$ *and* $M_{\mathbf{j}}$ *of adequate dimensions such that*

$$\sum_{\mathbf{ij}\in\mathcal{P}(\mathbf{kl})} A_{\mathbf{i}} X + B_{\mathbf{i}} M_{\mathbf{j}} + A_{\mathbf{i}}^T X + M_{\mathbf{j}}^T B_{\mathbf{i}}^T < 0,$$

where $\mathbf{kl} \in \mathbb{B}^{2p+}$, $\mathbb{B}^{2p+}$ *being the set of* $2p$*-bit binary numbers such that every bit is less than or equal to the following,* $\mathcal{P}(\mathbf{kl})$ *is the set of indices such that* $\mathbf{w_i}(\boldsymbol{x})\mathbf{w_j}(\boldsymbol{x}) = \mathbf{w_k}(\boldsymbol{x})\mathbf{w_l}(\boldsymbol{x})$.

References

1. Lancaster, P., Tismenetsky, M.: The Theory of Matrices: With Applications. Academic Press, San Diego (1985)
2. Boyd, S., El Ghaoui, L., Féron, E., Balakrishnan, V.: Linear Matrix Inequalities in System and Control Theory. Studies in Applied Mathematics, Society for Industrial and Applied Mathematics, Philadelphia, PA (1994)

3. Scherer, C., Weiland, S.: Linear Matrix Inequalities in Control. Delft University, Delft, The Netherlands (2005)
4. Strang, G.: Linear Algebra and Its Applications, 3rd edn. Thomson Publishing (1988)
5. Golub, G., van Loan, C.: Matrix Computations, 3rd edn. John Hopkins University Press, Baltimore (1996)
6. Meyer, C.: Matrix Analysis and Applied Linear Algebra. Society for Industrial and Applied Mathematics (SIAM), Reston (2001)
7. Scherer, C.W.: LPV control and full block multipliers. Automatica **37**(3), 361–375 (2001)
8. Skelton, R.E., Iwasaki, T., Grigoriadis, K.: A Unified Approach to Linear Control Design. Taylor & Francis, London (1998)
9. Apkarian, P., Gahinet, P.: A convex characterization of gain-scheduled h_∞ controllers. IEEE Trans. Autom. Control **40**(5), 853–864 (1995)
10. Tuan, H.D., Apkarian, P., Narikiyo, T., Kanota, M.: New fuzzy control model and dynamic output feedback parallel distributed compensation. IEEE Trans. Fuzzy Syst. **12**(1), 13–21 (2004)
11. Khalil, H.K.: Nonlinear Systems. Prentice-Hall, Upper Saddle River, NJ (2002)
12. Khalil, H.K.: Nonlinear Control. Pearson, New York (2014)
13. Vidyasagar, M.: Nonlinear systems analysis. SIAM, Reston (2002)
14. Löfberg, J.: YALMIP: a toolbox for modeling and optimization in MATLAB. In: Proceedings of the CACSD Conference, pp. 284–289 (2004)
15. Parrilo, P.A.: Structured semidefinite programs and semialgebraic geometry methods in robustness and optimization. PhD thesis, California Institute of Technology (2000)
16. Higham, N.: Analysis of the Cholesky decomposition of a semi-definite matrix. In: Cox, M., Hammarling, S.J. (eds.) Reliable Numerical Computation, pp. 161–185. Oxford University Press, Oxford (1990)
17. Reznick, B.: Some concrete aspects of Hilbert's 17th problem. Contemp. Math. **253**, 251–272 (2000)
18. Blekherman, G.: There are significantly more nonegative polynomials than sums of squares. Israel J. Math. **153**(1), 355–380 (2006)
19. Prajna, S., Papachristodoulou, A., Seiler, P., Parrilo, P.A.: SOSTOOLS: Sum of Squares Optimization Toolbox for MATLAB. California Inst. Technol, Pasadena, CA (2004)
20. Lofberg, J.: Pre-and post-processing sum-of-squares programs in practice. IEEE Trans. Autom. Control **54**(5), 1007–1011 (2009)
21. Seiler, P.: Sosopt: A toolbox for polynomial optimization. Preprint at arXiv:1308.1889 (2013)
22. Stengle, G.: A nullstellensatz and a positivstellensatz in semialgebraic geometry. Optimization Methods and Software **207**(2), 87–97 (1974)
23. Jarvis-Wloszek, Z., Feeley, R., Tan, W., Sun, K., Packard, A.: Control applications of sum of squares programming. In: Positive Polynomials in Control. Lecture Notes in Control and Information Science, vol. 312, pp. 3–22. Springer, Berlin (2005)
24. Bertsekas, D.: Nonlinear Programming, 2nd edn. Athena Scientific, Belmont (1999)
25. Scherer, C., Hol, C.: Matrix sum-of-squares relaxations for robust semi-definit programs. Math. Program. **107**, 189–211 (2006)
26. Li, J., Niemann, D., Wang, H., Tanaka, K.: Parallel distributed compensation for Takagi–Sugeno fuzzy models: multiobjective controller design. In: Proceedings of the 1999 American Control Conference, vol. 3, pp. 1832–1836 (1999)
27. Tanaka, K., Wang, H.O.: Fuzzy Control System Design and Analysis: A Linear Matrix Inequality Approach. Wiley, New York (2001)
28. Fang, C.-H., Liu, Y.-S., Kau, S.-W., Hong, L., Lee, C.-H.: A New LMI-Based Approach to Relaxed Quadratic Stabilization of T–S Fuzzy Control Systems. IEEE Trans. Fuzzy Syst. **14**, 286–397 (2006)
29. Chen, S.-S., Chang, Y.-C., Su, S.-F., Chung, S.-L., Lee, T.-T.: Robust static output-feedback stabilization for nonlinear discrete-time systems with time delay via fuzzy control approach. Fuzzy Systems, IEEE Transactions on **13**(2), 263–272 (2005)

30. Wang, H., Tanaka, K., Griffin, M.: An approach to fuzzy control of nonlinear systems: stability and design issues. IEEE Trans. Fuzzy Syst. **4**(1), 14–23 (1996)
31. Tuan, H., Apkarian, P., Narikiyo, T., Yamamoto, Y.: Parameterized linear matrix inequality techniques in fuzzy control system design. IEEE Trans. Fuzzy Syst. **9**(2), 324–332 (2001)
32. Liu, X., Zhang, Q.: New approaches to H_∞ controller designs based on fuzzy observers for T–S fuzzy systems via LMI. Automatica **39**(9), 1571–1582 (2003)
33. Pólya, G.: Uber positive Darstellung von Polynomen. Vierterterhartschrift d. Naturforschenden Gessellschaft in Zurich **73**, 141–145 (1928)
34. Powers, V., Reznick, B.: A new bound for Polya's theorem with applications to polynomials positive on polyhedra. J. Pure Appl. Algebra **164**(1–2), 221–229 (2001)
35. Scherer, C.: Relaxations for robust linear matrix inequality problems with verifications for exactness. SIAM J. Matrix Anal. Appl. **27**(2), 365–395 (2005)
36. Sala, A., Ariño, C.: Asymptotically necessary and sufficient conditions for stability and performance in fuzzy control: Applications of Polya's theorem. Fuzzy Sets Syst. **158**(24), 2671–2686 (2007)
37. Ariño, C., Sala, A.: Relaxed LMI conditions for closed-loop fuzzy systems with tensor-product structure. Eng. Appl. Artif. Intell. **20**(8), 1036–1046 (2007)
38. Gonzalez, T., Bernal, M., Sala, A., Aguiar, B.: Cancellation-based nonquadratic controller design for nonlinear systems via Takagi–Sugeno models. IEEE Transactions on Cybernetics **47**(9), 2628–2638 (2017)

Index

M. Bernal et al., *Analysis and Synthesis of Nonlinear Control Systems*, Studies in Systems, Decision and Control 408, https://doi.org/10.1007/978-3-030-90773-0